Mean square within

$$MSW = \frac{SSW}{dfw}$$

Mean square between

$$MSB = \frac{SSB}{dfb}$$

F ratio

$$F = \frac{MSB}{MSW}$$

CHAPTER 11

Chi square

$$\chi^2 \text{ (obtained)} = \Sigma \frac{(f_o - f_e)^2}{f_e}$$

CHAPTER 13

Phi

$$\phi = \sqrt{\frac{\chi^2}{N}}$$

Cramer's V

$$V = \sqrt{\frac{\chi^2}{(N)(\text{Minimum of } r - 1, c - 1)}}$$

Lambda

$$\lambda = \frac{E_1 - E_2}{E_1}$$

CHAPTER 14

Gamma

$$G = \frac{N_s - N_d}{N_s + N_d}$$

Spearman's rho

$$r_s = 1 - \frac{6\Sigma D^2}{N(N^2 - 1)}$$

CHAPTER 15

Least-squares regression line

$$Y = a + bX$$

Slope

$$b = \frac{N\Sigma\ XY - (\Sigma\ X)(\Sigma\ Y)}{N\Sigma\ X^2 - (\Sigma\ X)^2}$$

Y intercept

$$a = \overline{Y} - b\overline{X}$$

$$\frac{- (\Sigma\ X)(\Sigma\ Y)}{X)^2][N\Sigma\ Y^2 - (\Sigma\ Y)^2]}$$

Partial correlation coefficient

$$r_{yx.z} = \frac{r_{yx} - (r_{yz})(r_{xz})}{\sqrt{1 - r_{yz}^2}\ \sqrt{1 - r_{xz}^2}}$$

Least-squares multiple regression line

$$Y = a + b_1X_1 + b_2X_2$$

Partial slope for X_1

$$b_1 = \left(\frac{s_y}{s_1}\right)\left(\frac{r_{y1} - r_{y2}r_{12}}{1 - r_{12}^2}\right)$$

Partial slope for X_2

$$b_2 = \left(\frac{s_y}{s_2}\right)\left(\frac{r_{y2} - r_{y1}r_{12}}{1 - r_{12}^2}\right)$$

Y intercept

$$a = \overline{Y} - b_1\overline{X}_1 - b_2\overline{X}_2$$

Beta-weight for X_1

$$b_1^* = b_1\left(\frac{s_1}{s_y}\right)$$

Beta-weight for X_2

$$b_2^* = b_2\left(\frac{s_2}{s_y}\right)$$

Standardized least-squares regression line

$$Z_y = b_1^*Z_1 + b_2^*Z_2$$

Coefficient of multiple determination

$$R^2 = r_{y1}^2 + r_{y2.1}^2(1 - r_{y1}^2)$$

www.wadsworth.com

wadsworth.com is the World Wide Web site for Wadsworth Publishing Company and is your direct source to dozens of online resources.

At wadsworth.com you can find out about supplements, demonstration software, and student resources. You can also send e-mail to many of our authors and preview new publications and exciting new technologies.

wadsworth.com
Changing the way the world learns®

STATISTICS
A TOOL FOR
SOCIAL RESEARCH

Sixth Edition

Joseph F. Healey

Christopher Newport University

WADSWORTH

™

THOMSON LEARNING   Australia • Canada • Mexico • Singapore • Spain • United Kingdom • United States

Sociology Editor: Lin Marshall
Assistant Editor: Analie Barnett
Editorial Assistant: Reilly O'Neal
Marketing Manager: Matthew Wright
Project Manager, Editorial Production:
  Jerilyn Emori
Print/Media Buyer: Mary Noel
Permissions Editor: Stephanie Keough-Hedges
Production Service: Greg Hubit Bookworks

Copy Editor: Charles Cox
Illustrator: Lotus Art
Cover Designer: Denise Davidson
Cover Images: All images by PhotoDisc, except first image in second row and last image in third row, by Eyewire
Text and Cover Printer: R.R. Donnelley & Sons Company
Compositor: G&S Typesetters, Inc.

Printed in the United States of America
1 2 3 4 5 6 7 05 04 03 02 01

For permission to use material from this text, contact us by
Web: http://www.thomsonrights.com
Fax: 1-800-730-2215
Phone: 1-800-730-2214

Wadsworth/Thomson Learning
10 Davis Drive
Belmont, CA 94002-3098
USA

For more information about our products, contact us:
Thomson Learning Academic Resource Center
1-800-423-0563
http://www.wadsworth.com

International Headquarters
Thomson Learning
International Division
290 Harbor Drive, 2nd Floor
Stamford, CT 06902-7477
USA

UK/Europe/Middle East/South Africa
Thomson Learning
Berkshire House
168–173 High Holborn
London WC1V 7AA
United Kingdom

Asia
Thomson Learning
60 Albert Street, #15-01
Albert Complex
Singapore 189969

Canada
Nelson Thomson Learning
1120 Birchmount Road
Toronto, Ontario M1K 5G4
Canada

Library of Congress Cataloging-in-Publication Data
Healey, Joseph F.
    Statistics : a tool for social research / Joseph F. Healey.—6th ed.
        p. cm.
    Includes index.
    ISBN 0-534-55785-6
    1. Social sciences—Statistical methods. 2. Statistics. I. Title.
HA29.H39 2002
519.5—dc21   2001017976

BRIEF CONTENTS

DETAILED CONTENTS

PREFACE

Sociology and the other social sciences (including political science, social work, public administration, criminal justice, urban studies, and gerontology) are research-based disciplines, and statistics are part of their everyday language. To join the conversation, you must be literate in the vocabulary of research, data analysis, and scientific thinking. Fluency in statistics will help you understand the research reports you encounter in everyday life and the professional research literature of your discipline. You will also be able to conduct quantitative research, contribute to the growing body of social science knowledge, and reach your full potential as a social scientist.

Although essential, learning (and teaching) statistics can be a challenge. Students in statistics courses typically bring with them a wide range of mathematical backgrounds and an equally diverse set of career goals. They are often puzzled about the relevance of statistics for them and, not infrequently, there is some math anxiety to deal with.

This text introduces statistical analysis for the social sciences while addressing these challenges. The text makes minimal assumptions about mathematical background (the ability to read a simple formula is sufficient preparation for virtually all of the material in the text), and a variety of special features help students analyze data successfully. The text has been written especially for sociology and social work programs but is sufficiently flexible to be used in any program with a social science base.

The text is written at a level intermediate between a strictly mathematical approach and a mere "cookbook." I have not sacrificed comprehensive coverage or statistical correctness, but theoretical and mathematical explanations are kept at an elementary level, as is appropriate in a first exposure to social statistics. For example, I do not treat formal probability theory per se. Rather, the background necessary for an understanding of inferential statistics is introduced, informally and intuitively, in Chapters 5 and 6 while considering the concepts of the normal curve and the sampling distribution. The text makes no claim that statistics are "fun" or that the material can be mastered without considerable effort. At the same time, students are not overwhelmed with abstract proofs, formula derivations, and mathematical theory, which can needlessly frustrate the learning experience at this level.

GOAL OF THE TEXT AND CHANGES IN THIS EDITION

The goal of this text is to develop basic statistical literacy. The statistically literate person understands and appreciates the role of statistics in the research process, is competent to perform basic calculations, and can read and appreciate the professional research literature in their field as well as any research reports they may encounter in everyday life. This goal has not changed since the first edition of this text. However, in recognition of the fact that "mere computation" has become less of a challenge in this high-tech age, this edition places more stress on interpretation and computer applications and de-emphasizes computation. This change will be apparent in several ways. For example, a new feature called "Interpreting Statistics" has been added. These noncomputational sections are included in about half the chapters and present detailed examples of "what to say after the statistics have been calculated." They use real data and real research situations to illustrate the process of developing meaning and understanding, and they exemplify how statistics can be used to answer important questions. The issues addressed in these sections include surveillance in the workplace, the gender gap in income, and the deterrent effect of capital punishment.

Also, the end-of-chapter problem sets have been simplified (e.g., they use smaller data sets) in recognition of the fact that modern technology has rendered long, tedious hand calculation obsolete. More challenging problems (e.g., using larger data sets) are still available but have been moved to the companion Web site for this text. Also, for the first time this edition incorporates MicroCase—probably the most "user-friendly" statistical package on the market today—along with SPSS, the leading statistical software for social science research. Instructors can choose either statistical package for their course. The data set used in the computer exercises has been updated to the 1998 General Social Survey, and additional data sets and exercises are available at the Web site.

An additional major change in this edition is the creation of a Web site specifically for this text (*www.sociology.wadsworth.com*). The Web site offers, among other things, additional end-of-chapter problems, exercises using SPSS or MicroCase Explorit, additional data sets, flowcharts to help students select appropriate statistical procedures for various situations, a review of math (Appendix H in previous editions), and a variety of other features.

While the changes in this edition are significant, they continue to be focused on the goal of basic statistical literacy, the three aspects of which provide a framework for discussing the additional features of this text:

1. An Appreciation of Statistics. A statistically literate person understands the relevance of statistics for social research, can analyze and interpret the meaning of a statistical test, and can select an appropriate statistic for a given purpose and a given set of data. This textbook develops these qualities, within the constraints imposed by the introductory nature of the course, in the following ways:

- *The relevance of statistics.* Chapter 1 includes a discussion of the role of statistics in social research and stresses their usefulness as ways of analyzing and manipulating data and answering research questions. Each example problem is framed in the context of a research situation. A question is posed and then, with the aid of a statistic, answered. The relevance of statistics for answering questions is thus stressed throughout the text. This central theme of usefulness is further reinforced by a series of boxes labeled "Application," each of which illustrates some specific way statistics can be used to answer questions.

  Almost all end-of-chapter problems are labeled by the social science discipline or subdiscipline from which they are drawn: $\boxed{\text{SOC}}$ for sociology, $\boxed{\text{SW}}$ for social work, $\boxed{\text{PS}}$ for political science, $\boxed{\text{CJ}}$ for criminal justice, $\boxed{\text{PA}}$ for public administration, and $\boxed{\text{GER}}$ for gerontology. By identifying problems with specific disciplines, students can more easily see the relevance of statistics to their own academic interests. (Not incidentally, they will also see that the disciplines have a large subject matter in common.)

- *Interpreting statistics.* For most students, interpretation—saying what statistics mean—is a big challenge. The ability to interpret statistics can be developed only by exposure and experience. To provide exposure, I have been careful, in the example problems, to express the meaning of the statistic in terms of the original research question. To provide experience, the end-of-chapter problems almost always call for an interpretation of the statistic calculated. To provide examples, many of the Answers to Odd-Numbered Computational Problems in the back of the text are expressed in words as well as numbers. The new feature "Interpreting Statistics" provides additional, detailed examples of how to express the meaning of statistics.

- *Using Statistics: Ideas for Research Projects.* Appendix E, another new feature of this edition, offers ideas for independent data-analysis projects for students. The projects require students to use a computerized statistical package (SPSS or MicroCase) to analyze a data set. They can be assigned at intervals throughout the semester or at the end of the course. Each project provides an opportunity for students to practice and apply their statistical skills and, above all, to exercise their ability to understand and interpret the meaning of the statistics they produce. Additional ideas for research projects can be found in the *SPSS and MicroCase Companion for Statistics: A Tool for Social Research,* Sixth Edition, which can be ordered as a supplement to this text.

2. Computational Competence.   Students should emerge from their first course in statistics with the ability to perform elementary forms of data analysis—to execute a series of calculations and arrive at the correct answer. To be sure, computers and calculators have made computation less of an issue today. Yet, computation is inseparable from statistics and, since social science majors frequently do not have strong quantitative backgrounds, I have included a number of features to help students cope with these challenges:

- *Step-by-step computational algorithms* are provided for each statistic.

- *Extensive problem sets* are provided at the end of each chapter. For the most part, these problems use fictitious data and are designed for ease of computation.

- *Cumulative exercises* are included at the end of each part to provide practice in choosing, computing, and analyzing statistics. These exercises present only data sets and research questions. Students must choose appropriate statistics as part of the exercise. Many of these problems, like the end-of-chapter problems, have been simplified for ease of computation in this edition.

- *Solutions* to odd-numbered computational problems are provided so that students may check their answers.

- This edition continues to include *SPSS for Windows* to give students access to the computational power of the computer. However, an additional statistical package called *MicroCase* (using MicroCase Explorit) has been added to increase the flexibility of the text. Instructors may select either package. This feature is explained in more detail below.

3. The Ability to Read the Professional Social Science Literature. The statistically literate person can comprehend and critically appreciate research reports written by others. The development of this quality is a particular problem at the introductory level since (1) the vocabulary of professional researchers is so much more concise than the language of the textbook, and (2) the statistics featured in the literature are more advanced than those covered at the introductory level. To help bridge this gap, I have included a series of boxes labeled "Reading Statistics" beginning in Chapter 1. In each box, I briefly describe the reporting style typically used for the statistic in question and try to alert students about what to expect when they approach the professional literature. These inserts, many of which have been updated in this edition, include excerpts from the research literature and illustrate how statistics are actually applied and interpreted by social scientists.

Additional Features. A number of other features make the text more meaningful for students and more useful for instructors:

- *Readability and Clarity*. The writing style is informal and accessible to students without ignoring the traditional vocabulary of statistics. Problems and examples have been written to maximize student interest and to focus on issues of concern and significance. For the more difficult material (such as hypothesis testing), students are first walked through an example problem before being confronted by formal terminology and concepts. Each chapter ends with a summary of major points and formulas and a glossary of important concepts. A list of frequently used formulas inside the front cover and a glossary of symbols inside the back cover can be used for quick reference.

- *Organization and coverage.* The text is divided into four parts, with most of the coverage devoted to univariate descriptive statistics, inferential statistics, and bivariate measures of association. The distinction between description and inference is introduced in the first chapter and maintained throughout the text. In selecting statistics for inclusion, I have tried to strike a balance between the essential concepts with which students must be familiar and the amount of material students can reasonably be expected to learn in their first (and perhaps only) statistics course, while bearing in mind that different instructors will naturally wish to stress different aspects of the subject. Thus, the text covers a full gamut of the usual statistics, with each chapter broken into subsections so that instructors may choose the particular statistics they wish to include. In this edition, the text has been shortened and streamlined by moving some infrequently used techniques and statistical procedures to the companion Web site.

- *Learning objectives.* Learning objectives are stated at the beginning of each chapter. These are intended to serve as "study guides" and to help students identify and focus on the most important material.

- *Review of mathematical skills.* A comprehensive review of all of the mathematical skills that will be used in this text (formerly Appendix H) is available at the Web site for this text. Students who are inexperienced or out of practice with mathematics may want to study this review early in the course and/or refer to it as needed. A self-test is included so students may check their level of preparation for the course.

- *Statistical techniques and end-of-chapter problems are explicitly linked.* After a technique is introduced, students are directed to specific problems for practice and review. The "how-to-do-it" aspects of calculation are reinforced immediately and clearly.

- *End-of-chapter problems are organized progressively.* Simpler problems with small data sets are presented first. Often, explicit instructions or hints accompany the first several problems in a set. The problems gradually become more challenging and require more decision making by the student (e.g., choosing the most appropriate statistic for a certain situation). Thus, each problem set develops problem-solving abilities gradually and progressively. Some of the more challenging (e.g., using the largest) data sets have been moved to the Web site.

- *Computer applications.* To help students take advantage of the power of the computer, this text integrates two different statistical packages in parallel but separate sections at the end of chapters. Instructors may choose between SPSS and MicroCase. Appendix F provides an introduction to statistical packages in general and to SPSS and MicroCase in particular. The demonstrations at the ends of chapters explain how to use the statistical package to produce the statistics presented in the chapter. Student exercises analyzing data with SPSS or MicroCase are also included. Additional exercises with other data sets are also available at the Web site for this text. The student version of SPSS is available as a supplement to this text, and the student version of MicroCase is available online at the Web site.

- *Realistic, up-to-date data.* The database for computer applications in the text is a shortened version of the 1998 General Social Survey. This database will give students the opportunity to practice their statistical skills on "real-life" data. The database is described in Appendix G and is available in either SPSS or MicroCase format at the Web site for this text. Also available at the Web site are other data sets and additional exercises and projects.

- *Companion Web Site.* The Web site for this text includes additional material and some less frequently used techniques, a basic math review, additional end-of-chapter problems, data sets, a table of random numbers (formerly Appendix E), hypothesis tests for ordinal-level variables (formerly Chapter 11), "find-the-test" flowcharts, internet links, and a number of other features.

- *Instructor's Manual/Testbank.* The Instructor's Manual includes chapter summaries, a test item file of multiple-choice questions, answers to even-numbered computational problems, and step-by-step solutions to selected problems. In addition, the Instructor's Manual includes cumulative exercises (with answers) that can be used for testing purposes.

- *Study Guide.* The Study Guide, written by Professor Rebecca Davis, contains additional examples to illuminate basic principles, review problems with detailed answers, SPSS projects, and multiple-choice questions and answers that complement but do not duplicate the test item file.

Summary of Key Changes in this Edition.   The most important changes in this edition include:

- More emphasis on interpretation, less emphasis on hand computation
- A new feature called "Interpreting Statistics"
- A companion Web site offering a variety of features
- MicroCase Explorit has been added in addition to SPSS
- Learning objectives are stated at the beginning of each chapter

- The data set used in the text has been updated to the 1998 General Social Survey
- Less frequently used tests and techniques have been move to the Web site
- Many of the Reading Statistics inserts have been updated.

The text has been thoroughly reviewed for clarity and readability. As with previous editions, my goal is to provide a comprehensive, flexible, and student-oriented text that will provide a challenging first exposure to social statistics.

ACKNOWLEDGMENTS

This text has been in development, in one form or another, for almost twenty years. An enormous number of people have made contributions, both great and small, to this project and, at the risk of inadvertently omitting someone, I am bound to at least attempt to acknowledge my many debts.

This edition reflects the thoughtful guidance of both Lin Marshall and Eve Howard of Wadsworth, and I thank them for their contributions. Much of whatever integrity and quality this book has is a direct result of the very thorough (and often highly critical) reviews that have been conducted over the years. I am consistently impressed by the sensitivity of my colleagues to the needs of the students, and, for their assistance in preparing the sixth edition, I would like to thank Patricia R. Hoffman, University of Nebraska, Lincoln; Wen H. Kuo, University of Utah; Shawn McEntee, Salisbury State College; Mary Ann Powell, University of Nebraska, Omaha; and Suzan Waller, University of Central Oklahoma. Whatever failings are contained in the text are, of course, my responsibility and are probably the results of my occasional decisions not to follow the advice of my colleagues.

I would like to thank the instructors who made statistics understandable to me (Professors Satoshi Ito, Noelie Herzog, and Ed Erikson) and all of my colleagues at Christopher Newport University for their support and encouragement (especially Professors F. Samuel Bauer, Robert Durel, James Forte, Ruth Kernodle, Timothy Marshall, Cheryl Mathews, Lea Pellet, Virginia Purtle, and William Winter). I owe a special debt of gratitude to Professor Roy Barnes of the University of Michigan–Flint. He designed all of the new "Interpreting Statistics" features in this edition and wrote much of the narrative. I would be very remiss if I did not acknowledge the constant support and excellent assistance of Jessica Ledbetter, and I thank all of my students for their patience and thoughtful feedback. Also, I am grateful to the Literary Executor of the late Sir Ronald A. Fisher, F.R.S., to Dr. Frank Yates, F.R.S., and to Longman Group Ltd., London, for permission to reprint Appendixes B, C, and D, from their book *Statistical Tables for Biological, Agricultural and Medical Research* (6th edition, 1974).

Finally, I want to acknowledge the support of my family and rededicate this work to them. I have the extreme good fortune to be a member of an extended family that is remarkable in many ways and that continues to increase in size. Although I cannot list everyone, I would like to especially thank the older generation (my mother, Alice T. Healey), the next generation (my sons Kevin and Christopher, my daughter-in-law Jennifer), the new members (my wife Patricia Healey and Christopher, Katherine, and Jennifer Schroen), and the youngest generation (Benjamin and Caroline Healey).

1 INTRODUCTION

LEARNING OBJECTIVES: By the end of this chapter, you will be able to

1. Describe the limited but crucial role of statistics in social research.
2. Distinguish between three applications of statistics (univariate descriptive, bivariate descriptive, and inferential) and identify situations in which each is appropriate.
3. Distinguish between discrete and continuous variables and cite examples of each.
4. Identify and describe three levels of measurement and cite examples of variables from each.

1.1 WHY STUDY STATISTICS?

Students sometimes approach their first course in statistics with questions about the value of the subject matter. What, after all, do numbers and statistics have to do with understanding people and society? In a sense, this entire book will attempt to answer this question, and the value of statistics will become clear as we move from chapter to chapter. For now, the importance of statistics can be demonstrated, in a preliminary way, by briefly reviewing the research process as it operates in the social sciences. These disciplines are scientific in the sense that social scientists attempt to verify their ideas and theories through research. Broadly conceived, **research** is any process by which information is systematically and carefully gathered for the purpose of answering questions, examining ideas, or testing theories. Research is a disciplined inquiry that can take numerous forms. Statistical analysis is relevant only for those research projects in which the information collected is represented by numbers. Numerical information of this sort is called **data,** and the sole purpose of statistics is to manipulate and analyze data. **Statistics,** then, are a set of mathematical techniques used by social scientists to organize and manipulate data for the purpose of answering questions and testing theories.

What is so important about learning how to manipulate data? On one hand, some of the most important and enlightening works in the social sciences do not utilize any statistical techniques. There is nothing magical about data and statistics. The mere presence of numbers guarantees nothing about the quality of a scientific inquiry. On the other hand, data can be the most trustworthy information available to the researcher and, consequently, deserve special attention. Data that have been carefully collected and thoughtfully analyzed are the strongest, most objective foundations for building theory and enhancing understanding. Without a firm base in data, the social sciences would lose the right to the name *science* and would be of far less value.

Thus, the social sciences rely heavily on data analysis for the advancement of knowledge. Let us be very clear about one point: it is never enough merely to gather data (or, for that matter, any kind of information). Even the most objective and carefully collected numerical information does not and cannot speak for itself. The researcher must be able to use statistics effectively to organize, evaluate, and analyze the data. Without a good understanding of the principles of statistical analysis, the researcher will be unable to make sense of the data. Without the appropriate application of statistical techniques, the data will remain mute and useless.

Statistics are an indispensable tool for the social sciences. They provide the scientist with some of the most useful techniques for evaluating ideas, testing theory, and discovering the truth. The next section describes the relationships between theory, research, and statistics in more detail.

1.2 THE ROLE OF STATISTICS IN SCIENTIFIC INQUIRY

Figure 1.1 graphically represents the role of statistics in the research process. The diagram is based on the thinking of Walter Wallace and illustrates how the knowledge base of any scientific enterprise grows and develops. One point the diagram makes is that scientific theory and research continually shape each other. Statistics are one of the most important means by which research and theory interact. Let's take a closer look at the wheel.

FIGURE 1.1   THE WHEEL OF SCIENCE

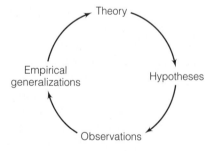

Source: Adapted from Walter Wallace, *The Logic of Science in Sociology* (Chicago: Aldine-Atherton, 1971).

The figure is circular and has no beginning or end, so we could begin our discussion at any point. For the sake of convenience, let's begin at the top and follow the arrows around the circle. A **theory** is an explanation of the relationships between phenomena. People naturally (and endlessly) wonder about problems in society (such as prejudice, poverty, child abuse, or serial murders) and, in their attempt to understand these phenomena, they develop explanations ("lack of education causes prejudice"). This kind of informal "theorizing" about society is no doubt very familiar to you. A major difference between our informal, everyday explanations of social phenomena and scientific theory is that the latter is subject to a rigorous testing process. Let's take the problem of racial prejudice as an example to illustrate how the research process works.

What causes racial prejudice? One possible explanation for this phenomenon is provided by a theory called the *contact hypothesis*. This theory was stated over 40 years ago by the social psychologist Gordon Allport, and it has been tested on a number of occasions since that time.[1] The theory links prejudice to the volume and nature of interaction between members of different racial groups. Specifically, the hypothesis asserts that contact situations in which different groups have equal status and are engaged in cooperative behavior will result in reductions of prejudice on all sides. The greater the extent to which contact is equal and cooperative, the more likely people will see each other as individuals and not as representatives of a particular group. For example, members of a racially mixed athletic team that cooperate with each other to achieve victory would tend to experience a decline in prejudice. On the other hand, when different groups compete for jobs, housing, or other valuable resources, prejudice would tend to increase.

The contact hypothesis is not a complete explanation of prejudice, but it will serve to illustrate the role of theory. We are trying to explain the relationship between two social phenomena: (1) prejudice and (2) equal-status, cooperative contact between members of different groups. People who have little contact will be more prejudiced, and those who experience more contact will be less prejudiced. Thus, we begin with a coherent theory that explains the relationship between two phenomena.

Before moving to the next area of Figure 1.1, let's take a moment to further examine theory. The contact hypothesis, like most theories, is stated in terms of causal relationships between variables. A **variable** is any trait that can change values from case to case. Examples of variables would be gender, age, income, or political party affiliation. In any specific theory, some

---

[1] Allport, Gordon, 1954. *The Nature of Prejudice*. Reading, Massachusetts: Addison-Wesley. For recent attempts to test this theory, see: Sigelman, Lee and Susan Welch, 1993. "The Contact Hypothesis Revisited: Black-White Interaction and Positive Racial Attitudes." *Social Forces*. 71:781–795 and Pettigrew, Thomas, 1997. "Generalized Intergroup Contact Effects on Prejudice." *Personality and Social Psychology Bulletin*. 23:173–185.

variables will be identified as causes, and others will be identified as effects or results. In the language of science, the causes are called **independent variables** and the effects or result variables are called **dependent variables.** In our theory, contact would be the independent variable (or the cause) and prejudice would be the dependent variable (the result or effect). In other words, we are arguing that equal-status contact is a cause of prejudice or that an individual's level of prejudice depends on the extent to which he or she participates in equal-status, cooperative contacts with other groups.

So far, we have a theory of prejudice, and an independent and a dependent variable. What we don't know yet is whether the theory is true or false. To find out, we need to compare our theory with the facts: we need to do some research. The next steps in the process would be to define our terms and ideas more specifically and exactly. One problem we often face in doing research is that scientific theories are too complex and abstract to be fully tested in a single research project. To conduct research, one or more hypotheses must be derived from the theory. A **hypothesis** is a statement about the relationship between variables that, while logically derived from the theory, is much more specific and exact.

For example, if we wished to test the contact hypothesis, we would have to say exactly what we mean by prejudice and we would need to describe "equal-status, cooperative contact" in great detail. There has been a great deal of research on the effect of contact on prejudice, and we would consult the research literature to develop and clarify our definitions of these concepts. As our definitions develop and the hypotheses take shape, we enter the next step of the research process. Now we will design the data-gathering phase of the project. We must decide how cases will be tested, how these cases will be selected, how exactly the variables will be measured, and a host of related matters. Ultimately, these plans will lead to the observation phase (the bottom of the wheel of science), where we actually measure social reality. Before we can do this, we must have a very clear idea of what we are looking for and a well-defined strategy for conducting the search.

To test the contact hypothesis, we would begin with people from different racial or ethnic groups. We might place some subjects in situations that required them to cooperate with members of other groups and other subjects in situations that feature intergroup competition. We would need to measure levels of prejudice before and after each type of contact. We might do this by administering a survey that asked subjects to agree or disagree with statements such as "Greater efforts must be made to racially integrate the public school system" or "Skin color is irrelevant and people are just people." Our goal would be to see if the people exposed to the cooperative contact situation actually experience a reduction in prejudice.

Now, finally, we come to statistics. As the observation phase of our research project comes to an end, we will be confronted with a large collection of numerical information or data. If our sample consisted of 100 people, we would have 200 completed surveys measuring prejudice: 100 completed before the contact situation and 100 filled out afterwards. Try to imagine dealing with 200 completed surveys. If we had asked each of the respondents just five questions to measure their prejudice, we would have a total of 1000 separate pieces of information to deal with. What do we do? We have to have some systematic way to organize and analyze this information and, at this point, statistics will become very valuable. Statistics will supply us with many ideas about "what to do" with the data; we will begin to look at some of the options in the next chapter. For now, let me stress two points about statistics.

First, statistics are crucial. Simply put, without statistics, quantitative research is impossible. Without quantitative research, the development of the social sciences would be severely impaired. Only by the application of statistical techniques can mere data help us shape and refine our theories and understand the social world better. Second, and somewhat paradoxically, the role of statistics is rather limited. As Figure 1.1 makes clear, scientific research proceeds through several mutually interdependent stages, and statistics become directly relevant only at the end of the observation stage. Before any statistical analysis can be legitimately applied, the preceding phases of the process must have been successfully completed. If the researcher has asked poorly conceived questions or has made serious errors of design or method, then even the most sophisticated statistical analysis is valueless. As useful as they can be, statistics cannot substitute for rigorous conceptualization, detailed and careful planning, or creative use of theory. Statistics cannot salvage a poorly conceived or designed research project. They cannot make sense out of garbage.

On the other hand, inappropriate statistical applications can limit the usefulness of an otherwise carefully done project. Only by successfully completing *all* phases of the process can a quantitative research project hope to contribute to understanding. A reasonable knowledge of the uses and limitations of statistics is as essential to the education of the social scientist as is training in theory and methodology.

As the statistical analysis comes to an end, we would begin to develop empirical generalizations. While we would be primarily focused on assessing our theory, we would also look for other trends in the data. Assuming that we found that equal-status, cooperative contact reduces prejudice in general, we might go on to ask if the pattern applies to males as well as females, to the well educated as well as the poorly educated, to older respondents as well as to the younger. As we probed the data, we might begin to develop some generalizations based on the empirical patterns we observe. For example, what if we found that contact reduced prejudice for younger respondents but not for older respondents? Could it be that younger people are less "set in their ways" and have attitudes and feelings that are more open to change? As we developed tentative explanations, we would begin to revise or elaborate our theory.

If we change the theory to take account of these findings, however, a new research project designed to test the revised theory is called for, and the wheel of science would begin to turn again. We (or perhaps some other researchers) would go through the entire process once again with this new—and, hopefully, improved—theory. This second project might result in further revisions and elaboration that would (you guessed it) require still more research projects, and the wheel of science would continue turning as long as scientists were able to suggest additional revisions or develop new insights. Every time the wheel turned, our understandings of the phenomena under consideration would (hopefully) improve.

This characterization of the research process does not include white-coated, clipboard-carrying scientists who, in a blinding flash of inspiration, discover some fundamental truth about reality and shout, "Eureka!" The truth is that, in the normal course of science, it is a rare occasion when we can say with absolute certainty that a given theory or idea is definitely true or false. Rather, evidence for (or against) a theory will gradually accumulate over time, and ultimate judgments of truth will likely be the result of many years of hard work, research, and debate.

Let's briefly review our imaginary research project. We began with an idea or theory about intergroup contact and racial prejudice. We imagined some of the steps we would have to take to test the theory and took a quick look at the various stages of the research project. We wound up back at the level of theory, ready to begin a new project guided by a revised theory. We saw how theory can motivate a research project and how our observations might cause us to revise the theory and, thus, motivate a new research project. Wallace's wheel of science illustrates how theory stimulates research and how research shapes theory. This constant interaction between theory and research is the lifeblood of science and the key to enhancing our understandings of the social world.

The dialog between theory and research occurs at many levels and in multiple forms. Statistics are one of the most important links between these two realms. Statistics permit us to analyze data, to identify and probe trends and relationships, to develop generalizations, and to revise and improve our theories. As you will see throughout this text, statistics are limited in many ways. They are also an indispensable part of the research enterprise. Without statistics, the interaction between theory and research would become extremely difficult, and the progress of our disciplines would be severely retarded. *(For practice in describing the relationship between theory and research and the role of statistics in research, see problems 1.1 and 1.2.)*

1.3 THE GOALS OF THIS TEXT

In the preceding section, I argued that statistics are a crucial part of the process by which scientific investigations are carried out and that, therefore, some training in statistical analysis is a crucial component in the education of every social scientist. In this section, we will address the questions of how much training is necessary and what the purposes of that training are. First, this textbook takes the point of view that statistics are tools. They can be a very useful means of increasing our knowledge of the social world, but they are not ends in themselves. Thus, we will not take a "mathematical" approach to the subject. The techniques will be presented as a set of tools that can be used to answer important questions. This emphasis does not mean that we will dispense with arithmetic entirely, of course. This text includes enough mathematical material so that you can develop a basic understanding of why statistics "do what they do." Our focus, however, will be on how these techniques are applied in the social sciences.

Second, all of you will soon become involved in advanced coursework in your major fields of study, and you will find that much of the literature used in these courses assumes at least basic statistical literacy. Furthermore, many of you, after graduation, will find yourselves in positions—either in a career or in graduate school—where some understanding of statistics will be very helpful or perhaps even required. Very few of you will become statisticians per se (and this text is not intended for the preprofessional statistician), but you must have a grasp of statistics in order to read and critically appreciate your own professional literature. As a student in the social sciences and in many careers related to the social sciences, you simply cannot realize your full potential without a background in statistics.

Within these constraints, this textbook is an introduction to statistics as they are utilized in the social sciences. The general goal of the text is to develop an appreciation—a "healthy respect"—for statistics and their place in the research process. You should emerge from this experience with the ability to use statistics intelligently and to know when other people have done so. You should be familiar with the advantages and limitations of the more commonly used statistical techniques, and you should know which techniques are appropriate for a given set of data and a given purpose. Lastly, you should develop sufficient statistical and computational skills and enough experience in the interpretation of statistics to be able to carry out some elementary forms of data analysis by yourself.

1.4 DESCRIPTIVE AND INFERENTIAL STATISTICS

As noted earlier, the general function of statistics is to manipulate data so that research question(s) can be answered. There are two general classes of statistical techniques that, depending on the research situation, are available to accomplish this task, and each is introduced in this section.

Descriptive Statistics. The first class of techniques is called **descriptive statistics** and is relevant (1) when the researcher needs to summarize or describe the distribution of a single variable and (2) when the researcher wishes to understand the relationship between two or more variables. If we are concerned with describing a single variable, then our goal will be to arrange the values or scores of that variable so that the relevant information can be quickly understood and appreciated. Many of the statistics that might be appropriate for this summarizing task are probably familiar to you. For example, percentages, graphs, and charts can all be used as single-variable descriptive statistics.

To illustrate briefly the usefulness of these kinds of statistics, consider the following problem: Suppose you wanted to summarize the distribution of the variable "family income" for a community of 10,000 families. How would you do it? Obviously, you couldn't simply list all incomes in the community and let it go at that. Presumably, you would want to develop some summary measures of the overall income distributions—perhaps an arithmetic average or the proportions of incomes that fall in various ranges (such as low, middle, and high). Or perhaps a graph or a chart would be more useful. Whatever specific method you choose, its function is the same: to reduce these thousands of individual items of information into a few easily understood numbers. The process of allowing a few numbers to summarize many numbers is called **data reduction** and is the basic goal of single-variable (or univariate) descriptive statistical procedures. The first part of this text is devoted to these statistics.

The second type of descriptive statistics is designed to help the investigator understand the relationship between two or more variables. These statistics, called **measures of association,** allow the researcher to quantify the strength and direction of a relationship. These statistics are very useful because they enable us to investigate two matters of central theoretical and practical importance to any science: causation and prediction. These techniques help us disentangle and uncover the connections between variables. They help us trace the ways in which some variables might have causal influences on others; and, depending on the strength of the relationship, they enable us to predict scores on one variable from the scores on another. Note that measures of association cannot, by themselves, prove that two variables are causally related. However, these techniques can provide valuable clues about causation and are therefore extremely important for theory testing and theory construction.

For example, suppose you were interested in the relationship between "time spent studying statistics" and "final grade in statistics" and had gathered the appropriate data from a group of college students. By calculating the appropriate measure of association, you could determine the strength of the relationship and its direction. Suppose you found a strong, positive relationship between these variables. You would infer that "study time" and "grade" were closely related (strength of the relationship) and that as one increased in value, the other also increased (direction of the relationship). You could make predictions from one variable to the other (for example, "the longer the study time, the higher the grade").

Now, as a result of finding this strong, positive relationship, you might be tempted to make causal inferences. That is, you might jump to such conclusions as "longer study time leads to (causes) higher grades." Such a conclusion might make a good deal of common sense and would certainly be supported by your statistical analysis. However, the causal nature of the relationship is in no way proven by the statistical analysis. Measures of association can be taken as important clues about causation, but the mere existence of a relationship should never be taken as conclusive proof of causation.

In fact, other variables might have an effect on the relationship. In the example above, we probably would not find a perfect relationship between "study time" and "final grade." That is, we will probably find some individuals who spend a great deal of time studying but receive low grades and some individuals who fit the opposite pattern. We know intuitively that other variables besides study time affect grades (such as efficiency of study techniques, amount of background in mathematics, and even random chance). Fortunately, researchers can incorporate these other variables into the analysis and measure their effects. Part III of this text is devoted to bivariate (two-variable) and Part IV to multivariate (more than two variables) descriptive statistics.

Inferential Statistics.   This second class of statistical techniques becomes relevant when we wish to generalize our findings from a **sample** to a **population.** A population is the total collection of all cases that the researcher wishes to understand better. Examples of possible populations would be adult male voters in the United States, all parliamentary democracies, unemployed Puerto Ricans in Atlanta, or sophomore college football players in the Midwest.

Populations can theoretically range from inconceivable in size ("all humanity") to quite small (all 35-year-old red-haired belly dancers currently residing in downtown Cleveland) but are usually fairly large. In fact, they are almost always too large to be measured. To put the problem another way, social scientists almost never have the resources or time to test every case in a population. Hence the need for **inferential statistics,** which involve using information from samples (carefully chosen subsets of the defined populations) to make inferences about populations. Samples are, of course, much cheaper to assemble, and—if the proper techniques are followed—generalizations based on these samples can be very accurate representations of the population.

Many of the concepts and procedures involved in inferential statistics may be unfamiliar. However, most of us are experienced consumers of inferential statistics—most familiarly, perhaps, in the form of public-opinion polls and election projections. When a public-opinion poll reports that 42% of the American electorate plans to vote for a certain presidential candidate, it is essentially reporting a generalization to a population ("the American electorate"—which numbers about 100 million people) from a carefully drawn sample (usually about 1500 respondents). Matters of inferential statistics will occupy our attention in Part II of this book. *(For practice in describing different statistical applications, see problems 1.3 and 1.7.)*

1.5 DISCRETE AND CONTINUOUS VARIABLES

In the next chapter, you will begin to encounter some of the broad array of statistics available to the social scientist. One aspect of using statistics that can be puzzling is deciding when to use which statistic. You will learn specific guidelines as you go along, but let me introduce some basic and general guidelines at this point. The first of these concerns discrete and continuous variables; the second, covered in the next section, concerns level of measurement.

A variable is said to be **discrete** if it has a basic unit of measurement that cannot be subdivided. The measurement process for discrete variables involves accurate counting of the number of units per case. For example, number of people per household is a discrete variable. The basic unit is people, and the fewest you can have is one. Note that the score of a given household on this variable will always be a whole number (you'll never find 2.7 people living in a specific household), and as long as we are counting accurately, the scores we report will always be exact.

A variable is **continuous** if the measurement of it can be subdivided infinitely—at least in a theoretical sense. A good example of such a variable would be time, which can be measured in nanoseconds (billionths of a second) or even smaller units. In a sense, when we measure a continuous variable, we are always approximating and rounding off the scores. We could report somebody's time in the 100-yard dash as 10.7 seconds or 10.732451 seconds, but, since time can be infinitely subdivided (if we have the technology to make the precise measurements), we will never be able to report the exact time elapsed. Since we cannot cite or work with infinitely long numbers, we must report the scores on continuous variables as if they were discrete. The distinction between the two types of variables relates more to measuring and processing the information than to the appearance of the data. This distinction between discrete and continuous variables is one of the most basic in statistics and will constitute one of the criteria by which we will choose among various statistics and graphic devices. *(For practice in distinguishing between discrete and continuous variables, see problems 1.4 through 1.8.)*

1.6 LEVEL OF MEASUREMENT

A second basic and general guideline for the selection of statistics is the **level of measurement.** Every statistical technique involves performing some mathematical operation, such as adding scores or ranking cases. Before you can properly use a technique, you must measure the variable being processed in a way that justifies the required mathematical operations. For example, many statistical techniques require that scores be added together. These techniques could be legitimately used only when the variable is measured in a way that permits addition. Thus, the researcher's choice of statistical techniques is heavily dependent on the way in which the variables have been measured.

The three levels of measurement are, in order of increasing sophistication, nominal, ordinal, and interval-ratio. Determining the level at which a variable has been measured is one of the first steps in any statistical analysis, and we will consider this matter at some length. I will make it a practice throughout this text to introduce level-of-measurement considerations for each statistical technique.

The Nominal Level of Measurement.   The most basic measurement procedure is to classify cases into the categories of a variable. All measurement involves classification as a minimum but, at the nominal level, classification into categories is the *only* measurement procedure permitted. The categories of the variable are not numerical and can be compared to each other only in terms of the number of cases classified in them. The categories cannot be thought of as "higher" or "lower" than each other along some numerical scale.

To illustrate, consider gender, a variable with only two categories, the smallest number a variable can have. Gender can be measured by simply placing subjects in the proper category and counting the numbers of males and females. When we are finished counting, we will be able to make statements about which category is larger or smaller (e.g., "there are more females than males in the sample"). We cannot, however, perform any mathematical operations on the categories (male and female) themselves. The categories are different from each other but cannot be placed on a scale, ranked, or ordered in any mathematical sense. Another example of a nominal variable is religious affiliation. There may be more or fewer Protestants or Catholics in a given sample, but the categories themselves (the religious affiliations) do not form a mathematical scale.

The measurement procedures for nominal variables are rudimentary but there are criteria and procedures that we need to observe in order to ensure adequate measurement. First, the categories of nominal-level variables must be mutually exclusive of each other so that no ambiguity exists concerning classification of any given case. Second, the categories must be exhaustive: a category—at least an "other" or miscellaneous category—must exist for every possible score that might be found.

Third, the categories of nominal variables should be relatively homogeneous. That is, our categories should include cases that are truly comparable or, to put it another way, we need to avoid categories that lump apples with oranges. There are no hard and fast guidelines for judging if a set of cate-

TABLE 1.1   FOUR SCALES FOR MEASURING RELIGIOUS PREFERENCE

Scale A (not mutually exclusive)	Scale B (not exhaustive)	Scale C (not homogeneous)	Scale D (an adequate scale)
Protestant	Protestant	Protestant	Protestant
Episcopalian	Catholic	Non-Protestant	Catholic
Catholic	Jew		Jew
Jew			None
None			Other
Other			

gories is appropriately homogeneous. The researcher must make that decision in terms of the specific purpose of the research, and categories that are too broad for some purposes may be perfectly adequate for others.

Table 1.1 demonstrates some errors of measurement in four different schemes for measuring the variable "religious preference." Scale A in the table violates the criterion of mutual exclusivity because of overlap between the categories Protestant and Episcopalian. Scale B is not exhaustive because it does not provide a category for people with no religious preference (None) or people who belong to religions other than the three listed. Scale C uses a category (Non-Protestant) that would be too broad for many research purposes. Scale D represents the way religious preference is often measured in North America, but note that these categories may be too general for some research projects and not comprehensive enough for others. For example, an investigation of issues that have strong moral and religious content (assisted suicide, abortion, or capital punishment) might need to make distinctions between the various Protestant denominations, and an effort to assess the diversity of American religions would need to add categories for Buddhists, Muslims, and so on.

As a final note, numerical labels are sometimes used to identify the categories of a variable measured at the nominal level. This practice is especially common when the data are being prepared for computer analysis. For example, the various religions might be labeled with a 1 indicating Protestant, a 2 signifying Catholic, and so on. You should understand that these numbers are merely labels or names and have no numerical quality to them. They cannot be added, subtracted, multiplied, or divided. The only mathematical operation permissible with nominal variables is counting the number of occurrences that have been classified into the various categories of the variable.

The Ordinal Level of Measurement.   In addition to classifying cases into categories, variables measured at the ordinal level allow the categories themselves to be ranked with respect to how much of the trait being measured they possess. The categories form a kind of numerical scale that can be ordered from "high" to "low." Thus, variables measured at the ordinal level are more sophisticated than nominal-level variables because, in addition to counting the number of cases in a category, we can rank the cases with respect to each other. Not only can we say that one case is different from another; we can also say that one case is higher or lower, more or less than another.

For example, the variable socioeconomic status (SES) is usually measured at the ordinal level in the social sciences. The categories of the variable are often ordered according to the following scheme:

4. Upper class
3. Middle class
2. Working class
1. Lower class

Individual cases can be compared in terms of the categories into which they are classified. Thus, an individual classified as a 4 (upper class) would be ranked higher than an individual classified as a 2 (working class). Other examples of variables measured at the ordinal level include attitude and opinion scales, such as those that measure prejudice, alienation, or political conservatism.

The major limitation of the ordinal level of measurement is that a particular score represents only position with respect to some other score. We can distinguish between high and low scores, but the distance between the scores cannot be described in precise terms. Although we know that a score of 4 is more than a score of 2, we do not know if it is twice as much as 2.

Since we don't know what the exact distances are from score to score on an ordinal scale, our options for statistical analysis are limited. For example, addition (and most other mathematical operations) assumes that the intervals between scores are exactly equal. If the distances from score to score are not equal, 2 + 2 might equal 3 or 5 or even 15. Thus, strictly speaking, statistics such as the average or mean (which requires that the scores be added together and then divided by the number of scores) are not permitted with ordinal-level variables. The most sophisticated mathematical operation fully justified with an ordinal variable is the ranking of categories and cases (although, as we will see, it is not unusual for social scientists to take some liberties with this strict criterion).

The Interval-Ratio Level of Measurement.[2] The categories of nominal-level variables have no numerical quality to them. Ordinal-level variables have categories that can be arrayed along a scale from high to low, but the

---

[2]Many statisticians distinguish between the interval level (equal intervals) and the ratio level (true zero point). I find the distinction unnecessarily cumbersome in an introductory text and will treat these two levels as one.

exact distances between categories or scores are undefined. Variables measured at the interval-ratio level not only permit classification and ranking but also allow the distance from category to category (or score to score) to be exactly defined.

Interval-ratio variables have two defining characteristics. First, they are measured in units that have equal intervals. For example, recording the ages of your respondents is a measurement procedure that would produce interval-ratio data because the unit of measurement (years) has equal intervals (the distance from year to year is 365 days). Similarly, if we ask people how many siblings they have, we would produce a variable with equal intervals: two siblings are one more than one and thirteen is one more than twelve.

The second characteristic of interval-ratio variables is that they have a true zero point. That is, the score of zero for these variables is not an arbitrary point: it indicates the absence or complete lack of whatever is being measured. For example, the variable "number of siblings" has a true zero point because it is possible to have no siblings at all. Similarly, it is possible to have zero years of education, to have no income at all, to score a zero on a multiple-choice test, and to be zero years old (although not for very long). Other examples of interval-ratio variables would be number of children, life expectancy, and years married. All mathematical operations are permitted for data measured at this level.

Table 1.2 summarizes this discussion by presenting the basic characteristics of the three levels of measurement. Most importantly, since different statistics require different mathematical operations, you need to remember that level of measurement is the first guideline to use in selecting a statistic. For example, computation of a mean or average requires addition and divi-

TABLE 1.2   BASIC CHARACTERISTICS OF THE THREE LEVELS OF MEASUREMENT

Levels	Examples	Measurement Procedures	Mathematical Operations Permitted
Nominal	Sex, race, religion, marital status	Classification into categories	Counting number of cases in each category of the variable; comparing sizes of categories
Ordinal	Social class (SES), attitude and opinion scales	Classification into categories plus ranking of categories with respect to each other	All above plus judgments of "greater than" and "less than"
Interval-Ratio	Age, number of children, income	All above plus description of distances between scores in terms of equal units	All above plus all other mathematical operations (addition, subtraction, multiplication, division, square roots, etc.)

READING STATISTICS 1: Introduction

By this point in your education you have developed an impressive array of skills for reading words. Although you may sometimes struggle with a difficult idea or stumble over an obscure meaning, you can comprehend virtually any written work that you are likely to encounter.

As you continue your education in the social sciences, you must develop an analogous set of skills for reading numbers and statistics. To help you reach a reasonable level of literacy in statistics, I have included a series of boxed inserts in this text labeled "Reading Statistics." These will appear in most chapters and will discuss how statistical results are typically presented in the professional literature. Each installment will include an extract or quotation from the professional literature so that we can analyze a realistic example.

As you will see, professional researchers use a reporting style that is quite different from the statistical language you will find in this text. Space in research journals and other media is expensive, and the typical research project requires the analysis of many variables. Thus, a large volume of information must be summarized in very few words. Researchers may express in a word or two

a result or an interpretation that will take us a paragraph or more to state.

Because this is an introductory textbook, I have been careful to break down the computation and logic of each statistic and to identify, even to the point of redundancy, what we are doing when we use statistics. In this text, we will never be concerned with more than a few variables at a time. We will have the luxury of analysis in detail and of being able to take pages or even entire chapters to develop a statistical idea or analyze a variable. Thus, a major theme of these boxed inserts will be to summarize how our comparatively long-winded (but more careful) vocabulary is translated into the concise language of the professional researcher.

When you have difficulty reading words, your tendency is (or, at least, should be) to consult reference books (especially dictionaries) to help you identify and analyze the elements (words) of the passage. When you have difficulty reading statistics, you should do exactly the same thing. I hope you will find this text a valuable reference book, but if you learn enough from this text to be able to use any source to help you read statistics, this text will have fulfilled one of its major goals.

---

sion, mathematical operations that are fully justified only when a variable is measured at the interval-ratio level.

Ideally, the researcher would utilize only those statistics that were fully justified by the level-of-measurement criteria. In this imperfect world, however, the most powerful and useful statistics (such as the mean) require interval-ratio variables, while most of the variables of interest to the social sciences are only nominal (race, sex, marital status) or ordinal (attitude scales). Relatively few concepts of interest to the social sciences are so precisely defined that they can be measured at the interval-ratio level. This disparity creates some very real difficulties in the research process. On one hand, the researcher should use the most sophisticated statistical procedures fully justified by the level-of-measurement criteria. Treating interval-ratio data as if they were only ordinal, for example, results in a significant loss of information and precision. Treated as an interval-ratio variable, the variable "age" can supply us with exact information regarding the differences between the cases (for example, "Individual A is three years and two months older than Individual B"). Treated only as an ordinal variable, however, the precision of our comparisons would suffer, and we could say only that "Individual A is older (or greater than) Individual B."

On the other hand, given the nature of the disparity, researchers are more likely to treat variables as if they were higher in level of measurement than they actually are. In particular, variables measured at the ordinal level, especially when they have many possible categories or scores, might be treated as if they were interval-ratio because the statistical procedures available at the higher level are more powerful, flexible, and interesting. This practice is common, but researchers should be cautious in assessing statistical results and developing interpretations when the level-of-measurement criterion has been violated. At any rate, level of measurement is a very basic characteristic of a variable, and we will always consider it when presenting statistical procedures. Level of measurement is also a major organizing principle for the material that follows, and you should make sure that you are familiar with these guidelines. *(For practice in determining the level of measurement of a variable, see problems 1.4 through 1.8.)*

SUMMARY

1. Within the context of social research, the purpose of statistics is to organize, manipulate, and analyze data so that the researcher can more easily answer his or her original question. Along with theory and methodology, statistics are a basic tool by which social scientists attempt to enhance their understanding of the social world.

2. There are two general classes of statistics. Descriptive statistics are used to summarize the distribution of a single variable and the relationships between two or more variables. Inferential statistics provide us with techniques by which we can generalize to populations from random samples.

3. Two basic guidelines for selecting statistical techniques were presented. Variables may be either discrete or continuous and may be measured at any of three different levels. At the nominal level, we can classify cases into categories of the variable and compare category sizes. At the ordinal level, categories and cases can be ranked with respect to each other. At the interval-ratio level, all mathematical operations are permitted.

GLOSSARY

Continuous variable. A variable with a unit of measurement that can be subdivided infinitely.

Data. Any information collected as part of a research project and expressed as numbers.

Data reduction. Summarizing many scores with a few statistics. A major goal of descriptive statistics.

Dependent variable. A variable that is identified as an effect, result, or outcome variable. The dependent variable is thought to be caused by the independent variable.

Descriptive statistics. The branch of statistics concerned with (1) summarizing the distribution of a single variable or (2) measuring the relationship between two or more variables.

Discrete variable. A variable with a basic unit of measurement that cannot be subdivided.

Hypothesis. A statement about the relationship between variables that is derived from a theory. Hypotheses are more specific than theories, and all terms and concepts are fully defined.

Independent variable. A variable that is identified as a causal variable. The independent variable is thought to cause the dependent variable.

Inferential statistics. The branch of statistics concerned with making generalizations from samples to populations.

Level of measurement. The mathematical characteristics of a variable as determined by the measurement process. A major criterion for selecting statistical techniques. Variables can be measured at any of three levels, each permitting certain mathematical operations and statistical techniques. The characteristics of the three levels are summarized in Table 1.2.

Measures of association. Statistics that summarize the strength and direction of the relationship between variables.

Population. The total collection of all cases in which the researcher is interested.

Research. Any process of gathering information systematically and carefully to answer questions or test theories. Statistics are useful for research projects in which the information is represented in numerical form or as data.

Sample. A carefully chosen subset of a population. In inferential statistics, information is gathered from a sample and then generalized to a population.

Statistics. A set of mathematical techniques for organizing and analyzing data.

Theory. A generalized explanation of the relationship between two or more variables.

Variable. Any trait that can change values from case to case.

MULTIMEDIA RESOURCES

The Wadsworth Sociology Resource Center: Virtual Society
http://sociology.wadsworth.com/

Visit the companion web site for the sixth edition of *Statistics: A Tool for Social Research* to access a wide range of student resources. Begin by clicking on the Student Resources section of the book's web site to access the following study tools:

- Basic math review
- Flash cards

- Additional chapter problems
- Statistics review
- Internet links
- Table of random numbers
- MicroCase and SPSS examples and exercises
- "Find the test" flowcharts
- Hypothesis testing for variables measured at the ordinal level

PROBLEMS

1.1 In your own words, describe the role of statistics in the research process. Using the "wheel of science" as a framework, explain how statistics link theory with research.

1.2 Find a research article in any social science journal. Choose an article on a subject of interest to you and don't worry about being able to understand all of the statistics that are reported.
  a. How much of the article is devoted to statistics per se (as distinct from theory, ideas, discussion, and so on)?
  b. Is the research based on a sample from some population? How large is the sample? How were subjects or cases selected? Can the findings be generalized to some population?
  c. What variables are used? Which are independent and which are dependent? For each variable, determine the level of measurement and whether the variable is discrete or continuous.

d. What statistical techniques are used? Try to follow the statistical analysis and see how much you can understand. Save the article and read it again after you finish this course and see if you do any better.

1.3 Distinguish between descriptive and inferential statistics. Describe a research situation in which each would be useful.

1.4 Below are some items from a public-opinion survey. For each item, indicate the level of measurement and whether the variable will be discrete or continuous.

a. What is your occupation? _____
b. How many years of school have you completed? _____
c. If you were asked to use one of these four names for your social class, which would you say you belonged in?
_____ Upper _____ Middle
_____ Working _____ Lower
d. What is your age? _____
e. In what country were you born? _____
f. What is your grade-point average? _____
g. What is your major? _____
h. The only way to deal with the drug problem is to legalize all drugs.
_____ Strongly agree
_____ Agree
_____ Undecided
_____ Disagree
_____ Strongly disagree
i. What is your astrological sign? _____
j. How many brothers and sisters do you have?
_____

1.5 Below are brief descriptions of how researchers measured a variable. For each situation, determine the level of measurement of the variable and whether it is continuous or discrete.

a. Race. Respondents were asked to select a category from the following list:
_____ Black _____ White _____ Other
b. Honesty. Subjects were observed as they passed by a spot on campus where an apparently lost wallet was lying. The wallet con-

tained money and complete identification. Subjects were classified into one of the following categories:
_____ Returned the wallet with money;
_____ Returned the wallet but kept the money;
_____ Did not return the wallet.
c. Social class. Subjects were asked about their family situation when they were 16 years old. Was their family
_____ Very well off compared to other families?
_____ About average?
_____ Not so well off?
d. Education. Subjects were asked how many years of schooling they and each parent had completed.
e. Racial integration on campus. Students were observed during lunchtime at the cafeteria for a month. The number of students sitting with students of other races was counted for each meal period.
f. Number of children. Subjects were asked: "How many children have you ever had? Please include any that may have passed away."
g. Student seating patterns in classrooms. On the first day of class, instructors noted where each student sat. Seating patterns were remeasured every two weeks until the end of the semester. Each student was classified as
_____ same seat as last measurement;
_____ adjacent seat;
_____ different seat, not adjacent;
_____ absent.
h. Physicians per capita. The number of practicing physicians was counted in each of 50 cities, and the researchers used population data to compute the number of physicians per capita.
i. Physical attractiveness. A panel of 10 judges rated each of 50 photos of a mixed-race sample of males and females for physical attractiveness on a scale from 0 to 20 with 20 being the highest score.
j. Number of accidents. The number of traffic accidents for each of 20 busy intersections in

a city was recorded. Also, each accident was rated as

_____ minor damage, no injuries;

_____ moderate damage, personal injury requiring hospitalization;

_____ severe damage and injury.

1.6 For each of the first 20 items in the General Social Survey (see Appendix G), indicate the level of measurement and whether the variable is continuous or discrete.

1.7 For each research situation summarized below, identify the level of measurement of all variables and indicate whether they are discrete or continuous. Also, decide which statistical applications are used: descriptive statistics (single variable), descriptive statistics (two or more variables), or inferential statistics. Remember that it is quite common for a given situation to require more than one type of application.

   a. The administration of your university is proposing a change in parking policy. You select a random sample of students and ask each one if he or she favors or opposes the change.

   b. You ask everyone in your social research class to tell you the highest grade he or she ever received in a math course and the grade on a recent statistics test. You then compare the two sets of scores to see if there is any relationship.

   c. Your aunt is running for mayor and hires you (for a huge fee, incidentally) to question a sample of voters about their concerns in local politics. Specifically, she wants a profile of the voters that will tell her what percent belong to each political party; what percent are male or female; and what percent favor or oppose the widening of the main street in town.

   d. Several years ago, a state reinstituted the death penalty for first-degree homicide. Supporters of capital punishment argued that this change would reduce the homicide rate. To investigate this claim, a researcher has gathered information on number of homicides in the state for the two-year periods before and after the change.

   e. A local automobile dealer is concerned about customer satisfaction. He wants to mail a sur-

vey form to all customers for the past year and ask them if they are satisfied, very satisfied, or not satisfied with their purchases.

1.8 For each research situation below, identify the independent and dependent variables. Classify each in terms of level of measurement and whether the variable is discrete or continuous.

   a. A graduate student is studying sexual harassment on college campuses and asks 500 female students if they personally have experienced any such incidents. Each student is asked to estimate the frequency of these incidents as either "often, sometimes, rarely, or never." The researcher also gathers data on age and major to see if there is any connection between these variables and frequency of sexual harassment.

   b. A supervisor in the Solid Waste Management Division of a city government is attempting to assess two different methods of trash collection. One area of the city is served by trucks with two-man crews who do "backyard" pickups, and the rest of the city is served by "hi-tech" single-person trucks with curbside pickup. The assessment measures include the number of complaints received from the two different areas over a six-month period, the amount of time per day required to service each area, and the cost per ton of trash collected.

   c. The adult bookstore near campus has been raided and closed by the police. Your social research class has decided to poll the student body and get their reactions and opinions. The class decides to ask each student if he or she supports or opposes the closing of the store, how many times each one has visited the store, and if he or she agrees or disagrees that "pornography is a direct cause of sexual assaults on women." The class also collects information on the sex, age, religious and political philosophy, and major of each student to see if opinions are related to these characteristics.

   d. For a research project in a political science course, a student has collected information about the quality of life and the degree of political democracy in 50 nations. Specifically, she used infant mortality rates to measure quality of life, and the percentage of all adults who are permitted to vote in national elections as a measure of democratization. Her hypothesis is that quality of life is higher in more democratic nations.

e. A highway engineer wonders if a planned increase in speed limit on a heavily traveled local avenue will result in any change in number of accidents. He plans to collect information on traffic volume, number of accidents, and number of fatalities for the six-month periods before and after the change.

f. Students are planning a program to promote "safe sex" and awareness of a variety of other health concerns for college students. To measure the effectiveness of the program, they plan to give a survey measuring knowledge about these matters to a random sample of the student body before and after the program.

g. Several states have drastically cut their budgets for mental health care. Will this increase the number of homeless people in these states? A researcher contacts a number of agencies serving the homeless in each state and develops an estimate of the size of the population before and after the cuts.

h. Does tolerance for diversity vary by race, ethnicity, or gender? A sample of white, black, Asian, Hispanic, and Native Americans have been given a survey that measures their interest in and appreciation of cultures and groups other than their own.

SPSS

Introduction to Computerized Statistical Packages and the General Social Survey

The problems at the end of chapters in this text have been written so that they can be solved with just a simple hand calculator. I've purposely kept the number of cases involved unrealistically low so that the tedium of mere calculation would not interfere unduly with the learning process. To provide a more realistic experience in the analysis of social science data, we will analyze a shortened version of the 1998 General Social Survey (GSS). This database is available on the compact disk that is included with some versions of this text, or it can be downloaded from our web site (***http://sociology.wadsworth.com***). The GSS is a public-opinion poll that has been conducted on nationally representative samples of citizens of the United States almost every year since 1972. The full survey includes hundreds of questions covering a broad range of social and political issues. The version supplied with this text has a limited number of variables and cases but is still actual, "real-life" data, so you have the opportunity to practice your statistical skills in a more realistic context.

One of the problems with reality, of course, is that it is often cumbersome and confusing. It's hard enough to do your homework with simplified problems, and you should be a little leery, in terms of your own time and effort, of promises of relevance and realism. This brings us to the second purpose of this section: computers and statistical packages. A statistical package is a set of computer programs for the analysis of data. The advantage of these packages is that, since the programs are already written, you can capitalize on the power of the computer with minimal computer literacy and virtually no programming experience.

This text utilizes two different statistical packages: the Statistical Package for the Social Sciences (SPSS) and MicroCase. Your instructor will specify which of the two you are using in this course. In these sections at the ends of chapters, I will explain how to use these packages to manipulate and analyze the GSS data, and I will illustrate and interpret the results. Obviously, you will read only the sections relevant to the specific statistical package being used in your course. Be sure to read Appendix F before attempting any data analysis.

2

BASIC DESCRIPTIVE STATISTICS
PERCENTAGES, RATIOS AND RATES, TABLES, CHARTS, AND GRAPHS

LEARNING OBJECTIVES   By the end of this chapter, you will be able to

1. Explain the purpose of descriptive statistics in making data comprehensible.
2. Compute and interpret percentages, proportions, ratios, and rates.
3. Construct and analyze frequency distributions for variables at each of the three levels of measurement.
4. Construct and analyze bar and pie charts, histograms, and line graphs.

Research results do not speak for themselves. They must be arranged in ways that allow the researcher (and his or her readers) to comprehend their meaning quickly. The primary function of descriptive statistics is to present research results clearly and concisely. Researchers use a process called *data reduction* to organize data into presentable form. Data reduction involves using a few numbers, a table, or a graphic device to summarize or stand for a larger array of data.

Little about this process is mysterious or difficult, but, as a researcher or a consumer of research, you should be aware of one problem. Data reduction inevitably loses information (precision and detail); and, therefore, summarizing statistics might present a misleading picture of research results. The issue is not whether to use descriptive statistics—they are always necessary. The issue is which descriptive statistics to use. The various summarizing techniques present results in different formats, are based on different mathematical operations, and characteristically lose information in different ways. Thus, in choosing among the various summarizers, several decisions must be made: how to present the data, what kind of information to lose, and how much detail can safely be obscured.

In this chapter, we will consider several commonly used techniques for presenting research results: percentages and proportions, ratios and rates, tables, charts, and graphs.

2.1 PERCENTAGES AND PROPORTIONS

Consider the following statement: "Of the 269 cases handled by the court, 167 resulted in prison sentences of five years or more." While there is nothing wrong with this statement, the same fact could have been more clearly conveyed if it had been reported as a percentage: "about 62% of all cases resulted in prison sentences of five or more years."

Percentages and proportions supply a frame of reference for reporting research results in the sense that they standardize the raw data: percentages to the base 100 and proportions to the base 1.00. The mathematical definitions of **proportions** and **percentages** are

FORMULA 2.1

$$\text{Proportion } (p) = \frac{f}{N}$$

FORMULA 2.2

$$\text{Percentage } (\%) = \left(\frac{f}{N}\right) \times 100$$

where f = frequency, the number of cases in any category
N = the number of cases in all categories

To illustrate the computation of percentages, consider the data presented in Table 2.1. To find the percentage of cases in the first category (sentences of five years or more), note that there are 167 cases in the category ($f = 167$) and a total of 269 cases in all ($N = 269$). So,

$$\text{Percentage } (\%) = \left(\frac{f}{N}\right) \times 100 = \left(\frac{167}{269}\right) \times 100 = (0.6208) \times 100 = 62.08\%$$

Using the same procedures, we can also find the percentage of cases in the second category:

$$\text{Percentage } (\%) = \left(\frac{f}{N}\right) \times 100 = \left(\frac{72}{269}\right) \times 100 = (0.2677) \times 100 = 26.77\%$$

TABLE 2.1   DISPOSITION OF 269 CRIMINAL CASES (fictitious data)*

Sentence	Frequency (f)	Proportion (p)	Percentage (%)
Five years or more	167	0.6208	62.08
Less than five years	72	0.2677	26.77
Suspended	20	0.0744	7.44
Acquitted	10	0.0372	3.72
	$N = 269$	1.0001	100.01

*The slight discrepancies in the totals of the proportion and percentage columns are due to rounding error.

Application 2.1

Not long ago, in a large social service agency, the following conversation took place between the executive director of the agency and a supervisor of one of the divisions.

> *Executive director:* Well, I don't want to seem abrupt, but I've only got a few minutes. Tell me, as briefly as you can, about this staffing problem you claim to be having.
>
> *Supervisor:* Ma'am, we just don't have enough people to handle our workload. Of the 177 full-time employees of the agency, only 50 are in my division. Yet, 6231 of the 16,722 cases handled by the agency last year were handled by my division.
>
> *Executive director (smothering a yawn):* Very interesting. I'll certainly get back to you on this matter.

How could the supervisor have presented his case more effectively? Because he wants to compare two sets of numbers (his staff versus the total staff and the workload of his division versus the total workload of the agency), proportions or percentages would be a more forceful way of presenting results. What if the supervisor had said, "Only 28.25% of the staff is assigned to my division, but we handle 37.26% of the total workload of the agency"? Is this a clearer message?

The first percentage is found by

$$\% = \left(\frac{f}{N}\right) \times 100 = \frac{50}{177} \times 100$$

$$= (.2825) \times 100 = 28.25\%$$

and the second percentage is found by

$$\% = \left(\frac{f}{N}\right) \times 100 = \left(\frac{6231}{16,722}\right) \times 100$$

$$= (.3726) \times 100 = 37.26\%$$

Both results could have been expressed as proportions. For example, the proportion of cases in the third category is 0.0744.

$$\text{Proportion } (p) = \frac{f}{N} = \frac{20}{269} = 0.0744$$

Percentages and proportions are easier to read and comprehend than frequencies. This advantage is particularly obvious when attempting to compare groups of different sizes. For example, based on the information presented in Table 2.2, which college has the higher relative number of social science

TABLE 2.2   DECLARED MAJOR FIELDS OF STUDY ON TWO COLLEGE CAMPUSES
(fictitious data)

Major	College A	College B
Business	103	312
Natural sciences	82	279
Social sciences	137	188
Humanities	93	217
	$N = 415$	$N = 996$

TABLE 2.3   DECLARED MAJOR FIELDS OF STUDY ON TWO COLLEGE CAMPUSES (fictitious data)*

Major	College A	College B
Business	24.82	31.33
Natural sciences	19.76	28.01
Social sciences	33.01	18.88
Humanities	22.41	21.79
	100.00%	100.01%
	(415)	(996)

*The slight discrepancy in total percentages for College B is due to rounding error.

majors? Because the total enrollments are so different, such comparisons are difficult to conceptualize from the raw frequencies. To make comparisons easier, the difference in size can be effectively eliminated by standardizing both distributions to the common base of 100 or, in other words, by computing percentages for both distributions. The same data are presented in percentages in Table 2.3.

The percentages in Table 2.3 make it easier to identify both differences and similarities between the two colleges. College A has a much higher percentage of social science majors (even though the absolute number of social science majors is less than at College B) and about the same percentage of humanities majors. How would you describe the differences in the remaining two major fields? *(For practice in computing and interpreting percentages and proportions, see problems 2.1 and 2.2.)*

Some further guidelines on the use of percentages and proportions:

1. When working with a small number of cases (say, fewer than 20), it is usually preferable to report the actual frequencies rather than percentages or proportions. With a small number of cases, the percentages can change drastically with relatively minor changes in the data. For example, if you begin with a data set that includes 10 males and 10 females (that is, 50% of each gender) and then add another female, the percentage distributions will change noticeably to 52.38% female and 47.62% male. Of course, as the number of observations increases, each additional case will have a smaller impact. If we started with 500 males and females and then added one more female, the percentage of females would change by only a tenth of a percent (from 50% to 50.10%).

2. Always report the number of observations along with proportions and percentages. This permits the reader to judge the adequacy of the sample size and, conversely, helps to prevent the researcher from lying with statistics. Statements like "two out of three people questioned prefer courses in statistics to any other course" might impress you, but the claim would lose its gloss if you learned that only three people were tested. *You should* *be extremely suspicious of reports that fail to report the number of cases that were tested.*

3. In spite of the fact that they require division, proportions and percentages can be calculated for variables at *any* level of measurement. This is not a violation of the level-of-measurement guideline (see Table 1.2). Percentages and proportions do not require the division of the *scores* of the variable (as would be the case in computing the average score on a test, for example) but rather the *number of cases* in a particular category of the variable. When we make a statement like "43% of the sample are female," we are merely expressing the relative size of a category (female) of the variable (gender) in a convenient way.

2.2 RATIOS AND RATES

Ratios and **rates** provide two additional ways in which the distribution of a variable can be simply and dramatically summarized. Ratios are especially useful for comparing categories in terms of relative frequency. Instead of standardizing the distribution of a variable to the base 100 or 1.00, as we did in computing percentages and proportions, we determine ratios by dividing the frequency of one category by the frequency in another. Mathematically, a ratio can be defined as

FORMULA 2.3

$$\text{Ratio} = \frac{f_1}{f_2}$$

where f_1 = the number of cases in the first category
f_2 = the number of cases in thè second category

To illustrate the use of ratios, suppose that you were interested in the relative sizes of the various religious denominations and found that a particular community included 1370 Protestant families and 930 Catholic families. To find the ratio of Protestants (f_1) to Catholics (f_2), divide 1370 by 930. The resultant ratio is 1.47, which means that for every Catholic family, there are 1.47 Protestant families.

Application 2.2

In Table 2.2, how many natural science majors are there compared to social science majors at College B? This question could be answered with frequencies, but a more easily understood way of expressing the answer would be with a ratio. The ratio of natural science to social science majors would be

$$\text{Ratio} = \frac{f_1}{f_2} = \frac{279}{188} = 1.48$$

For every social science major, there are 1.48 natural science majors at College B.

Note that ratios can be very economical ways of expressing the relative predominance of two categories. That Protestants outnumber Catholics in our example is obvious from the raw data. Percentages or proportions could have been used to summarize the overall distribution (that is, 59.56% of the families were Protestant, 40.44% were Catholic). Compared to these other methods, ratios are a precise measure of the relative frequency of one category per unit of the other category. They tell us in an exact way the extent to which one category outnumbers the other.

Ratios are often multiplied by some power of 10 to eliminate decimal points. For example, the ratio computed above might be reported as 147 instead of 1.47. This would mean that, for every 100 Catholic families, there are 147 Protestant families in the community. To ensure clarity, the comparison units for the ratio are often expressed as well. Based on a unit of ones, the ratio of Protestants to Catholics would be expressed as 1.47:1. Based on hundreds, the same statistic might be expressed as 147:100. *(For practice in computing and interpreting ratios, see problems 2.1 and 2.2.)*

Rates provide still another way of summarizing the distribution of a single variable. Rates are defined as the number of actual occurrences of some phenomenon divided by the number of possible occurrences per some unit of time. Rates are usually multiplied by some power of 10 to eliminate decimal points. For example, the crude death rate for a population is defined as the number of deaths in that population (actual occurrences) divided by the number of people in the population (possible occurrences) per year. This quantity is then multiplied by 1000. The formula for the crude death rate can be expressed as

$$\text{Crude death rate} = \frac{\text{Number of deaths in year}}{\text{Total population}} \times 1000$$

If there were 100 deaths during a given year in a town of 7000, the crude death rate for that year would be

$$\text{Crude death rate} = \frac{100}{7000} \times 1000 = (.01429) \times 1000 = 14.29$$

Or, for every 1000 people, there were 14.29 deaths during this particular year. By the same token, if a city of 237,000 people experienced 120 auto thefts during a particular year, the auto theft rate would be

$$\text{Auto theft rate} = \frac{120}{237,000} \times 100,000 = (0.0005063) \times 100,000 = 50.63$$

Or, for every 100,000 people, there were 50.63 auto thefts during the year in question. *(For practice in computing and interpreting rates, see problems 2.3 and 2.4.)*

Up to now, we have considered three techniques (proportions and percentages, ratios, and rates) for describing and summarizing data. All three techniques express, clearly and concisely, the distribution of a single variable.

Application 2.3

In 1995, in a city of 167,000, there were 2500 births. In 1985, when the population of the city was only 133,000, there were 2700 births. Is the birthrate rising or falling? Although this question can be answered from the preceding information, the trend in birthrates will be much more obvious if we compute birthrates for both years. Like crude death rates, crude birthrates are usually multiplied by 1000 to eliminate decimal points. For 1985:

$$\text{Crude birthrate} = \left(\frac{2700}{133,000}\right) \times 1000 = 20.30$$

In 1985, there were 20.30 births for every 1000 people in the city.

For 1995:

$$\text{Crude birthrate} = \left(\frac{2500}{167,000}\right) \times 1000 = 14.97$$

In 1995, there were 14.97 births for every 1000 people in the city. With the help of these statistics, the decline in the birthrate is clearly expressed.

They represent different ways of expressing information so that it can be quickly appreciated. All three techniques are also quite useful for purposes of comparison. For instance, the auto theft rate of 50.63 in our example might be difficult to interpret by itself (is this a high or a low rate?) but would take on much more meaning if you found that the rate for the preceding year had been 32.70. These statistics provide alternative ways of expressing relative frequency and are thus especially useful in making comparisons between different groups and/or different times.

2.3 FREQUENCY DISTRIBUTIONS: INTRODUCTION

Frequency distributions are tables that summarize the distribution of a variable by reporting the number of cases contained in each category of the variable. They are very helpful and commonly used ways of organizing and working with data. In fact, the construction of frequency distributions is almost always the first step in any statistical analysis.

To illustrate the usefulness of frequency distributions and to provide some data for examples, assume that the counseling center at a university is assessing the effectiveness of its services. Any realistic evaluation research would collect a variety of information from a large group of students but, for the sake of this example, we will confine our attention to just four variables and 20 students. The data are reported in Table 2.4.

Note that, even though the data in Table 2.4 represent an unrealistically low number of cases, it is difficult to discern any patterns or trends. For example, try to ascertain the general level of satisfaction of the students from Table 2.4. You may be able to do so with just 20 cases, but it will take some time and effort. Imagine the difficulty with 50 cases or 100 cases presented in this fashion. Clearly the data need to be organized in a format that allows the researcher (and his or her audience) to understand easily the distribution of the variables.

TABLE 2.4   DATA FROM COUNSELING CENTER SURVEY

Student	Sex	Marital Status	Satisfaction with Services*	Age
A	Male	Single	4	18
B	Male	Married	2	19
C	Female	Single	4	18
D	Female	Single	2	19
E	Male	Married	1	20
F	Male	Single	3	20
G	Female	Married	4	18
H	Female	Single	3	21
I	Male	Single	3	19
J	Female	Divorced	3	23
K	Female	Single	3	24
L	Male	Married	3	18
M	Female	Single	1	22
N	Female	Married	3	26
O	Male	Single	3	18
P	Male	Married	4	19
Q	Female	Married	2	19
R	Male	Divorced	1	19
S	Female	Divorced	3	21
T	Male	Single	2	20

*Key: (4) Very satisfied   (2) Dissatisfied
      (3) Satisfied   (1) Very dissatisfied

One general rule that applies to all frequency distributions is that *the categories of the frequency distribution must be exhaustive and mutually exclusive.* In other words, the categories must be stated in a way that permits each case to be counted in one and only one category. This basic principle applies to the construction of frequency distributions for variables measured at all three levels of measurement.

Beyond this rule, there are only guidelines to help you construct useful frequency distributions. As you will see, the researcher has a fair amount of discretion in stating the categories of the frequency distribution (especially with variables measured at the interval-ratio level). I will identify the issues to consider as you make decisions about the nature of any particular frequency distribution. Ultimately, however, the guidelines I state are aids for decision making, nothing more than helpful suggestions. As always, the researcher has the final responsibility for making sensible decisions and presenting his or her data in a meaningful way.

2.4 FREQUENCY DISTRIBUTIONS FOR VARIABLES MEASURED AT THE NOMINAL AND ORDINAL LEVELS

Nominal-Level Variables.   For nominal-level variables, construction of the frequency distribution is typically very straightforward. For each category of the variable being displayed, the occurrences are counted and the subtotals, along with the total number of cases (*N*), are reported. Table 2.5 displays a frequency distribution for the variable "sex" from the counseling center survey. For purposes of illustration, a column for tallies has been included in

TABLE 2.5   SEX OF RESPONDENTS, COUNSELING CENTER SURVEY

Sex	Tallies	Frequency(f)
Male	LHT LHT	10
Female	LHT LHT	10
		$N = 20$

this table to illustrate how the cases would be sorted into categories. (This column would not be included in the final form of the frequency distribution.) Take a moment to notice several other features of the table. Specifically, the table has a descriptive title, clearly labeled categories (male and female), and a report of the total number of cases at the bottom of the frequency column. *These items must be included in all tables regardless of the variable or level of measurement.*

The meaning of the table is quite clear. There are 10 males and 10 females in the sample, a fact that is much easier to comprehend from the frequency distribution than from the unorganized data presented in Table 2.4.

For some nominal variables, the researcher might have to make some choices about the number of categories he or she wishes to report. For example, the distribution of the variable "marital status" could be reported using the categories listed in Table 2.4. The resultant frequency distribution is presented in Table 2.6. Although this is a perfectly fine frequency distribution, it may be too detailed for some purposes. For example, the researcher might want to focus solely on "nonmarried" as distinct from "married" students. That is, the researcher might not be concerned with the difference between single and divorced respondents but may want to treat both as simply "not married." In that case, these categories could be grouped together and treated as a single entity, as in Table 2.7. Notice that, by this collapsing, information and detail have been lost. This latter version of the table would not allow the researcher to discriminate between the two unmarried states.

Ordinal-Level Variables.   Frequency distributions for ordinal-level variables are constructed following the same routines used for nominal-level variables. Table 2.8 reports the frequency distribution of the "satisfaction"

TABLE 2.6   MARITAL STATUS OF RESPONDENTS, COUNSELING CENTER SURVEY

Status	Frequency
Single	10
Married	7
Divorced	3
	$N = 20$

TABLE 2.7   MARITAL STATUS OF RESPONDENTS, COUNSELING CENTER SURVEY

Status	Frequency
Married	7
Not married	13
	$N = 20$

TABLE 2.8   SATISFACTION WITH SERVICES, COUNSELING CENTER SURVEY

Satisfaction	Frequency (f)	Percentage (%)
(4) Very satisfied	4	20
(3) Satisfied	9	45
(2) Dissatisfied	4	20
(1) Very dissatisfied	3	15
	N = 20	100%

TABLE 2.9   SATISFACTION WITH SERVICES, COUNSELING CENTER SURVEY

Satisfaction	Frequency (f)	Percentage (%)
Satisfied	13	65
Dissatisfied	7	35
	N = 20	100

variable from the counseling center survey. Note that a column of percentages by category has been added to this table. Such columns heighten the clarity of the table (especially with larger samples) and are common adjuncts to the basic frequency distribution for variables measured at all levels.

This table reports that most students were either satisfied or very satisfied with the services of the counseling center. The most common response (nearly half the sample) was "satisfied." If the researcher wanted to emphasize this major trend, the categories could be collapsed as in Table 2.9. Again, the price paid for this increased compactness is that some information (in this case, the exact breakdown of degrees of satisfaction and dissatisfaction) is lost. *(For practice in constructing and interpreting frequency distributions for nominal- and ordinal-level variables, see problem 2.5.)*

2.5 FREQUENCY DISTRIBUTIONS FOR VARIABLES MEASURED AT THE INTERVAL-RATIO LEVEL

Basic Considerations.   In general, the construction of frequency distributions for variables measured at the interval-ratio level is more complex than for nominal and ordinal variables. Interval-ratio variables usually have a large number of possible scores (that is, a wide range from the lowest to the highest score). The large number of scores requires some collapsing or grouping of categories to produce reasonably compact frequency distributions. To construct frequency distributions for interval-ratio-level variables, you must decide how many categories to use and how wide these categories should be.

For example, suppose you wished to report the distribution of the variable "age" for a sample drawn from a community. Unlike the college data reported in Table 2.4, a community sample would have a very broad range of ages. If you simply reported the number of times that each year of age (or

---

Application 2.4

The following list shows the ages of 50 prisoners enrolled in a work-release program. Is this group young or old? A frequency distribution will provide an accurate picture of the overall age structure.

18	60	57	27	19
20	32	62	26	20
25	35	75	25	21
30	45	67	41	30
37	47	65	42	25
18	51	22	52	30
22	18	27	53	38
27	23	32	35	42
32	37	32	40	45
55	42	45	50	47

We will use about 10 intervals to display these data. By inspection we see that the youngest prisoner is 18 and the oldest is 75. The range is thus 57. Interval size will be 57/10, or 5.7, which we can round off to either 5 or 6. Let's use a six-year interval beginning at 18. The limits of the lowest interval will be 18–23. Now we must state the limits of all other intervals, count the number of cases in each interval, and display these counts in a frequency distribution. Columns may be added for percentages, cumulative percentages, and/or cumulative frequency. The complete distribution, with a column added for percentages, is

Ages	Frequency	Percentages
18–23	10	20
24–29	7	14
30–35	9	18
36–41	5	10
42–47	8	16
48–53	4	8
54–59	2	4
60–65	3	6
66–71	1	2
72–77	1	2
	$N = 50$	100

The prisoners seem to be fairly evenly spread across the age groups up to the 48–53 interval. There is a noticeable lack of prisoners in the oldest age groups and a concentration of prisoners in their 20s and 30s.

---

score) occurred, you could easily wind up with a frequency distribution that contained 70, 80, or even more categories. Such a large frequency distribution would not present a concise picture. The scores (years) must be grouped into larger categories to heighten clarity and ease of comprehension. How large should these categories be? How many categories should be included in the table? Although there are no hard-and-fast rules for making these decisions, they always involve a trade-off between more detail (a greater number of narrow categories) or more compactness (a smaller number of wide categories).

Constructing the Frequency Distribution. To introduce the mechanics and decision-making processes involved, we will construct a frequency distribution to display the ages of the students in the counseling center survey. Because of the narrow age range of a group of college students, we can use categories of only one year (these categories are often called **class intervals** when working with interval-ratio data). The frequency distribution is con-

TABLE 2.10   AGE OF RESPONDENTS, COUNSELING CENTER SURVEY
(interval width = one year of age)

Class Intervals	Frequency (f)
18	5
19	6
20	3
21	2
22	1
23	1
24	1
25	0
26	1
	N = 20

structed by listing the ages from youngest to oldest, counting the number of times each score (year of age) occurs, and then totaling the number of scores for each category. Table 2.10 presents the information and reveals a concentration or clustering of scores in the 18 and 19 class intervals.

Even though the picture presented in this table is fairly clear, assume for the sake of illustration that you desire a more compact (less detailed) summary. To do this, you will have to group scores into wider class intervals. By increasing the interval width (say to two years), you can reduce the number of intervals and achieve a more compact expression. The grouping of scores in Table 2.11 clearly emphasizes the relative predominance of younger respondents. This trend in the data can be stressed even more by the addition of a column displaying the percentage of cases in each category.

Note that the class intervals in Table 2.11 have been stated with an apparent gap between them (that is, the class intervals are separated by a distance of one unit). At first glance, these gaps may appear to violate the principle of exhaustiveness; but, since age has been measured in whole numbers, the gaps actually pose no problem. Given the level of precision of the measurement (in years, as opposed to 10ths or 100ths of a year), no case could

TABLE 2.11   AGE OF RESPONDENTS, COUNSELING CENTER SURVEY
(interval width = two years)

Class Intervals	Frequency (f)	Percentage (%)
18–19	11	55
20–21	5	25
22–23	2	10
24–25	1	5
26–27	1	5
	N = 20	100

have a score falling between these class intervals. In fact, for these data, the set of class intervals contained in Table 2.11 constitutes a scale that is exhaustive and mutually exclusive. Each of the 20 respondents in the sample can be sorted into one and only one age category.

However, consider the difficulties that might have been encountered if age had been measured with greater precision. If age had been measured in 10ths of a year, into which class interval in Table 2.11 would a 19.4-year-old subject be placed? You can avoid this ambiguity by always stating the limits of the class intervals at the same level of precision as the data. Thus, if age were being measured in 10ths of a year, the limits of the class intervals in Table 2.11 would be stated in 10ths of a year. For example:

17.5–19.4
19.5–21.4
21.5–23.4
23.5–25.4
25.5–27.4

To maintain mutual exclusivity between categories, *no overlapping of class intervals is allowed.* If you state the limits of the class intervals at the same level of precision as the data and maintain a gap between intervals, you will always produce a frequency distribution where each case can be assigned to one and only one category.

Midpoints.[1]   On occasion, you will need to work with the **midpoints** of the class intervals, for example, when constructing or interpreting certain graphs. Midpoints are defined as the points exactly halfway between the upper and lower limits and can be found for any interval by dividing the sum of the upper and lower limits by two. Table 2.12 displays midpoints for two different sets of class intervals. *(For practice in finding midpoints, see problems 2.8b and 2.9b.)*

Cumulative Frequency and Cumulative Percentage.   Two commonly used adjuncts to the basic frequency distribution for interval-ratio data are the **cumulative frequency** and **cumulative percentage** columns. Their primary purpose is to allow the researcher (and his or her audience) to tell at a glance how many cases fall below a given score or class interval in the distribution.

To construct a cumulative frequency column, begin with the lowest class interval (i.e., the class interval with the lowest scores) in the distribution. The entry in the cumulative frequency columns for that interval will be the same as the number of cases in the interval. For the next higher interval, the cumulative frequency will be all cases in the interval plus all the cases in the

---

[1] Real class limits are discussed at the web site for this text.

TABLE 2.12   MIDPOINTS

Class Interval Width of Three Units	
Categories	Midpoints
0–2	1
3–5	4
6–8	7
9–11	10

Class Interval Width of Six Units	
Categories	Midpoints
100–105	102.5
106–111	108.5
112–117	114.5
118–123	120.5

TABLE 2.13   AGE OF RESPONDENTS, COUNSELING CENTER SURVEY

Class Interval	Frequency (f)	Cumulative Frequency
18–19	11	11
20–21	5	16
22–23	2	18
24–25	1	19
26–27	1	20
	N = 20	

first interval. For the third interval, the cumulative frequency will be all cases in the interval plus all cases in the first two intervals. Continue adding (or accumulating) cases until you reach the highest class interval, which will have a cumulative frequency of all the cases in the interval plus all cases in all other intervals. For the highest interval, cumulative frequency equals the total number of cases. Table 2.13 shows a cumulative frequency column added to Table 2.11. The arrows show the direction of addition.

The cumulative percentage column is quite similar to the cumulative frequency column. Begin by adding a column to the basic frequency distribution for percentages as in Table 2.11. This column shows the percentage of all cases in each class interval. To find cumulative percentages, follow the same addition pattern explained above for cumulative frequency. That is, the cumulative percentage for the lowest class interval will be the same as the percentage of cases in the interval. For the next higher interval, the cumulative percentage is the percentage of cases in the interval plus the percentage of cases in the first interval, and so on. Table 2.14 shows the age

TABLE 2.14   AGE OF RESPONDENTS, COUNSELING CENTER SURVEY

Class Interval	Frequency (f)	Cumulative Frequency	Percentage (%)	Cumulative Percentage (%)
18–19	11	11	55	55
20–21	5	16	25	80
22–23	2	18	10	90
24–25	1	19	5	95
26–27	1	20	5	100
	N = 20		100%	

data with a cumulative percentage column added. Again, the arrows show the direction of addition.

These cumulative columns are quite useful in situations where the researcher wants to make a point about how cases are spread across the range of scores. For example, Tables 2.13 and 2.14 show quite clearly that most students in the counseling center survey are less than 21 years of age. If the researcher wishes to impress this feature of the age distribution on his or her audience, then these cumulative columns are quite handy. Most realistic research situations will be concerned with many more than 20 cases and/or many more categories than our tables have. Since the cumulative percentage column is clearer and easier to interpret in such cases, it is normally preferred to the cumulative frequencies column.

Procedures for Constructing Frequency Distributions for Interval-Ratio Variables.   Guidelines for dealing with interval-ratio variables with many scores (wide ranges) can now be stated:

1. Decide how many class intervals you wish to use. One reasonable convention suggests that the number of intervals should be about 10. Many research situations may require fewer than 10 intervals, and it is common to find frequency distributions with as many as 15 intervals. Only rarely will more than 15 intervals be used, since the resultant frequency distribution would not be very concise.

2. Find the size of the class interval. Once you have decided how many intervals you will use, interval size can be found by dividing the range of the scores by the number of intervals and rounding to a convenient whole number.

3. State the lowest interval so that its lower limit is equal to or below the lowest score. By the same token, your highest interval will be the one that contains the highest score. All intervals must be equal in size.

4. State the limits of the class intervals at the same level of precision as you have used to measure the data. Do not overlap intervals. You will thereby

define the class intervals so that each case can be sorted into one and only one category.

5. Count the number of cases in each class interval and report these sub-totals in a column labeled "frequency." Report the total number of cases (N) at the bottom of this column. The table may also include a column for percentages, cumulative frequencies, and cumulative percentages.

6. Inspect the frequency distribution carefully. Has too much detail been lost? If so, reconstruct the table with a greater number of class intervals (or smaller interval size). Is the table too detailed? If so, reconstruct the table with fewer class intervals (or use wider intervals). Remember that the frequency distribution results from a number of decisions you make in a rather arbitrary manner. If the appearance of the table seems less than optimal given the purpose of the research, redo the table until you are satisfied that you have struck the best balance between detail and conciseness.

7. Remember to give your table a clear, concise title, and number the table if your report contains more than one. All categories and columns must also be clearly labeled. *(For practice in constructing and interpreting frequency distributions for interval-ratio level variables, see problems 2.5 to 2.9.)*

2.6 CONSTRUCTING FREQUENCY DISTRIBUTIONS FOR INTERVAL-RATIO LEVEL VARIABLES: A REVIEW

We covered a lot of ground in the preceding section, so let's pause and review these principles by considering a specific research situation. Below are the numbers of visits received over the past year by 90 residents of a retirement community.

0	52	21	20	21	24	1	12	16	12
16	50	40	28	36	12	47	1	20	7
9	26	46	52	27	10	3	0	24	50
24	19	22	26	26	50	23	12	22	26
23	51	18	22	17	24	17	8	28	52
20	50	25	50	18	52	46	47	27	0
32	0	24	12	0	35	48	50	27	12
28	20	30	0	16	49	42	6	28	2
16	24	33	12	15	23	18	6	16	50

Listed in this format, the data are a hopeless jumble from which no one could derive much meaning. The function of the frequency distribution is to arrange and organize these data so that their meanings will be made obvious.

First, we must decide how many class intervals to use in the frequency distribution. Following the guidelines established in the previous section,

TABLE 2.15 NUMBER OF VISITS PER YEAR, 90 RETIREMENT-COMMUNITY RESIDENTS

Class Interval	Frequency (f)	Cumulative Frequency	Percentage (%)	Cumulative Percentage (%)
0–4	10	10	11.11	11.11
5–9	5	15	5.56	16.67
10–14	8	23	8.89	25.56
15–19	12	35	13.33	38.89
20–24	18	53	20.00	58.89
25–29	12	65	13.33	72.22
30–34	3	68	3.33	75.55
35–39	2	70	2.22	77.77
40–44	2	72	2.22	79.99
45–49	6	78	6.67	86.66
50–54	12	90	13.33	99.99
	$N = 90$		99.99%*	

*Percentage columns will occasionally fail to total to 100% because of rounding error. If the total is between 99.90% and 100.10%, ignore the discrepancy. Discrepancies of greater than plus or minus 0.10% may indicate mathematical errors, and the entire column should be computed again.

let's use about 10 intervals. By inspecting the data, we can see that the lowest score is 0 and the highest is 52. The range of these scores is 52 − 0, or 52. To find the approximate interval size, divide the range (52) by the number of intervals (10). Since $52/10 = 5.2$, we can set the interval size at 5.

The lowest score is 0, so the lowest class interval will be 0–4. The highest class interval will be 50–54, which will include the high score of 52. All that remains is to state the intervals in table format, count the number of scores that fall into each interval, and report the totals in a frequency column. These steps have been taken in Table 2.15, which also includes columns for the percentages and cumulative percentages. Note that this table is the product of several relatively arbitrary decisions. The researcher should remain aware of this fact and inspect the frequency distribution carefully. If the table is unsatisfactory for any reason, it can be reconstructed with a different number of categories and interval sizes.

Now, with the aid of the frequency distribution, some patterns in the data can be discerned. There are three distinct clusterings of scores in the table. Ten residents were visited rarely, if at all (the 0–4 visits per year interval). The single largest interval, with 18 cases, is 20–24. Combined with the intervals immediately above and below, this represents quite a sizable grouping of cases (42 out of 90, or 46.66% of all cases) and suggests that the dominant visiting rate is about twice a month, or approximately 24 visits per year. The third grouping is in the 50–54 class interval with 12 cases, reflecting a visiting rate of about once a week. The cumulative percentage column indicates that the majority of the residents (58.89%) were visited 24 or fewer times a year.

You will often need to interpret and understand tables, charts, and graphs as you become more involved in the professional literature of the social sciences. These figures will be explained in the text of the article, but you should develop the habit of reading them for yourself and coming to your own conclusions. Here are some ideas to keep in mind when reading tables and charts.

First, there are many different formats for presenting results, and the tables and graphs you find in the research literature will not necessarily follow the conventions used in this text. Second, because of space limitations, tables and graphs will be presented with a minimum of detail. For example, the researcher may present a frequency distribution with only a percentage column.

Begin your analysis by first reading the title, all labels (that is, row and/or column headings), and any footnotes to the table. These will tell you exactly what information is contained in the figure or table. Inspect the body of the table or graph with the author's analysis in mind. See if you agree with the author's analysis. (You almost always will, but it never hurts to double-check and exercise your critical abilities.)

Finally, remember that most research projects analyze interrelationships among many variables. Because the tables, graphs, and charts covered in this chapter display variables one at a time, they are unlikely to be included in such research reports (or perhaps, included only as background information). Even when not reported, you can be sure that the research began with an inspection of frequency distributions or graphs for each variable. Univariate tables and graphs display a great deal of information about the variables in a compact, easily understood format and are almost universally used as descriptive devices.

Statistics in the Professional Literature

In this section, two brief excerpts from social science research are presented so that you can see how statistical information is actually interpreted and reported. The first excerpt presents a frequency distribution and the second a line chart.

Excerpt 1 Power and Gender

Are women who are unequal at home also at a disadvantage in their ability to participate in politics? Do women who have less control over family decision making and the family budget also have less impact on the political institution? Does powerlessness in one institution translate into powerlessness in another?

To find answers to these questions, researchers Nancy Burns, Kay Schlozman, and Sidney Verba interviewed 380 married couples. The interviews covered a variety of topics, including how each couple divided up the household chores. This information on domestic power relationships was then used to analyze political power. The table on page 40, reproduced verbatim, reports some of their findings.

Note that this table presents 10 different frequency distributions (i.e., two for each of five different variables). Also note that only percentages are reported, and the tables are horizontal rather than vertical as in this text. As always, the researchers report percentage totals and the actual number of cases.

As for interpretation, here's what the authors have to say about the table:

Table 2 shows the distribution of . . . responses for wives and husbands. Overall, a similar division of labor is reported, with wives doing more when it comes to cleaning, shopping, caring for children and, to a lesser extent, paying the bills, and with men taking primary responsibility for the car. . . . The proportion of couples in which the spouses gave contradictory answers —each claiming to do all or most of a particular chore—is relatively low. . . . The fewest contradictory responses are found in relation to traditional female chores: taking care of children and cleaning the house. Most wives— and few husbands—claim to do all or most of these tasks.

Source: Burns, Nancy, Kay Schlozman, and Sidney Verba. 1997. "The Public Consequences of Private Inequality: Family Life and Citizen Participation." *American Political Science Review.* 91: 373–387.

(continued)

TABLE 2   DIVISION OF HOUSEHOLD CHORES

	All	Most	Some	Little	None		N	Contradictory Responses[a]
Housecleaning								
Wives say they do	41	34	17	3	5	100%	374	7%
Husbands say they do	1	10	46	31	12	100%	349	9%
Grocery Shopping								
Wives say they do	47	34	14	2	4	101%	374	9%
Husbands say they do	9	13	38	25	15	100%	349	
Paying Bills								
Wives say they do	48	17	13	12	11	101%	374	9%
Husbands say they do	21	20	14	20	25	100%	348	
Repairs/Car								
Wives say they do	7	13	29	27	24	100%	374	17%
Husbands say they do	69	24	5	1	1	100%	349	
Taking Care of Children								
Wives say they do	11	63	22	1	4	101%	330	5%
Husbands say they do	—	7	75	10	9	101%	224	

[a]Both members of a couple claim to do all or most.

Want to find out what the researchers concluded? The citation is given on page 39, and the article is almost certainly in your library.

Excerpt 2 Racism in Children's Literature

How have African Americans been presented in children's books? Do these portrayals reflect patterns of prejudice, conflict, and exclusion in the larger society? To find out, researchers Bernice Pescosolido, Elizabeth Grauerholz, and Melissa Milkie analyzed thousands of children's books published in the United States between the 1930s and the 1990s. The line chart below shows some of their results.

Here's what the authors have to say about this chart:

Figure 1 shows that the visibility of Blacks in children's books varies systematically across

FIGURE 1   PERCENTAGE OF CHILDREN'S BOOKS THAT PORTRAY BLACK CHARACTERS, BY YEAR: ALL THREE BOOK SERIES, 1937 TO 1993

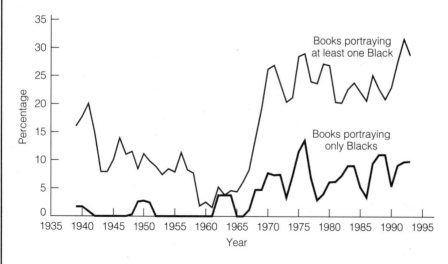

(continued)

READING STATISTICS 2: *(continued)*

the time period studied. The nonlinear time trend reveals roughly four phases of the dependent variable "visibility of Blacks" in these books. In the earliest phase, from 1938 to about 1957, Blacks are represented in modest proportion, with percentages decreasing significantly throughout this period. The second phase, 1958 through 1964, is marked by the virtual absence of Blacks. In the third phase, from the mid-1960s to the early 1970s, Blacks reappear. In the fourth phase, from 1975 to 1993, the percentage stabilizes, fluctuating between 20 and 30 percent for portrayal of at least one Black.

The authors go on to link these patterns to the changes in American race relations that occurred at the times in question. Want to find out what they concluded? The citation is given below.

Source: Pescosolido, Bernice, Elizabeth Grauerholz, and Melissa Milkie. 1997. "Culture and Conflict: The Portrayal of Blacks in U.S. Children's Picture Books Through the Mid- and Late-Twentieth Century." *American Sociological Review.* 62:443–464.

2.7 CHARTS AND GRAPHS

Researchers frequently use charts and graphs to present their data in ways that are visually more dramatic than frequency distributions. These devices are particularly useful for conveying an impression of the overall shape of a distribution and for highlighting any clustering of cases in a particular range of scores. Many graphing techniques are available, but we will examine just four. The first two, pie and bar charts, are appropriate for discrete variables at any level of measurement. The last two, histograms and line charts or frequency polygons, are used with interval-ratio variables.

The sections that follow explain how to construct graphs and charts "by hand." These days, however, computer programs are almost always used to produce graphic displays. Graphing software is sophisticated and flexible but also relatively easy to use and, if such programs are available to you, you should familiarize yourself with them. The effort required to learn these programs will be repaid in the quality of the final product. The section on computer applications at the end of this chapter includes a demonstration of how to produce bar charts and line charts.

Pie Charts.   To construct a **pie chart,** begin by computing the percentage of all cases that fall into each category of the variable. Then divide a circle (the pie) into segments (slices) proportional to the percentage distribution. Be sure that the chart and all segments are clearly labeled.

Figure 2.1 is a pie chart that displays the distribution of "marital status" from the counseling center survey. The frequency distribution (Table 2.6) is reproduced as Table 2.16, with a column added for the percentage distribution. Since a circle's circumference is 360°, we will apportion 180° (or 50%) for the first category, 126° (35%) for the second, and 54° (15%) for the last category. The pie chart visually reinforces the relative preponderance of single respondents and the relative absence of divorced students in the counseling center survey.

FIGURE 2.1 SAMPLE PIE CHART: MARITAL
STATUS OF RESPONDENTS
($N = 20$)

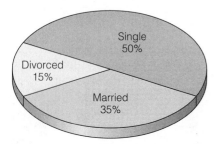

TABLE 2.16 MARITAL STATUS OF RESPONDENTS,
COUNSELING CENTER SURVEY

Status	Frequency (f)	Percentage (%)
Single	10	50
Married	7	35
Divorced	3	15
	$N = 20$	100

Bar Charts. Like pie charts, **bar charts** are relatively straightforward. Conventionally, the categories of the variable are arrayed along the horizontal axis (or abscissa) and frequencies, or percentages if you prefer, along the vertical axis (or ordinate). For each category of the variable, construct (or draw) a rectangle of constant width and with a height that corresponds to the number of cases in the category. The bar chart in Figure 2.2 reproduces the marital status data from Figure 2.1 and Table 2.16.

This chart would be interpreted in exactly the same way as the pie chart in Figure 2.1, and researchers are free to choose between these two methods of displaying data. However, if a variable has more than four or five categories, the bar chart would be preferred. With too many categories, the pie chart gets very crowded and loses its visual clarity. To illustrate, Figure 2.3 uses a bar chart to display the data on visiting rates for the retirement community presented in Table 2.15. A pie chart for this same data would have had 11 different "slices," a more complex or "busier" picture than that presented by the bar chart. In Figure 2.3, the clustering of scores in the "20 to 24" range (approximately two visits a month) is readily apparent, as are the groupings in the "0 to 4" and "50 to 54" ranges.

FIGURE 2.2 SAMPLE BAR CHART: MARITAL STATUS OF RESPONDENTS, COUNSELING CENTER SURVEY ($N = 20$)

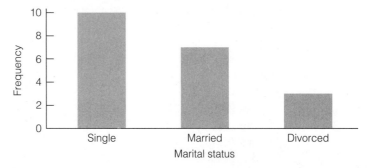

FIGURE 2.3   SAMPLE BAR CHART FOR VISITS PER YEAR, RETIREMENT COMMUNITY RESIDENTS ($N = 90$)

Bar charts are particularly effective ways to display the relative frequencies for two or more categories of a variable when you want to emphasize some comparisons. Suppose, for example, that you wished to make a point about changing rates of homicide victimization for white males and females since 1955. Figure 2.4 displays the data in a dramatic and easily comprehended way. The bar chart shows that rates for males are higher than rates for females, that rates for both sexes were highest in 1975, and that rates for males rose faster than rates for females between 1955 and 1975. *(For practice in constructing and interpreting pie and bar charts, see problems 2.5b and 2.10.)*

FIGURE 2.4   HOMICIDE VICTIMIZATION RATES, 1955–1996 (RATES PER 100,000 POPULATION, WHITES ONLY)

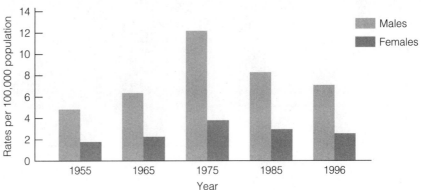

Histograms.   Histograms look a lot like bar charts and, in fact, are constructed in much the same way. They are most appropriate for continuous interval-ratio-level variables, but they are commonly used for discrete interval-ratio-level variables as well. To reflect the numerical quality of the scale of measurement, the bars are contiguous to each other. To construct a histogram from a frequency distribution, follow these steps:

1. Array the class intervals or scores along the horizontal axis (abscissa) using the class limits.
2. Array frequencies along the vertical axis (ordinate).
3. For each category in the frequency distribution, construct a bar with height corresponding to the number of cases in the category and with width corresponding to the limits of the class intervals.
4. Label each axis of the graph.
5. Title the graph.

As an example, Figure 2.5 uses a histogram to display the distribution of ages for a sample of respondents to a national public-opinion poll. The graph shows that the sample is concentrated in their 30s and 40s and that the number of respondents declines with age. Note also that there are no people in the sample younger than age 18, the usual cutoff point for respondents to public-opinion polls.

Line Charts.   Construction of a **line chart** or **frequency polygon** is similar to construction of a histogram. Instead of using bars to represent the frequencies, however, use a dot at the midpoint of each interval. Straight lines

FIGURE 2.5   AGE OF RESPONDENTS TO GENERAL SOCIAL SURVEY, 1998

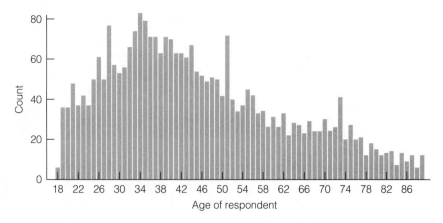

FIGURE 2.6   SAMPLE LINE CHART: NUMBER OF VISITS PER YEAR, RETIREMENT COMMUNITY RESIDENTS ($N = 90$)

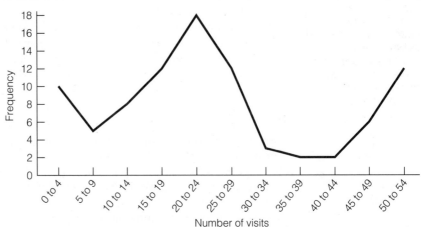

then connect the dots. Figure 2.6 displays a line chart for the visiting data previously displayed in the bar chart in Figure 2.3.

Line charts can also be used to display trends across time. Figure 2.7 shows both marriage and divorce rates per 100,000 population for the United States since 1972. Note that both rates were slowly falling since the early 1980s, but the marriage rate fell slightly faster.

Histograms and frequency polygons are alternative ways of displaying essentially the same message. Thus, the choice between the two techniques is left to the aesthetic pleasures of the researcher. *(For practice in constructing and interpreting histograms and line charts, see problems 2.7b, 2.8d, 2.9d, 2.11 and 2.12.)*

FIGURE 2.7   MARRIAGE AND DIVORCE RATES, UNITED STATES

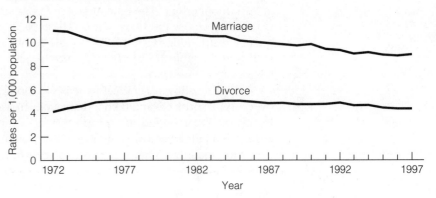

2.8 INTERPRETING STATISTICS: USING PERCENTAGES, FREQUENCY DISTRIBUTIONS, CHARTS, AND GRAPHS TO ANALYZE CHANGING PATTERNS OF WORKPLACE SURVEILLANCE

A sizeable volume of statistical material has been introduced in this chapter and it will be useful to conclude by focusing on meaning and interpretation. What can you say after you have calculated percentages, built a frequency-distribution, or constructed a graph or chart? Remember that statistics are tools to help us analyze information and answer questions. They never speak for themselves and they always have to be understood in the context of some research question or test of hypothesis. This section will provide an example of interpretation by posing and answering some questions from social science research. The interpretation (or words) will be explicitly linked to the statistical techniques and numbers so that you will be able to see how and why conclusions are developed.

Your New Job and Workplace Surveillance.   Congratulations! You have just landed a job with a major U.S. corporation and you now find yourself in the middle of the lunch hour in your cubicle. Should you log on to the D&D Web site and briefly engage in your favorite role-playing game? Are you considering sending a chain letter via the company's email system to your old college friends or using your desk phone to make an appointment to get your hair done? Before making a decision, consider a series of reports issued by the American Management Association (AMA). These reports suggest that the chances are growing that you may be the subject of workplace surveillance and that your email, telephone, and more recently, your Internet use may be monitored by your employer.

Overall Trends.   Since 1997, the AMA has surveyed companies about workplace surveillance and monitoring of employees.[2] The companies in the study are among the largest U.S. businesses, and the percentages in Table 2.17 suggest a relatively high level of employee surveillance and monitoring. Take a moment to note the components of Table 2.17: the table is numbered, titled, and completely labeled. Each column represents a specific year, and the rows represent responses (Yes or No) to the question "Does your company engage in monitoring?" The total number of companies surveyed in each year is reported across the bottom row. The table is actually four different frequency distributions (one for each year) combined into a single table. The entries in the table are percentages (not frequencies), and each column totals to 100%.

The table shows a gradual increase in monitoring and surveillance between 1997 and 1999 and then a dramatic increase between 1999 and 2000. In this one-year period, the proportion of companies reporting that they engaged in employee oversight increased from approximately two-thirds to over three-quarters. One possible explanation for the increase is that the AMA included Internet monitoring for the first time in the 2000 survey.

---

[2]American Management Association. 1998. Workplace Testing: Monitoring & Surveillance. 1999. Workplace Testing: Monitoring & Surveillance. 2000. AMA Research Reports: New York: AMA.

TABLE 2.17   WORKPLACE MONITORING AND SURVEILLANCE, 1997–2000

Does your company engage in monitoring?	Year			
	1997	1998	1999	2000
Yes	63.4	67.1	67.3	78.4
No	36.6	32.9	32.7	22.6
Total	100.0	100.0	100.0	100.0
$N =$	906	1,085	1,054	2,133

TABLE 2.18   MONITORING AND SURVEILLANCE IN 2000

Type of monitoring and surveillance:	Percent Indicating	
	Yes	No
Monitoring Internet connections	54.1%	45.9%
Telephone use (time spent, numbers called)	44.0%	66.0%
Storage and review of email messages	38.1%	61.9%
Video surveillance for security purposes	35.3%	64.7%
Storage and review of computer files	30.8%	69.2%
Computer use (time logged on, keystroke counts, etc.)	19.4%	80.6%
Video recording of employee job performance	14.6%	85.4%
Recording and review of telephone conversations	11.5%	88.5%
Storage and review of voice mail messages	6.8%	93.2%
N of Cases (2,133)		

Monitoring and Surveillance in 2000.   Although the computer has become an important component of the workday for more and more people, your employer may feel compelled to monitor its use. Table 2.18 reports the percentage of companies that indicated that they practiced a specific form of monitoring and surveillance.

Graphs are almost always a more effective method of information like that presented in Table 2.18. The variable (type of monitoring and surveillance) is nominal level (the "types" are different from each other but do not form a scale) and, with nine possible scores, a bar chart would be preferred to a pie chart. Figure 2.8 clearly shows that monitoring Internet connections was the most common form of surveillance with slightly over half of the companies (54.1%) practicing it. Telephone monitoring (44.0%) and storage and review of email messages (38.1%) were also very common. This graph clearly shows that it would be unwise to surf the net on company time, or use the phone or email for personal business.

Monitoring and Surveillance over Time.   What changes occurred in specific workplace monitoring practices between 1997 and 2000? Table 2.19

FIGURE 2.8 MONITORING AND SURVEILLANCE IN 2000

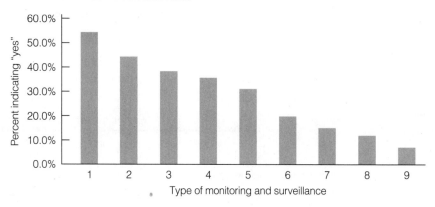

Key for types of monitoring and surveillance:
1. Monitoring Internet connections
2. Telephone use (time spent, numbers called)
3. Storage and review of email messages
4. Video surveillance for security purposes
5. Storage and review of computer files
6. Computer use (time logged on, keystroke counts, etc.)
7. Video recording of employee job performance
8. Recording and review of telephone conversations
9. Storage and review of voice mail messages

TABLE 2.19 MONITORING AND SURVEILLANCE, 1997 TO 2000

Type of Monitoring and Surveillance	Percent Indicating Yes			
	1997	1998	1999	2000
Recording and review of telephone conversations	10.4%	11.2%	10.6%	11.5%
Storage and review of voice mail messages	5.3%	5.3%	5.8%	6.8%
Storage and review of computer files	13.7%	19.6%	21.4%	30.8%
Storage and review of email messages	14.9%	20.2%	27.0%	38.1%
Video recording of employee job performance	15.7%	15.6%	16.1%	14.6%
Telephone use (time spent, numbers called)	34.4%	40.2%	38.6%	44.0%
Computer use (time logged on, keystroke counts, etc.)	16.1%	15.9%	15.2%	19.4%
Video surveillance for security purposes	33.7%	32.7%	32.8%	35.3%
Total Number of Companies	906	1,085	1,054	2,133

shows that in 1997 companies engaged in monitoring telephone use and video surveillance for security purposes at comparable rates (about 34%). Many forms of workplace monitoring and surveillance occurred among 10 and 15 percent of the companies surveyed. Only 5% of the companies engaged in the storage and review of voice mail messages in 1997.

By 1998, the percentage of companies monitoring the use of their telephones had increased to 40.2%. The percentage of companies reporting that they stored and reviewed email messages and computer files also increased. In 1999, four different forms of monitoring and surveillance were being used by over 20 percent of the companies: monitoring of telephone use, video surveillance, storage and review of email messages and computer files. By 2000, a higher percentage of companies were storing and reviewing email messages than were video taping for security purposes. Also noteworthy is the percentage of companies indicating that they store and review employee computer files, nearly one-third (30.8%) of the sample.

Once again, these trends and patterns would be more clearly presented and appreciated in the form of a graph. Since we are looking at changes in percentages over time, a line graph will be appropriate. Figure 2.9 illustrates relative stability in the rates of some forms of surveillance (e.g., video surveillance for security purposes) and sizeable increases in others (for example, monitoring of telephone use). The most obvious feature of the chart, however, is the steady increase in the percentage of companies that monitor employee email messages. In 1997, only about 15 percent of the companies engaged in this form of monitoring and surveillance. By 2000, this percentage had more than doubled to 38 percent.

Computers may be ubiquitous features of employment in our information-age economy, but these data suggest that the potential for workplace monitoring and surveillance is also increasing. The computer on your

FIGURE 2.9   MONITORING AND SURVEILLANCE, 1997–2000

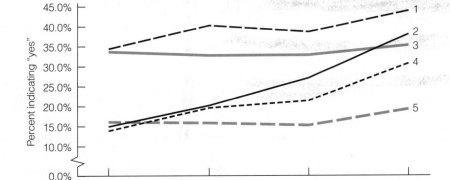

Key for types of monitoring and surveillance:
1. Telephone use (time spent, numbers called)
2. Storage and review of email messages
3. Video surveillance for security purposes
4. Storage and review of computer files
5. Computer use (time logged on, keystrokes)

desk is a double-edged sword. While it provides you with the necessary tools to do your job, employers are increasingly using the very same technology to watch you.

SUMMARY

1. We considered several different ways of summarizing the distribution of a single variable and, more generally, reporting the results of our research. Our emphasis throughout was on the need to communicate our results clearly and concisely. You will often find that, as you strive to communicate statistical information to others, the meanings of the information will become clearer to you as well.

2. Percentages and proportions, ratios, and rates represent several different techniques for enhancing clarity by expressing our results in terms of relative frequency. Percentages and proportions report the relative occurrence of some category of a variable compared with the distribution as a whole. Ratios compare two categories with each other, and rates report the actual occurrences of some phenomenon compared with the number of possible occurrences per some unit of time.

3. Frequency distributions are tables that summarize the entire distribution of some variable. It is very common to construct these tables for each variable of interest as the first step in a statistical analysis. Columns for percentages, cumulative frequency, and/or cumulative percentages often enhance the readability of frequency distributions.

4. Pie and bar charts, histograms, and line charts or frequency polygons are graphic devices used to express the basic information contained in the frequency distribution in a compact and visually dramatic way.

SUMMARY OF FORMULAS

Proportions	2.1	$p = \dfrac{f}{N}$
Percentage	2.2	$\% = \left(\dfrac{f}{N}\right) \times 100$
Ratios	2.3	$\text{Ratio} = \dfrac{f_1}{f_2}$

[handwritten: N = TOTAL]

[handwritten: Always Devide Bottom into Top]

GLOSSARY

Bar chart. A graphic display device for discrete variables. Categories are represented by bars of equal width, the height of each corresponding to the number (or percentage) of cases in the category.

Class intervals. The categories used in the frequency distributions for interval-ratio variables.

Cumulative frequency. An optional column in a frequency distribution that displays the number of cases within an interval and all preceding intervals.

Cumulative percentage. An optional column in a frequency distribution that displays the percentage of cases within an interval and all preceding intervals.

Frequency distribution. A table that displays the number of cases in each category of a variable.

Frequency polygon. A graphic display device for interval-ratio variables. Class intervals are represented by dots placed over the midpoints, the height of each corresponding to the number (or percentage) of cases in the interval. All dots are connected by straight lines. Same as a line chart.

Histogram. A graphic display device for interval-ratio variables. Class intervals are represented by contiguous bars of equal width (equal to the class limits), the height of each corresponding to the number (or percentage) of cases in the interval.

Line chart. See **Frequency polygon.**

Midpoint. The point exactly halfway between the upper and lower limits of a class interval.

Percentage. The number of cases in a category of a variable divided by the number of cases in all categories of the variable, the entire quantity multiplied by 100.

Pie chart. A graphic display device especially for discrete variables with only a few categories. A circle (the pie) is divided into segments proportional in size to the percentage of cases in each category of the variable.

Proportion. The number of cases in one category of a variable divided by the number of cases in all categories of the variable.

Rate. The number of actual occurrences of some phenomenon or trait divided by the number of possible occurrences per some unit of time.

Ratio. The number of cases in one category divided by the number of cases in some other category.

MULTIMEDIA RESOURCES

The Wadsworth Sociology Resource Center: Virtual Society
http://sociology.wadsworth.com/

Visit the companion web site for the sixth edition of *Statistics: A Tool for Social Research* to access a wide range of student resources. Begin by clicking on the Student Resources section of the book's web site to access the following study tools:

- Basic math review
- Flash cards

- Additional chapter problems
- Statistics review
- Internet links
- Table of random numbers
- MicroCase and SPSS examples and exercises
- "Find the test" flowcharts
- Hypothesis testing for variables measured at the ordinal level

PROBLEMS

2.1 SOC The tables that follow report the marital status of 20 respondents in two different apartment complexes. *(HINT: Make sure that you have the correct numbers in the numerator and denominator before solving the following problems. For example, problem 2.1a asks for "the percentage of respondents who are married in each complex," and the denominators will be 20 for these two fractions. Problem 2.1d, on the other hand, asks for the "percentage of the single respondents who live in Complex B," and the denominator for this fraction will be 4 + 6, or 10.)*

Status	Complex A	Complex B
Married	5	10
Unmarried ("living together")	8	2
Single	4	6
Separated	2	1
Widowed	0	1
Divorced	1	0
	20	20

a. What percentage of the respondents in each complex are married?

b. What is the ratio of single to married respondents at each complex?

c. What proportion of each sample are widowed?

d. What percentage of the single respondents live in Complex B?

e. What is the ratio of the "unmarried/living together" to the married at each complex?

2.2 At St. Algebra College, the numbers of males and females in the various major fields of study are as follows:

Major	Males	Females
Humanities	117	83
Social sciences	97	132
Natural sciences	72	20
Business	156	139
Nursing	3	35
Education	30	15

Read each of the following problems carefully before constructing the fraction and solving for the answer.

a. What percentage of social science majors are male?

b. What proportion of business majors are female?

c. For the humanities, what is the ratio of males to females?

d. What percentage of the total student body are males?

e. What is the ratio of males to females for the entire sample?

f. What proportion of the nursing majors are male?

g. What percentage of the sample are social science majors?

h. What is the ratio of humanities majors to business majors?

i. What is the ratio of female business majors to female nursing majors?

j. What proportion of the males are education majors?

2.3 CJ The town of Shinbone, Kansas, has a population of 211,732 and experienced 47 bank robberies, 13 murders, and 23 auto thefts during the past year. Compute a rate for each type of crime per 100,000 population. (*HINT: Make sure that*

you set up the fraction with size of population in the denominator.)

2.4 CJ The numbers of major felonies reported to the police in two different towns last year are listed below. Calculate the rate for each crime per 100,000 population. Relatively speaking, which town has the greater apparent crime problem? With which crimes? Write a paragraph describing the differences.

Crime	Town A (Population = 20,109)	Town B (Population = 764,213)
Homicide	13	78
Robbery	102	617
Auto theft	125	314
Rape	23	79
Burglary	178	537

2.5 SOC The scores of 15 respondents on four variables are reported below. These scores were taken from a public opinion survey called the General Social Survey, or the GSS. This data set, which is described in some detail in Appendix G, is used for the computer exercises in this text. Small subsamples from the GSS will be used throughout the text to provide "real" data for problems. For the actual questions and other details, see Appendix G. The numerical codes for the variables are as follows:

Sex	Support for Gun Control	Level of Education	Age
1 = Male	1 = In favor	0 = Less than HS	Actual years
2 = Female	2 = Opposed	1 = HS	
		2 = Jr. college	
		3 = Bachelor's	
		4 = Graduate	

Case No.	Sex	Support for Gun Control	Level of Education	Age
1	2	1	1	45
2	1	2	1	48
3	2	1	3	55
4	1	1	2	32
5	2	1	3	33
6	1	1	1	28
7	2	2	0	77

Case No.	Sex	Support for Gun Control	Level of Education	Age
8	1	1	1	50
9	1	2	0	43
10	2	1	1	48
11	1	1	4	33
12	1	1	4	35
13	1	1	0	39
14	2	1	1	25
15	1	1	1	23

a. Construct a frequency distribution for each variable. Include a column for percentages.

b. Construct pie and bar charts to display the distributions of sex, support for gun control, and level of education.

2.6 SW A local youth service agency has begun a sex education program for teenage girls who have been referred by the juvenile courts. The girls were given a 20-item test for general knowledge about sex, contraception, and anatomy and physiology upon admission to the program and again after completing the program. The scores of the first 15 girls to complete the program are listed below.

Case	Pretest	Posttest
A	8	12
B	7	13
C	10	12
D	15	19
E	10	8
F	10	17
G	3	12
H	10	11
I	5	7
J	15	12
K	13	20
L	4	5
M	10	15
N	8	11
O	12	20

Construct frequency distributions for the pretest and postest scores. Include a column for percentages. (HINT: *There were 20 items on the test so the maximum range for these scores is 20. If you use 10 class intervals to display these scores, the interval size will be 2. Since there are no scores of 0 or 1 for either test, you may state the first interval as 2–3. To make comparisons easier, both frequency distributions should have the same intervals.*)

2.7 SOC Sixteen high school students completed a class to prepare them for the College Boards. Their scores are reported below.

420	345	560	650
459	499	500	657
467	480	505	555
480	520	530	589

These same sixteen students were given a test of math and verbal ability to measure their readiness for college-level work. Scores are reported below in terms of the percentage of correct answers for each test.

Math Test

67	45	68	70
72	85	90	99
50	73	77	78
52	66	89	75

Verbal Test

89	90	78	77
75	70	56	60
77	78	80	92
98	72	77	82

a. Display each of these variables in a frequency distribution with columns for percentages and cumulative percentages.

b. Construct a histogram and frequency polygon for these data.

2.8 GER Following are reported the number of times 25 residents of a community for senior citizens left their homes for any reason during the past week.

0	2	1	7	3
7	0	2	3	17
14	15	5	0	7
5	21	4	7	6
2	0	10	5	7

a. Construct a frequency distribution to display these data.

b. What are the midpoints of the class intervals?

c. Add columns to the table to display the percentage distribution, cumulative frequency, and cumulative percentages.

d. Construct a histogram and a frequency polygon to display this distribution.

e. Write a paragraph summarizing this distribution of scores.

2.9 │SOC│ Twenty-five students completed a questionnaire that measured their attitudes toward interpersonal violence. Respondents who scored high believed that in many situations a person could legitimately use physical force against another person. Respondents who scored low believed that in no situation (or very few situations) could the use of violence be justified.

52	47	17	8	92
53	23	28	9	90
17	63	17	17	23
19	66	10	20	47
20	66	5	25	17

a. Construct a frequency distribution to display these data.
b. What are the midpoints of the class intervals?
c. Add columns to the table to display the percentage distribution, cumulative frequency, and cumulative percentage.
d. Construct a histogram and a frequency polygon to display these data.
e. Write a paragraph summarizing this distribution of scores.

2.10 │PA/CJ│ As part of an evaluation of the efficiency of your local police force, you have gathered the following data on police response time to calls for assistance during two different years. (Response times were rounded off to whole minutes.) Convert both frequency distributions into percentages and construct pie charts and bar charts to display the data. Write a paragraph comparing the changes in response time between the two years.

Response Time, 1986	f
21 minutes or more	35
16–20 minutes	75
11–15 minutes	180
6–10 minutes	375
Less than 6 minutes	210
	875

Response Time, 1996	f
21 minutes or more	45
16–20 minutes	95
11–15 minutes	155
6–10 minutes	350
Less than 6 minutes	250
	895

2.11 │SOC│ Figures 2.10 through 2.12 display trends in crime in the United States. Write a paragraph describing each of these graphs. What similarities and differences can you observe among the three graphs? (For example, do crime rates always change in the same direction?) Note the differences in the vertical axes from chart to chart

FIGURE 2.10   HOMICIDE RATES, 1984–1999

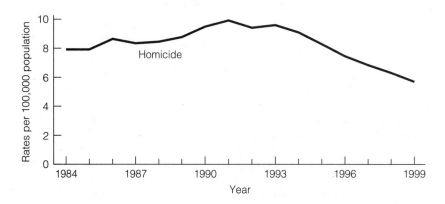

FIGURE 2.11   ROBBERY AND AGGRAVATED ASSAULT RATES, 1984–1999

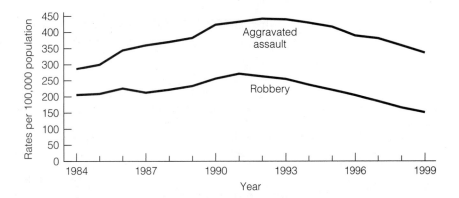

FIGURE 2.12   BURGLARY AND AUTO THEFT RATES, 1984–1999

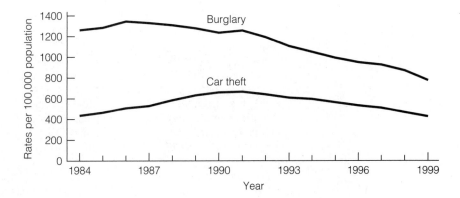

—for homicide the axis ranges from 0 to 12, while for burglary and auto theft the range is from 0 to 1600. The latter crimes are far more common, and a scale with smaller intervals is needed to display the rates.

2.12 PA The city's Department of Transportation has been keeping track of accidents on a particularly dangerous stretch of highway. Early in the year, the city lowered the speed limit on this highway and increased police patrols. Data on number of accidents before and after the changes are presented below. Did the changes work? Is the highway safer? Construct a line chart to display these two sets of data (use graphics software if available) and write a paragraph describing the changes.

	12 Months Before	12 Months After
January	23	25
February	25	21
March	20	18
April	19	12
May	15	9
June	17	10
July	24	11
August	28	15
September	23	17
October	20	14
November	21	18
December	22	20

Using *SPSS for Windows* to Produce Frequency Distributions and Graphs

Click the SPSS icon on your monitor screen to start *SPSS for Windows.* Load the 1998 GSS by clicking the file name on the first screen or by clicking **File, Open,** and **Data** on the **SPSS Data Editor** screen. You may have to change the drive specification to locate the 1998 GSS data supplied with this text (probably named **GSS98.sav**). Double click the file name to open the data set. When you see the message 'SPSS Processor is Ready' on the bottom of the screen, you are ready to proceed.

SPSS DEMONSTRATION 2.1 Frequency Distributions

We produced and examined a frequency distribution for the variable *sex* in Appendix F. Use the same procedures to produce frequency distributions for the variables *age* and *marital* (marital status). From the menu bar, click **Analyze.** From the menu that drops down, click **Descriptive Statistics** and **Frequencies.** The **Frequencies** window appears with the variables listed in alphabetical order in the left-hand box. The window may display variables by name (e.g. *abany, abhlth*) or by label (e.g., ABORTION IF WOMAN WANTS FOR ANY REASON). If labels are displayed, you may switch to variable names by clicking **Edit, Options,** and then making the appropriate selections on the "General" tab. You will have to restart SPSS before the change takes effect. See Appendix F and Table F.2 for further information. The variable *age* (AGE OF RESPONDENT) will be visible. Click on it to highlight it and then click the arrow button in the middle of the screen to move *age* to the right-hand window.

Find *marital* in the left-hand box by using the slider button or the arrow keys on the right-hand border to scroll through the variable list. As an alternative, type 'm', and the cursor will move to the first variable name in the list that begins with that letter. In this case, the variable is *marital,* the variable in which we are interested. Once *marital* is highlighted, click the arrow button in the center of the screen to move the variable name to the **Variables** box. There should now be two variable names in the box, *age* and *marital.* SPSS will process all variables listed in the right-hand box together. Click **OK** in the upper-right-hand corner, and SPSS will rush off to create the frequency distributions you requested.

The table will be in the **Output** window that will now be "closest" to you on the screen. The tables, along with other information, will be in the right-hand box of the **Output** window. To change the size of the output window, click the middle symbol (shaped like either a square or two intersecting squares) in the upper-right-hand corner of the **Output** window.

The frequency distribution for *marital* will look like this:

MARITAL STATUS

		Frequency	Percent	Valid Percent	Cumulative Percent
Valid	MARRIED	650	46.9	46.9	46.9
	WIDOWED	134	9.7	9.7	56.6
	DIVORCED	216	15.6	15.6	72.2
	SEPARATED	51	3.7	3.7	75.8
	NEVER MARRIED	335	24.2	24.2	100.0
	Total	1386	99.9	100.0	
Missing	NA	1	.1		
Total		1387	100.0		

Let's briefly examine the elements of this table. The variable description or label is printed at the top of the output ("MARITAL STATUS"). The various categories are printed on the left. Moving one column to the right, we find the actual frequencies or the number of times each score of the variable occurred. We see that 650 of the respondents were married, 134 were widowed, and so forth. Next are two columns that report percentages. The entries in the "Percent" column are based on *all* respondents who were asked this question. In this case, the denominator would include the respondent coded as "NA" (No Answer) for *marital*. The "Valid Percent" column eliminates all cases with missing values (e.g., cases coded as NA, DK for "Don't Know," or NAP for "Not Applicable"). Since we almost always ignore missing values, we will pay attention only to the "Valid Percent" column (even though there is little difference between the two columns in this case).

The final column is a cumulative percentage column (see Table 2.14). For nominal-level variables like *marital,* this information is not meaningful since the order in which the categories are stated is arbitrary. Finally, note that at the bottom of the table are totals for the number of cases, the two percentage columns, and the number of missing cases.

Turning to the frequency distribution for *age,* scroll the **Output** window until you get to the top of the table. Note that the table follows the same format as the table for *marital* but is much longer. This, of course, reflects the fact that *age* has many more possible scores than *marital*— so many scores, in fact, that the table is not easy to read or understand. The table needs to be made more compact by collapsing scores into fewer categories. For example, if ages were grouped into categories 10 years wide (10–19, 20–29, etc.), the number of categories would be reduced to about 8. We could collapse the categories by hand (see Section 2.5), or we could have SPSS do the work. The procedures for recoding or collapsing variables are explained in Demonstration 2.2.

SPSS DEMONSTRATION 2.2 Collapsing Categories with the Recode Command

The scores of interval-ratio level variables will often have to be collapsed in order to produce readable frequency distributions. *SPSS for Windows* provides a number of ways to change the scores of a variable, and one of the most useful of these is the **Recode** command. We will use this command to create a new version of *age* that has fewer categories and is more suitable for display in a frequency distribution. When we are finished, we will have two different versions of the same variable in the data set: the original interval-ratio version with age measured in years and a new ordinal-level version with collapsed categories. If we wish, the new version of *age* can be added to the permanent data file and used in the future.

We will collapse the values of *age* into categories of 10 years each. Since the youngest respondents are 18, let's (arbitrarily) begin with an interval of 10–19. The next interval will be 20–29, followed by 30–39, and so forth until the interval 80–89. As you will see, recoding requires many small steps, so please be patient and execute the commands as they are discussed.

1. In the **SPSS Data Editor** window, click **Transform** from the menu bar and then click **Recode.** A window will open that gives us two choices: **into same variable** or **into different variable.** If we choose the former ("into same variable"), the new version of the variable will *replace* the old version—the original version of *age* (with actual years) would disappear. We definitely do *not* want this to happen, so we will choose (click on) **into different variable.** This option will allow us to keep both the old and new versions of the variable.

2. The **Recode into Different Variable** window will open. A box containing an alphabetical list of variables will appear on the left. Use the cursor to highlight *age* and then click on the arrow button to move the variable to the **Input Variable → Output Variable** box. The input variable is the old version of *age,* and the output variable is the new, recoded version we will soon create.

3. In the **Output Variable** box on the right, click in the **Name** box and type a name for the new (output) variable. I suggest *ager* (*age r*ecoded) for the new variable, but you can assign any name as long as it does not duplicate the name of some other variable in the data set and is no longer than eight characters. Click the **Change** button and the expression *age → ager* will appear in the **Input Variable → Output Variable** box.

4. Click on the **Old and New Values** button in the middle of the screen, and a new dialog box will open. Read down the left-hand column until you find the **Range** button. Click on the button, and the cursor will move to the small box immediately below. In these boxes we will specify the low and high points of each interval of the new variable *ager.*

5. Starting with the interval 10–19, type 10 into the left-hand **Range** dialog box and then click on the right-hand box and type 19. In the **New Value** box in the upper-right-hand corner of the screen, click the **Value** button. Type 1 in the **Value** dialog box and then click the **Add** button directly below. The expression 10–19 → 1 will appear in the **Old → New** dialog box. This completes the first recode instruction to SPSS.

6. Continue recoding by returning to the **Range** dialog boxes on the left. Type 20 in the left-hand box and 29 in the right-hand box and then click the **Value** button in the **New Values** box. Type 2 in the **Value** dialog box and then click the **Add** button. The expression 20–29 → 2 appears in the **Old → New** dialog box.

7. Continue this sequence of operations until all old values of *age* have been recoded into new values. The last expression in the **Old → New** dialog box should be 80–89 → 8. Now click the **Continue** button at the bottom of the screen, and you will return to the **Recode into Different Variable** dialog box. Click **OK,** and SPSS will execute the transformation.

You now have a data set with one more variable named *ager* (or whatever name you gave the recoded variable). SPSS adds the new variable to the data set, and you can find it in the last column to the right in the data window. You can make the new variable a permanent part of the data set by saving the data file at the end of the session. If you do not wish to save the new, expanded data file, click **No** when you are asked if you want to save the data file. If you are using the student version of *SPSS for Windows,* remember that you are limited to a maximum of 50 variables.

To produce a frequency distribution for the new variable, click **Analyze, Descriptive Statistics,** and **Frequencies.** Find *ager* in the left-hand box and click the arrow key to move it to the right-hand box. Click **OK,** and you will soon be looking at the following table:

AGER

		Frequency	Percent	Valid Percent	Cumulative Percent
Valid	1.00	23	1.7	1.7	1.7
	2.00	247	17.8	17.8	19.5
	3.00	366	26.4	26.4	45.9
	4.00	266	19.2	19.2	65.1
	5.00	193	13.9	13.9	79.1
	6.00	121	8.7	8.7	87.8
	7.00	116	8.4	8.4	96.2
	8.00	53	3.8	3.8	100.0
	Total	1385	99.9	100.0	
Missing	System	2	.1		
Total		1387	100.0		

There are few respondents in their teens (category 1 or ages 10–19) and their 80s (category 8). The sample is clustered in their 30s and 40s (more than 45% of the sample is in the categories 3 and 4 or in their 30s and 40s). We are missing data on only 2 respondents for this variable. If you want to retain the data set with *ager* added, remember to save the data file at the end of your session.

SPSS DEMONSTRATION 2.3 Graphs and Charts

SPSS for Windows can produce a variety of graphs and charts and, in this demonstration, we will use the program to produce a bar chart and a line chart (frequency polygon). To conserve space, I will keep the choices as simple as possible, but you should explore the options for yourself. For any questions you might have that are not answered in this demonstration, click on **Help** on the main menu bar.

To produce a bar chart, first click on **Graphs** on the main menu bar and then click **Bar.** The **Bar Chart** dialog box appears with three choices for the type of graph we want. The **Simple** option is already highlighted, and this is the one we want, so just click **Define.** The **Define Simple Bar** dialog box appears with variable names listed on the left. Choose *degree* from the variable list by moving the cursor to highlight this variable name. Click the arrow button in the middle of the screen to move *degree* to the **Category Axis** box.

Note the **Bars Represent** box above the **Category Axis** box. This dialog box gives you control over the vertical axis of the graph, which can be calibrated in frequencies, percentages, or cumulative frequencies or percentages. Let's choose **N of cases or frequencies,** the option that is already selected.

Click the **Options** button in the lower-right-hand corner. In the **Options** dialog box, make sure that the button next to the **Display groups defined by missing values** option is *not* selected (or checked). Click **Continue** and then click **OK** in the **Define Simple Bar** dialog box, and the following bar chart will be produced:

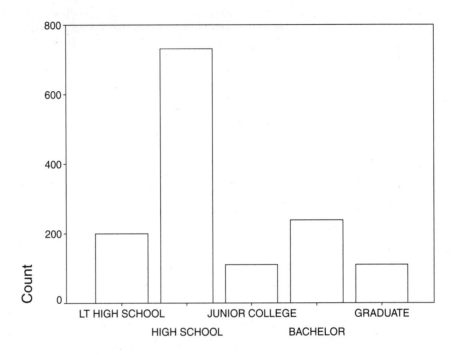

RS HIGHEST DEGREE

Click on any part of the graph and the **SPSS Chart Editor** window will appear. The menu bar across the top of the window gives you a wide array of options for the final appearance of the chart, including—under the **Format** command—the color of the bars and their borders. Explore these options at your leisure using the **Help** function as necessary.

Turning to the chart itself, by far the most common level of education for this sample was high school, with 'college' second and 'less than high school' a close third. When you are ready, the bar chart can be printed or saved by selecting the appropriate command from the **File** menu.

The procedures for producing a line chart are very similar to those for a bar chart. Click **Graphs** and then **Line.** The **Line Charts** dialog box will appear with the **Simple** option already chosen. This is the one we want, so click **Define,** and the **Define Simple Line** dialog box will appear. Your choices here are the same as in the **Define Simple Bar** dialog box. Line charts are appropriate for interval-ratio data, so let's use *educ* for this demonstration. You might think of this variable as a noncollapsed form of *degree,* and we should be able to compare the two variables quickly and easily.

Select *educ* from the variable list on the left and click the arrow button in the middle of the screen to move *educ* to the **Category Axis** box. Note that the **Lines Represent** box is the same as in the **Simple Bar** dialog box. We want the bars to represent *frequencies,* and this option is already selected. Click the **Options** button and make sure that the **Display Groups Defined by Missing Values** option is *not* selected (or checked). Click **Continue** and then click **OK** in the **Define Simple Line** dialog box, and the following line chart will be produced:

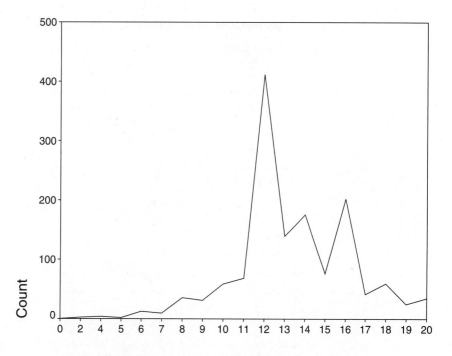

HIGHEST YEAR OF SCHOOL COMPLETED

The line chart can be edited by clicking on any part of the chart and activating the Chart Editor window. For example, to change the color of the line, click anywhere on the line in the Chart Editor window, then click **Format** and **Color.** Click on the new color and then click **Apply.** Don't forget to **Save** or **Print** the chart if you wish.

The chart has a very high peak at 12 years of education (high school) and another peak at 16 (college). This essentially duplicates the information in the bar chart for *degree.*

Using MicroCase to Produce Frequency Distributions and Graphs

Click the MicroCase icon on your monitor screen, and the **File and Data** screen (student version of MicroCase) or the **File Management** screen (full version) will open. Load the 1998 GSS by clicking the **Open File** command and clicking the file name. You may have to change the drive specification to locate the 1998 GSS data supplied with this text (probably named **GSS98.mc4** or something similar). Double click the file name to open the data set. The **File Settings** window will appear. Click **OK,** and you will be ready to proceed.

MICROCASE DEMONSTRATION 2.1 Frequency Distributions

We produced and examined a frequency distribution for the variable *sex* in Appendix F. Use the same procedures to produce frequency distributions for the variables *marital* (marital status) and *age.* Begin by clicking **Statistics Menu** (Student version of MicroCase) or **Basic Statistics** (full version) on the left of the screen and then click on **Univariate.** Look down the list until you find *marital* (variable #3) and highlight the variable name by clicking on it. Then, click the arrow pointing to the box labeled 'Primary variable.' Click **OK,** and the next screen you see will be a pie chart of the variable. To see the frequency distribution, click 'summary' in the **Statistics** box to the left of the screen. The frequency distribution for *marital* will look like this:

MARITAL—MARITAL STATUS

| Mean: 2.486 Std.Dev.: 1.647 N: 1386 |
| Median: 2.000 Variance: 2.714 Missing: 1 |
| 99% confidence interval +/- mean: 2.372 to 2.600 |
| 95% confidence interval +/- mean: 2.399 to 2.572 |

Category	Freq.	%	Cum.%	Z-Score
1) MARRIED	650	46.9	46.9	-0.902
2) WIDOWED	134	9.7	56.6	-0.295
3) DIVORCED	216	15.6	72.2	0.312
4) SEPARATED	51	3.7	75.8	0.919
5) NEVER MARR	335	24.2	100.0	1.526

Let's briefly examine the elements of this table. The variable name and description are printed at the top of the output (MARITAL—MARITAL STATUS) followed by some summary statistics. We will ignore these for the moment except to note that the total number of cases in the table is reported here as N: 1386 and that we are missing information on marital status for only one case (Missing: 1). Looking at the table itself, the categories are printed on the left. Moving one column to the right, we find the actual frequencies or the number of times each score of the variable occurred. We see that 650 of the respondents were married, 134 were widowed, and so forth. Next are columns for percentages, cumulative percentages (see

Table 2.14), and Z-scores. For nominal-level variables like *marital,* cumulative percentages are not meaningful since the order in which the categories are stated is arbitrary, and we will defer a discussion of Z-scores to Chapter 5.

To get a frequency distribution for *age,* begin by clicking the curled arrow button at the upper-left-hand side of the window and you will be returned to the **Univariate** screen. The variable name *marital* will still be in the "Primary Variable" window. You can erase it by clicking the **Clear All** button on the right and then moving *age* (variable #6) in place of *marital,* or you can simply double click *age* and it will automatically replace *marital* as the selected variable. Either way, click **OK,** and a histogram for *age* will be displayed. To view the frequency distribution, click 'summary' in the **Statistics** box. Note that the output for *age* follows the same format as for *marital* but that the table is much longer. This, of course, reflects the fact that *age* has many more possible scores than *marital*—so many scores, in fact, that the table is not easy to read or understand. The table needs to be made more compact by collapsing scores into fewer categories. For example, if ages were grouped into categories 10 years wide (10–19, 20–29, etc.), the number of categories would be reduced to 8. We could collapse the categories by hand (see Section 2.5), or we could have MicroCase do the work. The procedures for recoding or collapsing variables are explained in Demonstration 2.2.

MICROCASE DEMONSTRATION 2.2 Collapsing Categories

The scores of interval-ratio level variables will often have to be collapsed in order to produce readable frequency distributions. MicroCase provides a number of ways to change the scores of a variable, and one of the most useful of these is the **Collapse Categories** command. We will use this command to create a new version of *age* that has fewer categories and is more suitable for display in a frequency distribution. When we are finished, we will have two different versions of the same variable in the data set: the original interval-ratio version with age measured in years and a new ordinal-level version with collapsed categories. If we wish, the new version of *age* can be added to the permanent data file and used in the future.

We will collapse the values of *age* into categories of 10 years each. Since the youngest respondents are 18, let's (arbitrarily) begin with an interval of 10–19. The next interval will be 20–29, followed by 30–39, and so forth until the interval 80–89. As you will see, collapsing the scores of variables requires many small steps, so please be patient and execute the commands as they are discussed.

1. From the **File and Data** window (student version) or **Data Management** window (full version), click **Collapse Variables,** and a window will open that allows us to select the variable we want to transform. Highlight *age* (variable #6), then click on the arrow button to move the variable to the box labeled "variable to collapse" and click **OK.**

2. The next screen will show a histogram or bar chart for the selected variable along the top. To begin to collapse the categories of *age,* click on the top rectangle under "New Categories" in the middle of the screen. The rectangle will turn blank. Type 10–19 in the rectangle and then use the cursor to click on the histogram bars you want to be in the category. In other words, highlight the bar at the far left of the histogram (age 18) and drag the mouse to the right so that

the adjacent bar (age 19) turns the same color as the first bar. These are the only two categories that will be assigned to the 10–19 age group.

3. Click on the next rectangle under "New Categories" and type 20–29. Use the cursor to select the bar over age 20 and drag the mouse to the right until all bars between 20 and 29 turn the same color.

4. Continue clicking on the rectangles under "New Categories," entering the age limits for the intervals, and selecting categories by selecting bars from the chart until you reach the 80–89 category. Click the **Finish** button at the upper left and assign a name to this new variable. I suggest *ager* (*age re*coded), but you may choose your own name. Click **OK,** and the new variable will be added to the data set as variable #50. You now have a data set with one more variable named *ager* (or whatever name you gave the recoded variable).

To produce a frequency distribution for the new variable, return to the Statistics menu and click **Univariate,** select *ager* (variable #50), and click **OK.** You will soon be looking at the following table:

```
       AGER—Collapsed from AGE: AGE OF RESPONDENT
```

Mean: 3.048 Std.Dev.: 1.727 N: 1385				
Median: 3.000 Variance: 2.982 Missing: 2				
99% confidence interval +/- mean: 2.928 to 3.167				
95% confidence interval +/- mean: 2.957 to 3.139				

Category	Freq.	%	Cum.%	Z-Score
0) 10–19	23	1.7	1.7	-1.765
1) 20–29	247	17.8	19.5	-1.186
2) 30–39	366	26.4	45.9	-0.607
3) 40–49	266	19.2	65.1	-0.028
4) 50–59	193	13.9	79.1	0.552
5) 60–60	121	8.7	87.8	1.131
6) 70–79	116	8.4	96.2	1.710
7) 80–89	53	3.8	100.0	2.289

There are few respondents in their teens (category 0 or ages 10–19) and their 80s (category 7). The sample is clustered in the 30s and 40s (more than 45% of the sample is in the categories 2 and 3 or in their 30s and 40s). If you want to retain the data set with *ager* added, remember to save the data file at the end of your session.

MICROCASE DEMONSTRATION 2.3 Graphs and Charts

We have already seen that MicroCase produces pie and bar charts automatically when we use the **Univariate** command. In this demonstration, we will produce two more graphs, one for *degree* (highest level of education) and one for *educ* (years of education). Click the **Univariate** command on the **File and Data** menu and select *degree* (variable #9) as the "Primary Variable." Click **OK,** and a pie chart will be displayed. To get a bar chart, click the appropriate selection under "Graphs" in the upper-left-hand corner of the screen. The bar chart will look like this:

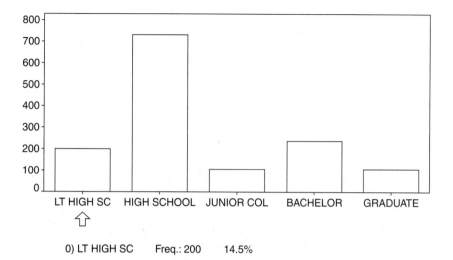

0) LT HIGH SC   Freq.: 200   14.5%

By far the most common level of education for this sample was high school, with 'college' second and 'less than high school' a close third. If you wish, the bar chart can be printed by selecting the appropriate command.

To get a bar chart for *educ,* click the curled arrow at the upper-left-hand side of the screen, and the **Univariate** screen will reappear. Select *educ* (variable #7) and move it to the "Primary Variable" window and then click **OK.** You might think of *educ* as a noncollapsed form of *degree,* and we should be able to compare the two variables quickly and easily.

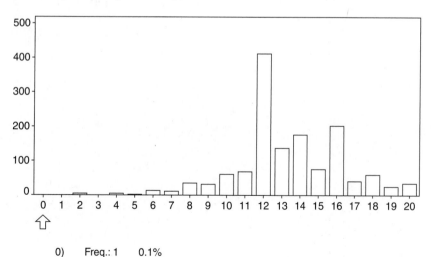

0)   Freq.: 1   0.1%

The chart has a very high peak at 12 years of education (high school) and another peak at 16 (college). This essentially duplicates the information in the bar chart for *degree.*

Exercises

2.1 Get frequency distributions for 5 or 10 nominal or ordinal variables, including *race* and *relig*. Get bar charts for each variable and write a sentence or two summarizing each frequency distribution. Your description should clearly identify the most and least common scores and any other noteworthy patterns you observe.

2.2 Get frequency distributions for *degree* and *educ*. Use the **Recode** command (SPSS) or the **Collapse Variables** command (MicroCase) to collapse *educ* into categories comparable to *degree*. That is, use the categories 0–11 for less than high school, 12 for high school graduates, 13–15 to approximate the number of junior college graduates, 16 for college graduates, and 17–20 for graduate degrees. Get a frequency distribution for recoded *educ* and compare it to the frequency distribution for *degree*. Describe the three frequency distributions in a few sentences.

2.3 Get a frequency distribution for *prestg80* (respondent's occupational prestige). Find the range of this variable and use this information to determine reasonable class intervals by following the procedures described in Section 2.5. Use the **Recode** command (SPSS) or the **Collapse Variables** command (MicroCase) to produce a final frequency distribution for the recoded variable and write a sentence or two of description and interpretation.

2.4 Choose five or six more variables of interest to you and produce bar charts or line charts for each. Remember that bar charts are more appropriate for nominal and ordinal variables and line charts for interval-ratio variables. Write a sentence or two of interpretation for each chart.

3

MEASURES OF CENTRAL TENDENCY

LEARNING OBJECTIVES   By the time you finish this chapter, you will be able to

1. Explain the purposes of measures of central tendency and interpret the information they convey.
2. Calculate, explain, and compare and contrast the mode, median, and mean.
3. Explain the mathematical characteristics of the mean.
4. Select an appropriate measure of central tendency according to level of measurement and skew.

3.1 INTRODUCTION   One clear benefit of frequency distributions, graphs, and charts is that they summarize the overall shape of a distribution of scores in a way that can be quickly comprehended. Often, however, you will need to report more detailed information about the distribution. Specifically, two additional kinds of statistics are almost always useful: some idea of the typical or average case in the distribution (for example, "the average starting salary for social workers is $26,000 per year") and some idea of how much variety or heterogeneity there is in the distribution ("starting salaries for social workers range from $19,000 per year to $28,000 per year"). The first kind of statistic, called **measures of central tendency,** will be the subject of this chapter. The second kind of statistic, measures of dispersion, will be presented in Chapter 4.

The three commonly used measures of central tendency—the mode, median, and mean—are all probably familiar to you. All three summarize an entire distribution of scores by describing the most common score (the mode), the middle case (the median), or the typical score of the cases (the mean) of that distribution. These statistics are powerful because they can reduce huge arrays of data to a single, easily understood number. Remember that the central purpose of descriptive statistics is to summarize or "reduce" data.

Even though they share a common purpose, the three measures of central tendency are quite different from each other. In fact, they will have the same value only under specific and limited conditions. As we shall see, they vary in terms of level-of-measurement considerations and, perhaps more importantly, they also vary in terms of how they define central tendency—they will not necessarily identify the same score or case as "typical." Thus, your choice of an appropriate measure of central tendency will depend in part on the way you measure the variable and in part on the purpose of the research.

3.2 THE MODE

The **mode** of any distribution is the value that occurs most frequently. For example, in the set of scores 58, 82, 82, 90, 98, the mode is 82 since it occurs twice and the other scores occur only once.

TABLE 3.1 RELIGIOUS PREFERENCE (fictitious data)

Denomination	Frequency
Protestant	128
Catholic	57
Jew	10
None	32
Other	15
	N = 242

The mode is, relatively speaking, a rather simple statistic, most useful when you want a "quick and easy" indicator of central tendency and when you are working with nominal-level variables. In fact, the mode is the only measure of central tendency that can be used with nominal-level variables. Such variables do not, of course, have numerical "scores" per se, and the mode of a nominally measured variable would be its largest category. For example, Table 3.1 reports the religious affiliations of a fictitious sample of 242 respondents. The mode of this distribution, the single largest category, is Protestant.

If a researcher desires to report only the most popular or common value of a distribution, or if the variable under consideration is nominal, then the mode is the appropriate measure of central tendency. However, keep in mind that the mode does have several limitations. First, some distributions have no mode at all (see Table 2.5) or so many modes that the statistic loses all meaning. Second, with ordinal and interval-ratio data, the modal score may not be central to the distribution as a whole. That is, *most common* does not necessarily mean "typical" in the sense of identifying the center of the distribution. For example, consider the following rather unusual (but not impossible) distribution of scores on a statistics test:

Scores (% correct)	Frequency
58	2
60	2
62	3
64	2
66	3
67	4
68	1
69	1
70	1
93	5
	N = 24

The mode of the distribution is 93. Is this score very close to the majority of the scores? If the instructor summarized this distribution by reporting only the modal score, would he or she be conveying an accurate picture of the distribution as a whole? *(For practice in finding and interpreting the mode, see problems 3.1 to 3.7.)*

3.3 THE MEDIAN

Unlike the mode, the **median (Md)** always represents the exact center of a distribution of scores. The median is the score of the case that is in the exact middle of a distribution: half the cases have scores higher and half the cases have scores lower than the case with the median score. Thus, if the median family income for a community is $30,000, half the families earn more than $30,000 and half earn less.

Before finding the median, the cases must be placed in order from the highest to the lowest score—or from the lowest to the highest. Once this is done, find the central or middle case. The median is the score associated with that case. If five students received grades of 93, 87, 80, 75, and 61 on a test, the median would be 80, the score that splits the distribution into two equal halves.

When the number of cases (N) is odd, the value of the median is unambiguous because there will always be a middle case. With an even number of cases, however, there will be two middle cases; and, in this situation, the median is defined as the score exactly halfway between the scores of the two middle cases.

To illustrate, assume that seven students were asked to indicate their level of support for the intercollegiate athletic program at their universities on a scale ranging from 10 (indicating great support) to 0 (no support). After arranging their responses from high to low, you can find the median by locating the case that divides the distribution into two equal halves. With a total of seven cases, the middle case would be the fourth case, since there will be three cases above and three cases below this case. If the seven scores were 10, 10, 8, 7, 5, 4, and 2, then the median is 7, the score of the fourth case. To find the middle case when N is odd, add 1 to N and divide by 2. With an N of 7, the median is the score associated with the $(7 + 1)/2$, or fourth, case. If N had been 25, the median would be the score associated with the $(25 + 1)/2$, or 13th, case.

Now, if we added a student to the sample whose support for athletics was measured as a 1 and thereby made N an even number (8), we would no longer have a single middle case. The ordered distribution of scores would now be 10, 10, 8, 7, 5, 4, 2, 1; any value between 7 and 5 would technically satisfy the definition of a median (that is, would split the distribution into two equal halves of four cases each). This ambiguity is resolved by defining the median as the average of the scores of the two middle cases. In the example above, the median would be defined as $(7 + 5)/2$, or 6.

To identify the two middle cases when N is an even number, divide N by 2 to find the first middle case and then increase that number by 1 to find the second middle case. In the example above with eight cases, the first middle case would be the fourth case ($N/2 = 4$) and the second middle case would be the ($N/2$) + 1, or fifth, case. If N had been 142, the first middle case would have been the 71st case and the second the 72nd case. Remember that the median is defined as the average of the scores associated with the two middle cases.*

Since the median requires that scores be ranked from high to low, it cannot be calculated for variables measured at the nominal level. Remember that the scores of nominal-level variables cannot be ordered: the scores are different from each other but do not form a mathematical scale of any sort. The median can be found for either ordinal or interval-ratio data but is generally more appropriate for the former. *(The median may be found for any problem at the end of this chapter.)*

3.4 OTHER MEASURES OF POSITION: PERCENTILES, DECILES, AND QUARTILES†

In addition to serving as a measure of central tendency, the median is also a member of a class of statistics that measure position or location. The median identifies the exact middle of a distribution, but it is sometimes useful to locate other points as well. We may want to know, for example, the scores that split the distribution into thirds or fourths or the point below which a given percentage of the cases fall. A familiar application of these measures would be scores on standardized tests, which are often reported in terms of location (for example, "a score of 476 is higher than 46% of the scores").

One commonly used statistic for reporting position is the **percentile.** A percentile identifies the point below which a specific percentage of cases fall. If a score of 476 is reported as the 46th percentile, this means that 46% of the cases had scores lower than 476. To find a percentile, first arrange the scores in order. Next, multiply the number of cases (N) by the proportional value of the percentile. For example, the proportional value for the 46th percentile would be 0.46. The resultant value identifies the number of the case that marks the percentile. To find the 37th percentile for a sample of 78 cases, multiply 78 by .37. The result is 28.86, and the 37th percentile would be 86/100 of the distance between the scores of the 28th and 29th cases. In most cases, we would probably round off 28.86 to 29, and call the score of the 29th case the 37th percentile. The slight inaccuracy would be worth the savings in time and calculational effort.

Note that, if we think in terms of percentiles, then the median is simply

*If the middle cases have the same score, that score is defined as the median. In the distribution 10, 10, 8, 6, 6, 4, 2, 1 the middle cases both have scores of 6 and, thus, the median would be defined as 6.
†This section is optional.

the 50th percentile and, in our example above, we would find the median by multiplying 78 by .50, finding the 39th case, and declaring the score of that case to be the 50th percentile. Notice again that we are cutting some corners here. Technically, the median would be the score halfway between the two middle cases (the 39th and 40th cases), but it is unlikely that this inaccuracy would be very significant.

Some other commonly used measures of position are **deciles** and **quartiles.** Deciles divide the distribution of scores into tenths. So, the first decile is the point below which 10% of the cases fall and is equivalent to the 10th percentile. The fifth decile is also the same as the 50th percentile, which is the same as (you guessed it) the median. Quartiles divide the distribution into quarters, and the first quartile is the same as the 25th percentile, the second quartile is the 50th percentile, and the third quartile is the 75th percentile. Any of these measures can be found by the method described above for percentiles. Remember that multiplying N by the proportional value of the percentile, decile, or quartile gives the number of the appropriate *case,* and it's the *score* of the case that actually marks the location. Also remember that this technique cuts some (probably minor) computational corners, and use it with caution. *(Quartiles, deciles, and percentiles may be found for any ordinal or interval-ratio variable in the problems at the end of this chapter.)*

3.5 THE MEAN

The **mean** ($\overline{X}$, read this as "ex-bar")*, or arithmetic average, is by far the most commonly used measure of central tendency. It reports the average score of a distribution, and its calculation is straightforward: to compute the mean, add the scores and then divide by the number of scores (N). To illustrate: a birth control clinic administered a 20-item test of general knowledge about contraception to 10 clients. The number of correct responses was 2, 10, 15, 11, 9, 16, 18, 10, 11, 7. To find the mean of this distribution, add the scores (total = 109) and divide by the number of scores (10). The result (10.9) is the average score on the test.

The mathematical formula for the mean is

FORMULA 3.1

$$\overline{X} = \frac{\Sigma(X_i)}{N}$$

where $\overline{X}$ = the mean;
$\Sigma(X_i)$ = the summation of the scores
N = the number of scores

Since this formula introduces some new symbols, let us take a moment to consider it. First, the symbol Σ (uppercase Greek letter **sigma**) is a mathematical operator just like the plus sign ($+$) or divide sign ($\div$). It stands for

---

*This is the symbol for the mean of a sample. The mean of a population is symbolized with the Greek letter mu (μ—read this symbol as "mew").

"the summation of" and directs us to add whatever quantities are stated immediately following it. The second new symbol is X_i ("X sub i"), which refers to any single score—the "ith" score. If we wished to refer to a particular score in the distribution, the specific number of the score could replace the subscript. Thus, X_1 would refer to the first score, X_2 to the second, X_{26} to the 26th, and so forth. The operation of adding all the scores is symbolized as $\Sigma(X_i)$.

This combination of symbols directs us to sum the scores, beginning with the first score and ending with the last score in the distribution. Thus, Formula 3.1 states in symbols what has already been stated in words (to calculate the mean, add the scores and divide by the number of scores), but in a very succinct and precise way.* *(For practice in computing the mean, see any of the problems at the end of this chapter.)*

Since the computation of a mean requires the mathematical procedures of addition and division, its use is fully justified only when working with interval-ratio data. However, researchers do calculate the mean for variables measured at the ordinal level, because the mean is much more flexible than the median and is a central feature of many interesting and powerful advanced statistical techniques. Thus, if the researcher plans to do any more than merely describe his or her data, the mean will probably be the preferable measure of central tendency even for ordinal-level variables.

3.6 SOME CHARACTERISTICS OF THE MEAN

Because the mean is the most commonly used measure of central tendency, let us consider its mathematical and statistical characteristics in some detail. First, the mean is always the center of any distribution of scores. The mean is the point around which all of the scores (X_i) cancel out. Symbolically:

$$\Sigma(X_i - \overline{X}) = 0$$

Or, if we take each score in a distribution, subtract the mean from it, and add all of the differences, the resultant sum will always be zero. To illustrate, consider the following set of test scores: 65, 73, 77, 85, and 90. The mean of these five scores is 390/5, or 78. So

X	$(X_i - \overline{X})$
65	$65 - 78 = -13$
73	$73 - 78 = -5$
77	$77 - 78 = -1$
85	$85 - 78 = 7$
90	$90 - 78 = 12$
$\Sigma X = 390$	$\Sigma(X_i - \overline{X}) = 0$

The total of the negative differences (-19) is exactly equal to the total of the positive differences $(+19)$ and these quantities will always be equal

*See the Basic Math Review at the web site for a further review of the summation sign.

regardless of the distribution. This algebraic relationship between the scores and the mean indicates that the mean is a good descriptive measure of the centrality of scores. You may think of the mean as a fulcrum that exactly balances all of the scores.

A second characteristic of the mean will only be mentioned at this point but is of great importance in later chapters. This characteristic is expressed in the statement:

$$\Sigma(X_i - \overline{X})^2 = \text{minimum}^*$$

If the differences between the scores and the mean are squared and then added, the resultant sum will be less than the sum of the squared differences between the scores and any other point in the distribution. That is, the mean is the point in a distribution around which the variation of the scores (as indicated by the squared differences) is minimized. In a sense, this algebraic property merely underlines the fact that the mean is closer to all of the scores than the other measures of central tendency. However, this characteristic of the mean is also important for the statistical techniques of correlation and regression, topics we will take up toward the end of this text.

The final important characteristic of the mean is that every score in the distribution affects it. The mode (which is only the most common score) and the median (which deals only with the score of the middle case or cases) are not so affected. This quality is both an advantage and a disadvantage. On one hand, the mean utilizes all the available information—every score in the distribution affects the mean. On the other hand, when a distribution has a few extreme cases (very high or low scores), the mean may become very misleading as a measure of centrality.

To illustrate, consider the following set of five scores: 15, 20, 25, 30, 35. Both the mean and median of this distribution are 25. ($\overline{X} = 125/5 = 25$. Md = score of third case = 25.) What will happen if we change the last score from 35 to 3500? Note that nothing happens to the median. It remains at exactly 25. The median is based only on the score of the middle case and is not affected by changes in the scores of other cases in the distribution. The mean, since it takes all scores into account, is very much affected. It becomes 3590/5, or 718. Clearly, the one extreme score in the data set disproportionately affects the mean. In this case, which of the two measures presents the more accurate or truer measure of central tendency for this distribution? *(For practice in dealing with the effects of extreme scores on means and medians, see problems 3.11, 3.13, 3.14, and 3.15)*

---

*To illustrate this principle, let us use the distribution of five test scores mentioned above. The differences between the scores and the mean have already been found. If we square and sum these differences, we would get $(-13)^2 + (-5)^2 + (-1)^2 + (7)^2 + (12)^2$, or $(169 + 25 + 1 + 49 + 144)$, or 388. So, $\Sigma(X_i - \overline{X})^2 = 388$. If we performed the same operation with any other number, the resultant sum would be greater than 388. For example, if we used 77 instead of the mean, the sum of the squared differences would be $(65 - 77)^2 + (73 - 77)^2 + (77 - 77)^2 + (85 - 77)^2 + (90 - 77)^2$, or $(-12)^2 + (-4)^2 + (0)^2 + (8)^2 + (13)^2$, or $(144 + 16 + 0 + 64 + 169)$, or 393.

Application 3.1

Ten students have been asked how many hours they spent in the college library during the past week. What is the average "library time" for these students? The hours are reported in the following list, and we will find the mode, the median, and the mean for these data.

$$X_i$$

X_i
0
2
5
5
7
10
14
14
20
30
107

By scanning the scores, we can see that two scores, 5 and 14, occurred twice, and no other score occurred more than once. This distribution has two modes, 5 and 14.

Because the number of cases is even, the median will be the average of the two middle cases

after all cases have been ranked in order. With 10 cases, the first middle case will be the ($N/2$), or (10/2), or fifth case. The second middle case is the ($N/2$) + 1, or (10/2) + 1, or sixth case. The median will be the score halfway between the scores of the fifth and sixth cases. Counting down from the top, we find that the score of the fifth case is 7 and the score of the sixth case is 10. The median for these data is (7 + 10)/2, or (17/2), or 8.5. The median, the score that divides this distribution in half, is 8.5.

The mean is found by first adding all the scores and then dividing by the number of scores. The sum of the scores is 107, so the mean is

$$\overline{X} = \Sigma\left(\frac{X_i}{N}\right) = \frac{107}{10} = 10.7$$

These 10 students spent an average of 10.7 hours in the library during the week in question.

Note that the mean is a higher value than the median. This indicates a positive skew in the distribution (a few extremely high scores). By inspection we can see that the positive skew is caused by the two students who spent many more hours (20 hours and 30 hours) in the library than the other eight students.

What general point can we derive from these considerations? Relative to the median, the mean is always pulled in the direction of extreme scores. The two measures of central tendency will have the same value when and only when a distribution is symmetrical. When a distribution has some extremely high scores (a positive **skew**), the mean will always have a greater numerical value than the median. If the distribution has some very low scores (a negative skew), the mean will be lower in value than the median. Figures 3.1 to 3.3 depict three different frequency polygons that demonstrate these relationships.

These relationships between medians and means also have a practical value. For one thing, a quick comparison of the median and mean will always tell you if a distribution is skewed and the direction of the skew. Second, they provide a simple and effective way to "lie" with statistics. For example, if you want to minimize the average score of a positively skewed distribution, report the median. Income data usually have a positive skew

FIGURE 3.1   A POSITIVELY SKEWED DISTRIBUTION (The mean is greater in value than the median.)

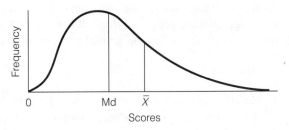

FIGURE 3.2   A NEGATIVELY SKEWED DISTRIBUTION (The mean is less than the median.)

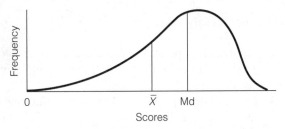

FIGURE 3.3   AN UNSKEWED, SYMMETRICAL DISTRIBUTION (The mean and median are equal.)

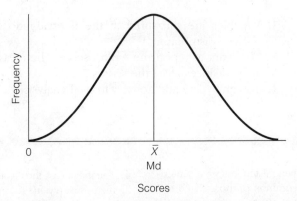

(there are only a few very wealthy people). If you want to impress someone with the general affluence of a mixed-income community, report the mean. If you want a lower figure, report the median. For the good and honest researcher, the selection of a measure of central tendency for a badly skewed distribution will hinge on what he or she wishes to show and, in most cases, either both statistics or the median alone should be reported.

3.7 CHOOSING A MEASURE OF CENTRAL TENDENCY   The selection of a measure of central tendency should, in general, be based on level-of-measurement considerations and on an evaluation of what each of the three statistics shows. Remember that the mode, median, and mean are three different statistics that will be the same value only under certain, specific conditions (that is, for symmetrical distributions with one mode). Each of the three has its own message to report and, in many circumstances, you might want to report all three. When choosing a single measure of central tendency, the following guidelines may be helpful:

Use the mode when

1. Variables are measured at the nominal level.
2. You want a quick and easy measure for ordinal and interval-ratio variables.
3. You want to report the most common score.

Use the median when

1. Variables are measured at the ordinal level.
2. Variables measured at the interval-ratio level have highly skewed distributions.
3. You want to report the central score. The median always lies at the exact center of a distribution.

Use the mean when

1. Variables are measured at the interval-ratio level (except for highly skewed distributions).
2. You want to report the typical score. The mean is "the fulcrum that exactly balances all of the scores."
3. You anticipate additional statistical analysis.

SUMMARY

1. The three measures of central tendency presented in this chapter share a common purpose. Each reports some information about the most typical or representative value in a distribution. Appropriate use of these statistics permits the researcher to report important information about an entire distribution of scores in a single, easily understood number.
2. The mode reports the most common score and is used most appropriately with nominally measured variables.
3. The median (Md) reports the score that is the exact center of the distribution. It is most appropriately used with variables measured at the ordinal level and with variables measured at the interval-ratio level when the distribution is skewed.

4. The mean ($\overline{X}$), the most frequently used of the three measures, reports the most typical score. It is used most appropriately with variables measured at the interval-ratio level (except when the distribution is highly skewed).
5. The mean has a number of mathematical characteristics that are significant for statisticians. First, it is the point in a distribution of scores around which all other scores cancel out. Second, the mean is the point of minimized variation. Last, as distinct from the mode or median, the mean is affected by every score in the distribution and is therefore pulled in the direction of extreme scores.

SUMMARY OF FORMULAS

Mean   3.1 $\overline{X} = \dfrac{\Sigma(X_i)}{N}$

GLOSSARY

Deciles. The points that divide a distribution of scores into 10ths.

Mean. The arithmetic average of the scores. $\overline{X}$ represents the mean of a sample, and μ, the mean of a population.

Measures of central tendency. Statistics that summarize a distribution of scores by reporting the most typical or representative value of the distribution.

Median (Md). The point in a distribution of scores above and below which exactly half of the cases fall.

Mode. The most common value in a distribution or the largest category of a variable.

Percentile. A point below which a specific percentage of the cases fall.

Quartiles. The points that divide a distribution into quarters.

Σ (uppercase Greek letter **sigma**). "The summation of."

Skew. The extent to which a distribution of scores has a few scores that are extremely high (positive skew) or extremely low (negative skew).

X_i ("X sub i"). Any score in a distribution.

MULTIMEDIA RESOURCES

The Wadsworth Sociology Resource Center: Virtual Society
http://sociology.wadsworth.com/

Visit the companion web site for the sixth edition of *Statistics: A Tool for Social Research* to access a wide range of student resources. Begin by clicking on the Student Resources section of the book's web site to access the following study tools:

- Basic math review
- Flash cards

- Additional chapter problems
- Statistics review
- Internet links
- Table of random numbers
- MicroCase and SPSS examples and exercises
- "Find the test" flowcharts
- Hypothesis testing for variables measured at the ordinal level

PROBLEMS

3.1 SOC A variety of information has been gathered from a sample of college freshmen and seniors including their region of birth, the extent to which they support legalization of marijuana (measured on a scale on which 7 = strong support, 4 = neutral, and 1 = strong opposition), the amount of money they spend each week out-of-pocket for food, drinks, and entertainment, how many movies they watched in their dorm rooms last week, their opinion of cafeteria food (10 = excellent, 0 = very bad), and their religious affiliation. Some results are presented below. Find the *most appropriate* measure of central tendency for each variable for freshmen and then for seniors. Report both the measure you selected as well as its value for each variable (e.g., "Mode = 3" or "Median = 3.5").

(HINT: Determine the level of measurement for each variable first. In general, this will tell you which measure of central tendency is appropriate. See Section 3.7 to review the relationship between

measure of central tendency and level of measurement. Also, remember that the mode is the most common score and, especially, remember to array scores from high to low before finding the median.)

FRESHMEN

Student	Region of Birth	Legali- zation	Out-of- pocket Expenses	Movies	Cafeteria Food	Religion
A	North	7	23	0	10	Protestant
B	North	4	29	14	7	Protestant
C	South	3	35	10	2	Catholic
D	Midwest	2	37	7	1	None
E	North	3	52	5	8	Protestant
F	North	5	38	1	6	Jew
G	South	1	42	0	10	Protestant
H	South	4	55	14	0	Other
I	Midwest	1	41	3	5	Other
J	West	2	33	4	6	Catholic

SENIORS

Student	Region of Birth	Legali- zation	Out-of- pocket Expenses	Movies	Cafeteria Food	Religion
K	North	7	55	0	1	None
L	Midwest	6	52	5	2	Protestant
M	North	7	50	11	8	Protestant
N	North	5	80	3	4	Catholic
O	South	1	52	4	3	Protestant
P	South	5	47	14	6	Protestant
Q	West	6	30	0	2	Catholic
R	West	7	39	7	9	None
S	North	3	35	5	4	None
T	West	5	75	3	7	Other
U	North	4	68	5	4	None

3.2 A variety of information has been collected for each of the nine high schools in a district. Find the most appropriate measure of central tendency for each variable and summarize this information in a paragraph. *(HINT: The level of measurement of the variable will generally tell you which measure of central tendency is appropriate. Remember to organize the scores from high to low before finding the median.)*

High School	Total Enrollment	Largest Racial/ Ethnic Group	Percent College Bound	Most Popular Sport	Condition of Physical Plant (scale of 1–10, 10 = high)
1	1400	White	25	Football	10
2	1223	White	77	Baseball	7
3	876	Black	52	Football	5
4	1567	Hispanic	29	Football	8
5	778	White	43	Basketball	4
6	1690	Black	35	Basketball	5
7	1250	White	66	Soccer	6
8	970	White	54	Football	9
9	1109	Hispanic	64	Soccer	3

3.3 PS You have been observing the local Democratic Party in a large city and have compiled some information about a small sample of party regulars. Find the appropriate measure of central tendency for each variable.

Respondent	Sex	Social Class	Number of Years in the Party	Education	Marital Status	Number of Children
A	M	High	32	High School	Married	5
B	M	Medium	17	High School	Married	0
C	M	Low	32	High School	Single	0
D	M	Low	50	Eighth grade	Widowed	7
E	M	Low	25	Fourth grade	Married	4
F	M	Medium	25	High School	Divorced	3
G	F	High	12	College	Divorced	3
H	F	High	10	College	Separated	2
I	F	Medium	21	College	Married	1
J	F	Medium	33	College	Married	5
K	M	Low	37	High School	Single	0
L	F	Low	15	High School	Divorced	0
M	F	Low	31	Eighth grade	Widowed	1

3.4 SOC You have compiled the information below on each of the graduates voted "most likely to succeed" by a local high school for a 10-year period. For each variable, find the appropriate measure of central tendency.

Case	Present Income	Marital Status	Owns a BMW?	Years of Schooling Completed after HS
A	24,000	Single	No	8
B	48,000	Divorced	No	4
C	54,000	Married	Yes	4
D	45,000	Married	No	4
E	30,000	Single	No	4
F	35,000	Separated	Yes	8
G	30,000	Married	No	3
H	17,000	Married	No	1
I	33,000	Married	Yes	6
J	48,000	Single	Yes	4

3.5 SOC For 15 respondents, data have been gathered on four variables (table below). Find and report the appropriate measure of central tendency for each variable.

Respondent	Racial or Marital Status	Ethnic Group	Age	Attitude on Abortion Scale*
A	Single	White	18	10
B	Single	Hispanic	20	9
C	Widowed	White	21	8

(continued next column)

(continued)

Respondent	Racial or Marital Status	Ethnic Group	Age	Attitude on Abortion Scale*
D	Married	White	30	10
E	Married	Hispanic	25	7
F	Married	White	26	7
G	Divorced	Black	19	9
H	Widowed	White	29	6
I	Divorced	White	31	10
J	Married	Black	55	5
K	Widowed	Asian American	32	4
L	Married	American Indian	28	3
M	Divorced	White	23	2
N	Married	White	24	1
O	Divorced	Black	32	9

*This scale is constructed so that a high score indicates strong opposition to abortion under any circumstances.

3.6 SOC Following are four variables for 30 cases from the General Social Survey. Age is reported in years. The variable "happy" consists of answers to the question "Taken all together, would you say that you are (1) very happy (2) pretty happy, or (3) not too happy?" Respondents were asked how many sex partners they had over the past five years. Responses were measured on the following scale: 0–4 = actual numbers; 5 = 5–10 partners; 6 = 11–20 partners; 7 = 21–100 partners; 8 = more than 100. For each variable, find the appropriate measure of central tendency and write a sentence reporting this statistical information as you would in a research report.

Respondent	Age	Happi-ness	Number of Partners	Religion
1	20	1	2	Protestant
2	32	1	1	Protestant
3	31	1	1	Catholic
4	34	2	5	Protestant
5	34	2	3	Protestant
6	31	3	0	Jew
7	35	1	4	None
8	42	1	3	Protestant
9	48	1	1	Catholic
10	27	2	1	None
11	41	1	1	Protestant
12	42	2	0	Other
13	29	1	8	None
14	28	1	1	Jew
15	47	2	1	Protestant
16	69	2	2	Catholic
17	44	1	4	Other
18	21	3	1	Protestant
19	33	2	1	None
20	56	1	2	Protestant
21	73	2	0	Catholic
22	31	1	1	Catholic
23	53	2	3	None
24	78	1	0	Protestant
25	47	2	3	Protestant
26	88	3	0	Catholic
27	43	1	2	Protestant
28	24	1	1	None
29	24	2	3	None
30	60	1	1	Protestant

3.7 Find the appropriate measure of central tendency for each variable displayed in Table 2.4. Report each statistic as you would in a formal research report.

3.8 |SOC| For problem 2.5, find the most appropriate measure of central tendency for each variable.

3.9 |SOC| The administration is considering a total ban on student automobiles. You have conducted a poll on this issue of fellow students and the neighbors who live around the campus and have calculated scores for your respondents. On the scale you used, a high score indicates strong opposition to the proposed ban. The scores are presented below for both groups. Calculate an appropriate measure of central tendency and compare the two groups in a sentence or two.

Students		Neighbors	
10	11	0	7
10	9	1	6
10	8	0	0
10	11	1	3
9	8	7	4
10	11	11	0
9	7	0	0
5	1	1	10
5	2	10	9
0	10	10	0

3.10 |SW| As the head of a social services agency, you believe that your staff of 20 social workers is very much overworked compared to 10 years ago. The case loads for each worker are reported below for each of the two years in question. Has the average case load increased? What measure of central tendency is most appropriate to answer this question? Why?

1990		2000	
52	55	42	82
50	49	75	50
57	50	69	52
49	52	65	50
45	59	58	55
65	60	64	65
60	65	69	60
55	68	60	60
42	60	50	60
50	42	60	60

3.11 |SOC| Compute the median and mean for each of the three variables presented in problem 2.7. Do any of these variables have positive or negative skews? How can you tell?

3.12 |SW| For the test scores presented in problem 2.6, compute a median and mean for both the pretest and posttest. Interpret these statistics.

3.13 |SOC| A sample of 25 freshmen at a major university completed a survey that measured their degree of racial prejudice (the higher the score, the greater the prejudice).
   a. Compute the median and mean scores for these data.

10	43	30	30	45
40	12	40	42	35
45	25	10	33	50
42	32	38	11	47
22	26	37	38	10

b. These same students completed the same survey during their senior year. Compute the median and mean for this second set of scores and compare them to the earlier set. What happened?

10	45	35	27	50
35	10	50	40	30
40	10	10	37	10
40	15	30	20	43
23	25	30	40	10

3.14 PA The data below represent the percentage of all workers in each city who use public transportation to commute to work.

City	Workers Using Public Transportation (%)
New York	62
Chicago	36
Los Angeles	9
Philadelphia	37
Detroit	18
Houston	8
Washington, D.C.	38
Dallas	11
San Francisco	36
Boston	39
Cleveland	22
Baltimore	27
St. Louis	21
Atlanta	21
San Diego	5
Milwaukee	19
Indianapolis	8

City	Workers Using Public Transportation (%)
Seattle	15
Pittsburgh	30
Denver	8
Kansas City	11
Minneapolis	19

Source: United States Bureau of the Census, *Statistical Abstracts of the United States: 1977* (98th edition). Washington, D.C., 1977.

a. Calculate the mean and median of this distribution.

b. Compare the mean and median. Which is the higher value? Why?

c. If you removed New York from this distribution and recalculated, what would happen to the mean? To the median? Why?

d. Report the mean and median as you would in a formal research report.

3.15 Professional athletes are threatening to strike because they claim that they are underpaid. The team owners have released a statement that says, in part, "the average salary for players was $1.2 million last year." The players counter by issuing their own statement that says, in part, "the average player earned only $753,000 last year." Is either side necessarily lying? If you were a sports reporter and had just read Chapter 3 of this text, what questions would you ask about these statistics?

SPSS for Windows

Using *SPSS for Windows* for Measures of Central Tendency and Percentiles

Start *SPSS for Windows* by clicking the SPSS icon on your monitor screen. Load the 1998 GSS and when you see the message "SPSS Processor is Ready" on the bottom of the "closest" screen, you are ready to proceed.

SPSS DEMONSTRATION 3.1 Producing Measures of Central Tendency

The only procedure in SPSS that will produce all three commonly used measures of central tendency (mode, median, and mean) is **Frequencies.** We used this procedure to produce frequency distributions in Demonstrations 2.1 and 2.2 and in Appendix F. Here we will use **Frequencies** to calculate measures of central tendency for three variables: *relig* (religious denomination), *attend* (frequency of attendance at religious services), and *age*.

The three variables vary in level of measurement, and we could request only the appropriate measure of central tendency for each variable. That is, we could request the mode for *relig* (nominal), the median for *attend* (ordinal), and the mean for *age* (interval-ratio). While this would be reasonable, it's actually more convenient to get all three measures for each variable and ignore the irrelevant output. Statistical packages typically generate more information than necessary, and it is common to disregard some of the output.

To produce modes, medians, and means, begin by clicking **Analyze** from the menu bar and then click **Descriptive Statistics** and **Frequencies.** In the **Frequencies** dialog box, find the variable names in the list on the left and click the arrow button in the middle of the screen to move the names (*relig, attend,* and *age*) to the **Variables** box on the right.

To request specific statistics, click the **Statistics** button at the bottom of the **Frequencies** dialog box, and the **Frequencies: Statistics** dialog box will open. Find the **Central Tendency** box on the right and click **Mean, Median,** and **Mode.** Click **Continue,** and you will be returned to the **Frequencies** dialog box, where you might want to click the **Display Frequency Tables** box. When this box is *not* checked, SPSS will not produce frequency distribution tables, and only the statistics we request (mode, median, and mean) will appear in the **Output** window. Click **OK,** and SPSS will produce the following output:

Statistics

		AGE OF RESPONDENT	HOW OFTEN R ATTENDS RELIGIOUS SERVICES	RS RELIGIOUS PREFERENCE
N	Valid	1385	1365	1372
	Missing	2	22	15
Mean		44.94	3.65	1.89
Median		41.00	3.00	1.00
Mode		34	0	1

Looking only at the most appropriate measures for each variable, the mode for *relig* ("RS RELIGIOUS PREFERENCE") is "1" (see the bottom line of output). What does the value mean? You can find out by consulting either the code book in Appendix G of the text or the online code book. To use the latter, click **Utilities** and then click **Variables** and find *relig* in the variable list on the left. Either way, you will find that a score of 1 indicates Protestant. This was the most common religious affiliation in the sample and, thus, is the mode.

The median for *attend* ("HOW OFTEN R ATTENDS RELIGIOUS SERVICES") is a value of "3.00." Again, use either Appendix G or the online code book, and you will see that the category associated with this score is "several times a year." This means that the middle case in this distribution of 1365 cases has a score of 3 (or that the middle case is in the interval "several times a year").

The output for *age* indicates that the respondents were, on the average,

44.94 years of age. Since age is an interval-ratio variable that has been measured in a defined unit (years), the value of the mean is numerically meaningful, and we do not need to consult the code book to interpret its meaning.

Note that SPSS did not hesitate to compute means and medians for the two variables that were not interval-ratio. The program cannot distinguish between numerical codes (such as the scores for *relig*) and actual numbers—to SPSS, all numbers are the same. Also, SPSS cannot screen your commands to see if they are statistically appropriate. If you request nonsensical statistics (average sex or median religious denomination), SPSS will carry out your instructions without hesitation. The blind willingness of SPSS to simply obey commands makes it easy for you, the user, to request statistics that are completely meaningless. Computers don't care about meaning; they just crunch the numbers.

In this case, it was more convenient for us to produce statistics indiscriminately and then ignore the ones that are nonsensical. This will not always be the case, and the point of all this, of course, is to caution you to use SPSS wisely. As the manager of your local computer center will be quick to remind you, computer resources (including paper) are not unlimited.

Before closing this demonstration, note that the mean for *age* is greater than the median. Recalling Section 3.6, this indicates that the variable has a positive skew or a few extremely high (old) cases. Verify this by doing a line chart for *age:* click **Graphs** and then click **Line** and then **Define.** The **Define Simple Line** dialog box appears with variable names listed on the left. Choose *age* from the variable list and click the arrow button in the middle of the screen to move the variable name to the **Category Axis** box. Click the **Options** button in the lower-left-hand corner and make sure that the button next to the **Display groups defined by missing values** option is *not* selected (or checked). Click **Continue** and then click **OK,** and the line chart for *age* will be produced.

Note how the chart peaks in the 30s and 40s age groups and then gradually declines into the 70s and 80s. Although this figure is not as smooth as Figure 3.1, it does indicate that *age* is skewed in a positive direction (i.e., has a few very high scores).

SPSS DEMONSTRATION 3.2 The Descriptives Command

The **Descriptives** command in *SPSS for Windows* is designed to provide summary statistics for continuous interval-ratio-level variables. By default (i.e., unless you tell it otherwise), **Descriptives** produces the mean, the minimum and maximum scores (i.e., the lowest and highest scores), and the standard deviation (which will be discussed in Chapter 4). Unlike **Frequencies,** this procedure will not produce frequency distributions.

To illustrate the use of **Descriptives,** let's run the procedure for *age, educ* (HIGHEST YEAR OF SCHOOL COMPLETED), and *tvhours* (HOURS PER DAY WATCHING TV). Click **Analyze, Descriptive Statistics,** and **Descriptives.** The **Descriptives** dialog box will open. This dialog box looks just like the **Frequencies** dialog box and works in the same way. Find the variable names in the list on the left and, once they are highlighted, click the arrow button in the middle of the screen to transfer them to the **Variables** box on the right. Click **OK,** and the following output will be produced:

Descriptive Statistics

	N	Minimum	Maximum	Mean	Std. Deviation
AGE OF RESPONDENT	1385	18	89	44.94	17.08
HIGHEST YEAR OF SCHOOL COMPLETED	1381	0	20	13.37	2.86
HOURS PER DAY WATCHING TV	1134	0	21	2.86	2.20
Valid N (listwise)	1126				

On the average, the sample is 44.94 years of age (this duplicates the **Frequencies** output in Demonstration 3.1), has over 13 years of education, and watches almost three hours of TV daily.

SPSS DEMONSTRATION 3.3 Finding Percentiles, Deciles, and Quartiles

The **Frequencies** procedure can be used to find percentiles, deciles, and quartiles for any variable (see Section 3.4). These statistics are especially useful for continuous variables with broad ranges of scores. One of the few variables that fits this description in the GSS data file is *age,* and we will once again use this variable for our illustrations.

Click **Analyze, Descriptive Statistics,** and **Frequencies** to get the **Frequencies** dialog box. Click the **Statistics** button to get the **Frequencies: Statistics** dialog box. For purposes of illustration, let's find quartiles, deciles, and the 23rd and 47th percentiles. In the **Percentile Values** box, check the boxes next to **Quartiles** and next to **Cut points for 10 equal groups.** The latter instruction will find deciles (ten equal groups) and could be changed to split the distribution in other ways (e.g., into thirds).

To get specific percentiles, check the box next to **Percentiles,** type the desired percentile values in the box to the right (23 and 47), and click the **Add** button after each value. Click **Continue,** and you will return to the **Frequencies** dialog box, where you might want to click the **Display Frequency Tables** box. Recall that when this box is *not* checked, SPSS will not produce frequency distribution tables. Click **OK,** and SPSS will produce the following output:

AGE OF RESPONDENT

N	Valid	1385
	Missing	2
Mean		44.94
Median		41.00
Mode		34
Percentiles	10	25.00
	20	30.00
	23	31.00
	25	32.00
	30	34.00
	40	37.00
	47	40.00
	50	41.00
	60	46.00
	70	52.00
	75	56.00
	80	60.00
	90	72.00

The output prints the deciles, quartiles, and the percentiles we requested in order. Find the 10th percentile (noted as 10.00 in the column labeled "Percentiles"). The score associated with this percentile is 25 (noted as a value of 25.00). This indicates that 10% of the sample were younger than 25. Similarly, 20% were younger than age 30, 23% (the percentile we requested) were less than 31, 25% (the first quartile) were younger than 32, and so forth. Note that the fifth decile, 2nd quartile, and 50th percentile are all associated with the age of 41, which is also the value of the median.

MicroCase

Using MicroCase for Measures of Central Tendency and Percentiles

Click the MicroCase icon on your monitor screen, and the **File and Data** screen (student version of MicroCase) or the **File Management** screen (full version) will open. Load the 1998 GSS by clicking the **Open File** command and clicking the file name. Double click the file name to open the data set. The **File Settings** window will appear. Click **OK,** and you will be ready to proceed.

MICROCASE DEMONSTRATION 3.1
Producing Measures of Central Tendency

We have already used the **Univariate** command to produce graphs and frequency distributions (see MicroCase Demonstrations at the end of Chapter 2 and Appendix F). As you may recall, MicroCase automatically reports the mean and the median above the frequency distribution, and the value of the mode can be found by inspecting the table. Here we will use this command to get measures of central tendency for *relig* (religious denomination, variable #22), *attend* (frequency of attendance at religious services, variable #23), and *age* (variable #6).

The three variables vary in level of measurement, and we will focus on the most appropriate measure of central tendency. That is, we will find the mode for *relig* (a nominal variable), the median for *attend* (an ordinal variable), and the mean for *age* (interval-ratio). To begin with *relig,* click the **Univariate** command, find *relig* (variable #22) in the list of variables, move it to the **Primary Variable** box, and click **OK.** The first screen you will see is a pie chart, and it will be immediately obvious from both the graph and the table to the left that "Protestant" is the modal (largest) category. This can also be confirmed by clicking **Summary** in the **Statistics** box on the left of the screen and producing the following frequency distribution, along with some other statistics:

RELIG -- RS RELIGIOUS PREFERENCE

| Mean: 1.889 | Std.Dev.: 1.218 | N: 1372 |
| Median: 1.000 | Variance: 1.484 | Missing: 15 |

99% confidence interval +/- mean: 1.804 to 1.974
95% confidence interval +/- mean: 1.825 to 1.954

Category	Freq.	%	Cum.%	Z-Score
1) Protestant	733	53.4	53.4	-0.730
2) Catholic	367	26.7	80.2	0.091
3) Jewish	23	1.7	81.9	0.912
4) None	189	13.8	95.6	1.732
5) OTHER	60	4.4	100.0	2.553

There are 733 Protestants in this sample of 1372, and this is by far the single largest category at 53.4%.

Click the curled arrow to return to the **Univariate** window and select *attend* as the primary variable and click **OK.** To get the median, click **Summary** in the **Statistics** box on the left of the screen, and the following output will be produced:

ATTEND -- HOW OFTEN R ATTENDS RELIGIOUS SERVICES

| Mean: 3.651 | Std.Dev.: 2.753 | N: 1365 |
| Median: 3.000 | Variance: 7.579 | Missing: 22 |

99% confidence interval +/- mean: 3.459 to 3.843
95% confidence interval +/- mean: 3.504 to 3.797

Category	Freq.	%	Cum.%	Z-Score
0) NEVER	260	19.0	19.0	-1.326
1) LT ONCE A	143	10.5	29.5	-0.963
2) ONCE A YEA	148	10.8	40.4	-0.600
3) SEVRL TIME	151	11.1	51.4	-0.236
4) ONCE A MON	106	7.8	59.2	0.127
5) 2-3X A MON	122	8.9	68.1	0.490
6) NRLY EVERY	97	7.1	75.2	0.853
7) EVERY WEEK	229	16.8	92.0	1.217
8) MORE THN O	109	8.0	100.0	1.580

The median for *attend* is reported in the statistics above the table as a value of "3.00", which—as you can see from the label to the left of the score—stands for "SEVRL TIME." The labels in the frequency distribution are highly abbreviated and, if you need some further translation, you can consult Appendix G of the textbook or check the **Variable Description** window on the **Univariate** dialog box. Either way, you'll see that the full label for the score of 3 is "several times a year."

To get measures of central tendency for *age,* click the curled arrow, select age as the primary variable, and click **OK.** To save space, we will reproduce only the statistics associated with this variable, not the frequency distribution:

```
           AGE  --  AGE OF RESPONDENT
Mean:    44.939    Std.Dev.: 17.080            N: 1385
Median: 41.000    Variance: 291.709   Missing: 2
99% confidence interval +/- mean: 43.756 to 46.121
95% confidence interval +/- mean: 44.039 to 45.838
```

The respondents were, on the average, 44.94 years of age. Since age is an interval-ratio variable that has been measured in a defined unit (years), the value of the mean is numerically meaningful, and we do not need to consult the code book to interpret its meaning.

Note that MicroCase did not hesitate to compute means and medians for the two variables that were not interval-ratio. The program cannot distinguish between numerical codes (such as the scores for *relig*) and actual numbers—to Micro-Case, all numbers are the same. Also, MicroCase cannot screen your commands to see if they are statistically appropriate. If you request nonsensical statistics (average sex or median religious denomination), MicroCase will carry out your instructions without hesitation. The blind willingness of MicroCase to simply obey commands makes it easy for you, the user, to request statistics that are completely meaningless. Computers don't care about meaning; they just crunch the numbers.

Before closing this demonstration, note that the mean for *age* is greater than the median. Recalling Section 3.6, this indicates that the variable has a positive skew or a few extremely high (old) cases. You can verify this by returning to the bar chart for *age:* click **bar freq** in the **Graphs** window. Note how the graph peaks in the 30s and 40s age groups and then gradually declines into the 70s and 80s. Although this figure is not as smooth as Figure 3.1, it does indicate that *age* is skewed in a positive direction (i.e., has a few very high scores).

MICROCASE DEMONSTRATION 3.2

In this demonstration, we'll examine the central tendency of several more variables using the **Univariate** command. With the 1998 GSS loaded, click **Univariate** and select *educ* as the **Primary Variable.** As we saw in Demonstration 2.3, *educ* is an interval-ratio variable with a peak at 12 years (high school) and a smaller peak at 16 years (college). To conserve space, we will not reproduce the frequency distribution or bar chart for this variable. The summary statistics are

```
EDUC -- HIGHEST YEAR OF SCHOOL COMPLETED
┌─────────────────────────────────────────────────────────┐
│ Mean:    13.366   Std.Dev.: 2.857        N: 1381         │
│ Median: 13.000    Variance: 8.160   Missing: 6           │
│ 99% confidence interval +/- mean: 13.168 to 13.564       │
│ 95% confidence interval +/- mean: 13.216 to 13.517       │
└─────────────────────────────────────────────────────────┘
```

The average number of years completed is 13.36, and the median is 13, a smaller value. This means that the variable has a slight positive skew or a few very low scores. You can confirm this by observing the bar chart.

Now let's look at *paeduc,* the years of education completed by the respondent's father. Once again, we reproduce only the summary statistics here:

```
PAEDUC -- HIGHEST YEAR SCHOOL COMPLETED, FATHER
┌─────────────────────────────────────────────────────────┐
│ Mean:    11.576   Std.Dev.: 4.099        N: 994          │
│ Median: 12.000    Variance: 16.798  Missing: 393         │
│ 99% confidence interval +/- mean: 11.241 to 11.912       │
│ 95% confidence interval +/- mean: 11.322 to 11.831       │
└─────────────────────────────────────────────────────────┘
```

The mean of 11.58 is almost two years lower than the mean of *educ,* and the median of 12 is a year lower, indicating a sharp change in levels of education from one generation to the next. Note also that almost 400 cases are missing for *paeduc* versus only 6 cases for *educ,* indicating that many respondents were not asked this question or simply did not have this information about their fathers.

While we are making comparisons, let's use another function of MicroCase and compare levels of education between males and females. Which gender has, on the average, completed more years of schooling? Return to the **Univariate** screen and click on *educ* to make it the primary variable. Next click on *sex* and then click the arrow to move this variable name to the **Optional Subset Variables** box in the middle of the **Univariate** dialog box. This box allows us to break the sample down into subgroups based on scores on the variable specified. In this case, we will create subgroups of males and then females. On the **Subset by Categories** dialog box, click the box next to **male** and then click **OK.** Click **OK** again on the **Univariate** dialog box, and the program will produce summary statistics and charts on *educ* for males only. The summary statistics will look like this:

```
EDUC -- HIGHEST YEAR OF SCHOOL COMPLETED
┌─────────────────────────────────────────────────────────┐
│ Subset: SEX Categories: MALE                             │
├─────────────────────────────────────────────────────────┤
│ Mean:    13.521   Std.Dev.: 2.841        N: 620          │
│ Median: 13.000    Variance: 8.072   Missing: 2           │
│ 99% confidence interval +/- mean: 13.227 to 13.815       │
│ 95% confidence interval +/- mean: 13.297 to 13.745       │
└─────────────────────────────────────────────────────────┘
```

Note that MicroCase reminds us that we are using a subset based on sex and that the statistics refer to males only.

Now return to the **Univariate** dialog box and click **Clear All.** Select *educ* once more as the primary variable and *sex* as the subset variable, but this time click the box next to **female** and click **OK.** The summary statistics for females are

EDUC -- HIGHEST YEAR OF SCHOOL COMPLETED

Subset: SEX Categories: FEMALE		
Mean: 13.240 Std.Dev.: 2.865 N: 761		
Median: 13.000 Variance: 8.207 Missing: 4		
99% confidence interval +/- mean: 12.973 to 13.508		
95% confidence interval +/- mean: 13.037 to 13.444		

Comparing means (13.52 for males, 13.24 for females) and medians (13 for both groups), we see that the subgroups are very nearly identical on this variable. Both males and females average more than a year higher than a high school education and the middle case for both subgroups has 13 years of education, a year beyond high school. The mean does show a higher level of education for males but, at this point, it is hard to know how meaningful or important this difference is. We will return to this comparison in a later chapter when we have some additional statistical tools to help us analyze the difference.

Exercises

3.1 Use **Frequencies** (SPSS) or **Univariate** (MicroCase) to get all three measures of central tendency for *marital, income98, race, region, polviews, cappun, prestg80,* and five more variables of your own choosing. Select the most appropriate measure for each of these variables and write a sentence or two reporting and summarizing the measure.

3.2 Use **Descriptives** (SPSS) or **Univariate** (MicroCase) to get means for *partnrs5* and *sexfreq*. Note that these variables are ordinal in level of measurement and that, therefore, calculation of the mean is not completely justified. Such violations of strict level-of-measurement considerations are common in social science research. Write a sentence or two reporting and summarizing the mean for each variable.

3.3 **(SPSS ONLY)** Revise the command you wrote in exercise 3.1 to also find the quartiles and the deciles for each of the variables listed.

3.4 (**MicroCase** only) Use the subset option on the **Univariate** dialog box to compare males and females on occupational prestige (*prestg80*) and church attendance (*attend*). Write up your results in a few sentences.

MEASURES OF DISPERSION

By the end of this chapter, you will be able to

1. Explain the purpose of measures of dispersion and the information they convey.
2. Compute and explain the index of qualitative variation (IQV), the range (R), the interquartile range (Q), the standard deviation (s), and the variance (s^2).
3. Select an appropriate measure of dispersion and correctly calculate and interpret the statistic.
4. Describe and explain the mathematical characteristics of the standard deviation.

4.1 INTRODUCTION

By themselves, measures of central tendency cannot summarize data completely. For a full description of a distribution of scores, measures of central tendency must be paired with **measures of dispersion.** Whereas the mean, median, and mode are designed to locate the typical and/or central scores, measures of dispersion provide information about the amount of heterogeneity or variety within a distribution of scores.

The importance of the concept of **dispersion** might be easier to grasp if we consider a brief example. Suppose that the director of public safety for a city has been given the task of evaluating two ambulance services that have contracted with the city to provide emergency medical aid. As a part of the investigation, she has collected data on the response time of both services to calls for assistance. Data collected for the past year show that the mean response time is 7.4 minutes for Service A and 7.6 minutes for Service B. The average response times are essentially the same and provide no basis for judging one service as more or less efficient than the other. Measures of dispersion, however, can reveal substantial differences in the underlying distributions even when the measures of central tendency are equivalent. For example, consider Figure 4.1, which displays the distribution of response times for the two services.

Compare the shapes of these two figures. Note that the frequency polygon for Service B is much flatter than that for Service A. The scores for Ser-

FIGURE 4.1   RESPONSE TIME FOR TWO AMBULANCE SERVICES

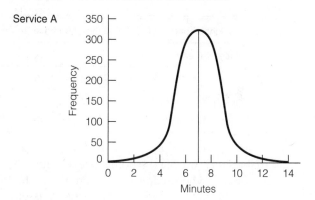

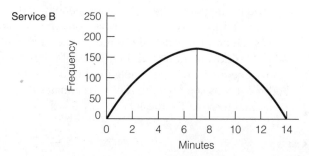

vice B are more spread out over the range of scores, while the scores for Service A are clustered or grouped around the mean. Comparatively speaking, Service B was much more variable in response time. Thus, even though both distributions have essentially the same mean, the distribution of scores for Service B is more heterogeneous. Had the dispersion of these distributions not been investigated, a possibly important difference in the performance of the two ambulance services might have gone unnoticed.

Although you can derive a general notion of what is meant by dispersion from Figure 4.1, the concept is not easily described in words alone. In this chapter we will introduce some of the more common measures of dispersion, each providing a precise, quantitative indication of heterogeneity. We will begin with the index of qualitative variation, mention two measures—the range and the interquartile range—quite briefly, and devote most of our attention to the standard deviation and the variance.

4.2 THE INDEX OF QUALITATIVE VARIATION (IQV)

The **index of qualitative variation (IQV)** is essentially the ratio of the amount of variation actually observed in a distribution of scores to the maximum variation that could exist in that distribution. The index varies from 0.00 (no variation) to 1.00 (maximum variation) and is used most commonly with

TABLE 4.1   RACIAL COMPOSITION OF THREE NEIGHBORHOODS

Neighborhood A		Neighborhood B		Neighborhood C	
Race	Frequency	Race	Frequency	Race	Frequency
White	90	White	60	White	30
Black	0	Black	20	Black	30
Other	0	Other	10	Other	30
	$N = 90$		$N = 90$		$N = 90$

variables measured at the nominal level. However, the IQV can be used with any variable when scores have been grouped into a frequency distribution.

To illustrate the logic of this statistic, assume that a researcher is interested in comparing the racial heterogeneity of the three small neighborhoods displayed in Table 4.1. By inspection, you see that neighborhood A is the least heterogeneous of the three. In fact, since all residents of A are white, there is no variation at all in this distribution. Neighborhood B is more heterogeneous than A, and neighborhood C is the most heterogeneous of the three. Let us see how the IQV substantiates these observations. The computational formula for the IQV is

FORMULA 4.1

$$IQV = \frac{k(N^2 - \Sigma f^2)}{N^2(k - 1)}$$

where k = the number of categories
N = the number of cases;
Σf^2 = the sum of the squared frequencies

To use this formula, the sum of the squared frequencies must first be computed. Do so by adding a column to the frequency distribution for the squared frequencies and then sum this column. This procedure is illustrated in Table 4.2.

The sum of the frequency column is, of course, N, and the sum of the squared frequency (Σf^2) is the total of the second column. Substituting these values into Formula 4.1 for neighborhood A, we would have an IQV of 0.00:

$$IQV = \frac{3(8100 - 8100)}{8100(2)}$$

$$IQV = \frac{3(0)}{16,200}$$

$$IQV = \frac{0}{16,200}$$

$$IQV = 0.00$$

TABLE 4.2   FINDING THE SUM OF THE SQUARED FREQUENCIES

	Neighborhood A	
Race	Frequency	Squared Frequency
White	90	8100
Black	0	0
Other	0	0
	$N = 90$	8100

	Neighborhood B	
Race	Frequency	Squared Frequency
White	60	3600
Black	20	400
Other	10	100
	$N = 90$	4100

	Neighborhood C	
Race	Frequency	Squared Frequency
White	30	900
Black	30	900
Other	30	900
	$N = 90$	2700

Since the values of k and N are the same for all three tables, the IQV for the other two neighborhoods can be found by changing the values for Σf^2. For neighborhood B,

$$\text{IQV} = \frac{3(8100 - 4100)}{8100(2)}$$

$$\text{IQV} = \frac{12,000}{16,200}$$

$$\text{IQV} = 0.74$$

and, similarly, for neighborhood C,

$$\text{IQV} = \frac{3(8100 - 2700)}{16,200}$$

$$\text{IQV} = \frac{16,200}{16,200}$$

$$\text{IQV} = 1.00$$

Thus, the IQV, in a quantitative and precise way, substantiates our impressions. Neighborhood A exhibits no variation on the variable (IQV = 0.00), neighborhood B has substantial variation (IQV = 0.74), and neighborhood C exhibits the maximum amount of variation possible (IQV = 1.00) for these data. *(For practice in calculating and interpreting the IQV, see problems 4.1 and 4.2.)*

4.3 THE RANGE (*R*) AND INTERQUARTILE RANGE (*Q*)

The **range (*R*)** is defined as the distance between the highest and lowest scores in a distribution. It is quite easy to calculate (high score minus low score) and is perhaps most useful to gain a quick and general notion of variability while scanning many distributions.

Unfortunately, the range is often deceptive as a measure of dispersion. It is based on only two scores in the distribution (the highest and lowest scores). Since almost any sizable distribution will contain some atypically high and low scores, the range might be quite misleading. Also, *R* yields no information about the nature of the scores between these two extremes.

The **interquartile range (*Q*)** is a kind of range. It avoids some of the problems associated with *R* by considering only the middle 50% of the cases in a distribution. To find *Q*, arrange the scores from highest to lowest and then divide the distribution into quarters (as distinct from halves, as in locating the median). The first quartile (Q_1) is the point below which 25% of the cases fall and above which 75% of the cases fall. The second quartile (Q_2) divides the distribution into halves (thus, Q_2 is equal in value to the median). The third quartile (Q_3) is the point below which 75% of the cases fall and above which 25% of the cases fall. Thus, if line *LH* represents a distribution of scores, the quartiles are located as shown:

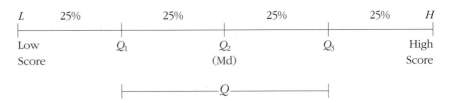

The interquartile range is defined as the distance from the third to the first quartile ($Q = Q_3 - Q_1$). Thus, *Q* essentially extracts the middle 50% of the cases and, like *R*, is based on only two scores. While *Q* avoids the problem of being based on the most extreme scores, it has all the other disadvantages associated with *R*. Most importantly, *Q* also fails to yield any information about the nature of the scores other than the two upon which it is based.

TABLE 4.3   PER CAPITA EXPENDITURES ON PUBLIC EDUCATION, 1998

State	Expenditures per Capita
New Jersey	1571
Michigan	1370
Wyoming	1347
Texas	1346
Maine	1258
Ohio	1206
Oregon	1185
Illinois	1147
New Hampshire	1142
Pennsylvania	1140
Virginia	1132
Idaho	1122
Nebraska	1117
Florida	1074
Arizona	1064
California	1061
North Carolina	1013
Alabama	1011
Louisiana	951
Mississippi	939

Source: United States Bureau of the Census, *Statistical Abstracts of the United States: 1999* (119th edition). Washington, D.C., 2000. p. 182.

4.4 COMPUTING THE RANGE AND INTER-QUARTILE RANGE*

Table 4.3 presents per capita school expenditures for 20 states. What are the range and interquartile range of these data?

Note that the scores have already been ordered from high to low. This makes the range easy to calculate and is necessary for finding the interquartile range. Of the 20 states listed in Table 4.3, New Jersey spent the most, per capita, on public education ($1571) and Mississippi spent the least ($939). The range is therefore 1571 − 939, or $632 ($R = \$632$).

To find Q, we must locate the first and third quartiles (Q_1 and Q_3). In a manner analogous to the technique for finding the median, we can define both of these points in terms of the scores associated with certain cases. Q_1 is determined by multiplying N by (.25). Since (20) $\times$ (.25) is 5, Q_1 is the score associated with the fifth case, counting up from the lowest score. The fifth case is California, with a score of 1061. So, $Q_1 = 1061$. The case that lies at the third quartile (Q_3) is given by multiplying N by (.75): (20) $\times$ (.75) = 15th case. The 15th case is Ohio, with a score of 1206 ($Q_3 = 1206$). Therefore:

$$Q = Q_3 - Q_1$$
$$Q = 1206 - 1061$$
$$Q = 145$$

*This section is optional.

In most situations, the locations of Q_1 and Q_3 will not be as obvious as they are when $N = 20$. For example, if N had been 157, then Q_1 would be $(157)(.25)$, or the score associated with the 39.25th case, and Q_3 would be $(157)(.75)$, or the score associated with the 117.75th case. Since fractions of cases are impossible, these numbers present some problems. The easy solution to this difficulty is to take the score of the closest case to the numbers that mark the quartiles. Thus, Q_1 would be defined as the score of the 39th case and Q_3 as the score of the 118th case.

The more accurate solution would be to take the fractions of cases into account. For example, Q_1 could be defined as the score that is one-quarter of the distance between the scores of the 39th and 40th cases, and Q_3 could be defined as the score that is three-quarters of the distance between the scores of the 117th and 118th cases. (This procedure could be analogous to defining the median—which is also Q_2—as halfway between the two middle scores when N is even.) In most cases, the differences in the values of Q for these methods would be quite small. *(For practice in finding and interpreting Q, see problem 4.5. The range may be found for any of the problems at the end of this chapter except 4.1 and 4.2.)*

4.5 THE STANDARD DEVIATION

The basic limitation of both Q and R is their failure to use all the scores in the distribution—they do not capitalize on all the available information. Also, neither Q nor R yields any information about the average or typical deviation of the scores. How can we design a measure of dispersion that would correct these faults? We can begin with some specifications: a good measure of dispersion should

1. Use all the scores in the distribution.
2. Describe the average or typical deviation of the scores.
3. Increase in value as the distribution of scores becomes more heterogeneous.

One way to develop a statistic to meet these criteria would be to start with the distances between each score and some central point (that is, the mean). The distances between the scores and the mean ($X_i - \overline{X}$) are called **deviations,** and, obviously, the greater the variety or heterogeneity of the scores, the greater the deviations. If the scores were clustered around the mean, the deviations would be small, but they would increase as the scores became more spread out or more varied. How can we use the deviations of the scores around the mean to develop a useful statistic?

One course of action would be to use the sum of the deviations— $\Sigma(X_i - \overline{X})$—as the basis for a statistic, but, as we saw in Section 3.6, the sum of deviations will always be zero. To illustrate, consider a distribution of five scores: 10, 20, 30, 40, and 50. If the deviations of the scores from the mean were summed, we would have

Scores (X_i)	Deviations $(X_i - \overline{X})$
10	$(10 - 30) = -20$
20	$(20 - 30) = -10$
30	$(30 - 30) = 0$
40	$(40 - 30) = 10$
50	$(50 - 30) = 20$
$\Sigma(X_i) = 150$	$\Sigma(X_i - \overline{X}) = 0$

$$\overline{X} = 150/5 = 30$$

Still, the sum of the deviations are a logical basis for a statistic that measures the amount of variety in a set of scores, and statisticians have developed two ways around the fact that the positive deviations always equal the negative deviations. Both solutions eliminate the negative signs. The first does so by using the absolute values or by ignoring signs when summing the deviations. In the example above, this would produce a sum of (20 + 10 + 0 + 10 + 20) or 60.

The second solution squares the deviations. This makes all values positive because negative numbers produce positive numbers when multiplied by themselves. In the example above, the sum of the squared deviations would be (400 + 100 + 0 + 100 + 400) or 1000.

We now have two strategies for designing a good measure of dispersion, but we need to make one last adjustment. Whether using absolute or squared values, the sum of the deviations will increase with sample size (that is, the larger the number of scores, the greater the value of the measure). This would make it very difficult to compare the relative variability of distributions of different sizes. We can solve this problem by dividing the sum of the deviations—whether squared or absolute—by N (sample size) and thus standardizing for samples of different sizes.

The measure of dispersion based on the sum of the absolute values is called the **average deviation (AD)** and is defined in Formula 4.2.

FORMULA 4.2

$$AD = \frac{\Sigma|X_i - \overline{X}|}{N}$$

where $|X_i - \overline{X}|$ directs us to add up the deviations without regard to sign

For the sample problem above, AD would be

$$AD = \frac{|20| + |10| + |0| + |10| + |20|}{5}$$

$$AD = \frac{60}{5} = 12$$

The average deviation indicates that, on the average, the scores lie 12 units from the mean.

The average deviation fulfills all three of our criteria for a good measure of dispersion, but absolute values are difficult to manipulate algebraically and, for this reason, the average deviation is not often used in the social sciences.

The measure of dispersion that eliminates negative signs by squaring all deviations yields a statistic known as the **variance,** which is symbolized as s^2 when referring to a sample and σ^2 (lowercase Greek letter sigma squared) when referring to a population. The variance is used primarily in inferential statistics, although it is a central concept in the design of some measures of association. For purposes of describing the dispersion of a distribution, a closely related statistic called the **standard deviation** (symbolized as s for samples and σ for populations) is typically used, and this statistic will be our focus for the remainder of the chapter. The formulas for the sample variance and standard deviation are

FORMULA 4.3

$$s^2 = \frac{\Sigma(X_i - \overline{X})^2}{N}$$

FORMULA 4.4

$$s = \sqrt{\frac{\Sigma(X_i - \overline{X})^2}{N}}$$

Strictly speaking, Formulas 4.3 and 4.4 are for the variance and standard deviation of a population. Slightly different formulas, with $N - 1$ instead of N in the denominator, should be used when we are working with random samples rather than entire populations. This is important because many of the electronic calculators and statistical software packages that you might be using use "$N - 1$" in their calculations and will produce results that are at least slightly different from Formulas 4.3 and 4.4. Some calculators offer the choice of "$N - 1$" or "N" in the denominator. If you use the latter, the values calculated for the standard deviation should match the values in this text.

At any rate, for the problem above, the standard deviation is

$$s = \sqrt{\frac{\Sigma(X_i - \overline{X})^2}{N}}$$

$$s = \sqrt{\frac{(-20)^2 + (-10)^2 + (0)^2 + (10)^2 + (20)^2}{5}}$$

$$s = \sqrt{\frac{400 + 100 + 0 + 100 + 400}{5}}$$

$$s = \sqrt{\frac{1000}{5}}$$

$$s = \sqrt{200}$$

$$s = 14.14$$

TABLE 4.4   COMPUTING THE STANDARD DEVIATION BY THE DEFINITIONAL FORMULA (Formula 4.4)

Test Score (X_i)	$(X_i - \overline{X})$	$(X_i - \overline{X})^2$
57	-19	361
60	-16	256
65	-11	121
70	-6	36
78	2	4
80	4	16
82	6	36
85	9	81
90	14	196
93	17	289
$\Sigma(X_i) = 760$	$\Sigma(X_i - \overline{X}) = 0$	$\Sigma(X_i - \overline{X})^2 = 1396$

$$\overline{X} = 760/10 = 76.0$$

To find the variance, square the standard deviation. For this problem, the variance is $s^2 = (14.14)^2 = 200$.

4.6 COMPUTING THE STANDARD DEVIATION*

Formula 4.4 is the definitional formula for the standard deviation, but it is tedious to use in actual computations. For greater speed and ease of calculation, a number of computational formulas can be derived from Formula 4.4, one of the most commonly used of which is Formula 4.5.

FORMULA 4.5

$$s = \frac{1}{N}\sqrt{N\Sigma X_i^2 - (\Sigma X_i)^2}$$

This formula is easier to use than Formula 4.4 since it eliminates computational steps. The formula also introduces two new symbols, which we should consider before proceeding. The first new symbol is ΣX_i^2 and the second is $(\Sigma X_i)^2$. Although these symbols might seem quite similar, they refer to two very different operations. The first—ΣX_i^2—is read as "the sum of the squared scores" and is found by first squaring and then summing the scores. On the other hand, $(\Sigma X_i)^2$ is read as "the sum of the scores, squared" and is found by first summing the scores and then squaring the sum. See the Basic Math Review at the web site for this text for a more comprehensive treatment of the differences between these symbols.

To illustrate, we will work through the computation of the standard deviation using the definitional and computational formulas. The data in Table 4.4 represent the scores of 10 students on a statistics test. Using the figures calculated in Table 4.4 and Formula 4.4, we get

---

*This section is optional.

$$s = \sqrt{\frac{\Sigma(X_i - \overline{X})^2}{N}}$$

$$s = \sqrt{\frac{1396}{10}}$$

$$s = \sqrt{139.6}$$

$$s = 11.82$$

TABLE 4.5 COMPUTING THE STANDARD DEVIATION BY THE COMPUTATIONAL FORMULA (Formula 4.6)

Test Score (X_i)	X_i²
57	3249
60	3600
65	4225
70	4900
78	6084
80	6400
82	6724
85	7225
90	8100
93	8649
$\Sigma X_i = 760$	$\Sigma X_i^2 = 59,156$

Formula 4.5 eliminates the need to subtract the mean from each score and, thus, is a significantly faster computational routine. Using data from Table 4.5 and Formula 4.5, we get

$$s = \frac{1}{N}\sqrt{N\Sigma X_i^2 - (\Sigma X_i)^2}$$

$$s = \frac{1}{10}\sqrt{(10)(59{,}156) - (760)^2}$$

$$s = \frac{1}{10}\sqrt{591{,}560 - 577{,}600}$$

$$s = \frac{1}{10}\sqrt{13{,}960}$$

$$s = \frac{1}{10}(118.15)$$

$$s = 11.82$$

Note that the value of the standard deviation is exactly the same as that computed using the definitional formula. Formula 4.5 is equivalent to, but significantly easier to use than, Formula 4.4. Also note that the variance can be found by squaring the value of the standard deviation. In the example above, the variance would be $(11.82)^2$, or 139.71. *(For practice in computing and interpreting s and s², see any of the problems at the end of this chapter except 4.1 and 4.2. Problems with smaller data sets, such as 4.3 to 4.5, are recommended for practicing computations until you are comfortable with these procedures.)*

4.7 INTERPRETING THE STANDARD DEVIATION

It is very possible that the meaning of the standard deviation (i.e., why we calculate it) is not obvious to you at this point. You might be asking: "Once I've gone to the trouble of calculating the standard deviation, what do I have?" The meaning of this measure of dispersion can be expressed in three ways. The first and most important involves the normal curve, and we will defer this interpretation until the next chapter.

A second way of thinking about the standard deviation is as an index of variability: the standard deviation increases in value as the distribution becomes more variable. The reverse relationship also holds: the less the variability in the distribution, the lower the value of the standard deviation. The limiting case in this direction would be a distribution with no dispersion—

Application 4.1

At a local preschool, 10 children were observed for 1 hour, and the number of aggressive acts committed by each was recorded in the following list. What is the standard deviation of this distribution? We will use Formula 4.5 to compute the standard deviation.

Number of Aggressive Acts (X_i)	X_i^2
0	0
3	9
5	25
2	4
7	49
10	100
0	0
8	64
2	4
0	0
$\sum X_i = 37$	$\sum X_i^2 = 255$

Substituting these values into Formula 4.5, we have

$$s = \frac{1}{N} \sqrt{N \sum X_i^2 - (\sum X_i)^2}$$

$$s = \frac{1}{10} \sqrt{(10)(255) - (37)^2}$$

$$s = \frac{1}{10} \sqrt{2550 - 1369}$$

$$s = \frac{1}{10} \sqrt{1181}$$

$$s = \frac{1}{10} (34.37)$$

$$s = 3.44$$

The standard deviation for these data is 3.44.

Application 4.2

You have just won the state lottery and are, overnight, fabulously wealthy. Even after a brief visit with the IRS you still have enough money to live wherever you want and completely ignore such mundane considerations as work. One of your first decisions is to find the "nicest place to live" in all the world. Because you are somewhat eccentric, your only criterion for "nicest place" is climate. Specifically, you want to locate a city where the temperature is exactly 78°F. After much research you find three cities where the average daily temperature is exactly 78°F. Which of the three cities will you choose as your permanent residence?

The preceding information reports central tendency (the average temperature) and does not provide you with a basis for deciding among the three cities. What if you also discovered that the standard deviation and the range of the daily temperature were, respectively, .7 and 3 degrees for City A; 10.3 and 30 degrees in City B; and 25.8 and 103 degrees in City C? Can you choose a permanent residence now? Which city would you choose? Why?

a distribution where every case had exactly the same score. In such a case, the standard deviation would be 0. Thus, 0 is the lowest value possible for the standard deviation (although there is no upper limit).

A third way to get a feel for the meaning of the standard deviation is by comparing one distribution with another. Assume we had the following information on the distribution of household income in two cities:

City A	City B
$\overline{X} = 31{,}233$	$\overline{X} = 31{,}233$
$s = 1729$	$s = 5268$

The two cities are equal in central tendency, but income is much more variable from household to household in City B. This would be the case, for example, if City B has more families in both the higher and lower income brackets, and families in City A were more clustered around the mean income. The standard deviation is extremely useful for making comparisons of this sort between distributions of scores.

4.8 INTERPRETING STATISTICS: USING MEASURES OF CENTRAL TENDENCY AND DISPERSION TO ANALYZE DIVERSITY IN THE UNITED STATES

As the United States moves into the 21st century, "diversity" is no longer simply a matter of political correctness: it has become a social and political reality. Using data from the United States Census Bureau,[1] this installment of "interpreting statistics" describes the racial and ethnic composition of the fifty states and District of Columbia to the year 2025.

Trends in Racial and Ethnic Group Percentages.   The growing diversity of the nation is reflected in the changing percentages of the five racial and ethnic groups included in Table 4.6. The table presents the mean and median percentages of the groups in the fifty states plus the District of Columbia (central tendency) along with the range and standard deviation (dispersion) for each of six years. The statistics for the first three years are based on actual data, while those for 2005, 2015, and 2025 are projections. Race and ethnicity are, of course, nominal level variables. If we worked with the actual categories of the variable (black, white, etc.), we could use only the mode and the IQV as summary statistics. The *percentage* of each group in a state, on the other hand, can be treated as an interval-ratio variable since it has a true zero point (when a state has no members of a group among its residents) and equal intervals.

Take a few minutes to inspect Table 4.6 and see if you can detect the basic trends. This may take some effort on your part since the table is admit-

---

[1]Estimates for 1990 are from "Population Estimates for States by Age, Race, Sex, and Hispanic Origin: July 1, 1990," Population Estimates Program, Population Division, U.S. Census Bureau, Washington, D.C. Projections for 1995–2025 have been constructed from "Projected State Population, by Sex, Race, and Hispanic Origin: 1995–2025," Resident Population, Series A projections.

As was the case with frequency distributions, measures of central tendency and dispersion may not be presented in research reports in the professional literature. Given the large number of variables included in a typical research project and the space limitations in the journals and other media, there may not be room for the researcher to describe each variable fully. Furthermore, the great majority of research reports focus on relationships between variables rather than the distribution of single variables. In this sense, univariate descriptive statistics will be irrelevant to the main focus of the report.

This does not mean, of course, that univariate descriptive statistics are irrelevant to the research project; nor do I mean to imply that researchers do not calculate and interpret these statistics. Measures of central tendency and dispersion will be calculated and interpreted for virtually every variable, in virtually every research project. However, these statistics are less likely to be included in final research reports than the more analytical statistical techniques to be presented in the remainder of this text. Furthermore, some of these statistics (for example, the mean and standard deviation) serve a dual function. They not only are valuable descriptive statistics but also form the basis for many analytical techniques. Thus, they may be reported in the latter role if not in the former.

When included in research reports, measures of central tendency and dispersion will most often be presented in some summary form for all relevant variables—often in the form of a table. Means and standard deviations for many variables might, for example, be presented in the following table format.

Variable	$\overline{X}$	s	N
Age	33.2	1.3	1078
Number of children	2.3	.7	1078
Years married	7.8	1.5	1052
Income	30,078	982	978
.	.	.	.
.	.	.	.
.	.	.	.
.	.	.	.

These tables describe the overall characteristics of the sample succinctly and clearly. If you inspect the table carefully, you will have a good sense of the nature of the sample on the traits relevant to

the project. Note that the number of cases varies from variable to variable in the preceding table. This is normal and is caused by missing data or incomplete information on some of the cases.

Statistics in the Professional Literature

This section presents two brief excerpts from social science research. In both cases, the descriptive statistics are not the focus of the article but, rather, the basis for a statistical test that we will consider in Chapter 9. The first article presents means and standard deviations and the second presents means only.

Excerpt 1 Gender Differences in the Productivity of Scientists

Professor J. Scott Long notes that "The average female scientist publishes fewer articles and receives fewer citations than the average male scientist. This finding persists across disciplines and through time." (p. 1297) What factors account for the lower productivity of female scientists? In an attempt to find out, Professor Long assembled a sample of 556 males and 603 females that earned doctoral degrees in biochemistry between 1950 and 1967. He gathered information about their scholarly careers, their research productivity, and their personal lives. Some of his results are presented in the table below. Inspect the table carefully to see if you can distill the essential meanings of the information presented.

Sex Differences in the Levels of Major Variables

	Sex	Mean	Std. Dev.
Age at BS degree	♀	22.089	2.339
	♂	22.728	2.220
Age at Ph.D.	♀	31.299	5.709
	♂	29.651	3.821
Articles by student— three-year count	♀	1.529	1.627
	♂	1.932	2.232
Baccalaureate selectivity	♀	2.958	1.819
	♂	3.066	1.694
Number of children at Ph.D	♀	0.286	0.710
	♂	0.844	1.084
Interruption between BS and Ph.D. (in years)	♀	0.527	0.500
	♂	0.344	0.476

(continued)

READING STATISTICS 3: *(continued)*

Professor Long uses means and standard deviations to summarize and describe the sample and also uses these two statistics as the basis for a test that helps to identify important or consequential differences by gender. Here's what the author has to say about the differences:

Males and females come from undergraduate institutions of nearly identical selectivity. While females are about three-fourths of a year younger than males at completion of the baccalaureate, they are more than a year and one-half older than males upon completion of the Ph.D. The older age of females upon completion of the Ph.D. is the result of females interrupting their education between the baccalaureate and the Ph.D. more frequently and longer than males. While males are more likely to be married and to have children before receiving the Ph.D., females are more likely to have had their educations interrupted, possibly as a result of family obligations.

The author concludes that the key factor accounting for the productivity difference is that females have fewer and weaker relationships with faculty mentors because they more often interrupt their education for child bearing and child rearing. Professor Long concludes: "For females, having young children strongly decreases the odds of collaborating with the mentor; this in turn decreases predoctoral productivity." Curious and want to learn more? Complete bibliographic information is given below.

J. Scott Long: 1990. "The Origins of Sex Differences in Science." *Social Forces,* Vol. 68 (4): 1297–1315. Copyright © The University of North Carolina Press. Reprinted by permission.

Excerpt 2 The Adjustment to the United States of Immigrants from Africa: Implications for Research and for Social Work Practice

Professor Hugo Kamya notes that researchers have largely ignored immigrant groups from Africa. As a result, little information is available to social workers and other service providers to help them understand the specific problems, stresses, and coping mechanisms of these groups.

Professor Kamya tested a sample of 52 African immigrants on their levels of stress, hardiness (or ability to cope with stress), spiritual well-being, and level of self-esteem. He was particularly interested in the role of religion in the process of adjusting to U.S. society and divided the sample into two groups in terms of whether they saw religion as a minor or major influence. Some of his results are reported in the following table:

Mean Scores for Influence of Religion in Relation to Hardiness, Stress, Self-Esteem, Spiritual Well-Being, and Coping Resources

	Religion Is	
	Minor Influence (N = 15)	Major Influence (N = 37)
Hardiness overall	45.2	45.4
Stress	37.2	41.7
Self-esteem	66.8	76.1
Spiritual well-being overall	88.8	101.8
Coping resources total	50.8	54.9

Here's what Professor Kamya has to say about these results:

Immigrants for whom religion was a major influence reported higher scores on the spiritual well-being scale and the coping resources scale. This finding supports the literature that suggests that religion is a major coping resource for immigrants. Any threat, therefore, to a person's cultural or religious beliefs poses a threat to the person's ability to cope.

Would you like to learn more? Bibliographic information is given below.

Hugo A. Kamya. 1997. "African Immigrants in the United States: The Challenge for Research and Practice." *Social Work.* 42: 154–163.

TABLE 4.6   SUMMARY MEASURES OF RACIAL AND ETHNIC DIVERSITY, 1990–2025

Estimate for Year	1990	1995	2000	2005	2015	2025
Percentage White Population						
Mean	84.76	84.13	83.47	82.82	81.55	80.32
Median	88.24	87.56	87.09	86.31	84.93	83.64
Standard deviation	13.74	13.72	13.52	13.43	13.53	13.83
Range	66.80	65.54	64.73	64.80	65.21	66.41
Percentage African American Population						
Mean	10.76	11.04	11.30	11.56	12.10	12.63
Median	7.18	7.12	7.37	7.30	8.35	8.74
Standard deviation	12.11	12.01	11.91	11.83	11.94	12.17
Range	65.81	63.31	61.42	59.46	58.27	58.49
Percentage Native American Population						
Mean	1.63	1.62	1.64	1.66	1.76	1.86
Median	0.43	0.45	0.45	0.46	0.50	0.49
Standard deviation	2.94	2.88	2.84	2.80	2.86	2.99
Range	15.62	15.07	14.22	13.14	11.63	10.73
Percentage Asian American Population						
Mean	2.86	3.21	3.59	3.96	4.59	5.20
Median	1.16	1.43	1.75	1.95	2.26	2.54
Standard deviation	8.73	9.81	8.82	8.90	9.15	9.58
Range	62.59	63.06	62.78	62.91	63.42	65.02
Percentage Hispanic Origin Population						
Mean	5.40	6.06	6.76	7.44	8.76	10.11
Median	2.39	3.05	3.58	4.08	4.81	5.75
Standard deviation	7.49	7.95	8.42	8.95	9.99	11.03
Range	37.98	38.53	38.95	39.99	42.97	46.29
Total Number of Cases	51	51	51	51	51	51

tedly "busy" and presents a lot of information. As you read the table, keep the definitions of the statistics in mind: the mean reports the average score, and the median reports the score of the middle case. The range is the distance from the highest to lowest score, and the standard deviation, like the range, measures the amount of dispersion or variety in the scores. Don't try to absorb the information in the table all at once. Rather, scan the results and look for general trends.

For example, let's begin with the measures of central tendency. With the aid of these statistics, we can identify two patterns. Perhaps the most significant trend revealed by the table is that both the mean and median percentage of the population that is white decreases steadily across the years. Thus, white Americans will become less of a numerical majority as the 21st century progresses. Secondly, for the same variable, note that the mean is lower in value than the median for all six years. This means that the percentage white has a negative skew (see Section 3.6) or a few cases with very low scores. If you looked at the raw data (not reproduced here), you would see that the skew is produced almost entirely by two cases: the District of Columbia and Hawaii. While most states are more than 75% or 80% white, these two cases are only about a third white.

In contrast to the steady decline in the two measures of central tendency, the measures of dispersion for percent white remain fairly constant, hovering around 13.6 for the standard deviation and 66 for the range. The range is fairly easy to interpret—there is about a 66 percentage point difference between the states that have the highest and lowest scores on percent white. Some states (e.g., New Hampshire and Iowa) are as much as 98% white while, as we noted earlier, the District of Columbia and Hawaii are both about one-third white. The stability of the range means that this pattern of some states with very high scores and some with very low scores will remain the same.

Although the interpretation of the standard deviation is less intuitive than that of the range, it makes the same point. The stability in this statistic shows that the variation in percent white from state to state will not change much in the near future. Remember that the standard deviation can be interpreted as an index of variation: the higher the value of s, the greater the variety in the scores. If the states all had exactly the same percentage of whites in their populations, the value of s would be zero. The more the states vary in percent white, the higher the value of s. The stability in the standard deviation for percent white means that the amount of diversity between the states will be about the same in 2025 as it was in 1990.

What does the table say about the other racial and ethnic groups? Note that in 1990, the cases averaged about 10% African Americans in their population and that the averages for the other groups (especially Native and Asian Americans) were quite small. Over the years, however, the percentage of all of the minority groups increases, again underscoring the growing diversity of U.S. society. Increases are particularly pronounced for Americans of Asian and Hispanic origin. By 2025, the average percentage of Asian Americans across the fifty states and the District of Columbia is projected to increase from less than 3% to more than 5%, and the mean percentage of the population of Hispanic Origin will almost double to over 10%.

The median percentages across time for the fifty states and the District of Columbia also exhibit steady increases. For all four "nonwhite" groups, the mean is greater in value than the median, indicating a *positive* skew for these variables. This is exactly the reverse of the pattern for the percentage white and is especially apparent for Native and Asian Americans. The positive skew means that while most states have a low percentage of Native and Asian American populations, there are a few states with relatively high proportions and these states "pull" the mean upwards relative to the median.

Finally, consider dispersion. We do not have the space to analyze all groups but consider the trends for the percentage of population of Hispanic Origin as an illustration. Both the standard deviation and the range show a clear trend towards greater variability. The standard deviation steadily increases from 7.49 in 1990 to 11.03 by 2025. Similarly, the range increases from 37.98 to 46.29. In other words, the fifty states and the District of Columbia will become more variable in their percentage of population of Hispanic Origin. How could this happen? Some states will become "more Hispanic" (increase their percentage of Hispanic Origin population) while others stay the same. In other words, some states that were low on this variable in 1990 will not change much while the relative size of the Hispanic population in other states (e.g., California, Arizona, Florida) will grow. The "high end" of the variable will increase while the "low end" will remain stable, thus increasing the diversity of the scores and, as a result, the value of the range and standard deviation. How would you summarize the measures of dispersion for the other groups?

What can we say to summarize the information presented in Table 4.6 briefly and clearly? The United States is facing profound changes. While the society will remain majority white, the relative numerical advantage of the majority group will shrink. Other groups (especially Hispanic and Asian Americans), traditions, languages, and life styles will play a larger role in American life in the years to come. On the other hand, the amount of diversity among the fifty states and the District of Columbia will remain about the same through 2025. Some states will remain "lily white" (95% or more white) while others will continue to have a low percentage of whites and a growing percentage of racial and ethnic minority groups.

Trends in Racial and Ethnic Group Percentages by Region.   The closing observation to the last section suggests that not all states will experience the same pattern of increasing diversity. What can we learn by grouping the states by region and calculating means for each of the five racial and ethnic categories over time? Table 4.7 presents patterns for three time periods: 1990, 2005, and 2025.

As you work with descriptive statistics, one skill that you will develop is the ability to see the proverbial forest through the trees. That is, you need to be able to look at a large number of results, such as those in Table 4.7, and identify general trends. What patterns stand out in this table in terms of central tendency and dispersion, and how would you express these patterns? This statistical information could be approached in a variety of ways, but some points seem most obvious. First, the regions vary considerably in their ethnic and racial makeup. States in the South have the lowest percentage of whites and the highest percentage of African Americans; the Northeast and the Midwest are the "whitest" regions; and the West is and will continue to be the most Asian and Hispanic region through the year 2025. Second, the average percentage white declines steadily in all four regions—a trend that reinforces the message of Table 4.6. Third, as Table 4.6 demonstrated for the nation as a whole, Table 4.7 exhibits dramatic increases in the mean percentage of Asian American and Hispanic Origin population. While the Midwest exhibits consistently low average percentages for these two racial

TABLE 4.7   MEAN PERCENTAGES BY REGION AND BY YEAR

Geographical Region	1990		2005		2025	
	Mean	Std Dev	Mean	Std Dev	Mean	Std Dev
Northeast						
% White	91.2	7.1	88.5	8.7	85.2	10.5
% African Am.	6.7	6.1	7.8	6.7	9.3	7.4
% Native Am.	0.3	0.1	0.3	0.1	0.4	0.2
% Asian Am.	1.9	1.2	3.3	2.3	5.1	3.4
% Hispanic Origin	4.7	4.2	7.2	6.0	10.7	8.0
Total N of Cases	9		9		9	
Midwest						
% White	91.0	4.5	89.0	4.5	86.7	4.6
% African Am.	6.5	5.1	7.4	5.3	8.4	5.7
% Native Am.	1.4	2.1	1.7	2.5	2.2	3.3
% Asian Am.	1.1	0.6	1.9	1.0	2.7	1.4
% Hispanic Origin	2.2	2.0	3.4	3.0	4.9	4.3
Total N of Cases	12		12		12	
South						
% White	76.0	14.3	74.5	13.5	72.3	13.6
% African Am.	22.0	14.5	22.8	13.4	24.3	13.4
% Native Am.	0.8	1.9	0.8	2.0	0.9	2.1
% Asian Am.	1.2	0.8	1.9	1.3	2.6	1.7
% Hispanic Origin	3.7	6.3	5.2	7.8	7.0	9.7
Total N of Cases	17		17		17	
West						
% White	86.0	16.8	84.0	16.7	81.5	17.5
% African Am.	2.9	2.3	3.3	2.3	3.6	2.5
% Native Am.	3.8	4.4	3.7	3.9	3.8	3.7
% Asian Am.	7.4	16.9	9.0	16.9	11.1	17.7
% Hispanic Origin	11.0	10.8	14.3	12.1	18.5	14.3
Total N of Cases	13		13		13	

and ethnic groups, note the high average percentages for these groups across the Northeast and especially the West. What other trends and patterns can you identify?

Application of the Index of Qualitative Variation.   Table 4.7 presents information about dispersion with the standard deviation but it is also possible to capture the increasing racial and ethnic diversity of the population in the United States with the Index of Qualitative Variation (or IQV). Recall that the IQV is an appropriate measure of dispersion for nominal-level variables. Since people of Hispanic Origin can be of any racial group, we can treat the variable of race as a four-category nominal-level variable. Table 4.8 presents the actual frequency of the various racial groups for 1990 through 2000 and the estimated frequencies for three future years. Recall that the higher the IQV, the more variable the distribution. Table 4.8 clearly demonstrates that between 1990 and 2025, the United States will steadily increase in variability.

TABLE 4.8   ESTIMATED RACIAL DISTRIBUTION IN THE UNITED STATES POPULATION, 1990–2025

	Frequency (in 1,000s) by Year					
	1990	1995	2000	2005	2015	2025
White	209,196	218,031	225,534	232,464	247,197	262,222
African American	30,627	33,128	35,441	37,741	42,587	47,539
Native American	2,076	2,236	2,401	2,561	2,932	3,317
Asian American	7,565	9,349	11,245	13,212	17,411	21,971
Total	249,464	262,744	274,621	285,978	310,127	335,049
Index of Qualitative Variation (IQV)	0.3743	0.3922	0.4095	0.4261	0.4567	0.4839

As Bob Dylan immortally sang, "times are changing," and this statement certainly applies to the racial and ethnic composition of the United States. Based on the statistics presented here, racial and ethnic diversity will increase in every region, and the United States will have a very different complexion by the year 2025.

A Note on Racial and Ethnic Categories.   The racial categories and labels used in this analysis reflect standard usage for the U.S. Bureau of the Census, other governmental agencies, and social science research. It is important to realize, however, that the categories are largely arbitrary and do not reflect the totality of the growing racial complexity of the United States, let alone the world. For example, the categories do not provide a place for the growing number of "mixed race" Americans, many of whom could fit in three, four, or all five of these racial/ethnic categories. Many people—scientists as well as "ordinary" people—already regard these categories as invalid, overly rigid representations of racial and ethnic heritage. The increasingly problematic nature of these categories suggests that the true extent of racial and ethnic diversity is even greater than that reflected in this analysis.

SUMMARY

1. Measures of dispersion summarize information about the heterogeneity or variety in a distribution of scores. When combined with an appropriate measure of central tendency, these statistics convey a large volume of information in just a few numbers. While measures of central tendency locate the central points of the distribution, measures of dispersion indicate the amount of diversity in the distribution.

2. The index of qualitative variation (IQV) can be computed for any variable that has been organized into a frequency distribution. It is the ratio of the amount of variation observed in the distribution to the maximum variation possible in the distribution. The IQV is most appropriate for variables measured at the nominal level.

3. The range (R) is the distance from the highest to the lowest score in the distribution. The interquartile range (Q) is the distance from the third to the first quartile (the "range" of the middle 50% of the scores). These two ranges can be used with variables measured at either the ordinal or interval-ratio level.

4. The standard deviation (s) is the most important

measure of dispersion because of its central role in many more advanced statistical applications. The standard deviation has a minimum value of zero (indicating no variation in the distribution) and increases in value as the variability of the distribution increases. It is used most appropriately with variables measured at the interval-ratio level.

5. The variance (s^2) is used primarily in inferential statistics and in the design of some measures of association.

SUMMARY OF FORMULAS

Index of Qualitative Variation	4.1	$IQV = \dfrac{k(N^2 - \Sigma f^2)}{N^2(k - 1)}$
Average Deviation	4.2	$AD = \dfrac{\Sigma |X_i - \overline{X}|}{N}$
Variance	4.3	$s^2 = \dfrac{\Sigma(X_i - \overline{X})^2}{N}$
Standard Deviation (definitional)	4.4	$s = \sqrt{\dfrac{\Sigma(X_i - \overline{X})^2}{N}}$
Standard Deviation (computational)	4.5	$s = \dfrac{1}{N}\sqrt{N\Sigma X_i^2 - (\Sigma X_i)^2}$

GLOSSARY

Average deviation (AD).   The average of the absolute deviations of the scores around the mean.

Deviations.   The distances between the scores and the mean.

Dispersion.   The amount of variety or heterogeneity in a distribution of scores.

Index of qualitative variation (IQV).   A measure of dispersion for variables that have been organized into frequency distributions.

Interquartile range (Q).   The distance from the third quartile to the first quartile.

Measures of dispersion.   Statistics that indicate the amount of variety or heterogeneity in a distribution of scores.

Range (R).   The highest score minus the lowest score.

Standard deviation.   The square root of the squared deviations of the scores around the mean, divided by N. The most important and useful descriptive measure of dispersion; s represents the standard deviation of a sample; σ, the standard deviation of a population.

Variance.   The squared deviations of the scores around the mean divided by N. A measure of dispersion used primarily in inferential statistics and also in correlation and regression techniques; s^2 represents the variance of a sample; σ^2, the variance of a population.

MULTIMEDIA RESOURCES

The Wadsworth Sociology Resource Center: Virtual Society
http://sociology.wadsworth.com/

Visit the companion web site for the sixth edition of *Statistics: A Tool for Social Research* to access a wide range of student resources. Begin by clicking on the Student Resources section of the book's web site to access the following study tools:

- Basic math review
- Flash cards
- Additional chapter problems
- Statistics review
- Internet links

- Table of random numbers
- MicroCase and SPSS examples and exercises
- "Find the test" flowcharts

- Hypothesis testing for variables measured at the ordinal level

PROBLEMS

4.1 SOC The marital status of residents of four apartment complexes is reported below. Compute the index of qualitative variation (IQV) for each neighborhood. Which is the most heterogeneous of the four? Which is the least? *(HINT: It may be helpful to organize your computations as in Table 4.2.)*

Complex A		Complex B	
Marital Status	Frequency	Marital Status	Frequency
Single	26	Single	10
Married	31	Married	12
Divorced	12	Divorced	8
Widowed	5	Widowed	7
	N = 74		N = 37

Complex C		Complex D	
Marital Status	Frequency	Marital Status	Frequency
Single	20	Single	52
Married	30	Married	3
Divorced	2	Divorced	20
Widowed	1	Widowed	10
	N = 53		N = 85

4.2 SOC/CJ Compute the Index of Qualitative Variation for the distributions presented in problems 2.1 and 2.4.

4.3 Compute the range and standard deviation of 10 scores reported below. Use both the definitional and the computational formulas for computing s. *(HINT: It may be helpful to organize your computations as in Tables 4.4 and 4.5.)*

10, 12, 15, 20, 25, 30, 32, 35, 40, 50

4.4 For the 10 test scores below, compute the standard deviation using both the definitional and computational formulas.

77, 83, 69, 72, 85, 90, 95, 75, 55, 45

4.5 GER Compute the range, interquartile range, standard deviation, and variance for the data presented in problem 2.8.

4.6 For problem 3.2, find the standard deviation for enrollment, percent college bound, and condition of physical plant. What information does the standard deviation add to what you already knew about these variables?

4.7 SOC Labor force participation rates (percent employed), percent high school graduates, and mean income for males and females in 10 states are reported below. Calculate a mean and a standard deviation for both groups for each variable and describe the differences. Are males and females unequal on any of these variables? How great is the gender inequality?

	Labor Force Participation		% High School Graduates		Mean Income	
State	Male	Female	Male	Female	Male	Female
A	74	54	65	67	35,623	27,345
B	81	63	57	60	32,345	28,134
C	81	59	72	76	35,789	30,546
D	77	60	77	75	38,907	31,788
E	80	61	75	74	42,023	35,560
F	74	52	70	72	34,000	35,980
G	74	51	68	66	25,800	19,001
H	78	55	70	71	29,000	26,603
I	77	54	66	66	31,145	30,550
J	80	75	72	75	34,334	29,117

4.8 SW Compute the standard deviation for the pretest and posttest scores presented in problem 2.6. Considering the means you calculated in problem 3.12, briefly describe how the sample changed from test to test. What does the standard deviation add to the information you already had?

4.9 SOC Compute a standard deviation for the two sets of scores presented in problem 3.9. What does the standard deviation add to the information you already had?

4.10 [CJ] Per capita expenditures for police protection for 20 cities are reported below for 1995 and 2000. Compute a mean and standard deviation for each year, and describe the differences in expenditures for the five-year period.

1995		2000	
180	167	210	225
95	101	110	209
87	120	124	201
101	78	131	141
52	107	197	94
117	55	200	248
115	78	119	140
88	92	87	131
85	99	125	152
100	103	150	178

4.11 [PA] Compute the range and standard deviation for the data presented in problem 3.14. What would happen to the value of the standard deviation if you removed New York City from this distribution and recalculated? Why?

4.12 [SOC] Below are listed the rates of abortion per 100,000 women for 20 states in 1973 and 1975. Describe what happened to these distributions over the two-year period. Did the average rate increase or decrease? What happened to the dispersion of this distribution? What happened between 1973 and 1975 that might explain these changes in central tendency and dispersion? *(HINT: It was a Supreme Court decision.)*

State	1973	1975
Maine	3.5	9.5
Massachusetts	10.0	25.7
New York	53.5	40.7
Pennsylvania	12.1	18.5
Ohio	7.3	17.9
Michigan	18.7	20.3
Iowa	8.8	14.7
Nebraska	7.3	14.3
Virginia	7.8	18.0
South Carolina	3.8	10.3
Florida	15.8	30.5
Tennessee	4.2	19.2
Mississippi	0.2	0.6
Arkansas	2.9	6.3

State	1973	1975
Texas	6.8	19.1
Montana	3.1	9.9
Colorado	14.4	24.6
Arizona	6.9	15.8
California	30.8	33.6
Hawaii	26.3	31.6

Source: United States Bureau of the Census, *Statistical Abstracts of the United States: 1977* (98th edition). Washington, D.C., 1977.

4.13 [SW] One of your goals as the new chief administrator of a large social service bureau is to equalize work loads within the various divisions of the agency. You have gathered data on case loads per worker within each division. Which division comes closest to the ideal of an equalized work load? Which is farthest away?

A	B	C	D
50	60	60	75
51	59	61	80
55	58	58	74
60	55	59	70
68	56	59	69
59	61	60	82
60	62	61	85
57	63	60	83
50	60	59	65
55	59	58	60

4.14 [SW] Compute the standard deviation for both data sets presented in problem 3.10. Compare the standard deviations for the two time periods along with the changes in the mean. What happened? Why?

4.15 [SOC] Compute the standard deviation for both sets of data presented in problem 3.13. Compare the standard deviation computed for freshmen with the standard deviation computed for seniors. What happened? Why? Does this change relate at all to what happened to the mean over the four-year period? How? What happened to the shapes of the underlying distributions?

4.16 At St. Algebra College, the math department ran some special sections of the freshman math course using a variety of innovative teaching techniques. Students were randomly assigned to either the traditional sections or the experimental sections, and all students were given the same final exam. The results of the final are summarized below. What was the effect of the experimental course?

Traditional	Experimental
$\overline{X} = 77.8$	$\overline{X} = 76.8$
$s = 12.3$	$s = 6.2$
$N = 478$	$N = 465$

4.17 CJ You're the governor of the state and must decide which of four metropolitan police departments will win the annual award for efficiency. The performance of each department is summarized in monthly arrest statistics as reported below. Which department will win the award? Why?

	Departments		
A	B	C	D
$\overline{X} = 601.30$	633.17	592.70	599.99
$s = 2.30$	27.32	40.17	60.23

SPSS for Windows

Using *SPSS for Windows* to Produce Measures of Dispersion

Start *SPSS for Windows* by clicking the SPSS icon on your monitor screen and load the 1998 GSS data set.

SPSS DEMONSTRATION 4.1 Producing the Range and the Standard Deviation

Most of the statistics discussed in this chapter are available from either the **Frequencies** or **Descriptives** procedures that you are already familiar with. In this demonstration, we will use **Descriptives** to find the range and standard deviation for *age* (YEARS OF AGE), *educ* (HIGHEST YEAR OF SCHOOL COMPLETED), and *tvhours* (HOURS PER DAY WATCHING TV). These three variables were also used in SPSS Demonstration 3.2.

From the main menu, click **Analyze, Descriptive Statistics,** and **Descriptives.** The **Descriptives** dialog box will open. Use the cursor to find the names of the three variables in the list on the left and click the right arrow button to transfer them to the **Variables** box. Click **OK,** and SPSS will produce the same output we analyzed in Demonstration 3.2. Now, however, we will consider dispersion rather than central tendency. The output looks like this:

Descriptive Statistics

	N	Minimum	Maximum	Mean	Std. Deviation
AGE OF RESPONDENT	1385	18	89	44.94	17.08
HIGHEST YEAR OF SCHOOL COMPLETED	1381	0	20	13.37	2.86
HOURS PER DAY WATCHING TV	1134	0	21	2.86	2.20
Valid N (listwise)	1126				

The standard deviation for each variable is reported in the column labeled "Std. Deviation," and the range can be computed from the values given in the Minimum and Maximum columns. The standard deviation for *age* was slightly more than

17 years, and the youngest and oldest respondents were 18 and 89, respectively. For *educ,* the standard deviation was 2.86, and scores ranged from zero to 20. Respondents with scores of 20 have completed four years of formal education beyond the bachelor's level. The standard deviation is 2.20 for *tvhours,* and scores ranged from 0 hours of television watching to 21 (nearly the maximum possible in a single day).

At this point, the range is probably easier to understand and interpret than the standard deviation. As we saw in Section 4.8, the latter is more meaningful when we have a point of comparison. For example, suppose we were interested in the variable *tvhours* and how television-viewing habits have changed over the years. The **Descriptives** output for 1998 shows that people watched an average of 2.86 hours a day with a standard deviation of 2.20. Suppose that a sample from 1978 showed an average of 3.70 hours of television viewing a day with a standard deviation of 1.1. You could conclude that television watching had, on the average, decreased over the 20-year period but that Americans had also become much more diverse in their viewing habits.

SPSS DEMONSTRATION 4.2 Using the COMPUTE Command to Create an "Attitude Toward Abortion" Scale

SPSS provides a variety of ways to transform and manipulate variables. In the SPSS exercises at the end of Chapter 2, the **Recode** command was introduced as a way of changing the values associated with a variable. In this demonstration, we will use the **Compute** command to create new variables and summary scales.

Let's begin by considering the two questions from the 1998 GSS that measure attitudes toward abortion, *abany* and *abhlth.* The items present two different situations under which an abortion might be desired and ask the respondent to react to each situation independently. *Abany* asks if an abortion should be possible for "any reason," and *abhlth* asks if a legal abortion should be available when the woman's health is in danger. Since these two situations are distinct, each item should be analyzed in its own right. Suppose, however, that you wanted to create a summary scale that indicated a person's *overall* feelings about abortion.

One way to do this would be to add the scores on the two variables together. This would create a new variable, which we will call *abscale,* with three possible scores. If a respondent was consistently "pro-abortion" and answered "yes" (coded as "1") to both items, the respondent's score on the summary variable would be 2. A score of 3 would occur when a respondent answered "yes" to one item and "no" to the other. This might be labeled an "intermediate" or "moderate" position. The final possibility would be a score of 4, if the respondent answered "no" to both items. This would be a consistent "anti-abortion" position.

If Response on *abany* Is	and	Response on *abhlth* Is	Score on *abscale* Will Be
1 (Pro)		1 (Pro)	2 (Pro)
1 (Pro)		2 (Anti)	3 (Moderate)
2 (Anti)		1 (Pro)	3 (Moderate)
2 (Anti)		2 (Anti)	4 (Anti)

The new variable, *abscale,* summarizes each respondent's overall position on the issue. Once created, *abscale* could be analyzed, transformed, and manipulated exactly like a variable actually recorded in the data file.

To use the **Compute** command, click **Transform** and then **Compute** from the main menu. The **Compute Variable** window will appear. Find the **Target Variable** box in the upper-left-hand corner of this window. The first thing we need to do is assign a name to the new variable we are about to compute (*abscale*) and type that name in this box. Next, we need to tell SPSS how to compute the new variable. In this case, *abscale* will be computed by adding the scores of *abany* and *abhlth.* Find *abany* in the variable list on the left and click the arrow button in the middle of the screen to transfer the variable name to the **Numeric Expression** box. Next, click the plus sign (+) on the calculator pad under the **Numeric Expression** box, and the sign will appear next to *abany.* Finally, highlight *abhlth* in the variable list and click the arrow button to transfer the variable name to the **Numeric Expression** box.

The expression in the **Numeric Expression** box should now read

abany + abhlth

Click **OK,** and *abscale* will be created and added to the data set. If you want to keep this new variable permanently, click **Save** from the **File** menu, and the updated data set with *abscale* added will be saved to disk. If you are using the student version of SPSS, remember that your data set is limited to 50 variables.

We now have three variables that measure attitudes towards abortion—two items referring to specific situations, and a more general, summary item. It is always a good idea to check the frequency distribution for computed and recoded variables to make sure that the computations were carried out correctly. Use the **Frequencies** procedure (click **Analyze, Descriptive Statistics,** and **Frequencies**) to get tables for *abany, abhlth,* and *abscale.* Your output will look like this:

Statistics

		ABORTION IF WOMAN WANTS FOR ANY REASON	WOMANS HEALTH SERIOUSLY ENDANGERED	ABSCALE
N	Valid	887	895	855
	Missing	500	492	532

ABORTION IF WOMAN WANTS FOR ANY REASON

		Frequency	Percent	Valid Percent	Cumulative Percent
Valid	YES	372	26.8	41.9	41.9
	NO	515	37.1	58.1	100.0
	Total	887	64.0	100.0	
Missing	NAP	449	32.4		
	DK	49	3.5		
	NA	2	.1		
	Total	500	36.0		
Total		1387	100.0		

WOMANS HEALTH SERIOUSLY ENDANGERED

		Frequency	Percent	Valid Percent	Cumulative Percent
Valid	YES	791	57.0	88.4	88.4
	NO	104	7.5	11.6	100.0
	Total	895	64.5	100.0	
Missing	NAP	449	32.4		
	DK	42	3.0		
	NA	1	.1		
	Total	492	35.5		
Total		1387	100.0		

ABSCALE

		Frequency	Percent	Valid Percent	Cumulative Percent
Valid	2.00	370	26.7	43.3	43.3
	3.00	383	27.6	44.8	88.1
	4.00	102	7.4	11.9	100.0
	Total	855	61.6	100.0	
Missing	System	532	38.4		
Total		1387	100.0		

Note that the level of approval for the two specific situations varies sharply. While an overwhelming majority (88%) of the sample approved of the legal right to an abortion when the woman's health was in danger, a majority (58%) disapproved when the woman wants an abortion for "any reason." Looking at the combined scores on *abscale,* we see that about 43% of the sample approved of the legal right to an abortion in both cases (scored 2), and less than 12% disapproved in both situations (scored 4). A very large percentage (45%) had a score of 3, which means that they approved in one situation (when health is at stake) but not in the other.

Note that fewer than 900 respondents answered the original two abortion items. Remember that no respondent is given the entire GSS, and the vast majority of the "missing cases" received a form of the GSS that did not include these two items. Now look at *abscale,* the summary scale, and note that even fewer cases (855) are included in the summary scale than in either of the two original items. When SPSS executes a **Compute** statement, it automatically eliminates any cases that are missing scores on any of the constituent items. If these cases were not eliminated, a variety of errors and misclassifications could result. For example, if cases with missing scores were included, a person who scored a 2 ("anti-abortion") on *abany* and then failed to respond to *abhlth* would have a total score of 2 on *abscale.* Thus, this case would be treated as "pro-abortion" when, actually, the only information we have indicates that this respondent is "anti-abortion." To eliminate this kind of error, cases with missing scores on any of the constituent variables are deleted from calculations.

MicroCase

Using MicroCase to Produce Measures of Dispersion

Start MicroCase by clicking the icon on your monitor screen and open the 1998 GSS data set.

MICROCASE DEMONSTRATION 4.1
Producing the Range and the Standard Deviation

In this demonstration, we will use the **Univariate** command, which you are already familiar with, to find the range and standard deviation for *age, educ,* and *tvhours.* These three variables were also used in MicroCase Demonstration 3.2.

From the Statistics menu, click **Univariate** and a dialog box will open. Select *age* (variable #6), *educ* (variable #7), and then *tvhours* (variable #42) one at a time and click **Summary** in the "Statistics" box on the left to get frequency distributions and descriptive statistics. The latter are presented below for each of the three variables.

```
              AGE -- AGE OF RESPONDENT
Mean:    44.939   Std.Dev.: 17.080          N: 1385
Median: 41.000    Variance: 291.709   Missing: 2
99% confidence interval +/- mean: 43.756 to 46.121
95% confidence interval +/- mean: 44.039 to 45.838
```

```
         EDUC -- HIGHEST YEAR OF SCHOOL COMPLETED
Mean:    13.366   Std.Dev.: 2.857          N: 1381
Median: 13.000    Variance: 8.160   Missing: 6
99% confidence interval +/- mean: 13.168 to 13.564
95% confidence interval +/- mean: 13.216 to 13.517
```

```
          TVHOURS -- HOURS PER DAY WATCHING TV
Mean:    2.857    Std.Dev.: 2.197          N: 1134
Median: 2.000     Variance: 4.825   Missing: 253
99% confidence interval +/- mean: 2.689 to 3.025
95% confidence interval +/- mean: 2.729 to 2.985
```

The standard deviation ("Std. Dev.") for each variable is reported in the top row and the range can be found from the scores in the frequency distribution. The standard deviation for *age* was slightly more than 17 years, and the youngest and oldest respondents were 18 and 89, respectively. For *educ,* the standard deviation was 2.86, and scores ranged from zero to 20. Respondents with scores of 20 have completed four years of formal education beyond the bachelor's level. The standard deviation is 2.20 for *tvhours,* and scores ranged from 0 hours of television watching to 21 (nearly the maximum possible in a single day).

At this point, the range is probably easier to understand and interpret than the standard deviation. As we saw in Section 4.8, the latter is more meaningful when we have a point of comparison. For example, suppose we were interested in the variable *tvhours* and how television-viewing habits have changed over the years. The **Univariate** output for 1998 shows that people watched an average of 2.86 hours a day with a standard deviation of 2.20. Suppose that a sample from 1978 showed an average of 3.70 hours of television viewing a day with a standard deviation of 1.1. You could conclude that television watching had, on the average, decreased over the 20-year period but that Americans had also become much more diverse in their viewing habits.

MICROCASE DEMONSTRATION 4.2 Using the SUM RECODE Command to Create an "Attitude Toward Abortion" Scale

MicroCase provides a variety of ways to transform and manipulate variables. In the MicroCase exercises at the end of Chapter 2, the **Collapse Categories** command was introduced as a way of changing the values associated with a variable. In this demonstration, we will use the **Sum Recode** command to create new variables and summary scales.

Let's begin by considering the two questions from the 1998 GSS that measure attitudes toward abortion, *abhlth* (variable #32) and *abany* (variable # 33). The items present two different situations under which an abortion might be desired and ask the respondent to react to each situation independently. *Abany* asks if an abortion should be possible for "any reason," and *abhlth* asks if a legal abortion should be available when the woman's health is in danger. Since these two situations are distinct, each item should be analyzed in its own right. Suppose, however, that you wanted to create a summary scale that indicated a person's *overall* feelings about abortion.

One way to do this would be to add the scores on the two variables together. This would create a new variable, which we will call *abscale,* with three possible scores. If a respondent was consistently "pro-abortion" and answered "yes" (coded as "1") to both items, the respondent's score on the summary variable would be 2. A score of 3 would occur when a respondent answered "yes" to one item and "no" to the other. This might be labeled an "intermediate" or "moderate" position. The final possibility would be a score of 4, if the respondent answered "no" to both items. This would be a consistent "anti-abortion" position.

If Response on *abany* Is	and	Response on *abhlth* Is	Score on *abscale* Will Be
1 (Pro)		1 (Pro)	2 (Pro)
1 (Pro)		2 (Anti)	3 (Moderate)
2 (Anti)		1 (Pro)	3 (Moderate)
2 (Anti)		2 (Anti)	4 (Anti)

The new variable, *abscale,* summarizes each respondent's overall position on the issue. Once created, *abscale* could be analyzed, transformed, and manipulated exactly like a variable actually recorded in the data file.

To use the **Sum Recode** command, click on **Recode Variables** on the "Data Management" window (full version of MicroCase) or **Define and Edit Variables** on the "File and Data Menu" (student version) and then click on the **Sum/Index** button. The "Sum Recode: Step 1" window will appear and present a number of options for creating the scale. By default, the buttons next to "Sum the values of the variables" and "Assign no value (e.g., missing data) to the case" should be checked. These are the options we want: if they are not checked, do so now. The first option will add the scores on *abany* and *abhlth* together, and the second will eliminate any case that is missing a score for either variable. This is desirable because, if these cases were not eliminated, a variety of errors and misclassifications could result. For example, if cases with missing scores were included, a person who scored a 2 ("anti-abortion") on *abany* and then failed to respond to *abhlth* would have a total score of 2 on *abscale*. Thus, this case would be treated as "pro-abortion" when, actually, the only information we have indicates that this respondent is "anti-abortion." To eliminate this kind of error, cases with missing scores on any of the constituent variables are deleted from calculations.

Click **Next,** and the "Sum Recode: Step 1" window will appear. Select *abhlth* and *abany* and click **Next** to get the "Sum Recode: Step 3" window. This window gives us an opportunity to change our choices and provides other information that we will ignore for now. Click **Next** once more, and the "Sum Recode: Step 4" window will appear. Enter *abscale* as the name of the variable and click **Next** and then **Finish** at the next window. You will be returned to the "Sum Recode" window. Click **OK** and the new variable (*abscale*) will be added to the data set. If you want to keep this new variable permanently, remember to save the data file. If you are using the student version of MicroCase, remember that your data set is limited to 50 variables.

We now have three variables that measure attitudes towards abortion—two items referring to specific situations, and a more general, summary item. It is always a good idea to check the frequency distribution for computed and recoded variables to make sure that the computations were carried out correctly. Use the **Univariate** command to get tables for *abany, abhlth,* and *abscale*. Your output will look like this:

```
ABANY -- ABORTION IF WOMAN WANTS FOR ANY REASON
```

Mean: 1.581   Std.Dev.: 0.494   N: 887
Median: 2.000   Variance: 0.244   Missing: 500
99% confidence interval +/- mean: 1.538 to 1.623
95% confidence interval +/- mean: 1.548 to 1.613

Category	Freq.	%	Cum.%	Z-Score
1) YES	372	41.9	41.9	-1.176
2) NO	515	58.1	100.0	0.849

```
ABHLTH -- WOMANS HEALTH SERIOUSLY ENDANGERED
```

Mean: 1.116   Std.Dev.: 0.321   N: 895
Median: 1.000   Variance: 0.103   Missing: 492
99% confidence interval +/- mean: 1.089 to 1.144
95% confidence interval +/- mean: 1.095 to 1.137

Category	Freq.	%	Cum.%	Z-Score
1) YES	791	88.4	88.4	-0.362
2) NO	104	11.6	100.0	2.756

```
    abscale -- Sum of the following variables:
    ABHLTH, ABANY -- Cronbach's alpha is 0.45
  (listwise deletion). There are 532 missing cases.
```

Mean: 2.687	Std.Dev.: 0.674		N: 855	
Median: 3.000	Variance: 0.454		Missing: 532	
99% confidence interval +/- mean: 2.627 to 2.746				
95% confidence interval +/- mean: 2.641 to 2.732				

Value	Freq.	%	Cum.%	Z-Score
2	370	43.3	43.3	-1.019
3	383	44.8	88.1	0.465
4	102	11.9	100.0	1.949

Note that the level of approval for the two specific situations varies sharply. While an overwhelming majority (88%) of the sample approved of the legal right to an abortion when the woman's health was in danger, a majority disapproved when the woman wants an abortion "for any reason." Looking at the combined scores on *abscale,* we see that about 43% of the sample approved of the legal right to an abortion in both cases (scored 2), and less than 12% disapproved in both situations (scored 4). A very large percentage (45%) had a score of 3, which means that they approved in one situation (when health is at stake) but not in the other.

Note that fewer than 900 respondents answered the original two abortion items. Remember that no respondent is given the entire GSS, and the vast majority of the "missing cases" received a form of the GSS that did not include these two items. Now look at *abscale,* the summary scale, and note that even fewer cases (855) are included in the summary scale than in either of the two original items. This is because we eliminated any cases that were missing scores for either *abany* or *abhlth.*

Exercises

4.1 Use **Descriptives** (SPSS) or **Univariate** (MicroCase) to produce univariate descriptive statistics for *paeduc* (respondent's father's years of education), *prestg80* (respondent's occupational prestige), and *papres80* (respondent's father's occupational prestige).

    a. Compare the statistics for *prestg80* and *papres80.* Describe the difference in occupation prestige between the two generations. Are the respondents higher or lower in prestige than their fathers? Is the sample more or less homogeneous on this variable than their fathers?

    b. Compare the statistics for *educ* (produced in Demonstration 4.1) and *paeduc.* Describe the differences between the two generations. Are the respondents more or less educated than their fathers? Is the sample more or less homogeneous on this variable than their fathers?

4.2 Use the **Compute** (SPSS) or **Sum Recode** (MicroCase) command to create a summary scale for *fehelp* and *fefam.* Get univariate descriptive statistics for the summary scale and for *fehelp* and *fefam.* Write a few sentences summarizing these tables using our description of the distributions for *abany, abhlth,* and *abindex* as a model.

5 THE NORMAL CURVE

By the end of this chapter, you will be able to

1. Define and explain the concept of the normal curve.
2. Convert empirical scores to *Z* scores and use *Z* scores and the normal curve table (Appendix A) to find areas above, below, and between points on the curve.
3. Express areas under the curve in terms of probabilities.

5.1 INTRODUCTION

The **normal curve** is a concept of great importance in statistics. As we shall see in Part II, it is central to the theory that underlies inferential statistics. Also, in combination with the standard deviation, the normal curve can be used to construct precise descriptive statements about empirical distributions. This chapter will conclude our treatment of statistics as descriptive devices in Part I and lay important groundwork for Part II.

The normal curve is a theoretical model, a kind of frequency polygon that is perfectly symmetrical and smooth. It is bell shaped and unimodal, with its tails extending infinitely in both directions. Of course, no empirical distribution has a shape that perfectly matches this ideal model, but many are close enough to permit the assumption of normality. In turn, this assumption makes possible one of the most important uses of the normal curve—description of empirical distributions based on our knowledge of the theoretical normal curve.

The crucial point about the normal curve is that distances along the abscissa (horizontal axis) of the distribution, when measured in standard deviations from the mean, always encompass the same proportion of the total area under the curve. In other words, regardless of the nature of the underlying distribution, the distance from any given point to the mean (when measured in standard deviations) will cut off exactly the same proportion of the total area.

To illustrate, Figures 5.1 and 5.2 present two hypothetical distributions

FIGURE 5.1 IQ SCORES FOR A SAMPLE OF MALES

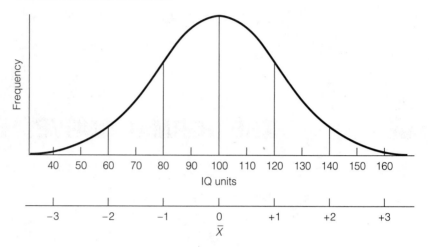

FIGURE 5.2 IQ SCORES FOR A SAMPLE OF FEMALES

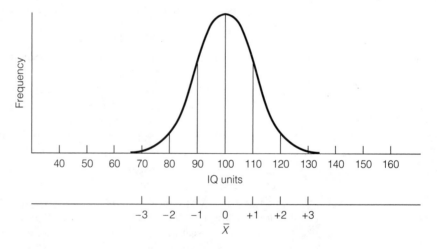

of IQ scores, one for a group of males and one for a group of females, both normally distributed (or nearly so), such that

Males	Females
$\overline{X} = 100$	$\overline{X} = 100$
$s = 20$	$s = 10$
$N = 1000$	$N = 1000$

Figures 5.1 and 5.2 are drawn with two scales on the horizontal axis (or abscissa). The first scale is stated in "IQ units" and the second in standard deviations from the mean. These scales should help you to visualize the relation-

ships explained below between distances from the mean as measured in units of the standard deviation and areas under the curve.

On any normal curve, the distance between ±1 standard deviation encompasses exactly 68.26% of the total area under the curve. The standard deviation for males is 20, so in Figure 5.1, 68.26% of the total area lies between the score of 80 (−1 standard deviation) and 120 (+1 standard deviation). The standard deviation for females is 10, so the same percentage of the area lies between the scores of 90 and 110. As long as an empirical distribution is normal, 68.26% of the total area will always be encompassed between ±1 standard deviation—regardless of the trait being measured and the number values of the mean and standard deviation.

It will be useful to familiarize yourself with the following relationships between distances from the mean and areas under the curve:

> between ±1 standard deviation lies 68.26% of the area
> between ±2 standard deviations lies 95.44% of the area
> between ±3 standard deviations lies 99.72% of the area

These relationships are displayed graphically in Figure 5.3.

The relationship between distance from the mean and area allows us to describe empirical distributions that are at least approximately normal. The position of individual scores can be described with respect to the mean, the distribution as a whole, or any other score in the distribution.

The areas between scores can also be expressed, if desired, in numbers of cases rather than percentage of total area. For example, a normal distribution of 1000 cases will contain about 683 cases (68.26% of 1000 cases) between ±1 standard deviation of the mean, about 954 between ±2 standard deviations, and about 997 between ±3 standard deviations. Thus, for any normal distribution, only a few cases will be farther away from the mean than ±3 standard deviations.

FIGURE 5.3   AREAS UNDER THE THEORETICAL NORMAL CURVE

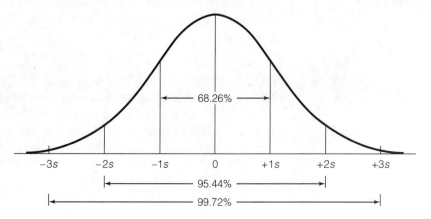

5.2 COMPUTING Z SCORES To find the percentage of the total area (or number of cases) above, below, or between scores in an empirical distribution, the original scores must first be expressed in units of the standard deviation or converted into **Z scores.** The original scores could be in any unit of measurement (feet, IQ, dollars), but Z scores always have the same values for their mean and standard deviation. Think of converting the original scores into Z scores as a process of changing value scales—similar to changing from meters to yards, kilometers to miles, or gallons to liters. These units are different but equally valid ways of expressing distance or volume. The original (or "raw") scores and Z scores are two equally valid but different ways of measuring distances under the normal curve.

The standardized theoretical normal curve has a mean of 0 and a standard deviation of 1. When an empirical normal distribution is standardized, its mean will be converted to 0, its standard deviation to 1, and all values will be expressed in Z-score form. The formula for computing Z scores is

FORMULA 5.1
$$Z = \frac{X_i - \overline{X}}{S}$$

This formula will convert any score (X_i) from an empirical normal distribution into the equivalent Z score. To illustrate with the men's IQ data (Figure 5.1), the Z-score equivalent of a raw score of 120 would be

$$Z = \frac{120 - 100}{20} = +1.00$$

The equivalent Z score of positive 1.00 indicates that this score lies one standard deviation unit above (to the right of) the mean. A negative score would fall below (to the left of) the mean. *(For practice in computing Z scores, see any of the problems at the end of this chapter.)*

5.3 THE NORMAL CURVE TABLE The theoretical normal curve has been very thoroughly analyzed and described by statisticians. The areas related to any Z score have been precisely determined and organized into a table format. This **normal curve table** or Z-score table is presented as Appendix A in this text. The table consists of three columns, with Z scores in the left-hand column, areas between the Z score and the mean of the normal curve in the middle column, and areas beyond the Z score in the right-hand column. To find the area between any Z score and the mean, go down the column labeled "Z" until you find the Z score. For our example, go down the column until you find the score 1.00. The entry in the column labeled "Area Between Mean and Z" is 0.3413. The table presents all areas in the form of proportions, but we can easily translate these into percentages by multiplying them by 100 (see Chapter 2). We could say either that "a proportion of 0.3413 of the total area under the curve lies between a Z score of 1.00 and the mean," or "34.13% of the total area lies between a score of 1.00 and the mean."

The third column in the table presents "Areas Beyond Z." These are areas above positive scores or below negative scores. This column will be used when we want to find an area above or below certain Z scores, an application that will be explained in Section 5.4.

To conserve space, the table includes only positive Z scores. Since the normal curve is perfectly symmetrical, however, the areas related to any negative score will be exactly the same as those of the corresponding positive score. For example, the area between a Z score of -1.00 and the mean will also be 34.13%, exactly the same as the area we found previously for a score of $+1.00$. As will be repeatedly demonstrated below, however, the sign of the Z score is extremely important and should be carefully noted.

For practice in using Appendix A to describe areas under an empirical normal curve, verify that the Z scores and areas given below are correct for the men's IQ distribution. For each IQ score, the equivalent Z score is computed using Formula 5.1, and then Appendix A is used to find areas between the score and the mean. ($\overline{X} = 100$, $s = 20$ throughout.)

IQ Score	Z Score	Area Between Z and the Mean
110	+0.50	19.15%
125	+1.25	39.44%
133	+1.65	45.05%
138	+1.90	47.13%

The same procedures apply when the Z-score equivalent of an actual score happens to be a minus value (that is, when the raw score lies below the mean).

IQ Score	Z Score	Area Between Z and the Mean
93	−0.35	13.68%
85	−0.75	27.34%
67	−1.65	45.05%
62	−1.90	47.13%

Remember that the areas in Appendix A will be the same for Z scores of the same numerical value regardless of sign. The area between the score of 138 ($+1.90$) and the mean is the same as the area between 62 (-1.90) and the mean. *(For practice in using the normal curve table, see any of the problems at the end of this chapter.)*

5.4 FINDING TOTAL AREA ABOVE AND BELOW A SCORE

To this point, we have seen how the normal curve table can be used to find areas between a Z score and the mean. The information presented in the table can also be used to find other kinds of areas in empirical distributions that are at least approximately normal in shape. For example, suppose you need to determine the total area below the scores of two male subjects in the distribution described in Figure 5.1. The first subject has a score of 117 ($X_1 = 117$), which is equivalent to a Z score of $+0.85$:

$$Z_1 = \frac{X_1 - \overline{X}}{s} = \frac{117 - 100}{20} = \frac{17}{20} = +0.85$$

The plus sign of the Z score indicates that the score should be placed above (to the right of) the mean. To find the area below a positive Z score, the area between the score and the mean must be added to the area below the mean. Since the normal curve is symmetrical (unskewed), the mean will be equal to the median, and the area below the mean will therefore be 50%. Study Figure 5.4 carefully. We are interested in the shaded area.

By consulting the normal curve table, we find that the area between the score and the mean is 30.23% of the total area. The area below a Z score of $+0.85$ is therefore 80.23% (50% + 30.23%). This subject scored higher than 80.23% of the persons tested.

The second subject has an IQ score of 73 ($X_2 = 73$), which is equivalent to a Z score of -1.35:

$$Z_2 = \frac{X_2 - \overline{X}}{s} = \frac{73 - 100}{20} = -\frac{27}{20} = -1.35$$

To find the area below a negative score, we use the column labeled "Area Beyond Z." The area of interest is depicted in Figure 5.5, and we must determine the size of the shaded area. The area beyond a score of -1.35 is given as 0.0885, which we can express as 8.85%. The second subject ($X_2 = 73$) scored higher than 8.85% of the tested group.

In the examples above, we use the techniques for finding the area below a score. Essentially the same techniques are used to find the area above a score. If we need to determine the area above an IQ score of 108, for example, we would first convert to a Z score,

$$Z = \frac{X - \overline{X}}{s} = \frac{108 - 100}{20} = \frac{8}{20} = 0.40$$

FIGURE 5.4 FINDING THE AREA BELOW A POSITIVE Z SCORE

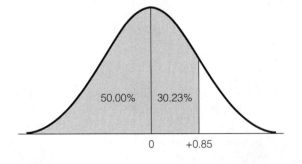

FIGURE 5.5 FINDING THE AREA BELOW A NEGATIVE Z SCORE

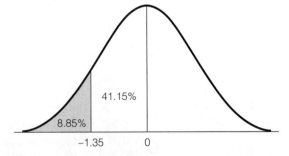

FIGURE 5.6   FINDING THE AREA ABOVE A POSITIVE Z SCORE

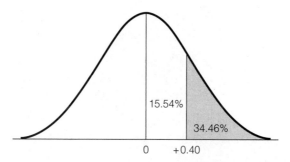

15.54%

34.46%

0   +0.40

TABLE 5.1   FINDING AREAS ABOVE AND BELOW POSITIVE AND NEGATIVE Z SCORES

	When the Z Score Is	
To Find Area:	Positive	Negative
Above Z	Look in Column C	Add Column B Area to .5000 or 50.00%
Below Z	Add Column B Area to .5000 or 50.00%	Look in Column C

and then proceed to Appendix A. The shaded area in Figure 5.6 represents the area in which we are interested. The area above a positive score is found in the "Area Beyond Z" column, and, in this case, the area is 0.3446, or 34.46%.

These procedures are summarized in Table 5.1. To find the total area above a positive Z score or below a negative Z score, go down the "Z" column until you find the score. The area you are seeking will be in the "Area Beyond Z" column. To find the total area below a positive Z score or above a negative score, locate the score and then add the area in the "Area Between Mean and Z" to either .5000 (for proportions) or 50.00 (for percentages). These techniques might be confusing at first, and you will find it helpful to draw the curve and shade in the areas in which you are interested. *(For practice in finding areas above or below Z scores, see problems 5.1 to 5.5 and 5.7 to 5.9.)*

5.5 FINDING AREAS BETWEEN TWO SCORES

On occasion, you will need to determine the area between two scores rather than the total area above or below one score. In the case where the scores are on opposite sides of the mean, the area between the scores can be found by adding the areas between each score and the mean. Using the men's IQ data as an example, if we wished to know the area between the IQ scores of 93 and 112, we would convert both scores to Z scores, find the area between

FIGURE 5.7   FINDING THE AREA BETWEEN TWO SCORES

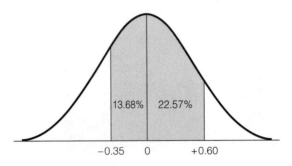

each score and the mean from Appendix A, and add these two areas together. The first IQ score of 93 converts to a Z score of -0.35:

$$Z_1 = \frac{X_1 - \overline{X}}{s} = \frac{93 - 100}{20} = -\frac{7}{20} = -0.35$$

The second IQ score (112) converts to $+0.60$:

$$Z_2 = \frac{X_2 - \overline{X}}{s} = \frac{112 - 100}{20} = \frac{12}{20} = 0.60$$

Both scores are placed on Figure 5.7. We are interested in the total shaded area. The total area between these two scores is 13.68% + 22.57%, or 36.25%. Therefore, 36.25% of the total area (or about 363 of the 1000 cases) lies between the IQ scores of 93 and 112.

Application 5.1

You have just received your score on a test of intelligence. If your score was 78 and you know that the mean score on the test was 67 with a standard deviation of 5, how does your score compare with the distribution of all test scores?

If you can assume that the test scores are normally distributed, you can compute a Z score and find the area below or above your score. The Z-score equivalent of your raw score would be

$$Z = \frac{X_i - \overline{X}}{s}$$

$$Z = \frac{78 - 67}{5}$$

$$Z = \frac{11}{5}$$

$$Z = 2.20$$

Turning to Appendix A, we find that the "Area Between Mean and Z" is 0.4861, which could also be expressed as 48.61%. Since this is a positive Z score, we need to add this area to 50.00% to find the total area below. Your score is higher than (48.61 + 50.00), or 98.61%, of all the test scores. You did pretty well!

FIGURE 5.8   FINDING THE AREA BETWEEN TWO SCORES

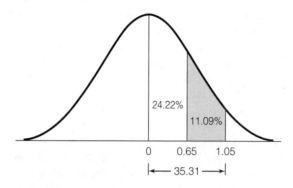

When the scores of interest are on the same side of the mean, a different procedure must be followed to determine the area between them. For example, if we were interested in the area between the scores of 113 and 121, we would begin by converting these scores into Z scores:

$$Z_1 = \frac{X_1 - \overline{X}}{s} = \frac{113 - 100}{20} = \frac{13}{20} = +0.65$$

$$Z_2 = \frac{X_2 - \overline{X}}{s} = \frac{121 - 100}{20} = \frac{21}{20} = +1.05$$

The scores are noted in Figure 5.8; we are interested in the lined area. To find the area between two scores on the same side of the mean, find the area between each score and the mean (given in Appendix A) and then subtract the smaller area from the larger. Between the Z score of $+0.65$ and the mean lies 24.22% of the total area. Between $+1.05$ and the mean lies 35.31% of the total area. Therefore, the area between these two scores is 35.31% $-$ 24.22%, or 11.09% of the total area. The same technique would be followed if both scores had been below the mean. The procedures for finding areas between two scores are summarized in Table 5.2. *(For practice in finding areas between two scores, see problems 5.3, 5.4, 5.6 to 5.9.)*

TABLE 5.2   FINDING AREAS BETWEEN SCORES

Situation:	Procedure:
Scores are on the SAME side of the mean:	Find areas between each score and the mean in column b. Subtract the smaller area from the larger area.
Scores are on OPPOSITE sides of the mean:	Find areas between each score and the mean in column b. Add the two areas together.

5.6 USING THE NORMAL CURVE TO ESTIMATE PROBABILITIES

To this point, we have thought of the theoretical normal curve as a way of describing the percentage of total area above, below, and between scores in an empirical distribution. We have also seen that these areas can be converted into the number of cases above, below, and between scores. The theoretical normal curve may also be thought of as a distribution of probabilities. Specifically, we may use the properties of the theoretical normal curve (Appendix A) to estimate the probability that a case randomly selected from an empirical normal distribution will have a score that falls in a certain range. In terms of techniques, these probabilities will be found in exactly the same way as areas were found. Before we consider these mechanics, however, let us examine what is meant by the concept of probability.

Although we are rarely systematic or rigorous about it, we all attempt to deal with probabilities every day and, indeed, we base our behavior on our estimates of the likelihood that certain events will occur. We constantly ask (and answer) questions such as, What is the probability of precipitation? Of drawing to an inside straight in poker? Of the cheap retreads on my car going flat? Of passing the final exam in Introduction to Nuclear Physics if I don't study?

To estimate the probability of events such as these, we must first be able to define what would constitute a "success." The examples above contain several different definitions of a success (that is, precipitation, drawing a certain card, flat tires, and passing grades). To determine a probability per se, a fraction must be established, with the numerator equaling the number of events that would constitute a success and the denominator equaling the total number of possible events where a success could theoretically occur.

To illustrate, assume that we wish to know the probability of selecting a specific card—say, the king of hearts—in one draw from a well-shuffled deck of cards. Our definition of a success is quite specific (drawing the king of hearts); and with the information given, we can establish a fraction. Only one card satisfies our definition of success, so the number of events that would constitute a success is 1; this value will be the numerator of the fraction. There are 52 possible events (that is, 52 cards in the deck), so the denominator will be 52. The fraction is thus 1/52, which represents the probability of selecting the king of hearts on one draw from a well-shuffled deck of cards. Our probability of success is 1 out of 52.

We can leave the fraction established above as it is, or we can express it in several other ways. For example, we can express it as an odds ratio by inverting the fraction, showing that the odds of selecting the king of hearts on a single draw are 52:1 (or fifty-two to one). We can express the fraction as a proportion by dividing the numerator by the denominator. For our example above, the corresponding proportion is .0192, which is the proportion of all possible events that would satisfy our definition of a success. In the social sciences, probabilities are usually expressed as proportions, and we will follow this convention throughout the remainder of this section.

As conceptualized here, probabilities have an exact meaning: over the long run, the events that we define as successes will bear a certain proportional relationship to the total number of events. The probability of .0192 for selecting the king of hearts in a single draw really means that, over thousands of selections of one card at a time from a full deck of 52 cards, the proportion of successful draws would be .0192. Or, for every 10,000 draws, 192 would be the king of hearts, and the remaining 9808 selections would be other cards. Thus, when we say that the probability of drawing the king of hearts in one draw is .0192, we are essentially applying our knowledge of what would happen over thousands of draws to a single draw.

Like proportions, probabilities range from 0.00 (meaning that the event has absolutely no chance of occurrence) to 1.00 (a certainty). As the value of the probability increases, the likelihood that the defined event will occur also increases. A probability of .0192 is close to zero, and this means that the event (drawing the king of hearts) is unlikely or improbable.

Combining this way of thinking about probability with our knowledge of the theoretical normal curve allows us to estimate the likelihood of selecting a case that has a score within a certain range. For example, suppose we wished to estimate the probability that a randomly chosen subject from the distribution of men's IQ scores would have an IQ score between 95 and the mean score of 100. Our definition of a success here would be the selection of any subject with a score in the specified range. Normally, we would next establish a fraction with the numerator equal to the number of subjects with scores in the defined range and the denominator equal to the total number of subjects. However, if the empirical distribution is normal in form, we can skip this step since the probabilities, in proportion form, are already stated in Appendix A. That is, the areas in Appendix A can be interpreted as probabilities.

To determine the probability that a randomly selected case will have a score between 95 and the mean, we would convert the original score to a Z score:

$$Z = \frac{X - \overline{X}}{s} = \frac{95 - 100}{20} = -\frac{5}{20} = -0.25$$

The area between this score and the mean (0.0987) is the probability we are seeking. The probability that a randomly selected case will have a score between 95 and 100 is 0.0987 (or, rounded off, 0.1, or one out of 10). In the same fashion, the probability of selecting a subject from any range of scores can be estimated. Note that the techniques for estimating probabilities are exactly the same as those for finding areas. The only new information introduced in this section is the idea that the areas in the normal curve table can also be thought of as probabilities.

Let us close by stressing a very important point about probabilities and the normal curve. The probability is very high that any case randomly selected from a normal distribution will have a score close in value to that of the mean. The shape of the normal curve is such that most cases are clustered around the mean and decline in frequency as we move farther away— either to the right or to the left—from the mean value. In fact, given what we know about the normal curve, the probability that a randomly selected case will have a score within ±1 standard deviation of the mean is 0.6826, whereas the probability of the case having a score beyond 3 standard deviations from the mean is 0.0028 (1.000 − 0.9972). Thus, if we randomly select a number of cases from a normal distribution, we will often select cases that have scores close to the mean but rarely select cases that have scores far above or below the mean. *(For practice in using the normal curve table to find probabilities, see problems 5.8 to 5.10 and 5.13.)*

SUMMARY

1. The normal curve, in combination with the mean and standard deviation, can be used to construct precise descriptive statements about empirical distributions that are normally distributed. This chapter also lays some important groundwork for Part II.
2. To work with the theoretical normal curve, raw scores must be transformed into their equivalent Z scores. Z scores allow us to find areas under the theoretical normal curve (Appendix A).

3. We considered three uses of the theoretical normal curve: finding total areas above and below a score, finding areas between two scores, and expressing these areas as probabilities. This last use of the normal curve is especially germane because inferential statistics are centrally concerned with estimating the probabilities of defined events in a fashion very similar to the process introduced in Section 5.6.

SUMMARY OF FORMULAS

Z scores	5.1	$Z_i = \dfrac{X_i - \overline{X}}{s}$

GLOSSARY

Normal curve. A theoretical distribution of scores that is symmetrical, unimodal, and bell shaped. The standard normal curve always has a mean of 0 and a standard deviation of 1.

Normal curve table. Appendix A; a detailed de-

scription of the area between a Z score and the mean of any standardized normal distribution.

Z scores. Standard scores; the way scores are expressed after they have been standardized to the theoretical normal curve.

MULTIMEDIA RESOURCES

The Wadsworth Sociology Resource Center: Virtual Society
http://sociology.wadsworth.com/

Visit the companion web site for the sixth edition of *Statistics: A Tool for Social Research* to access a wide range of student resources. Begin by clicking on the Student Resources section of the book's web site to access the following study tools:

- Basic math review
- Flash cards

- Additional chapter problems
- Statistics review
- Internet links
- Table of random numbers
- MicroCase and SPSS examples and exercises
- "Find the test" flowcharts
- Hypothesis testing for variables measured at the ordinal level

PROBLEMS

5.1 Scores on a quiz were normally distributed and had a mean of 10 and a standard deviation of 3. For each score below, find the Z score and the percentage of area above and below the score.

X_i	Z Score	% Area Above	% Area Below
5			
6			
7			
8			
9			
11			
12			
14			
15			
16			
18			

5.2 Assume that the distribution of a college entrance exam is normal with a mean of 500 and a standard deviation of 100. For each score below, find the equivalent Z score, the percentage of the area above the score, and the percentage of the area below the score.

X_i	Z Score	% Area Above	% Area Below
650			
400			
375			
586			
437			
526			
621			
498			
517			
398			

5.3 The senior class has been given a comprehensive examination to assess their educational experi-

ence. The mean on the test was 74, and the standard deviation was 10. What percentage of the students had scores

a. between 75 and 85? _____
b. between 80 and 85? _____
c. above 80? _____
d. above 83? _____
e. between 80 and 70? _____
f. between 75 and 70? _____
g. below 75? _____
h. below 77? _____
i. below 80? _____
j. below 85? _____

5.4 For a normal distribution where the mean is 50 and the standard deviation is 10, what percentage of the area is

a. between the scores of 40 and 47? _____
b. above a score of 47? _____
c. below a score of 53? _____
d. between the scores of 35 and 65? _____
e. above a score of 72? _____
f. below a score of 31 and above a score of 69? _____
g. between the scores of 55 and 62? _____
h. between the scores of 32 and 47? _____

5.5 At St. Algebra College, the 200 freshmen enrolled in Introductory Biology took a final exam on which their mean score was 72 and their standard deviation was 6. On page 134 are the grades of 10 students. Convert each into a Z score and determine the *number of people* who scored higher or lower than each of the 10 students. (*HINT: Multiply the appropriate proportion by N and round the result.*)

X_i	Z Score	Number of Students Above	Number of Students Below
60			
57			
55			
67			
70			
72			
78			
82			
90			
95			

5.6 If a distribution of test scores is normal with a mean of 78 and a standard deviation of 11, what percentage of the area lies

a. below 60? _____
b. below 70? _____
c. below 80? _____
d. below 90? _____
e. between 60 and 65? _____
f. between 65 and 79? _____
g. between 70 and 95? _____
h. between 80 and 90? _____
i. above 99? _____
j. above 89? _____
k. above 75? _____
l. above 65? _____

5.7 SOC A scale measuring prejudice has been administered to a large sample of respondents. The distribution of scores is approximately normal with a mean of 31 and a standard deviation of 5. What percentage of the sample had scores

a. below 20? _____
b. below 40? _____
c. between 30 and 40? _____
d. between 35 and 45? _____
e. above 25? _____
f. above 35? _____

5.8 CJ The average burglary rate for a jurisdiction has been 311 per year with a standard deviation of 50. What is the likelihood that next year the number of burglaries will be

a. less than 250? _____
b. less than 300? _____
c. more than 350? _____
d. more than 400? _____

e. between 250 and 350? _____
f. between 300 and 350? _____
g. between 350 and 375? _____

5.9 For a math test on which the mean was 59 and the standard deviation was 4, what is the probability that a student randomly selected from this class will have a score

a. between 55 and 65? _____
b. between 60 and 65? _____
c. above 65? _____
d. between 60 and 50? _____
e. between 55 and 50? _____
f. below 55? _____

5.10 SOC On the scale mentioned in problem 5.7, if a score of 40 or more is considered "highly prejudiced," what is the probability that a person selected at random will have a score in that range?

5.11 The local police force in Shinbone, Kansas, gives all applicants an entrance exam and accepts only those applicants who score in the top 15% on this test. If the mean score this year is 87 and the standard deviation is 8, would an individual with a score of 110 be accepted?

5.12 After taking the state merit examinations for the positions of social worker and employment counselor, you receive the following information on the tests and on your performance. On which of the tests did you do better?

Social Worker	Employment Counselor
$\overline{X} = 118$	$\overline{X} = 27$
$s = 17$	$s = 3$
Your score = 127	Your score = 29

5.13 In a distribution of scores with a mean of 35 and a standard deviation of 4, which event is more likely: that a randomly selected score will be between 29 and 31 or that a randomly selected score will be between 40 and 42?

5.14 To be accepted into an honor society, students must have GPAs in the top 10% of the school. If the mean GPA is 2.78 and the standard deviation is .33, which of the following GPAs would qualify?

3.20   3.21   3.25   3.30   3.35

Using *SPSS for Windows* to Transform Raw Scores into *Z* Scores

DEMONSTRATION 5.1   Computing *Z* Scores

The **Descriptives** program introduced at the end of Chapter 4 can also be used to compute *Z* scores for any variable. These *Z* scores are then available for further operations and may be used in other tasks. SPSS will create a new variable consisting of the transformed scores of the original variable. The program uses the letter *Z* and the first seven letters of the variable name to designate the normalized scores of a variable.

In this demonstration, we will have SPSS compute *Z* scores for *age.* First, load the 1998 GSS data set and then click **Analyze, Descriptive Statistics,** and **Descriptives.** Find *age* in the variable list and click the arrow to move the variable to the **Variable(s):** window. Find the "Save standardized values as variables" option below the variable list and click the box next to it. With this option selected for **Descriptives,** SPSS will compute *Z* scores for all variables listed in the **Variable(s):** window. Click **OK,** and SPSS will produce the usual set of descriptive statistics for *age.* It will also add the new variable (called *zage*), which contains the standardized scores for *age,* to the data set. To verify this, run the **Descriptives** command again, and you will find *zage* in the variable list. Transfer *zage* to the **Variable(s):** window with *age,* click **OK,** and the following output will be produced:

Descriptive Statistics

	N	Minimum	Maximum	Mean	Std. Deviation
AGE OF RESPONDENT	1385	18	89	44.94	17.08
Zscore: AGE OF RESPONDENT	1385	-1.57725	2.57978	-2.8593122E-15	1.0000000
Valid N (listwise)	1385				

Note that the mean for *zage* is reported as $-2.86E\text{-}15$. This is a form of scientific notation and can be read by moving the decimal point to the left the number of places that follows "E-" ($-.00000000000000286$). In other words, like any set of *Z* scores, *zage* effectively has a mean of zero and a standard deviation of one. The new variable *zage* can be treated just like any other variable and used in any SPSS procedure.

If you would like to inspect the scores of *zage,* use the **Case Summaries** procedure. Click **Analyze, Reports,** and then **Case Summaries.** Move both *age* and *zage* to the **Variable(s):** window. Find the **Limit cases to first** option at the bottom of the window. This option can be used to set the number of cases included in the output. For this exercise, let's set a limit of 10 cases. Make sure the box to the left of the option is checked and type 10 in the box to the right. Click **OK,** and the following output will be produced:

Case Summaries[a]

	AGE OF RESPONDENT	Zscore: AGE OF RESPONDENT
1	60	.88184
2	27	-1.05030
3	21	-1.40160
4	35	-.58190
5	43	-.11351
6	29	-.93320
7	39	-.34770
8	45	.00359
9	29	-.93320
10	41	-.23061
Total N	10	10

[a]Limited to first 10 cases.

Scan the list of scores and note that the scores that are close in value to the mean of *age* (44.94) are very close to the mean of *zage* (0.00), and the further away the score is from 44.94, the greater the numerical value of the *Z* score. Also note that, of course, scores below the mean (less than 44.94) have negative signs and scores above the mean (greater than 44.94) have positive signs.

MicroCase

Using MicroCase to Transform Raw Scores into *Z* Scores

DEMONSTRATION 5.1 Computing *Z* Scores

The **Sum Recode** command program introduced at the end of Chapter 4 can be used to compute *Z* scores for any variable. MicroCase can create a new variable consisting of the transformed scores of the original variable, and these *Z* scores are then available for further operations and may be used in other tasks.

In this demonstration, we will compute *Z* scores for *age*. First, load the 1998 GSS data set and then click **Recode Variables** on the **Data Management** window (full version of MicroCase) or **Define and Edit Variables** on the **File and Data Menu** (student version). Next click **Transform Variables** and the **Transform Variables: Step 1** window will appear. Select *age* from the variable list on the left and select **Z score** in the **Select Transformation** box. Click **Next** and enter *zage* ("*Z* scores for *age*") as the name of the new variable. Click **Next** and then **Finish** and *zage* will be computed and added to the data set.

The transformed variable can be used just like any other variable. To illustrate, get basic descriptive statistics for *zage* using the Univariate procedure. Your output should look like this:

```
            zage -- Z-Score of AGE
      (mean=44.94 standard deviation=17.07)
```

```
Mean:     0.001   Std.Dev.: 1.000          N: 1385
Median: -0.230    Variance: 1.001  Missing: 2
99% confidence interval +/- mean: -0.069 to 0.070
95% confidence interval +/- mean: -0.052 to 0.053
```

Note that, like any other normal curve, *zage* has a mean of zero (in this case, the mean is virtually zero at .001) and a standard deviation of 1.000. Inspect the frequency distribution for *zage* (not reproduced here to conserve space). The *Z* scores are listed in the far right-hand column of the table beginning with −1.58, equivalent to a raw score of 18. Scan the list of scores and note that the further away the score is from 44.94, the greater the numerical value of the *Z* score. The closer the score is to the center of the distribution (the closer it is to the mean), the lower the *Z* score. Also note that scores below the mean (less than 44.94) have negative signs and scores above the mean (greater than 44.94) have positive signs.

Exercises

5.1 Compute *Z* scores for *prestg80, papres80,* and *tvhours.*
   a. (SPSS) Use the **Case Summaries** procedure to display the normalized and "raw" scores for each variable for 10 cases.
   b. (MicroCase) Use the **Univariate** command to get descriptive statistics and frequency distributions for the transformed variables.
5.2 (SPSS only) Use the **Graphs** command to get simple line charts of *age* and then *zage.* How close are these charts to smooth, bell-shaped, normal curves?

1. To what extent do people apply religion to the problems of everyday living? Fifteen people have responded to a series of questions including the following:

   1. What is your religion?
      1. Protestant
      2. Catholic
      3. Jewish
      4. None
      5. Other (Muslim, Hindu, etc.)
   2. On a scale of 1 to 10 (with 10 being the highest), how strong would you say your faith is?
   3. How many times a day do you pray?
   4. When things aren't going well, my religion is my major source of comfort.
      1. Strongly agree
      2. Slightly agree
      3. Neither agree nor disagree
      4. Slightly disagree
      5. Strongly disagree
   5. How old are you?

Case	Religion	Strength	Pray	Comfort	Age
1	1	2	0	2	30
2	2	8	1	1	67
3	1	8	3	1	45
4	1	6	0	1	43
5	3	9	3	1	32
6	3	3	0	3	18
7	4	0	0	5	52
8	5	9	6	1	37
9	1	5	0	2	54
10	2	8	2	1	55
11	2	3	0	5	33
12	1	6	1	3	45
13	1	8	2	2	37
14	1	7	2	1	50
15	2	9	1	1	25

   For each variable, construct a frequency distribution and calculate appropriate measures of central tendency and dispersion. Write a sentence summarizing each variable.

2. A survey measuring attitudes toward interracial dating was administered to 1000 people. The survey asked the following questions:

   1. What is your age?
   2. What is you sex?
      1. Male
      2. Female

3. Marriages between people of different racial groups just don't work out and should be banned by law.
   1. Strongly agree
   2. Agree
   3. Undecided
   4. Disagree
   5. Strongly disagree
4. How many years of schooling have you completed?
5. Which category below best describes the place where you grew up?
   1. Large city
   2. Medium-size city
   3. Suburbs of a city
   4. Small town
   5. Rural area
6. What is your marital status?
   1. Married
   2. Separated or divorced
   3. Widowed
   4. Never married

The scores of 20 respondents are reproduced below.

Case	Age	Sex	Attitudes on Interracial Dating	Years of School	Area	Marital Status
1	17	1	5	12	1	4
2	25	2	3	12	2	1
3	55	2	3	14	2	1
4	45	1	1	12	3	1
5	38	2	1	10	3	1
6	21	1	1	16	5	1
7	29	2	2	16	2	2
8	30	2	1	12	4	1
9	37	1	1	12	2	1
10	42	2	3	18	5	4
11	57	2	4	12	2	3
12	24	2	2	12	4	1
13	27	1	2	18	3	2
14	44	1	1	15	1	1
15	37	1	1	10	5	4
16	35	1	1	12	4	1
17	41	2	2	15	3	1
18	42	2	1	10	2	4
19	20	2	1	16	1	4
20	21	2	1	16	1	4

a. For each variable, construct a frequency distribution and select and calculate an appropriate measure of central tendency and a measure of dispersion. Summarize each variable in a sentence.

b. For all 1000 respondents, the mean age was 34.70 with a standard deviation of 3.4 years. Assuming the distribution of age is approximately normal, compute Z scores for each of the first 10 respondents above and determine the percentage of the area below (younger than) each respondent.

3. The data set below is taken from the General Social Survey. Abbreviated versions of the questions along with the meanings of the codes are also presented. See Appendix G for the codes and the complete question wordings. For each variable, construct a frequency distribution and select and calculate an appropriate measure of central tendency and a measure of dispersion. Summarize each variable in a sentence.

1. How many children have you ever had? (Values are actual numbers.)
2. Respondent's educational level:
   0. Less than HS
   1. HS
   2. Jr. college
   3. Bachelor's degree
   4. Graduate school
3. Race:
   1. White
   2. Black
   3. Other
4. "... methods of birth control should be available to teenagers ..."
   1. Strongly agree
   2. Agree
   3. Disagree
   4. Strongly disagree
5. Number of hours of TV watched per day. (Values are actual numbers of hours.)
6. What is your religious preference?
   1. Protestant
   2. Catholic
   3. Jewish
   4. None
   5. Other

Case	Number of Children	Years of School	Race	Birth Control	TV	Religion
1	3	1	1	3	3	1
2	2	0	1	4	1	1
3	4	2	1	2	3	1
4	0	3	1	1	2	1
5	5	1	1	3	2	1
6	1	1	1	3	3	1
7	9	0	1	1	6	1
8	6	1	2	3	4	1
9	4	3	1	1	2	4
10	2	1	3	1	1	1
11	2	0	1	2	4	1

Case	Number of Children	Years of School	Race	Birth Control	TV	Religion
12	4	1	2	1	5	2
13	0	1	1	3	2	2
14	2	1	1	4	2	1
15	3	1	2	3	4	1
16	2	0	1	2	2	1
17	2	1	1	2	2	1
18	0	3	1	3	2	1
19	3	0	1	3	5	2
20	2	1	2	1	10	1
21	2	1	1	3	4	1
22	1	0	1	3	5	1
23	0	2	1	1	2	2
24	0	1	1	2	0	4
25	2	4	1	1	1	2

6

INTRODUCTION TO
INFERENTIAL STATISTICS
SAMPLING AND THE
SAMPLING DISTRIBUTION

LEARNING OBJECTIVES

By the end of this chapter, you will be able to

1. Explain the purpose of inferential statistics in terms of generalizing from a sample to a population.
2. Define and explain the basic techniques of random sampling.
3. Explain and define these key terms: population, sample, parameter, statistic, representative, EPSEM.
4. Differentiate between the sampling distribution, the sample, and the population.
5. Explain the two theorems presented.

6.1 INTRODUCTION

One of the goals of social science research is to test our theories and hypotheses using many different types of people drawn from a broad cross section of society. Obviously, we can place the most confidence in theories that have been tested with many different groups and in a variety of social settings. The problem we often face in social science research is that the populations in which we are interested are too large to test. For example, a theory concerning political party preference among U.S. citizens would be most suitably tested using the entire electorate, but it is clearly impossible to interview every member of this group (almost 100 million people). Indeed, even for theories that could be reasonably tested with smaller populations—such as a local community or the student body at a university—the logistics of gathering data from every single case (entire populations) are staggering to contemplate.

If it is too difficult or expensive to do research with entire populations, how can we reasonably test our theories? To deal with this problem, social scientists select samples, or subsets of cases, from the populations of interest. Our goal in inferential statistics is to learn about the characteristics of a population (these are often called **parameters**), based on what we can learn from our samples. Two applications of inferential statistics are covered in this text. In estimation procedures, covered in Chapter 7, a "guess" of the population parameter is made, based on what is known about the sample. In hypothesis testing, covered in Chapters 8 through 11, the validity of a hypothesis about the population is tested against sample outcomes. In this chapter, we will look briefly at sampling (the techniques for selecting cases for a sample) and then introduce a key concept in inferential statistics: the sampling distribution.

6.2 TECHNIQUES FOR PROBABILITY SAMPLING

Social scientists have developed a variety of techniques for selecting samples from populations. In this chapter, we will review the basic procedures for selecting probability samples, the only type of sample that fully supports the use of inferential statistical techniques to generalize to populations. These types of samples are often described as "random," and you may be more familiar with this terminology. Because of its greater familiarity, I will often use the phrase "random sample" in the following chapters. The term "probability sample" is preferred, however, because, in everyday language, "random" is often used to mean "by coincidence" or to give a connotation of unpredictability. As you will see, probability samples are selected by techniques that are careful and methodical and leave no room for haphazardness. Interviewing the people you happen to meet in a mall one afternoon may be "random" in some sense, but this technique will not result in a sample that could support inferential statistics.

Before considering probability sampling, let me point out that social scientists often use nonprobability samples. For example, social scientists studying small group dynamics or the structure of attitudes or personal values might use the students enrolled in their classes as subjects. Such "convenience" samples are very useful for a number of purposes (e.g., exploring ideas or pretesting survey forms before embarking on a more ambitious project) and are typically less costly and easier to assemble. The major limitation of these samples is that results cannot be generalized beyond the group being tested. If a theory of prejudice, for example, has been tested only on college students, we cannot conclude that the theory would be true for other types of people. Therefore, even when the evidence is very strong, we cannot place a lot of confidence in theories tested on nonprobability samples only.

When constructing a probability sample, our goal is to select cases so that the final sample is **representative** of the population from which it was drawn. A sample is representative if it reproduces the important characteristics of the population. For example, if the population consists of 60% females and 40% males, the sample should contain essentially the same proportions. In other words, a representative sample is very much like the population

—only smaller. It is crucial for inferential statistics that samples be representative: if they are not, generalizing to the population becomes, at best, extremely hazardous.

The fundamental principle of probability sampling is that a sample is very likely to be representative if it is selected by a principle called **EPSEM,** which stands for the "**E**qual **P**robability of **SE**lection **M**ethod." The EPSEM principle is that every element or case in the population must have an equal probability of being selected for the sample. Our goal is to select a representative sample, and the technique we use to achieve that goal is to follow the rule of EPSEM.

Remember that the EPSEM selection technique and the representativeness of the final sample are two different things. In other words, the fact that a sample is selected according to EPSEM does not guarantee that it will be an exact representation or microcosm of the population. The probability is very high that an EPSEM sample will be representative but, just as a perfectly honest coin will sometimes show 10 heads in a row when flipped, an EPSEM sample will occasionally present an inaccurate picture of the population. One great strength of inferential statistics is that they allow the researcher to estimate the probability of this type of error and interpret results accordingly.

6.3 EPSEM SAMPLING TECHNIQUES

The most basic EPSEM sampling technique produces a **simple random sample.** To implement the EPSEM principle, we need a list of all elements or cases in the population and a system for selecting cases from the list that will guarantee that every case has an equal chance of being selected for the sample. The selection process could be based on a number of different kinds of operations (for example, drawing cards from a well-shuffled deck, flipping coins, throwing dice, drawing numbers from a hat, and so on). Cases are often selected by using tables of random numbers. These tables are lists of numbers that have no pattern to them (that is, they are random), and an example of such a table is available at the web site for this text.

To use the table of random numbers, first assign each case on the population list a unique identification number. Then, select cases for the sample when their identification number corresponds to the number chosen from the table. Since the numbers in the table are in random order and any number is just as likely as any other number, we will meet the EPSEM criteria by following this procedure. Stop selecting cases when you have reached your desired sample size and, if an identification number is selected more than once, ignore the repeats.*

---

*Ignoring identification numbers when they are repeated is called "sampling without replacement." Technically, this practice compromises the randomness of the selection process. However, if the sample is a small fraction of the total population, we will be unlikely to select the same case twice, and ignoring repeats will not bias our conclusions.

Rigidly following the procedures above will generate random samples as long as you select from a complete list of the population. However, these procedures can be very cumbersome when there is a long list of cases. Consider the situation when your population numbers 10,000 cases. It is perfectly possible that the first case you select will come from the front of the list, the second from the back, the third from the front again, and so on—leading to a great deal of paper shuffling and a fair amount of confusion. To save time and money in such a situation, researchers often use a technique called **systematic sampling,** where only the first case is randomly selected. Thereafter, every *k*th case is selected. For example, if you are drawing from a list of 10,000 and desire a sample of 200, select the first case randomly and every 10,000/200th, or 50th, case thereafter. If you randomly start with case #13, then your second case will be #63, your third #113, and so on, until you reach the end of the list.

Note that systematic sampling does not strictly conform to the criterion of EPSEM. That is, once the first case has been selected, the other cases no longer have an equal probability of being chosen. In our example above, cases other than the 13th, the 63rd, the 113th, and so on will not be selected for the sample. In general, this increased probability of error is very slight as long as the list from which cases are chosen is itself random, or at least noncyclical with respect to the traits you wish to measure. For example, if you are concerned with ethnicity and are drawing your sample from an alphabetical list, you might encounter difficulties because there is a tendency for certain ethnic names to begin with the same letter (for example, the Irish prefix *O*). Therefore, when using systematic sampling, pay careful attention to the nature of the population list as well as your sampling technique.

A third type of EPSEM sampling produces the **stratified sample.** This technique is very desirable because it guarantees that the sample will be representative on the selected traits. To apply this technique, you first stratify (or divide) the population list into sublists according to some relevant trait and then sample from the sublists. If you select a number of cases from each sublist proportional to the numbers for that characteristic in the population, the sample will be representative of the population.

For example, suppose that you are drawing a sample of 300 of your classmates and you wish to have proportional representation from every major field on campus. If only 10% of the student body is majoring in zoology, the first two sampling techniques we discussed could result in a sample with very few (or even no) zoologists. If, however, you first divide the population into sublists by major, you can use EPSEM to select exactly 30 zoologists from the appropriate sublist. Following the same procedure with other majors will create a sample that is, by definition, representative of the population on this characteristic. Thus, stratified samples are guaranteed to meet the all-important criterion of representativeness (at least for the traits that are used to stratify the samples).

A major limitation of stratified random sampling is that the exact composition of the population is often unknown. If we have no information about the population, we will be unable to establish a scheme for stratification and determine how many cases should be taken from each sublist.

To this point, sampling techniques have been presented as straightforward processes of randomly selecting cases from a list or sublists of the population. However, sampling is rarely so uncomplicated, and the major difficulty almost always centers on what might appear, at first glance, to be the easiest part: establishing the list of the population. For many of the populations of interest to the social sciences, there are no complete, up-to-date lists. There is no list of United States citizens, no list of the residents of any given state, and no complete, up-to-date list of residents of your local community. Devices such as telephone books or city directories might appear to contain complete lists of local residents. However, the former will omit unlisted numbers and a disproportionate number of low-income households, and the latter is very likely to be outdated.

Social scientists have devised several ways of dealing with the limitations imposed by the scarcity of lists. Probably the most significant of these is **cluster sampling,** which involves selecting groups of cases (clusters) rather than single cases. The clusters are often based on geography, and the selection of clusters often proceeds in stages. For example, you might draw a cluster sample of your city or town by first numbering all of the voting precincts within the political boundaries. Next, you would use EPSEM to select a sample of precincts. The second stage of selection would involve numbering the blocks within each of the selected precincts and, following EPSEM, selecting a sample of blocks. A third stage might involve the selection of households within each selected block. When these stages are completed, you would have a sample that had a very high probability of being representative of the entire city without ever using a list of residents of the city.

Cluster sampling is less trustworthy than the other techniques summarized above. A cluster sample is a less accurate representation of the population than a simple random sample of comparable size. In part, this decreased accuracy is a result of the multiple selection stages described above. With a simple random sample, the sample is drawn in one selection from the list of the population. In a multistage cluster sample, each stage in the selection process (e.g., first the precincts, then the blocks, and then the households) has a probability of error. That is, each time we sample, we run the risk of selecting an unrepresentative sample. In simple random sampling, we run this risk once; with cluster sampling we will run the risk anew at each stage.

Although we have to treat inferences to populations based on cluster samples with some additional caution, we often have no alternative method of sampling. While it may be extremely difficult (or even impossible) to construct an accurate list of an entire city population, all you need to compile a cluster sample is a map (or a list of voting precincts, census tracts, and so forth).

By way of summary, let me return to a major point. The purpose of inferential statistics is to acquire knowledge about populations, based on the information derived from samples drawn from that population. Each of the statistics to be presented in the following chapters requires that samples be selected according to EPSEM. While even the most painstaking and sophisticated sampling techniques will not guarantee representativeness, the probability is high that EPSEM samples will be representative of the populations from which they are selected.

6.4 THE SAMPLING DISTRIBUTION

Once we have selected a probability sample according to some EPSEM procedure, what do we know? On one hand, we can gather a great deal of information from the cases in the sample. On the other hand, we know nothing about the population. Indeed, if we had information about the population, we probably wouldn't need the sample. Remember that we use inferential statistics to learn more about populations, and information from the sample is important primarily insofar as it allows us to generalize to the population.

When we use inferential statistics, we generally measure some variable (e.g., age, political party preference, or opinions about abortion) in the sample and then use the information from the sample to learn more about that variable in the population. In Part I of this text, you learned that three types of information are generally necessary to adequately characterize a variable: (1) the shape of its distribution, (2) some measure of central tendency, and (3) some measure of dispersion. Clearly, all three kinds of information can be gathered (or computed) on a variable from the cases in the sample. Just as clearly, none of the information can be gathered for the population. Except in rare situations (for example, IQ and height are thought to be approximately normal in distribution), nothing can be known about the exact shape of the distribution of a variable in the population. The means and standard deviations of variables in the population are also unknown. Let me remind you that if we had this information for the population, inferential statistics would be unnecessary.

In statistics, we link information from the sample to the population with a device known as the **sampling distribution:** the theoretical, probabilistic distribution of a statistic for all possible samples of a certain sample size (N). That is, the sampling distribution includes statistics that represent every conceivable combination of cases from the population. A crucial point about the sampling distribution is that its characteristics are based on the laws of probability, not on empirical information, and are very well known. In fact, the sampling distribution is the central concept in inferential statistics, and a prolonged examination of its characteristics is certainly in order.

FIGURE 6.1 THE RELATION-
SHIPS BETWEEN THE SAMPLE,
SAMPLING DISTRIBUTION, AND
POPULATION

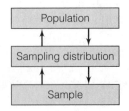

As illustrated by Figure 6.1, the general strategy of all applications of inferential statistics is to move between the sample and the population via the sampling distribution. Thus, three separate and distinct distributions are involved in every application of inferential statistics:

1. The sample distribution, which is empirical (i.e., it exists in reality) and known in the sense that the shape, central tendency, and dispersion of any variable can be ascertained for the sample. Remember that the information from the sample is important primarily insofar as it allows the researcher to learn about the population.

2. The population distribution, which, while empirical, is unknown. Amassing information about or making inferences to the population is the sole purpose of inferential statistics.

3. The sampling distribution, which is nonempirical or theoretical. Because of the laws of probability, a great deal is known about this distribution. Specifically, the shape, central tendency, and dispersion of the distribution can be deduced and, therefore, the distribution can be adequately characterized.

The utility of the sampling distribution is implied by its definition. Because it encompasses all possible sample outcomes, the sampling distribution enables us to estimate the probability of any particular sample outcome, a process that will occupy our attention for the next five chapters.

The sampling distribution is theoretical, which means that it is never obtained in reality by the researcher. However, to understand better the structure and function of the distribution, let's consider an example of how one might be constructed. Suppose that we wanted to gather some information about the age of a particular community of 10,000 individuals. We draw an EPSEM sample of 100 residents, ask all 100 respondents their age, and use those individual scores to compute a mean age of 27. This score is noted on the graph in Figure 6.2. Note that this sample is one of countless possible combinations of 100 people taken from this population of 10,000, and the mean of 27 is one of millions of possible sample outcomes.

Now, replace the 100 respondents in the first sample and draw another

FIGURE 6.2 CONSTRUCTING A SAMPLE DISTRIBUTION

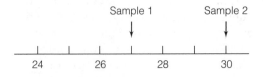

sample of the same size ($N = 100$) and again compute the average age. Assume that the mean for the second sample is 30 and note this sample outcome on Figure 6.2. This second sample is another of the countless possible combinations of 100 people taken from this population of 10,000, and the sample mean of 30 is another of the millions of possible sample outcomes. Replace the respondents from the second sample and draw still another sample, calculate and note the mean, replace this third sample, and draw a fourth sample, continuing these operations an infinite number of times, calculating and noting the mean of each sample. Now, try to imagine what Figure 6.2 would look like after thousands of individual samples had been collected and the mean had been computed for each sample. What shape, mean, and standard deviation would this distribution of sample means have after we had collected all possible combinations of 100 respondents from the population of 10,000?

For one thing, we know that each sample will be at least slightly different from every other sample, since we will never sample exactly the same 100 people twice. Hence, because each sample is a unique combination of individuals, each sample mean will be at least slightly different in value. We also know that even though the samples are chosen according to EPSEM, they will not be representative of the population in every single case. For example, if we continue taking samples of 100 people long enough, we will eventually choose a sample that includes only the very youngest residents. Such a sample would have a mean much lower than the true population mean. Likewise, some of our samples will include only senior citizens and will have means that are much higher than the population mean. Common sense suggests, however, that such nonrepresentative samples will be rare and that most sample means will cluster around the true population value.

To illustrate further, assume that we somehow come to know that the true mean age of the population is 30. As we have seen above, most of the sample means will also be approximately 30. Thus, the sampling distribution of these sample means should peak at 30. Some of the sample means will "miss the mark," but the frequency of such misses should decline as we get farther away from 30. That is, the distribution should slope to the base as we get farther away from the population value (sample means of 29 or 31 should be common; means of 20 or 40 should be rare). Since the samples are random, the means should miss an equal number of times on either side of the population value, and the distribution itself should therefore be roughly symmetrical. In other words, the sampling distribution of all possible sample means should be approximately normal and will resemble the distribution presented in Figure 6.3.

These common-sense notions about the shape of the sampling distribution and other very important information about central tendency and dispersion are stated in two theorems. The first of these theorems states that

FIGURE 6.3   A SAMPLING DISTRIBUTION OF SAMPLE MEANS

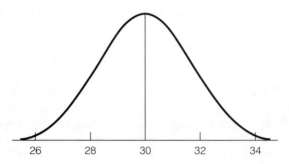

| 26 | 28 | 30 | 32 | 34 |

> If repeated random samples of size N are drawn from a normal population with mean μ and standard deviation σ, then the sampling distribution of sample means will be normal with a mean μ and a standard deviation of $\sigma/\sqrt{N}$.

To translate: if we begin with a trait that is normally distributed across a population (IQ, height, or weight, for example) and take an infinite number of equally sized random samples from that population, then the sampling distribution of sample means will be normal. If it is known that the variable is distributed normally in the population, it can be assumed that the sampling distribution will be normal.

The theorem tells us more than the shape of the sampling distribution of all possible sample means, however. It also defines its mean and standard deviation. In fact, it says that the mean of the sampling distribution will be exactly the same value as the mean of the population. That is, if we know that the mean IQ of the entire population is 100, then we know that the mean of any sampling distribution of sample mean IQ's will also be 100. Exactly why this should be so is not a matter that can be fully explained at this level. Recall, however, that most sample means will cluster around the population value over the long run. Thus, the fact that these two values are equal should have intuitive appeal. As for dispersion, the theorem says that the standard deviation of the sampling distribution, also called the **standard error of the mean,** will be equal to the standard deviation of the population divided by the square root of N (symbolically: $\sigma\sqrt{N}$).

If the mean and standard deviation of a normally distributed population are known, the theorem allows us to compute the mean and standard deviation of the sampling distribution.* Thus, we will know exactly as much

---

*In the typical research situation, the values of the population mean and standard deviation are, of course, unknown. However, these values can be estimated from sample statistics, as we shall see in the chapters that follow.

about the sampling distribution (shape, central tendency, and dispersion) as we ever knew about any empirical distribution. But what if the population distribution is not normal? This eventuality (very common, in fact) is covered by a second theorem, called the **Central Limit Theorem:**

> If repeated random samples of size N are drawn from any population, with mean μ and standard deviation σ, then, as N becomes large, the sampling distribution of sample means will approach normality, with mean μ and standard deviation $\sigma/\sqrt{N}$.

The importance of the Central Limit Theorem is that it removes the constraint of normality in the population. Whenever sample size is large, it can be assumed that the sampling distribution is normal, with a mean equal to the population mean and a standard deviation equal to $\sigma/\sqrt{N}$ regardless of the shape of the population. Thus, even if we are working with a variable that has a skewed distribution (such as income, which almost always has a positive skew), we can still assume a normal sampling distribution.

The issue remaining, of course, is to define what is meant by a large sample. A good rule of thumb is that if sample size (N) is 100 or more, the Central Limit Theorem applies, and you can assume that the sampling distribution is normal in shape. When N is less than 100, you must have good evidence of a normal population distribution before you can assume that the sampling distribution is normal. Thus, a normal sampling distribution can be ensured by the expedient of using fairly large samples.

6.5 SYMBOLS AND TERMINOLOGY

In the following chapters, we will be working with three entirely different distributions. The purpose of inferential statistics is to acquire knowledge of the population from information gathered from a sample by means of the sampling distribution. Furthermore, we will be concerned with several different kinds of sampling distributions—including the sampling distribution of sample means and the sampling distribution of sample proportions.*

To distinguish clearly among these various distributions, we will often use symbols. The symbols used for the means and standard deviations of samples and populations have already been introduced in Chapters 3 and 4. Table 6.1 introduces some of the symbols that will be used for the sampling distribution in summary form for quick reference. Basically, the sampling distribution is denoted with Greek letter symbols that are subscripted according to the sample statistic of interest.

---

*Symbols for the sampling distributions of statistics other than means and proportions will be introduced at the appropriate time.

TABLE 6.1 SYMBOLS FOR MEANS AND STANDARD DEVIATIONS OF THREE DISTRIBUTIONS

	Mean	Standard Deviation	Proportion
1. Samples	$\overline{X}$	s	P_s
2. Populations	μ	σ	P_u
3. Sampling distributions			
of means	$\mu_{\overline{X}}$	$\sigma_{\overline{X}}$	
of proportions	μ_p	σ_p	

SUMMARY

1. Since populations are almost always too large to test, a fundamental strategy of social science research is to select a sample from the defined population and then use information from the sample to generalize to the population. This is done either by estimation or by hypothesis testing.

2. Several techniques are commonly used for selecting random samples. Each of these techniques involves selecting cases for the sample according to EPSEM. Even the most rigorous technique, however, cannot guarantee representativeness. One of the great strengths of inferential statistics is that the probability of this kind of error (nonrepresentativeness) can be estimated.

3. The sampling distribution, the central concept in inferential statistics, is a theoretical distribution of all possible sample outcomes. Since its overall shape, mean, and standard deviation are known (under the conditions specified in the two theorems), the sampling distribution can be adequately characterized and utilized by researchers.

4. The two theorems that were introduced in this chapter state that when the variable of interest is normally distributed in the population or when sample size is large, the sampling distribution will be normal in shape, its mean will be equal to the population mean, and its standard deviation (or standard error) will be equal to the population standard deviation divided by the square root of N.

5. All applications of inferential statistics involve generalizing from the sample to the population by means of the sampling distribution. Both estimation procedures and hypothesis testing incorporate the three distributions, and it is crucial that you develop a clear understanding of each distribution and its role in inferential statistics.

GLOSSARY

Central Limit Theorem. A theorem that specifies the mean, standard deviation, and shape of the sampling distribution, given that the sample is large.

Cluster sample. A method of sampling by which geographical units are randomly selected and all cases within each selected unit are tested.

EPSEM. The **E**qual **P**robability of **SE**lection **M**ethod for selecting samples. Every element or case in the population must have an equal probability of selection for the sample.

μ. The mean of a population.

$\mu_{\overline{X}}$. The mean of a sampling distribution of sample means.

μ_p. The mean of a sampling distribution of sample proportions.

Parameter. A characteristic of a population.

P_s. (P-sub-s) Any sample proportion.

P_u. (P-sub-u) Any population proportion.

Representative. The quality a sample is said to have if it reproduces the major characteristics of the population from which it was drawn.

Sampling distribution. The distribution of all possible sample outcomes of a given statistic. Under conditions specified in two theorems, the sampling distribution will be normal in shape with a mean equal to the population value and a standard deviation equal to the population standard deviation divided by the square root of N.

Simple random sample. A method for choosing cases from a population by which every case and every combination of cases has an equal chance of being included.

Standard error of the mean. The standard deviation of a sampling distribution of sample means.

Stratified sample. A method of sampling by which cases are selected from sublists of the population.

Systematic sampling. A method of sampling by which the first case from a list of the population is randomly selected. Thereafter, every kth case is selected.

MULTIMEDIA RESOURCES

The Wadsworth Sociology Resource Center: Virtual Society
http://sociology.wadsworth.com/

Visit the companion web site for the sixth edition of *Statistics: A Tool for Social Research* to access a wide range of student resources. Begin by clicking on the Student Resources section of the book's web site to access the following study tools:

- Basic math review
- Statistics review

- Flash cards
- Internet links
- Additional chapter problems
- Table of random numbers
- MicroCase and SPSS examples and exercises
- "Find the text" flowcharts
- Hypothesis testing for variables measured at the ordinal level

PROBLEMS

6.1 Imagine that you had to gather a random sample ($N = 300$) of the student body at your school. How would you acquire a list of the population? Would the list be complete and accurate? What procedure would you follow in selecting cases (that is, simple or systematic random sampling)? Would cluster sampling be an appropriate technique (assuming that no list was available)? Describe in detail how you would construct a cluster sample.

6.2 This exercise is extremely tedious and hardly ever works out the way it ought to (mostly because not many people have the patience to draw an "infinite" number of even very small samples). However, if you want a more concrete and tangible understanding of sampling distributions and the two theorems presented in this chapter, then this exercise may have a significant payoff. Below are listed the ages of a population of college students ($N = 50$). By a random method (such as a table of random numbers), draw at least 50 samples of size 2 (that is, 50 pairs of cases), compute a mean for each sample, and plot the means on a frequency polygon. (Incidentally, this exercise will work better if you draw 100 or 200 samples and/or use larger samples than $N = 2$.)

a. The curve you've just produced is a sampling distribution. Observe its shape; after 50 samples, it should be approaching normality. What is your estimate of the population mean (μ) based on the shape of the curve?

b. Calculate the mean of the sampling distribution ($\mu_{\overline{X}}$). Be careful to do this by summing the sample means (not the scores) and dividing by the number of samples you've drawn. Now compute the population mean (μ). These two means should be very close in value because $\mu = \mu_{\overline{X}}$ by the Central Limit Theorem.

c. Calculate the standard deviation of the sampling distribution (use the means as scores) and the standard deviation of the population. Compare these two values. You should find that $\sigma_{\overline{X}} = \sigma / \sqrt{N}$.

d. If none of the above exercises turned out as they should have, it is for one or more of the following reasons:

1. You didn't take enough samples. You may need as many as 100 or 200 (or more) samples to see the curve begin to look "normal."

2. Sample size (2) is too small. An N of 5 or 10 would work much better.

3. Your sampling method is not truly random and/or the population is not arranged in random fashion.

17	20	20	19	20
18	21	19	20	19
19	22	19	23	19
20	23	18	20	20
22	19	19	20	20
23	17	18	21	20
20	18	20	19	20
22	17	21	21	21
21	20	20	20	22
18	21	20	22	21

SPSS for Windows

Using *SPSS for Windows* to Draw Random Samples

DEMONSTRATION 6.1 Estimating Average Age

SPSS for Windows includes a procedure for drawing random samples from a database. We can use this procedure to illustrate some points about sampling and to convince the skeptics in the crowd that properly selected samples will produce statistics that are close approximations of the corresponding population values or parameters. In this demonstration, the 1998 GSS sample will be treated as a population, and its characteristics will be treated as parameters.

The instructions below will calculate a mean for *age* for three random samples of different sizes drawn from the 1998 GSS sample. The actual average age of the 1998 GSS sample (which will be the parameter or μ) is 44.94 (see Demonstration 5.1). The samples are roughly 10%, 25%, and 50% of the population size, and the program selects them by a process that is quite similar to a table of random numbers. Therefore, these samples may be considered "simple random samples."

As a part of this procedure we also request the "standard error of the mean" or S.E. MEAN. This is the standard deviation of the sampling distribution ($\sigma / \sqrt{N}$) for a sample of this size. This statistic will be of interest because we can expect our sample means to be within this distance of the population value or parameter.

With the 1998 GSS loaded, click **Data** from the menu bar of the Data Editor and then click **Select Cases.** The **Select Cases** window appears and presents a number of different options. To select random samples, click the button next to "Random sample of cases" and then click on the **Sample** button. The **Select Cases: Random Sample** window will open. We can specify the size of the sample in two different ways. If we use the first option, we can specify that the sample will include a certain percentage of cases in the database. The second option allows us to specify the exact number of cases in the sample. Let's use the first option and request a 10% sample by typing 10 into the box on the first line. Click **Continue,** and then click **OK** on the **Select Cases** window. The sample will be selected and can now be processed.

To find the mean age for the 10% sample, click **Analyze, Descriptive Statistics,** and then **Descriptives.** Find *age* in the variable list and transfer it to the **Variable(s):** window. On the **Descriptives** menu, click the **Options** button and select S.E. MEAN in addition to the usual statistics. Click **Continue** and then **OK,** and the requested statistics will appear in the output window.

Now, to produce a 25% sample, return to the **Select Cases** window by clicking **Data** and **Select Cases.** Click the **Reset** button at the bottom of the window and then click **OK** and the full data set ($N = 1385$) will be restored. Repeat the procedure we followed for selecting the 10% sample. Click the button next to "Random sample of cases" and then click on the **Sample** button. The **Select Cases: Random Sample** window will open. Request a 25% sample by typing 25 in the box, click **Continue** and **OK,** and the new sample will be selected.

Run the **Descriptives** procedure for the 25% sample (don't forget S.E. MEAN) and note the results. Finally, repeat these steps for a 50% sample. Results are summarized below:

Sample %	Sample Size	Sample Mean	Standard Error	Sample Mean Plus and Minus Standard Error
10%	143	43.65	1.41	42.24–45.06
25%	319	44.42	.95	43.47–45.37
50%	685	44.93	.65	44.28–45.58

Note that standard error (or the standard deviation of the sampling distribution) decreases as sample size increases. This should reinforce the common-sense notion that larger samples will provide more accurate estimates of population values. All three samples produced estimates (sample means) that are quite close in value to the population value of 44.94. All three sample means are within a standard error of the population mean.

Furthermore, the smallest sample is the least accurate (43.65 is 1.29 years below the true population value of 44.94), the 25% sample is closer (.52 of a year too low), and the largest sample is very close (only .01 year too low). This relationship between accuracy and size of sample should make sense intuitively, but remember that random samples will not always be so cooperative, and it is perfectly possible that your samples will behave quite differently.

This demonstration should reinforce one of the main points of this chapter: statistics calculated on samples that have been selected according to the principle of EPSEM will (almost always) be reasonable approximations of their population counterparts.

Exercise

6.1 Following the procedures in Demonstration 6.1, select three samples from the 1998 GSS database: 15%, 30%, and 60%. Get descriptive statistics for *age* (don't forget to get the standard error), and use the results to complete the following table:

Sample %	Sample Size	Sample Mean	Standard Error	Sample Mean Plus and Minus Standard Error
15%	_____	_____	_____	_____
30%	_____	_____	_____	_____
60%	_____	_____	_____	_____

Summarize these results. What happens to standard error as sample size increases? Why? How accurate are the estimates (sample means)? Are all sample means within a standard error of 44.94? How does the accuracy of the estimates change as sample size changes?

7 ESTIMATION PROCEDURES

By the end of this chapter, you will be able to

1. Explain the logic of estimation and the role of the sample, sampling distribution, and population.
2. Define and explain the concepts of bias and efficiency.
3. Construct and interpret confidence intervals for sample means and sample proportions.
4. Explain the relationships between confidence level, sample size, and the width of the confidence interval.

7.1 INTRODUCTION

The object of this branch of inferential statistics is to estimate population values or parameters from statistics computed from samples. Although the techniques presented in this chapter may be new to you, you are certainly familiar with their most common applications: public-opinion polls and election projections. Polls and surveys on every conceivable issue—from the sublime to the petty—have become a staple of the mass media and popular culture. The techniques you will learn in this chapter are essentially the same as those used by the most reputable and scientific pollsters.

There are two kinds of estimation procedures. First, a point estimate is a sample statistic that is used to estimate a population value. For example, a newspaper story that reports that, based on a public opinion poll, 74% of Americans support capital punishment, is reporting a **point estimate.** The second kind of estimation procedure involves **confidence intervals,** which consist of a range of values (an interval) instead of a single point. Rather than estimating a specific figure as in a point estimate, an interval estimate might be phrased as "between 39% and 45% of the electorate will vote for the candidate." In this latter estimate, we are essentially guessing that the population value falls between 39% and 45%, but we do not specify where.

7.2 BIAS AND EFFICIENCY

In both point and interval estimation, sample statistics serve as the estimators. Which of the many available sample statistics should be used? Estimators can be selected according to two criteria: **bias** and **efficiency.** An estimator is unbiased if and only if the mean of its sampling distribution is equal to the population value of interest. We know from the theorems presented in Chapter 6 that sample means conform to this criterion. The mean of the sampling distribution of sample means (which we will note symbolically as $\mu_{\bar{x}}$) is the same as the population mean (μ).

Sample proportions (P_s) are also unbiased. That is, if we calculate sample proportions from repeated random samples of size N and then array them on a graph, the sampling distribution of sample proportions will have a mean (μ_p) equal to the population proportion (P_u). Thus, if we are concerned with coin flips and sample honest coins 10 at a time ($N = 10$), the sampling distribution will have a mean equal to 0.5, which is the probability that an honest coin will be heads (or tails) when flipped. All statistics other than sample means and sample proportions are biased (that is, have sampling distributions with means not equal to the population value).*

Since we know that sample means and proportions are unbiased, we can determine the probability that they lie within a given distance of the population values we are trying to estimate. To illustrate, consider a specific problem. Assume that we wish to estimate the average income of a population. A random sample ($N = 500$) is taken, and a sample mean of $35,000 is computed. Note that we have no idea what the population value (μ) is (if we did, we wouldn't need the sample), but it is μ that we are interested in. The sample mean of $35,000 is important and interesting primarily insofar as it can give us information about the population.

The two theorems presented in Chapter 6 give us a great deal of information about the sampling distribution of all possible sample means. Because N is large, we know that the sampling distribution is normal and that its mean is equal to the population mean. We also know that all normal curves contain about 68% of the cases (the cases here are sample means) within ± 1 Z, 95% of the cases within ± 2 Z's, and more than 99% of the cases within ± 3 Z's of the mean. Remember that we are discussing the sampling distribution here—the distribution of all possible sample outcomes or, in this instance, sample means. Thus, the probabilities are very good (approximately 68 out of 100 chances) that our sample mean of $35,000 is within ± 1 Z, excellent (95 out of 100) that it is within ± 2 Z's, and overwhelming (99 out of 100) that it is within ± 3 Z's of the mean of the sampling distribution (which is the same value as the population mean). These relationships are graphically depicted in Figure 7.1.

If an estimator is unbiased, it is probably an accurate estimate of the

---

*In particular, the sample standard deviation (s) is a biased estimator of the population standard deviation (σ). As you might expect, there is less dispersion in a sample than in a population and, as a consequence, s will underestimate σ. As we shall see, however, sample standard deviation can be corrected for this bias and still serve as an estimate of the population standard deviation for large samples.

FIGURE 7.1   AREAS UNDER THE SAMPLING DISTRIBUTION OF SAMPLE MEANS

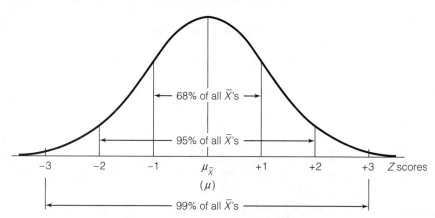

population parameter (μ in this case). Note that in less than 1% of the cases, a sample mean will be more than ± 3 Z's away from the mean of the sampling distribution (very inaccurate) by random chance alone. We literally have no idea if our sample mean of \$35,000 is in this small minority. We do know, however, that the odds are high that our sample mean is considerably closer than ± 3 Z's to the mean of the sampling distribution and, thus, to the population mean.

The second desirable characteristic of an estimator is efficiency, which is the extent to which the sampling distribution is clustered about its mean. Efficiency or clustering is essentially a matter of dispersion. The smaller the standard deviation of a sampling distribution, the greater the clustering and the higher the efficiency. Remember that the standard deviation of the sampling distribution of sample means, or the standard error of the mean, is equal to the population standard deviation divided by the square root of $N (\sigma_{\bar{x}} = \sigma/\sqrt{N})$. Therefore, the standard deviation of the sampling distribution is an inverse function of N. As sample size increases, $\sigma_{\bar{x}}$ will decrease. We can improve the efficiency (or decrease the standard deviation of the sampling distribution) for any estimator by increasing sample size.

An example should make this clearer. Consider two samples of different sizes

Sample 1	Sample 2
$\bar{X}_1 = \$35,000$	$\bar{X}_2 = \$35,000$
$N_1 = 100$	$N_2 = 1000$

Both sample means are unbiased, but which is the more efficient estimator? Consider sample 1 and assume, for the sake of illustration, that the population standard deviation (σ) is \$500.* In this case, the standard deviation of

---

*In reality, of course, the value of σ would be unknown.

the sampling distribution of all possible sample means with an N of 100 would be $\sigma/\sqrt{N}$, or $500/\sqrt{100}$, or $50.00. For sample 2, the standard deviation of all possible sample means with an N of 1000 would be much smaller. Specifically, it would be equal to $500/\sqrt{1000}$ or $15.81.

Sampling distribution 2 is much more clustered than sampling distribution 1. In fact, distribution 2 contains 68% of all possible sample means within ±15.81 of the μ while distribution 1 requires a much broader interval of ±50.00 to do the same. The estimate based on a sample with 1000 cases is much more likely to be close in value to the population parameter than is an estimate based on a sample of 100 cases. Figures 7.2 and 7.3 illustrate these relationships graphically.

To summarize, since the standard deviation of all sampling distributions is an inverse function of N, the larger the sample, the greater the clustering and the higher the efficiency. In part, these relationships between sample size and the standard deviation of the sampling distribution do nothing more than underscore our common-sense notion that much more confidence can be placed in large samples than in small (as long as both have been randomly selected).

FIGURE 7.2 A SAMPLING DISTRIBUTION WITH $N = 100$ AND $\sigma_{\bar{x}} = \$50.00$

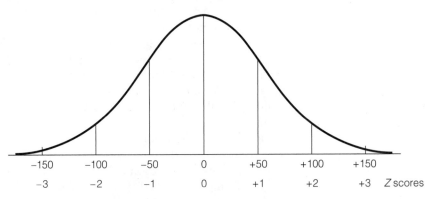

FIGURE 7.3 A SAMPLING DISTRIBUTION WITH $N = 1000$ AND $\sigma_{\bar{x}} = \$15.81$

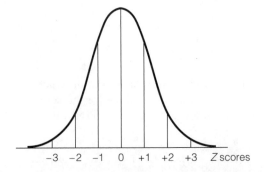

7.3 ESTIMATION PROCEDURES: INTRODUCTION

The procedure for constructing a point estimate is straightforward. Draw an EPSEM sample, calculate either a proportion or a mean, and estimate that the population parameter is the same as the sample statistic. Remember that the larger the sample, the greater the efficiency and the more likely that the estimator is approximately the same as the population value. Also remember that, no matter how rigid the sampling procedure or how large the sample, there is always some chance that the estimator is very inaccurate.

Compared to point estimates, interval estimates are more complicated but safer because when we guess a range of values, we are more likely to include the population parameter. The first step in constructing an interval estimate is to decide on the risk that you are willing to take of being wrong. An interval estimate is wrong if it does not include the population parameter. This probability of error is called **alpha** (symbolized α). The exact value of alpha will depend on the nature of the research situation, but a 0.05 probability is commonly used. Setting alpha equal to 0.05, also called using the 95% **confidence level,** means that over the long run the researcher is willing to be wrong only 5% of the time. Or, to put it another way, if an infinite number of intervals were constructed at this alpha level (and with all other things being equal), 95% of them would contain the population value and 5% would not. In reality, of course, only one interval is constructed and, by setting the probability of error very low, we are setting the odds in our favor that the interval will include the population value.

The second step is to picture the sampling distribution, divide the probability of error equally into the upper and lower tails of the distribution, and then find the corresponding Z score. For example, if we decided to set alpha equal to 0.05, we would place half (0.025) of this probability in the lower tail and half in the upper tail of the distribution. The sampling distribution would thus be divided as illustrated in Figure 7.4.

We need to find the Z score that marks the beginnings of the shaded areas. In Chapter 5, we learned how to calculate Z scores and find areas under the normal curve. Here, we will reverse that process. We need to find

FIGURE 7.4   THE SAMPLING DISTRIBUTION WITH ALPHA (α) EQUAL TO 0.05

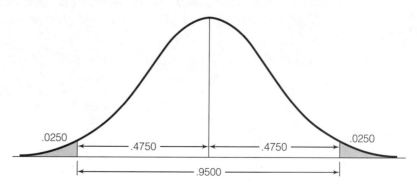

FIGURE 7.5 FINDING THE *Z* SCORE THAT CORRESPONDS TO AN ALPHA (α) OF 0.05

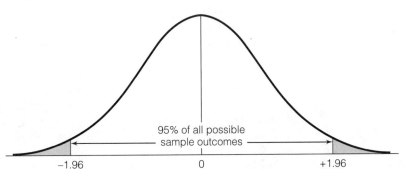

95% of all possible sample outcomes

−1.96   0   +1.96

the *Z* score beyond which lies a proportion of .0250 of the total area. To do this, go down column C of Appendix A until you find this proportion (.0250). The associated *Z* score is 1.96. Since the curve is symmetrical and we are interested in both the upper and lower tails, we designate the *Z* score that corresponds to an alpha of .05 as ±1.96 (see Figure 7.5).

We now know that 95% of all possible sample outcomes fall within ±1.96 *Z*-score units of the population value. In reality, of course, there is only one sample outcome but, if we construct an interval estimate based on ±1.96 *Z*'s, the probabilities are that 95% of all such intervals will trap the population value. Thus, we can be 95% confident that our interval contains the population value.

Besides the 95% level, there are three other commonly used confidence levels: the 90% level (α = .10), the 99% level (α = 0.01), and the 99.9% level (α = .001). To find the corresponding *Z* scores for these levels, follow the procedures outlined above for an alpha of 0.05. Table 7.1 summarizes all the information you will need.

You should turn to Appendix A and confirm for yourself that the *Z* scores in Table 7.1 do indeed correspond to these alpha levels. As you do, note that, in the cases where alpha is set at 0.10 and 0.01, the precise areas we seek do not appear in the table. For example, with an alpha of 0.10, we would look in column C ("Area beyond") for the area .0500. Instead we find an area of .0505 (*Z* = ±1.64) and an area of .0495 (*Z* = ±1.65). The *Z* score we are

TABLE 7.1 *Z* SCORES FOR VARIOUS LEVELS OF ALPHA (α)

Confidence Level	Alpha	α/2	*Z* Score
90%	.10	.0500	±1.65
95%	.05	.0250	±1.96
99%	.01	.0050	±2.58
99.9%	.001	.0005	±3.29

seeking is somewhere between these two other scores. When this condition occurs, take the larger of the two scores as Z. This will make the interval as wide as possible under the circumstances and is thus the most conservative course of action. In the case of an alpha of 0.01, we encounter the same problem (the exact area .0050 is not in the table), resolve it the same way, and take the larger score as Z. Finally, note that in the case where alpha is set at .001, we can choose from several Z scores. Although our table is not detailed enough to show it, the closest Z score to the exact area we want is ±3.291, which we can round off to ±3.29. *(For practice in finding Z scores for various levels of confidence, see problem 7.3.)*

The third step is to actually construct the confidence interval. In the sections that follow, we illustrate how to construct an interval estimate first with sample means and then with sample proportions.

7.4 INTERVAL ESTIMATION PROCEDURES FOR SAMPLE MEANS

The formula for constructing a confidence interval based on sample means is given in Formula 7.1:

FORMULA 7.1

$$\text{c.i.} = \overline{X} \pm Z \left(\frac{\sigma}{\sqrt{N}} \right)$$

Where: c.i. = confidence interval

$\overline{X}$ = the sample mean

Z = the Z score as determined by the alpha level

$\dfrac{\sigma}{\sqrt{N}}$ = the standard deviation of the sampling distribution or the standard error of the mean

As an example, suppose you wanted to estimate the average IQ of a community and had randomly selected a sample of 200 residents, with a sample mean IQ of 105. Assume that the population standard deviation for IQ scores is about 15, so we can set σ equal to 15. If we are willing to run a 5% chance of being wrong and set alpha at 0.05, the corresponding Z score will be 1.96. These values can be directly substituted into Formula 7.1, and an interval can be constructed

$$\text{c.i.} = \overline{X} \pm Z \left(\frac{\sigma}{\sqrt{N}} \right)$$

$$\text{c.i.} = 105 \pm 1.96 \left(\frac{15}{\sqrt{200}} \right)$$

$$\text{c.i.} = 105 \pm 1.96 \left(\frac{15}{14.14} \right)$$

$$\text{c.i.} = 105 \pm (1.96)(1.06)$$

$$\text{c.i.} = 105 \pm 2.08$$

That is, our estimate is that the average IQ for the population in question is somewhere between 102.92 (105 − 2.08) and 107.08 (105 + 2.08). Since 95% of all possible sample means are within ±1.96 Z's (or 2.08 IQ units in this case) of the mean of the sampling distribution, the odds are very high that our interval will contain the population mean. In fact, even if the sample mean is as far off as ± 1.96 Z's (which is unlikely), our interval will still contain $\mu_{\bar{x}}$ and, thus, μ. Only if our sample mean is one of the few that is more than ± 1.96 Z's from the mean of the sampling distribution will we have failed to include the population mean.

Note that in the example above, the value of the population standard deviation was supplied. Needless to say, it is unusual to have such information about a population. In the great majority of cases, we will have no knowledge of σ. In such cases, however, we can estimate σ with s, the sample standard deviation. Unfortunately, s is a biased estimator of σ, and the formula must be changed slightly to correct for the bias. For larger samples, the bias of s will not affect the interval very much. The revised formula for cases in which σ is unknown is

FORMULA 7.2

$$\text{c.i.} = \overline{X} \pm Z \left(\frac{s}{\sqrt{N-1}} \right)$$

In comparing this formula with 7.1, note that there are two changes. First, σ is replaced by s and, second, the denominator of the last term is the square root of $N - 1$ rather than the square root of N. The latter change is the correction for the fact that s is biased.

Let me stress here that the substitution of s for σ is permitted only for large samples (that is, samples with 100 or more cases). For smaller samples, when the value of the population standard deviation is unknown, the standardized normal distribution summarized in Appendix A cannot be used in the estimation process. It is perfectly possible to construct meaningful interval estimates for samples smaller than 100 but, to do so, we must use a different theoretical distribution, called the Student's t distribution, to find areas under the sampling distribution. We will defer the presentation of the t distribution until Chapter 8 and confine our attention here to estimation procedures for large samples only.

Let us close this section by working through a sample problem with Formula 7.2. Average income for a random sample of a particular community is $35,000, with a standard deviation of $200. What is the 95% interval estimate of the population mean, μ?

Given that

$$\overline{X} = \$35{,}000$$

$$s = \$200$$

$$N = 500$$

and using an alpha of 0.05, the interval can be constructed.

$$\text{c.i.} = \overline{X} \pm Z\left(\frac{s}{\sqrt{N-1}}\right)$$

$$\text{c.i.} = 35{,}000 \pm 1.96\left(\frac{200}{\sqrt{499}}\right)$$

$$\text{c.i.} = 35{,}000 \pm 17.55$$

The average income for the community as a whole is between \$34,982.45 (35,000 − 17.55) and \$35,017.55 (35,000 + 17.55). Remember that this interval has only a 5% chance of being wrong (that is, of not containing the population mean). *(For practice with confidence intervals for sample means, see problems 7.1, 7.4–7.7, 7.18 and 7.19a–7.19c.)*

7.5 INTERVAL ESTIMATION PROCEDURES FOR SAMPLE PROPORTIONS (LARGE SAMPLES)

Estimation procedures for sample proportions are essentially the same as those for sample means. The major difference is that, since proportions are different statistics, we must use a different sampling distribution. In fact, again based on the Central Limit Theorem, we know that sample proportions have sampling distributions that are normal in shape with means (μ_p) equal to the population value (P_u) and standard deviations (σ_p) equal to $\sqrt{P_u(1 - P_u)/N}$. The formula for constructing confidence intervals based on sample proportions is

FORMULA 7.3

$$\text{c.i.} = P_s \pm Z\sqrt{\frac{P_u(1 - P_u)}{N}}$$

The values for P_s and N come directly from the sample, and the value of Z is determined by the confidence level, as was the case with sample means. This leaves one unknown in the formula, P_u—the same value we are trying to estimate. This dilemma can be resolved by setting the value of P_u at 0.5. Since the second term in the numerator under the radical $(1 - P_u)$ is the reciprocal of P_u, the entire expression will always have a value of 0.5 × 0.5, or 0.25, which is the maximum value this expression can attain. That is, if we set P_u at any value other than 0.5, the expression $P_u(1 - P_u)$ will decrease in value. If we set P_u at 0.4, for example, the second term $(1 - P_u)$ would be 0.6, and the value of the entire expression would decrease to 0.24. Setting P_u at 0.5 ensures that the expression $P_u(1 - P_u)$ will be at its maximum possible value and, consequently, the interval will be at maximum width. This is the most conservative solution possible to the dilemma posed by having to assign a value to P_u in the estimation equation.

To illustrate these procedures, assume that you wish to estimate the proportion of students at your university who missed at least one day of classes because of illness last semester. Out of a random sample of 200 students, 60 reported that they had been sick enough to miss classes at least once during

---

Application 7.1

A study of the leisure activities of Americans was conducted on a sample of 1000 households. The respondents identified television viewing as a major form of leisure activity. If the sample reported an average of 6.2 hours of television viewing a day, what is the estimate of the population mean? The information from the sample is

$$\overline{X} = 6.2$$
$$s = 0.7$$
$$N = 1000$$

If we set alpha at 0.05, the corresponding Z score will be ± 1.96, and the 95% confidence interval will be

$$\text{c.i.} = \overline{X} \pm Z\left(\frac{s}{\sqrt{N-1}}\right)$$

$$\text{c.i.} = 6.2 \pm 1.96\left(\frac{0.7}{\sqrt{1000-1}}\right)$$

$$\text{c.i.} = 6.2 \pm 1.96\left(\frac{0.7}{31.61}\right)$$

$$\text{c.i.} = 6.2 \pm (1.96)(.022)$$

$$\text{c.i.} = 6.2 \pm .04$$

Based on this result, we would estimate that the population spends an average of $6.2 \pm .04$ hours per day viewing television. The lower limit of our interval estimate $(6.2 - .04)$ is 6.16, and the upper limit $(6.2 + .04)$ is 6.24. Thus, another way to state the interval would be

$$6.16 \leq \mu \leq 6.24$$

The population mean is greater than or equal to 6.16 and less than or equal to 6.24. This estimate has a 5% chance of being wrong (that is, of not containing the population mean).

---

the previous semester. The sample proportion upon which we will base our estimate is thus 60/200, or 0.30. At the 95% level, the interval estimate will be

$$\text{c.i.} = P_s \pm Z\sqrt{\frac{P_u(1-P_u)}{N}}$$

$$\text{c.i.} = 0.30 \pm 1.96\sqrt{\frac{(0.5)(0.5)}{200}}$$

$$\text{c.i.} = 0.30 \pm 1.96\sqrt{\frac{.25}{200}}$$

$$\text{c.i.} = 0.30 \pm (1.96)(0.035)$$

$$\text{c.i.} = 0.30 \pm 0.07$$

Based on this sample proportion of 0.30, you would estimate that the proportion of students who missed at least one day of classes because of illness was between 0.23 and 0.37. The estimate could, of course, also be phrased in percentages by reporting that between 23% and 37% of the student body was affected by illness at least once during the past semester. (*For*

You are most likely to encounter estimates to the population in the mass media in the form of public-opinion polls, election projections, and the like. Professional polling firms use interval estimates, and responsible reporting by the media will usually emphasize the estimate itself (for example, "In a recently conducted survey of the American public, 45% of the respondents said that they approve of the president's performance") but also will report the width of the interval ("This estimate is accurate to within ±3%," or "Figures from this poll are subject to a sampling error of ±3%"), the alpha level (usually as the confidence level of 95%), and the size of the sample ("1458 households were surveyed").

Election projections and voter analyses, at least at the presidential level, have been common for several decades. More recently, public opinion polling has been increasingly used to gauge reactions to everything from the newest movies to the hottest gossip to the president's conduct during the latest crisis. Newsmagazines routinely report poll results as an adjunct to news stories, and similar stories are regular features of TV news and newspapers. I would include an example or two of these applications here, but polls have become so pervasive that you can do your own example. Just pick up a newsmagazine or newspaper, leaf through it casually, and I bet that you'll find at least one poll. Read the story and see if you can identify the population, the confidence interval width, the sample size, and the confidence level. Bring the news item to class and dazzle your instructor.

As a citizen, as well as a social scientist, you should be extremely suspicious of estimates that do not include such vital information as sample size or interval width. You should also check to see how the sample was assembled. If the sample was selected in a nonrandom fashion (for example, a magazine polling only its subscribers), it cannot be regarded as representative of the American public. You should, of course, read all polls and surveys critically and analytically, but you should place confidence only in polls that are based on samples selected according to the rule of EPSEM (see Chapter 6) from some defined population.

Advertisements, commercials, and reports published by partisan groups sometimes report statistics that seem to be estimates to the population. Often, however, such estimates are based on woefully inadequate sampling sizes and biased sampling procedures, and the data are collected under circumstances that evoke a desired response. "Person in the street" (or shopper in the grocery store) interviews have a certain folksy appeal but must not be accorded the same credibility as surveys conducted by reputable polling firms.

Public Opinion Surveys in the Professional Literature

For the social sciences, probably the single most important consequence of the growth in opinion polling is that many nationally representative databases are now available for research purposes. These high-quality databases are often available for a nominal fee, and they make it possible to conduct "state-of-the-art" research without the expense and difficulty of collecting data yourself. This is an important development because we can now easily test our theories against very high-quality data, and our conclusions will thus have a stronger empirical basis. Our research efforts will have greater credibility with our colleagues, with policy makers, and with the public at large.

One of the more important and widely used databases of this sort is called the General Social Survey, or the GSS. In virtually every year since 1972, the National Opinion Research Council has questioned a nationally representative sample of Americans about a wide variety of issues and concerns. Since many of the questions are asked every year, the GSS offers a longitudinal record of American sentiment and opinion about a large variety of topics. Each year, new topics of current concern are added and explored, and the variety of information available thus continues to expand. Like other nationally representative samples, the GSS sample is chosen by a complex probability design that resembles cluster sampling (see Chapter 6). Sample size varies from 1400 to over 3000, and estimates based on samples this large will be accurate to within ±3% (see Table 7.4 and Section 7.6). The computer exercises in this text are based on the 1998 GSS, and this database is described more fully in Appendix F.

practice with confidence intervals for sample proportions, see problems 7.2, 7.8–7.12, and 7.19d–7.19g.)

7.6 CONTROLLING THE WIDTH OF INTERVAL ESTIMATES

The width of a confidence interval for either sample means or sample proportions can be partly controlled by manipulating two terms in the equation. First, the confidence level can be raised or lowered and, second, the interval can be widened or narrowed by gathering samples of different size. The researcher alone determines the risk he or she is willing to take of being wrong (that is, of not including the population value in the interval estimate). The exact confidence level (or alpha level) will depend, in part, on the purpose of the research. For example, if potentially harmful drugs were being tested, the researcher would naturally demand very high levels of confidence (99.99% or even 99.999%). On the other hand, if intervals are being constructed only for loose "guesstimates," then much lower confidence levels can be tolerated (such as 90%).

The relationship between interval size and confidence level is that intervals widen as confidence levels increase. This should make intuitive sense. Wider intervals are more likely to trap the population value; hence, more confidence can be placed in them.

To illustrate this relationship, let us return to the example where we estimated the average income for a community. In this problem, we were working with a sample of 500 residents, and the average income for this sample was $35,000, with a standard deviation of $200. We constructed the 95% confidence interval and found that it extended 17.55 around the sample mean (that is, the interval was $35,000 ± 17.55).

If we had constructed the 90% confidence interval for these sample data (a lower confidence level), the Z score in the formula would have decreased to ± 1.65, and the interval would have been narrower

$$\text{c.i.} = 35,000 \pm 1.65\left(\frac{200}{\sqrt{499}}\right)$$

$$\text{c.i.} = 35,000 \pm (1.65)(8.95)$$

$$\text{c.i.} = 35,000 \pm 14.77$$

On the other hand, if we had constructed the 99% confidence interval, the Z score would have increased to ± 2.58, and the interval would have been wider:

$$\text{c.i.} = 35,000 \pm 2.58\left(\frac{200}{\sqrt{499}}\right)$$

$$\text{c.i.} = 35,000 \pm (2.58)(8.95)$$

$$\text{c.i.} = 35,000 \pm 23.09$$

TABLE 7.2   INTERVAL ESTIMATES FOR FOUR CONFIDENCE LEVELS ($\overline{X}$ = $35,000, s = $200, N = 500 throughout)

Alpha	Confidence Level	Interval	Interval Width
.10	90%	$35,000 ± 14.77	$29.54
.05	95%	$35,000 ± 17.55	$35.10
.01	99%	$35,000 ± 23.09	$46.18
.001	99.9%	$35,000 ± 29.45	$58.90

At the 99.9% confidence level, the Z score would be ±3.29, and the interval would be wider still

$$\text{c.i.} = 35{,}000 \pm 3.29\left(\frac{200}{\sqrt{499}}\right)$$

$$\text{c.i.} = 35{,}000 \pm (3.29)(8.95)$$

$$\text{c.i.} = 35{,}000 \pm 29.45$$

These four intervals are grouped together in Table 7.2, and the increase in interval size can be readily observed. Although sample means have been used to illustrate the relationship between interval width and confidence level, exactly the same relationships apply to sample proportions. *(To further explore the relationship between alpha and interval width, see problem 7.13.)*

Sample size bears the opposite relationship to interval width. As sample size increases, interval width decreases. Larger samples give more precise (narrower) estimates. Again, an example should make this clearer. In Table 7.3, confidence intervals for four samples of various sizes are constructed and then grouped together for purposes of comparison. The sample

TABLE 7.3   INTERVAL ESTIMATES FOR FOUR DIFFERENT SAMPLES
($\overline{X}$ = $35,000, s = $200, alpha = 0.05 throughout)

Sample 1 (N = 100)		Sample 2 (N = 500)	
c.i. = $35,000 ± 1.96(200/$\sqrt{99}$) c.i. = $35,000 ± 39.40		c.i. = $35,000 ± 1.96(200/$\sqrt{499}$) c.i. = $35,000 ± 17.55	
Sample 3 (N = 1000)		**Sample 4 (N = 10,000)**	
c.i. = $35,000 ± 1.96(200/$\sqrt{999}$) c.i. = $35,000 ± 12.40		c.i. = $35,000 ± 1.96(200/$\sqrt{9999}$) c.i. = $35,000 ± 3.92	

Sample	N	Interval Width
1	100	$78.80
2	500	$35.10
3	1,000	$24.80
4	10,000	$ 7.84

Application 7.2

If 45% of a random sample of 1000 Americans reports that walking is their major physical activity, what is the estimate of the population value? The sample information is

$$P_s = 0.45$$
$$N = 1000$$

Note that the percentage of "walkers" has been stated as a proportion. If we set alpha at 0.05, the corresponding Z score will be ± 1.96, and the interval estimate of the population proportion will be

$$\text{c.i.} = P_s \pm Z\sqrt{\frac{P_u(1 - P_u)}{N}}$$

$$\text{c.i.} = 0.45 \pm 1.96\sqrt{\frac{(0.5)(0.5)}{1000}}$$

$$\text{c.i.} = 0.45 \pm 1.96\sqrt{0.00025}$$
$$\text{c.i.} = 0.45 \pm (1.96)(0.016)$$
$$\text{c.i.} = 0.45 \pm 0.03$$

We can now estimate that the proportion of the population for which walking is the major form of physical exercise is between 0.42 and 0.48. That is, the lower limit of the interval estimate is (0.45 − 0.03) or 0.42, and the upper limit is (0.45 + 0.03) or 0.48. We may also express this result in percentages and say that between 42% and 48% of the population walk as their major form of physical exercise. This interval has a 5% chance of not containing the population value.

---

data are the same as in Table 7.2, and the confidence level is 95% throughout. The relationships illustrated in Table 7.3 also hold true, of course, for sample proportions. *(To further explore the relationship between sample size and interval width, see problem 7.14.)*

Notice that the decrease in interval width (or, increase in precision) does not bear a constant or linear relationship with sample size. With sample 2 as compared to sample 1, the sample size was increased by a factor of 5, but the interval is not five times as narrow. This is an important relationship because it means that N might have to be increased many times over to appreciably improve the accuracy of an estimate. Since the cost of a research project is a direct function of sample size, this relationship implies a point of diminishing returns in estimation procedures. A sample of 10,000 will cost about twice as much as a sample of 5000, but estimates based on the larger sample will not be twice as precise.

7.7 INTERPRETING STATISTICS: CONFIDENCE INTERVALS, PUBLIC OPINION POLLS, AND THE 2000 PRESIDENTIAL ELECTION

The presidential election of 2000 between Governor George W. Bush and Vice-President Al Gore was the closest in recent history. From Labor Day to the time voters entered the voting booth, the pollsters consistently declared the race "too close to call," and "a statistical dead heat." In this installment of "Interpreting Statistics," we will use our newly acquired knowledge of confidence intervals to explore these statements.

READING STATISTICS 5: Using Representative Samples to Track National Trends

In Reading Statistics 4, I mentioned nationally representative databases such as the General Social Survey (GSS) and the Gallup poll. Among many other uses, this information can be used to track national trends and shifts in public opinion over time (at least since 1972, the first year the GSS was administered). In a recent article, Professor Mark Warr used these polls to assess the idea that Americans are "fed up with crime" and "paralyzed by fear" about their own vulnerability to victimization. Although such characterizations are frequently found in the mass media, do they accurately describe the level of fear in society?

The Gallup poll and the GSS have, since the mid-1960s, repeatedly included an item that measures fear of victimization: "is there an area right around here—that is, within a mile—where you would be afraid to walk at night?" Professor Warr constructed a line chart (see Chapter 2) that allowed him to observe changes in the level of fear along with changes in the actual rate of violent crime.

Remember that, although the data on trends in "fear" were collected from random samples of about 1500 respondents, they can be used to infer the characteristics of the population to within about three percentage points. Here's what Professor Warr has to say about the chart:

> The most striking feature of [the responses to the survey items] is the relative constancy of fear over the past 2 decades. After rising moderately during the 1960s, fear has remained relatively constant since the 1970s, varying no more than about 10 percent. This stability stands in sharp contrast to recurring media accounts of "skyrocketing" or "epidemic" fear in the United States.*

Professor Warr goes on to review other pieces of evidence and concludes that American public opinion about crime—although harsh and punitive in some regards—is "often more informed and accurate than one might surmise from journalistic accounts."

*Mark Warr. 1995. "Public Opinion on Crime and Punishment." *Public Opinion Quarterly.* 59:296–310. Copyright 1995 by the University of Chicago Press. Reprinted by permisssion of the author and publisher.

FIGURE 1   PERCENTAGE OF RESPONDENTS AFRAID TO WALK ALONE AT NIGHT, 1965–93, AND NATIONAL CRIME SURVEY VIOLENT CRIME RATE, 1973–92

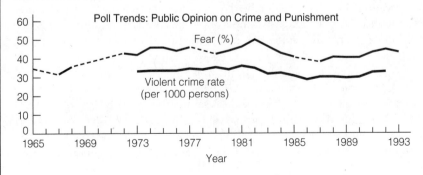

TABLE 7.4 GALLUP POLLING RESULTS, FEBRUARY TO NOVEMBER 2000

Polling Dates		Size of Sample (N)	Percentage of Sample for	
Start	End		Gore	Bush
Feb. 25	Feb. 27	1,004*	43	52
Mar. 10	Mar. 12	1,006*	43	49
Mar. 30	Apr. 2	998*	45	46
Apr. 7	Apr. 9	502	41	50
Apr. 28	Apr. 30	499	44	49
May 18	May 21	1,011*	38	45
June 6	June 7	528	44	48
June 23	June 25	1,020*	39	52
July 14	July 16	635	46	48
Aug. 4	Aug. 5	653	37	54
Aug. 18	Aug. 19	1,043*	47	46
Sep. 8	Sep. 10	727	49	42
Sep. 21	Sep. 23	693	46	47
Sep. 28	Sep. 30	691	45	45
Oct. 20	Oct. 22	769	44	46
Oct. 30	Nov. 1	2,123	43	47
Nov. 5	Nov. 6	2,350	46	48

*Size of the sample is based on adults, 18 years or older, not a smaller subsample of "likely" voters.

The Data. Polling has become a big business in the United States, and many different agencies attempt to forecast election results. Rather than compare all of these polls, we will examine results for the Gallup Organization only.* Drawing results from a single source has several advantages. First, we can be sure that the data were collected in a consistent manner over time. Second, the Gallup Organization is independent—not beholden to either political party—and has a long history of collecting national survey data.

Table 7.4 presents the results of seventeen Gallup surveys conducted between February 2000 and the day prior to the election. The right-hand columns list the percentages of the sample that said, at the time the poll was conducted, that they intended to vote for Bush or Gore. These are point estimates of the population parameters (the actual percentage of voters in the entire national electorate for Mr. Bush or Mr. Gore at that specific time). Note that, for ease of communication, Table 7.4 expresses results in percentages. Actual estimates would be based on sample proportions.

The table shows three distinct phases in the election. The first and longest was the early days of the campaign when Bush and Gore were establishing themselves as candidates for their respective parties. Between February

---

*For a complete listing of the Gallup Organization's archived polling results, see www.gallup.com/poll/releases.

FIGURE 7.6   GALLUP TRACKING POLLS, FEBRUARY–NOVEMBER, 2000

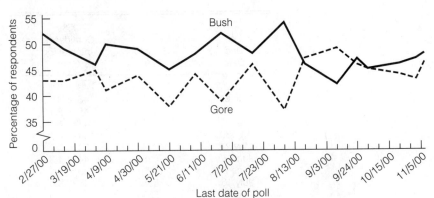

and early August, Bush was the consistent front runner, though his lead dropped to just a few points at times. The second phase occurred after the Republican and Democratic National Conventions in the late summer. In this briefer period between mid-August and September, the advantage narrowly shifted to Gore. The final period, running from mid-September to the election itself, shows Bush retaking the lead but by only the narrowest of margins. As we saw in Chapter 2, graphs are generally more effective ways to communicate results than tables. Figure 7.6 reproduces the data in Table 7.4 and visually emphasizes the three phases of the presidential contest.

Confidence Intervals.   The point estimates in Table 7.4 and Figure 7.6 are very unlikely to match their respective parameters exactly. As we have seen, confidence intervals are safer because they use ranges of values rather than single points to estimate population values. Table 7.5 displays the 95% confidence intervals for each candidate over the course of the election campaign. Note once again that results are expressed as percentages rather than proportions.

Let's analyze the information presented in this table piece by piece. Beginning at the left, we have the dates of the polls and the sample size (N) for each poll. Next are the 95% confidence intervals for each candidate and, finally, the width of the interval estimates. For example, the poll ending February 27 showed that 43% of the sample intended to vote for Gore. Changing the estimate to a proportion and substituting the proper values into Formula 7.3, we would have a confidence interval of

$$.43 \pm 1.96\left(\sqrt{\frac{(.5)(.5)}{1004}}\right)$$

$$.43 \pm .031$$

TABLE 7.5 SAMPLE SIZE, CONFIDENCE INTERVALS, AND LIKELY WINNERS

Poll End Date	Sample Size	95% Confidence Interval for Gore	95% Confidence Interval for Bush	Interval Width
Feb. 27*	1004	43 ± 3.1	52 ± 3.1	6.2
Mar.12	1006	43 ± 3.1	49 ± 3.1	6.2
Apr. 2	998	45 ± 3.1	46 ± 3.1	6.2
Apr. 9*	502	41 ± 4.4	50 ± 4.4	8.8
Apr. 30	499	44 ± 4.4	49 ± 4.4	8.8
May 21*	1011	38 ± 3.1	45 ± 3.1	6.2
June 7	528	44 ± 4.3	48 ± 4.3	8.6
June 25*	1020	39 ± 3.1	52 ± 3.1	6.2
July 16	635	46 ± 3.9	48 ± 3.9	7.8
Aug. 5*	653	37 ± 3.8	54 ± 3.8	7.6
Aug. 19	1043	47 ± 3.0	46 ± 3.0	6.0
Sep. 10	727	49 ± 3.6	42 ± 3.6	7.2
Sep. 23	693	46 ± 3.7	46 ± 3.7	7.4
Sep. 30	691	45 ± 3.7	45 ± 3.7	7.4
Oct. 22	769	44 ± 3.5	46 ± 3.5	7.0
Nov. 1	2123	43 ± 2.1	47 ± 2.1	4.2
Nov. 6	2350	46 ± 2.0	48 ± 2.0	4.0

*Clear advantage to one of the candidates (no overlap in confidence intervals).

Translated back to percentages, the interval would be

$$43\% \pm 3.1\%$$

This interval extends from 39.9% to 46.1%, a width of 6.2 percentage points.

Table 7.5 underscores a very important point about estimation: as sample size increases, the width of the interval becomes narrower (see also Table 7.3). For example, several of the polls were based on relatively small samples of approximately 500 or 600 likely voters. In these instances, intervals were as much as 8.8 percentage points wide. In contrast, the last two polls that were taken in November were based on much larger samples and, as a consequence, the intervals are about ±4% wide, less than half the width for the smaller samples. Note also that although the November samples are about four times larger than the smaller samples, they are only about twice as narrow. Remember that interval width does not bear a constant or linear relationship with sample size (see Section 7.6).

The most important information in Table 7.5 is, of course, the intervals themselves. As Election Day drew near, why was there so much uncertainty about who was leading? Why were the results too close to call? Remember that projections of the "likely winner" should be based on a comparison of the confidence intervals (not the point estimates) and, if the intervals overlap, it is possible that the apparent losing candidate is actually ahead.

For example, the very last poll in November showed that support for Gore was between 44% (46% − 2%) and 48% (46% + 2%), and support for

Bush was between 46% and 50%. Remember that while we can be 95% sure that the parameter is in the interval, it may be *anywhere* between the upper and lower limits. Thus, it was just as likely that Gore would win (e.g., support for Gore could have been as high as 48% vs. 46% for Bush) or that Bush would be the victor (e.g., if support for Gore was as low as 44% vs. 50% for Bush). In fact, the race was "too close to call" for the majority of polls taken since February. As Table 7.5 shows, there were only five polls (indicated with an asterisk) that showed a clear lead (at the 95% confidence level) for one of the candidates.

When a race is this close, the polls, with their margins of error of 2 or 3 percent, cannot identify the winner. When the ballots were finally counted (and recounted), the candidates indeed wound up in a virtual tie, separated by only a comparatively few thousand votes out of more than 100 million votes cast. Gore received 48.3% of the popular vote and, slightly higher than the upper limit of the November 6th estimate, Bush garnered 48.1%, a parameter within the range of the last November interval estimate.

SUMMARY

1. Population values can be estimated with sample values. With point estimates, a single sample statistic is used to estimate the corresponding population value. With confidence intervals, we estimate that the population value falls within a certain range of values.

2. Estimates based on sample statistics must be unbiased and relatively efficient. Of all the sample statistics, only means and proportions are unbiased. The means of the sampling distributions of these statistics are equal to the respective population values. Efficiency is largely a matter of sample size. The greater the sample size, the lower the value of the standard deviation of the sampling distribution,

the more tightly clustered the sample outcomes will be around the mean of the sampling distribution, and the more efficient the estimate.

3. With point estimates, we estimate that the population value is the same as the sample statistic (either a mean or proportion). With interval estimates, we construct a confidence interval, a range of values into which we estimate that the population value falls. The width of the interval is a function of the risk we are willing to take of being wrong (the alpha level) and the sample size. The interval widens as our probability of being wrong decreases and as sample size decreases.

SUMMARY OF FORMULAS

Confidence interval for a sample mean, population standard deviation known:

$$7.1 \quad \text{c.i.} = \overline{X} \pm Z\left(\frac{\sigma}{\sqrt{N}}\right)$$

Confidence interval for a sample mean, population standard deviation unknown:

$$7.2 \quad \text{c.i.} = \overline{X} \pm Z\left(\frac{s}{\sqrt{N-1}}\right)$$

Confidence interval for a sample proportion, large samples:

$$7.3 \quad \text{c.i.} = P_s \pm Z\sqrt{\frac{P_u(1 - P_u)}{N}}$$

GLOSSARY

Alpha (α). The probability of error or the probability that a confidence interval does not contain the population value. Alpha levels are usually set at 0.10, 0.05, 0.01, or 0.001.

Bias. A criterion used to select sample statistics as estimators. A statistic is unbiased if the mean of its sampling distribution is equal to the population value of interest.

Confidence interval. An estimate of a population value in which a range of values is specified.

Confidence level. A frequently used alternate way of expressing alpha, the probability that an interval estimate will not contain the population value. Confidence levels of 90%, 95%, 99%, and 99.9% correspond to alphas of 0.10, 0.05, 0.01, and 0.001, respectively.

Efficiency. The extent to which the sample outcomes are clustered around the mean of the sampling distribution.

Point estimate. An estimate of a population value where a single value is specified.

MULTIMEDIA RESOURCES

The Wadsworth Sociology Resource Center: Virtual Society
http://sociology.wadsworth.com/

Visit the companion web site for the sixth edition of *Statistics: A Tool for Social Research* to access a wide range of student resources. Begin by clicking on the Student Resources section of the book's web site to access the following study tools:

- Basic math review
- Flash cards

- Additional chapter problems
- Statistics review
- Internet links
- Table of random numbers
- MicroCase and SPSS examples and exercises
- "Find the test" flowcharts
- Hypothesis testing for variables measured at the ordinal level

PROBLEMS

7.1 For each set of sample outcomes below, construct the 95% confidence interval for estimating μ, the population mean.

a. $\overline{X} = 5.2$	b. $\overline{X} = 100$	c. $\overline{X} = 20$
$s = .7$	$s = 9$	$s = 3$
$N = 157$	$N = 620$	$N = 220$
d. $\overline{X} = 1020$	e. $\overline{X} = 7.3$	f. $\overline{X} = 33$
$s = 50$	$s = 1.2$	$s = 6$
$N = 329$	$N = 105$	$N = 220$

7.2 For each set of sample outcomes below, construct the 99% confidence interval for estimating P_u.

a. $P_s = .14$	b. $P_s = .37$	c. $P_s = .79$
$N = 100$	$N = 522$	$N = 121$
d. $P_s = .43$	e. $P_s = .40$	f. $P_s = .63$
$N = 1049$	$N = 548$	$N = 300$

7.3 For each confidence level below, determine the corresponding Z score.

Confidence Level	Alpha	Area Beyond Z	Z score
95%	.05	.0250	±1.96
94%			
92%			
97%			
98%			
99.9%			

7.4 [SW] You have developed a series of questions to measure "burnout" in social workers. A random sample of 100 social workers working in greater metropolitan Shinbone, Kansas, has an average score of 10.3, with a standard deviation of 2.7.

What is your estimate of the average burnout score for the population as a whole? Use the 95% confidence level.

7.5 SOC A researcher has gathered information from a random sample of 178 households. For each variable below, construct confidence intervals to estimate the population mean. Use the 90% level.
  a. An average of 2.3 people reside in each household. Standard deviation is .35.
  b. There was an average of 2.1 television sets ($s = .10$) and .78 telephones ($s = .55$) per household.
  c. The households averaged 6.0 hours of television viewing per day ($s = 3.0$).

7.6 SOC A random sample of 100 television programs contained an average of 2.37 acts of physical violence per program. At the 99% level, what is your estimate of the population value?

$$\overline{X} = 2.37$$
$$s = .30$$
$$N = 100$$

7.7 SOC A random sample of 429 college students was interviewed about a number of matters.
  a. They reported that they had spent an average of $178.23 on textbooks during the previous semester. If the sample standard deviation for these data is $15.78, construct an estimate of the population mean at the 99% level.
  b. They also reported that they had visited the health clinic an average of 1.5 times a semester. If the sample standard deviation is 0.3, construct an estimate of the population mean at the 99% level.
  c. On the average, the sample had missed 2.8 days of classes per semester because of illness. If the sample standard deviation is 1.0, construct an estimate of the population mean at the 99% level.
  d. On the average, the sample had missed 3.5 days of classes per semester for reasons other than illness. If the sample standard deviation is 1.5, construct an estimate of the population mean at the 99% level.

7.8 CJ A random sample of 500 residents of Shinbone, Kansas, shows that exactly 50 of the re-

spondents had been the victims of violent crime over the past year. Estimate the proportion of victims for the population as a whole, using the 90% confidence level. *(HINT: Calculate the sample proportion P_s before using Formula 7.3. Remember that proportions are equal to frequency divided by N.)*

7.9 SOC The survey mentioned in problem 7.5 found that 25 of the 178 households consisted of unmarried couples who were living together. What is your estimate of the population proportion? Use the 95% level.

7.10 PA A random sample of 324 residents of a community revealed that 30% were very satisfied with the quality of trash collection. At the 99% level, what is your estimate of the population value?

7.11 SOC A random sample of 1496 respondents was questioned about their sexual attitudes and behaviors. Construct estimates to the population at the 90% level for each of the results reported below. Express the final confidence interval in percentages (e.g. "between 40 and 45% agreed that premarital sex was always wrong").
  a. When asked to agree or disagree with the statement "Explicit sexual books and magazines lead to rape and other sex crimes," 823 agreed.
  b. When asked to agree or disagree with the statement "There is far too much explicit sex on television these days," 650 agreed.
  c. 375 of the sample said that they had sex with someone other than their spouse at least once while married.
  d. 1023 of the sample said that they had sex before they were first married.
  e. 800 agreed that premarital sex was "always wrong."

7.12 SW A random sample of 100 patients treated in a program for alcoholism and drug dependency over the past 10 years was selected. It was determined that 53 of the patients had been readmitted to the program at least once. At the 95% level, construct an estimate to the population proportion.

7.13 For the sample data below, construct four different interval estimates of the population mean, one each for the 90%, 95%, 99%, and 99.9% level. What happens to the interval width as confidence level increases? Why?

$$\overline{X} = 100$$
$$s = 10$$
$$N = 500$$

7.14 For each of the three sample sizes below, construct the 95% confidence interval. Use a sample proportion of 0.40 throughout. What happens to interval width as sample size increases? Why?

$$P_s = 0.40$$
Sample A: $N = 100$
Sample B: $N = 1000$
Sample C: $N = 10,000$

7.15 |PS| Two individuals are running for mayor of Shinbone. You conduct an election survey a week before the election and find that 51% of the respondents prefer candidate A. Can you predict a winner? Use the 99% level. *(HINT: in a two-candidate race, what percentage of the vote would the winner need? Does the confidence interval indicate that candidate A has a sure margin of victory? Remember that while the population parameter is probably ($\alpha = .01$) in the confidence interval, it may be anywhere in the interval.)*

$$P_s = 0.51$$
$$N = 578$$

7.16 |SOC| A random sample of 1000 registered voters in a state were asked if they favored a proposal to restrict the sale of handguns. If 510 approved of the legislation, is it safe, at the 95% level, to conclude that "a majority" of the population approves?

7.17 |SOC| The fraternities and sororities at St. Algebra College have been plagued by declining membership over the past several years and want to know if the incoming freshman class will be a fertile recruiting ground. Not having enough money to survey all 1600 freshmen, they commission you to survey the interests of a random

sample. You find that 35 of your 150 respondents are "extremely" interested in social clubs. At the 95% level, what is your estimate of the number of freshmen who would be extremely interested? *(HINT: the high and low values of your final confidence interval are proportions. How can proportions also be expressed as numbers?)*

7.18 |SOC| You are the consumer-affairs reporter for a daily newspaper. Part of your job is to investigate the claims of manufacturers, and you are particularly suspicious of a new economy car that the manufacturer claims will get 78 miles per gallon. After checking the mileage figures for a random sample of 120 owners of this car, you find an average miles per gallon of 75.5 with a standard deviation of 3.7. At the 99% level, do your results tend to confirm or refute the manufacturer's claims?

7.19 |SOC| The General Social Survey (GSS) summarized in Appendix G and used for computer exercises in this text was given to an EPSEM sample that should be representative of American society. The entire sample consists of about 3000 respondents, but the version used here has fewer than 1500 respondents.

For each sample statistic reported below, construct the 95% confidence interval estimate of the population value. Note that we will be using a "sample of a sample" as the basis for the estimates, and we should correct for the additional source of error. As you will see, however, the sample still produces reasonable estimates in those cases where the population parameter is known.

To judge the accuracy of your estimates, information on population parameters is given where possible in the Answers to Odd-Numbered Problems. See Appendix G for the codes and the actual wording of the questions

a. The average occupational prestige score for 1318 respondents was 43.87, with a standard deviation of 13.52.

b. The respondents ($N = 1134$) reported watching an average of 2.86 hours of TV per day with a standard deviation of 2.20.

c. The average number of children for 1385 respondents was 1.81, with a standard deviation of 1.67.

d. Of the 1372 respondents who answered the question, 367 identified themselves as Catholic.

e. Three hundred thirty-five of the 1386 respondents said that they had never married.

f. Of the 830 respondents for whom we have information, 441 said that they voted for Clinton in the 1996 presidential election.

g. When asked about capital punishment, 925 of the 1266 respondents said that they favored the death penalty for murder.

SPSS for Windows

Using *SPSS for Windows* Procedures to Produce Confidence Intervals

SPSS DEMONSTRATION 7.1 Generating Statistical Information for Use in Constructing Confidence Intervals

SPSS does not provide any programs specifically for constructing confidence intervals, although some of the procedures we'll cover in future chapters do include confidence intervals as part of the output. Rather than make use of these programs, I want to use this section to confirm something you may already suspect: fancy computer programs are not always helpful or even particularly useful. The arithmetic required by estimation procedures (let's face it) is not particularly difficult, and you could probably complete the formulas faster by hand than by SPSS.

On the other hand, who wants to do all the calculations it would take to get the summary statistics on which the estimates will be based? Calculating the mean age for the GSS sample would require the addition of almost 1500 numbers. The dimensions of that task should suggest the proper role of the computer. Once you know the mean age, the rest of the calculations for confidence intervals are not very formidable. So, how do you get the sample statistics?

Take a look at problem 7.19. I used a number of variables from the GSS data set for this problem and got the information I needed by using **Descriptives** for the interval-ratio variables and **Frequencies** for the nominal variables. **Descriptives** gives means and standard deviations, and **Frequencies** produces frequency distributions with a column for percentages.

MicroCase

Using MicroCase to Produce Confidence Intervals

MICROCASE DEMONSTRATION 7.1 Confidence Intervals

MicroCase automatically reports the 95% and 99% confidence intervals for sample means whenever we use the **Univariate** command. Thus, for variables for which computation of means are justified (interval-ratio variables and ordinal-level variables with broad ranges), the statistical package will do all the work. Remember that MicroCase cannot identify nominal level variables for which the computation of a mean is not justified, and it will happily compute estimates of average gender or marital status when told to do so.

Unfortunately, MicroCase has no comparable feature for nominal-level variables and sample proportions, and this may be a good time to confirm something you may already suspect: fancy computer programs are not always helpful or even particularly useful. The arithmetic required by estimation procedures (let's face it) is not particularly difficult and, once you have the necessary sample information, you could probably complete the formulas as fast by hand as by issuing commands to a computer. MicroCase will do the hard work of computing sample proportions, and we will then substitute these values into Formula 7.3 to construct the confidence intervals.

Exercises

7.1 a. SPSS: Use the **Descriptives** command to calculate means and standard deviations for *papres80, age,* and *income98.* Use the output to formulate confidence interval estimates for the population values.

MicroCase: Use the **Univariate** command to get the 95% confidence interval for *papres80, age,* and *income98.*

b. Express the confidence intervals in words, as if you were reporting results in a newspaper story.

7.2 a. SPSS: Use the **Frequencies** command to get sample proportions (convert the percentages in the frequency distributions) to estimate the population parameter for each of the following: proportion with college degree (*degree*), proportion black (*race*), and proportion favoring gun control (*gunlaw*). Use the 95% confidence level.

MicroCase: Use the **Univariate** command to get sample proportions (convert the percentages in the frequency distributions) to estimate the population parameter for each of the following: proportion with college degree (*degree*), proportion black (*race*), and proportion favoring gun control (*gunlaw*). Use the 95% confidence level.

b. Express the confidence intervals in words, as if you were reporting results in a newspaper story.

(NOTE: your estimate based on age will not reflect the total American population. The GSS sample is restricted to people age 18 and older, and the mean of the sample is much higher than the mean of the population as a whole.)

8

HYPOTHESIS TESTING I
THE ONE-SAMPLE CASE

LEARNING OBJECTIVES   By the end of this chapter, you will be able to

1. Explain the logic of hypothesis testing.
2. Define and explain the conceptual elements involved in hypothesis testing, especially the null hypothesis, the sampling distribution, the alpha level, and the test statistic.
3. Explain what it means to "reject the null hypothesis" or "fail to reject the null hypothesis."
4. Identify and cite examples of situations in which one-sample tests of hypothesis are appropriate.
5. Test the significance of single-sample means and proportions using the five-step model and correctly interpret the results.
6. Explain the difference between one- and two-tailed tests and specify when each is appropriate.
7. Define and explain Type I and Type II errors and relate each to the selection of an alpha level.

8.1 INTRODUCTION   Chapter 7 introduced the techniques for estimating population parameters from sample statistics. In chapters 8 through 11, we will investigate a second application of inferential statistics called **hypothesis testing** or **significance testing.** In this chapter, the techniques for hypothesis testing in the one-sample case will be introduced. These procedures could be used in situations such as the following:

1. A researcher has selected a sample of 789 older citizens who live in a particular state and also has information on the percentage of the entire population of the state that was victimized by crime during the past year. Are older citizens, as represented by this sample, more or less likely to be victimized?

2. Are the GPAs of college athletes different from the GPAs of the general student body? To investigate, the academic records of a random sample

of 235 student athletes from a large state university are compared with the overall GPA of all students.

3. A sociologist has been hired to assess the effectiveness of a rehabilitation program for alcoholics in her city. The program serves a large area, and she does not have the resources to test every single client. Instead, she draws a random sample of 127 people from the list of all clients and questions them on a variety of issues. She notices that, on the average, her sample misses fewer days of work each year than the city as a whole. Are alcoholics treated by the program more reliable workers?

In each of these situations, we have randomly selected samples (of senior citizens, athletes, or treated alcoholics) that we want to compare to a population (the entire state, student body, or city). Note that we are not interested in the sample per se but in the larger group from which it was selected (*all* senior citizens in the state, athletes on this campus, or people who have completed the treatment program). Specifically, we want to know if the groups represented by the samples are different from the populations on a specific trait or variable (victimization rates, GPAs, or absenteeism).

Of course, it would be better if we could test all senior citizens, athletes, or treated alcoholics rather than these smaller samples. However, as we have seen, researchers usually do not have the resources necessary to test everyone in a large group and must use random samples instead. In these situations, conclusions will be based on a comparison of a single sample (which represents a larger group) and the population. For example, if we found that the rate of victimization for the sample of senior citizens was higher than the state rate, we might conclude that "senior citizens are significantly more likely to be crime victims." The word "significantly" is a key word: it means that the difference between the sample's victimization rate and the population's rate is very unlikely to be caused by random chance alone. In other words, *all* senior citizens (not just the 789 people in the sample) have a higher victimization rate than the state as a whole. On the other hand, if we found little difference between the GPAs of the sample of athletes and the student body as a whole, we might conclude that athletes (*all* athletes on this campus) are essentially the same as other students in terms of academic achievement.

Thus, we can use samples to represent larger groups (senior citizens, athletes, or treated alcoholics) and compare and contrast the characteristics of the sample with the population and be extremely confident in our conclusions. Remember, however, that the EPSEM procedure for drawing random samples does not guarantee representativeness and, thus, there will always be a small amount of uncertainty in our conclusions. One of the great advantages of inferential statistics, however, is that we will be able to estimate the probability of error and evaluate our decisions accordingly.

8.2 AN OVERVIEW OF HYPOTHESIS TESTING

We'll begin with a general overview of hypothesis testing, using our third research situation above as an example, and introduce the more technical considerations and proper terminology throughout the remainder of the chapter. Absentee rates for the sample of clients and the city as a whole are

Community	Sample
$\mu = 7.2$ days per year	$\overline{X} = 6.8$ days per year
$\sigma = 1.43$	$N = 127$

We can see that there is a difference in rates of absenteeism and that the sample has an average absentee rate almost half a day lower than the population. Can we conclude that ALL treated alcoholics, as represented by this sample of 127, have lower absentee rates? Since we are working with a random sample (not all alcoholics treated in the program), there are two possible explanations for the difference in absentee rates.

The first explanation, which we will label explanation A, is that the difference between 7.2 days and 6.8 days reflects a real difference in absentee rates between treated alcoholics and the community. The difference is statistically significant and very unlikely to have occurred by random chance alone. If explanation A is true, treated alcoholics are different from everyone else, and the sample did *not* come from a population with a mean absentee rate of 7.2 days.

The second explanation, or explanation B, is that the observed difference was caused by random chance. In other words, there is no important difference between treated alcoholics and the community as a whole, and the difference between 6.8 days and 7.2 days of absenteeism is trivial and due to the effects of random chance. If explanation B is true, treated alcoholics are just like everyone else, and the sample *did* come from a population with a mean absentee rate of 7.2 days.

Which explanation for the observed difference is correct? As long as we are working with a sample rather than the entire group, we cannot be absolutely (100%) sure about the answer to this question. However, we can set up a decision-making procedure so conservative that one of the two explanations can be chosen, with the knowledge that the probability of making an incorrect choice is very low.

This decision-making process, in broad outline, begins with the assumption that explanation B is correct. Symbolically, the assumption that the mean absentee rate for all treated alcoholics is the same as the rate for the community as a whole can be stated as

$$\mu = 7.2 \text{ days per year}$$

Remember that this μ refers to the mean for all treated alcoholics. This assumption, $\mu = 7.2$, can be tested statistically.

If explanation B (treated alcoholics are not different from the community as a whole) is true, then the probability of getting the observed sample outcome ($\overline{X}$ = 6.8) can be found. Let us add an objective decision rule in advance. If the odds of getting the observed difference are less than .05 (5 out of 100, or 1 in 20), we will reject explanation B. If this explanation were true, a difference of this size (7.2 days vs. 6.8 days) would be a very rare event, and in hypothesis testing, we always bet *against* rare events.

The remaining problem is to determine the probability of the observed sample outcome if explanation B is correct. We can determine this probability with precision because of what we know about the sampling distribution of all possible sample outcomes. Looking back at the information we have and applying the Central Limit Theorem (see Chapter 6), we may assume that the sampling distribution is normal in shape, has a mean of 7.2 ($\mu_{\overline{x}}$ = μ), and a standard deviation of $1.43/\sqrt{127}$ because $\sigma_{\overline{x}} = \sigma/\sqrt{N}$. We also know that the standard normal distribution can be interpreted as a distribution of probabilities (see Chapter 5) and that the particular sample outcome noted above ($\overline{X}$ = 6.8) is one of thousands of possible sample outcomes. The sampling distribution, with the sample outcome noted, is depicted in Figure 8.1.

Using our knowledge of the standardized normal distribution, we can add further useful information to this sampling distribution of sample means. Specifically, with Z scores, we can depict the decision rule stated previously: any sample outcome with probability less than 0.05 (assuming that explanation B is true) will cause us to reject explanation B. The probability of 0.05 can be translated into an area and divided equally into the upper and lower tails of the sampling distribution. Using Appendix A, we find that the Z-score equivalent of this area is ±1.96. The areas and Z scores are depicted in Figure 8.2.

The decision rule can now be rephrased. Any sample outcome falling in the shaded areas depicted in Figure 8.2 by definition has a probability of occurrence of less than 0.05. Such an outcome would be a rare event and would cause us to reject explanation B. All that remains is to translate our

FIGURE 8.1 THE SAMPLING DISTRIBUTION OF ALL POSSIBLE SAMPLE MEANS

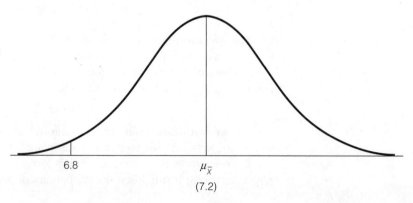

6.8 $\mu_{\overline{X}}$
        (7.2)

FIGURE 8.2   THE SAMPLING DISTRIBUTION OF ALL POSSIBLE SAMPLE MEANS

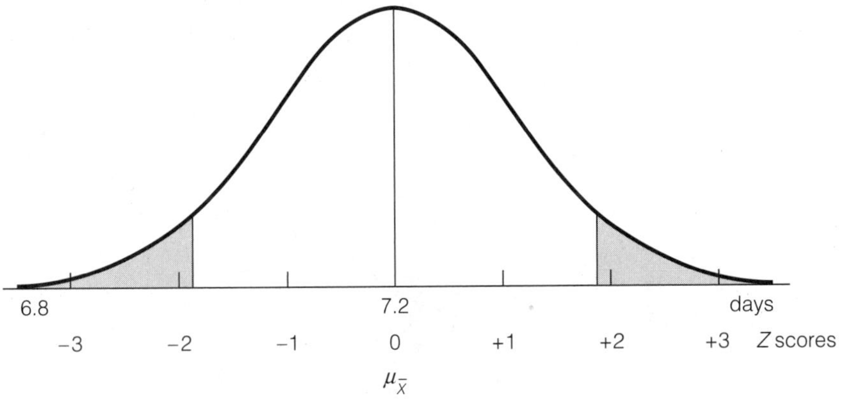

sample outcome into a Z score so we can see where it falls on the curve. To do this, we use the standard formula for locating any particular raw score under a normal distribution. When we use known or empirical distributions, this formula is expressed as

$$Z = \frac{X_i - \overline{X}}{s}$$

Or, to find the equivalent Z score for any raw score, subtract the mean of the distribution from the raw score and divide by the standard deviation of the distribution. Since we are now concerned with the sampling distribution of all sample means rather than an empirical distribution, the symbols in the formula will change, but the form remains exactly the same:

FORMULA 8.1
$$Z = \frac{\overline{X} - \mu}{\sigma / \sqrt{N}}$$

Or, to find the equivalent Z score for any sample mean, subtract the mean of the sampling distribution, which is equal to the population mean or μ, from the sample mean and divide by the standard deviation of the sampling distribution.

Recalling the data given on this problem, we can now find the Z-score equivalent of the sample mean.

$$Z = \frac{\overline{X} - \mu}{\sigma / \sqrt{N}}$$

$$Z = \frac{6.8 - 7.2}{1.43 / \sqrt{127}}$$

$$Z = \frac{-.40}{.127}$$

$$Z = -3.15$$

FIGURE 8.3 THE SAMPLING DISTRIBUTION OF SAMPLE MEANS WITH THE SAMPLE OUTCOME ($\overline{X} = 6.8$) NOTED IN Z SCORES

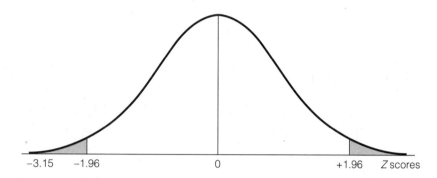

In Figure 8.3, this Z score of -3.15 is noted on the distribution of all possible sample means, and we see that the sample outcome does fall in the shaded area. If explanation B is true, this particular sample outcome has a probability of occurrence of less than 0.05 (how much less than 0.05 is irrelevant, since the decision rule was established in advance). The sample outcome ($\overline{X} = 6.8$ or $Z = -3.15$) would therefore be rare if explanation B were true, and the researcher may reject explanation B. If explanation B were true, this sample outcome would be extremely unlikely. It appears that our sample of 127 treated alcoholics comes from a population that is significantly different from the community on the trait of absenteeism.

Keep in mind that our decisions in significance testing are based on information gathered from random samples. On rare occasions, a sample may not be representative of the population from which it was selected. Let me repeat that the decision-making process outlined above has a very high probability of resulting in correct decisions, but, as long as we must work with samples rather than populations, we face an element of risk in the process. That is, the decision with respect to explanation B might be incorrect if this sample happens to be one of the few that is unrepresentative of the population of alcoholics treated in this program. One important strength of hypothesis testing is that the probability of making an incorrect decision can be estimated. In the example at hand, explanation B was rejected and the probability of this decision being incorrect is 0.05—the decision rule established at the beginning of the process. To say that the probability of rejecting explanation B incorrectly is 0.05 means that, if we repeated this same test an infinite number of times, we would incorrectly reject explanation B only 5 times out of every 100.

8.3 THE FIVE-STEP MODEL FOR HYPOTHESIS TESTING All the formal elements and concepts used in hypothesis testing were surreptitiously sneaked into the preceding discussion. This section presents their proper names and introduces a **five-step model** for organizing all hypothesis testing:

Step 1. Making assumptions

Step 2. Stating the null hypothesis

Step 3. Selecting the sampling distribution and establishing the critical region

Step 4. Computing the test statistic

Step 5. Making a decision

We will look at each step individually, using the problem from Section 8.2 as an example throughout.

Step 1. Making Assumptions. Any application of a statistical technique requires that certain assumptions be made about the data. When conducting a test of hypothesis with a single sample mean, we must assume that the sample has been randomly selected from a defined population. Also, to justify computation of a mean, we must assume that the variable is interval-ratio in level of measurement. Finally, we must assume that the sampling distribution of all possible sample means is normal in shape so that we may use the standardized normal distribution to find areas under the sampling distribution. We will refer to these assumptions about the sampling process, the level of measurement, and the shape of the sampling distribution as the "mathematical model." You should be very certain of the correctness of these model assumptions.

Usually, we will state these assumptions in abbreviated form. For example:

> Model: Random sampling
> Level of measurement is interval-ratio
> Sampling distribution is normal

Step 2. Stating the Null Hypothesis (H_0). The **null hypothesis** is the formal name for explanation B and is always a statement of "no difference." The exact form of the null hypothesis will vary depending on the test being conducted. In the single-sample case, the null hypothesis states that the sample comes from a population with a certain characteristic. In our example, the null is that treated alcoholics are "no different" from the community as a whole, that their average days of absenteeism is also 7.2, and that the difference between 7.2 and the sample mean of 6.8 is caused by random chance. Symbolically, the null would be stated as

$$H_0: \mu = 7.2$$

where μ refers to the mean of the population of treated alcoholics. The null hypothesis is the central element in any test of hypothesis because the entire process is aimed at rejecting or failing to reject the H_0.

Usually, the researcher believes that there is a significant difference and desires to reject the null hypothesis. At this point in the five-step model, the researcher's belief is stated in a **research hypothesis (H_1),** a statement that directly contradicts the null hypothesis. Thus, the researcher's goal in hypothesis testing is often to gather evidence for the research hypothesis by rejecting the null hypothesis. The research hypothesis can be stated in several ways. One form would simply assert that the sample was selected from a population that did not have a certain characteristic or, in terms of our example, had a mean that was not equal to a specific value

$$(H_1: \mu \neq 7.2)$$

where $\neq$ means "not equal to"

Symbolically, this statement asserts that the sample does not come from a population with a mean of 7.2, or that treated alcoholics are different from the community as a whole. The research hypothesis is enclosed in parentheses to emphasize that it has no formal standing or role in the hypothesis-testing process (except, as we shall see in the next section, in choosing between one-tailed and two-tailed tests). It serves as a reminder of what the researcher believes to be the truth.

Step 3. Selecting the Sampling Distribution and Establishing the Critical Region. The sampling distribution is, as always, the probabilistic yardstick against which a particular sample outcome is measured. By assuming that the null hypothesis is true (and *only* by this assumption), we can attach values to the mean and standard deviation of the sampling distribution and thus measure the probability of any specific sample outcome. There are several different sampling distributions, but for now we will confine our attention to the sampling distribution of sample means. We can measure areas under this distribution by using the standard normal curve as summarized in Appendix A.

The **critical region** consists of the areas under the sampling distribution that include unlikely sample outcomes. Prior to the test of hypothesis, we must define what is meant by *unlikely*. That is, we must specify in advance those sample outcomes so unlikely that they will lead us to reject the H_0. This decision rule will establish the critical region or **region of rejection.** The word *region* is used because, essentially, we are describing those areas under the sampling distribution that contain unlikely sample outcomes. In the example above, this area corresponded to a Z score of ± 1.96, called **Z (critical),** that was graphically displayed in Figure 8.2. The shaded area was the critical region. Any sample outcome for which the Z-score equivalent fell in this area (that is, below -1.96 or above $+1.96$) would have caused us to reject the null hypothesis.

By convention, the size of the critical region is reported as alpha (α), the proportion of all of the area included in the critical region. In the example above, our **alpha level** was 0.05. Other commonly used alphas are 0.10, 0.01, and 0.001.

In abbreviated form, all the decisions made in this step are noted below. The critical region is noted by the Z scores that mark its beginnings.

$$\text{Sampling distribution} = Z \text{ distribution}$$
$$\alpha = 0.05$$
$$Z\,(\text{critical}) = \pm 1.96$$

(For practice in finding Z (critical) scores, see problem 8.1a.)

Step 4. Computing the Test Statistic. To evaluate the probability of any given sample outcome, the sample value must be converted into a Z score. Solving the equation for Z-score equivalents is called computing the **test statistic,** and the resultant value will be referred to as **Z (obtained)** in order to differentiate the test statistic from the critical region. In our example above, we found a Z (obtained) of -3.15. *(For practice in computing obtained Z scores for means, see problems 8.1c, 8.2 to 8.7, and 8.15e and f)*

Step 5. Making a Decision. As the last step in the hypothesis-testing process, the test statistic is compared with the critical region. If the test statistic falls into the critical region, our decision will be to reject the null hypothesis. If the test statistic does not fall into the critical region, we fail to reject the null hypothesis. In our example, the two values were

$$Z\,(\text{critical}) = \pm 1.96$$
$$Z\,(\text{obtained}) = -3.15$$

and we saw that the Z (obtained) fell in the critical region (see Figure 8.3). Our decision was to reject the null hypothesis and conclude that the difference observed with respect to absenteeism was unlikely to have occurred by chance alone. On the trait of absenteeism, treated alcoholics are different from the community as a whole.

This five-step model will serve us as a framework for decision making throughout the hypothesis-testing chapters. The exact nature and method of expression for our decisions will be different for different situations. However, familiarity with the 5-step model will assist you in mastering this material by providing a common frame of reference for all significance testing.

8.4 ONE-TAILED AND TWO-TAILED TESTS OF HYPOTHESIS

The five-step model for hypothesis testing is fairly rigid, and the researcher has little room for making choices. Follow the steps one by one as specified above, and you cannot make a mistake. Nonetheless, the researcher must still make two crucial choices. First, he or she must decide between a one-tailed and a two-tailed test. Second, an alpha level must be selected. In this section, we will discuss the former decision and, in Section 8.5, the latter.

The choice between a one- and two-tailed test is based on the researcher's expectations about the population from which the sample was selected. These expectations are reflected in the research hypothesis (H_1), which is contradictory to the null hypothesis and usually states what the researcher believes to be "the truth." In most situations, the researcher will wish to support the research hypothesis by rejecting the null hypothesis.

The format for the research hypothesis may take either of two forms, depending on the relationship between what the null hypothesis states and what the researcher believes to be the truth. The null hypothesis states that the population has a specific characteristic. In the example that has served us throughout this chapter, the null stated, in symbols, that "all treated alcoholics have the *same* absentee rate (7.2 days) as the community." The researcher might believe that the population of treated alcoholics actually has *less* absenteeism (their population mean is *lower than* the value stated in the null hypothesis), *more* absenteeism (their population mean is *greater than* the value stated in the null hypothesis), or he or she might be unsure about the direction of the difference.

If the researcher is unsure about the direction, the research hypothesis would state only that the population mean is "not equal" to the value stated in the null hypothesis. The research hypothesis stated in Section 8.3 ($\mu \neq 7.2$) was in this format. This is called a **two-tailed test** of significance. The researcher will be equally concerned with the possibility that the true population value is greater than the value specified in the null hypothesis and the possibility that the true population value is less than the value specified in the null hypothesis.

In other situations, the researcher might be concerned only with differences in a specific direction. If the direction of the difference can be predicted, or if the researcher is concerned only with differences in one direction, a **one-tailed test** can be used. A one-tailed test may take one of two forms, depending on the researcher's expectations about the direction of the difference. If the researcher believes that the true population value is greater than the value specified in the null, the research hypothesis would reflect that belief. In our example, if we had predicted that treated alcoholics had higher absentee rates than the community (or, averaged *more* days of absenteeism than 7.2), our research hypothesis would have been

$$(H_1: \mu > 7.2)$$

where $>$ signifies "greater than"

If we predicted that treated alcoholics had lower absentee rates than the community (or, averaged *fewer* days of absenteeism than 7.2), our research hypothesis would have been

$$(H_1: \mu < 7.2)$$

where $<$ signifies "less than"

One-tailed tests are often appropriate when programs designed to solve a problem or improve a situation are being evaluated. If the program for treating alcoholics made them *less* reliable workers, for example, the program would be considered a failure, at least on that one criterion. Or, consider the evaluation of a program designed to reduce unemployment. The evaluators would be concerned only with outcomes that show a decrease in the unemployment rate. If the rate shows no change, or if unemployment increases, the program is a failure, and both of these outcomes might be considered equally negative by the researchers.

In terms of the five-step model, the choice of a one-tailed or two-tailed test determines what we do with the critical region under the sampling distribution in step 3. As you recall, in a two-tailed test, we split the critical region equally into the upper and lower tails of the sampling distribution. In a one-tailed test, we place the entire critical area in one tail of the sampling distribution. If we believe that the population characteristic is greater than the value stated in the null hypothesis (if the H_1 includes the $>$ symbol), we place the entire critical region in the upper tail. If we believe that the characteristic is less than the value stated in the null hypothesis (if the H_1 includes the $<$ symbol), the entire critical region goes in the lower tail.

For example, in a two-tailed test with alpha equal to 0.05, the critical region begins at Z (critical) $= \pm 1.96$. In a one-tailed test at the same alpha level, the Z (critical) is $+1.65$ if the upper tail is specified and -1.65 if the lower tail is specified. The contrast in the placement of the critical region is graphically summarized in Figure 8.4, and the critical Z scores for the most common alpha levels are given in Table 8.1 for both one- and two-tailed tests.

Note that the critical Z values for one-tailed tests are closer to the mean of the sampling distribution. Thus, a one-tailed test is more likely to reject the H_0 without affecting the alpha level (assuming that we have specified the correct tail). One-tailed tests are a way of statistically both having and eating your cake and should be used whenever (1) the direction of the difference can be confidently predicted, or (2) the researcher is concerned only with differences in one tail of the sampling distribution.

An example will clarify how the one-tailed test is used. After many years of work, a sociologist has noted that sociology majors seem more sophisticated, charming, and cosmopolitan than the rest of the student body. A "Sophistication Scale" test has been administered to the entire student body and to a random sample of 100 sociology majors, and these results have been obtained

Student Body	Sociology Majors
$\mu = 17.3$	$\overline{X} = 19.2$
$\sigma = 7.4$	$N = 100$

FIGURE 8.4 ESTABLISHING THE CRITICAL REGION, ONE-TAILED TESTS VERSUS TWO-TAILED TESTS (alpha = 0.05)

A. The two-tailed test, Z (critical) = ±1.96

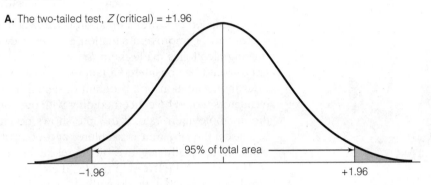

B. The one-tailed test for upper tail, Z (critical) = +1.65

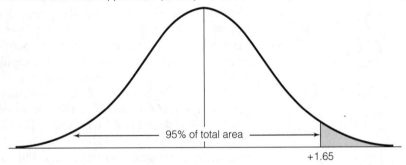

C. The one-tailed test for lower tail, Z (critical) = −1.65

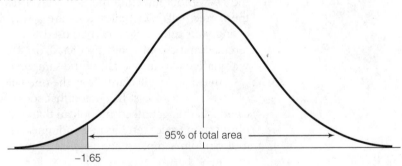

TABLE 8.1   FINDING CRITICAL Z SCORES FOR ONE-TAILED TESTS
(Single Sample Means)

Alpha	Two-tailed Value	One-tailed Value	
		Upper Tail	Lower Tail
.10	±1.65	+1.29	−1.29
.05	±1.96	+1.65	−1.65
.01	±2.58	+2.33	−2.33
.001	±3.29	+3.10	−3.10

We will use the five-step model to test the H_0 of no difference between sociology majors and the general student body.

Step 1. Making Assumptions. Since we are using a mean to summarize the sample outcome, we must assume that the Sophistication Scale generates interval-ratio-level data. With a sample size of 100, the Central Limit Theorem applies, and we can assume that the sampling distribution is normal in shape.

Model: Random sampling
Level of measurement is interval-ratio
Sampling distribution is normal

Step 2. Stating the Null Hypothesis (H_0). The null hypothesis states that there is no difference between sociology majors and the general student body. The research hypothesis (H_1) will also be stated at this point. The researcher has predicted a direction for the difference ("Sociology majors are *more* sophisticated"), so a one-tailed test is justified. The two hypotheses may be stated as

$$H_0: \mu = 17.3$$
$$(H_1: \mu > 17.3)$$

Step 3. Selecting the Sampling Distribution and Establishing the Critical Region. We will use the standardized normal distribution (Appendix A) to find areas under the sampling distribution. If alpha is set at 0.05, the critical region will begin at the Z score +1.65. That is, the researcher has predicted that sociology majors are *more* sophisticated and that this sample comes from a population that has a mean *greater than* 17.3, so he will be concerned only with sample outcomes in the upper tail of the sampling distribution. If sociology majors are *the same as* other students in terms of sophistication (if the H_0 is true), or if they are *less* sophisticated (and come from a population with a mean less than 17.3), the theory is disproved. These decisions may be summarized as

$$\text{Sampling distribution} = Z \text{ distribution}$$
$$\alpha = 0.05 \text{ (one-tailed)}$$
$$Z \text{(critical)} = +1.65$$

FIGURE 8.5 *Z* (OBTAINED) VERSUS *Z* (CRITICAL) (alpha = 0.05, one-tailed test)

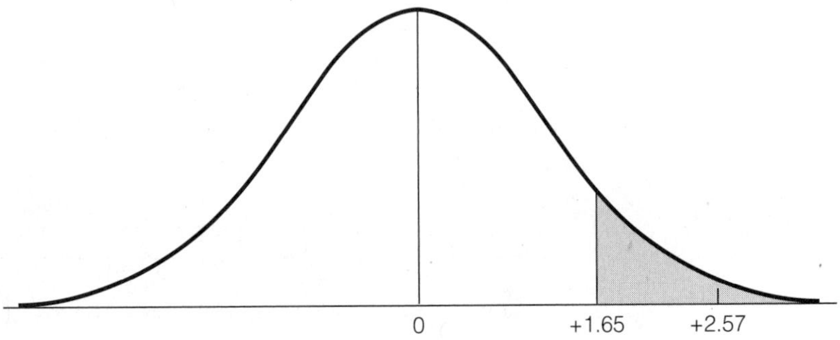

$$0 \qquad +1.65 \qquad +2.57$$

Step 4. Computing the Test Statistic.

$$Z\,(\text{obtained}) = \frac{\overline{X} - \mu}{\sigma/\sqrt{N}}$$

$$Z\,(\text{obtained}) = \frac{19.2 - 17.3}{7.4/\sqrt{100}}$$

$$Z\,(\text{obtained}) = +2.57$$

Step 5. **Making a Decision.** Comparing the *Z* (obtained) with the *Z* (critical):

$$Z\,(\text{critical}) = +1.65$$
$$Z\,(\text{obtained}) = +2.57$$

We see that the test statistic falls into the critical region. This outcome is depicted graphically in Figure 8.5. We will reject the null hypothesis because, if the H_0 were true, a difference of this size would be very unlikely. There is a significant difference between sociology majors and the general student body in terms of sophistication. Since the null hypothesis has been rejected, the research hypothesis (sociology majors are more sophisticated) is supported. *(For practice in dealing with tests of significance for means that may call for one-tailed tests, see problems 8.2, 8.3, 8.6, 8.8, and 8.17).*

8.5 SELECTING AN ALPHA LEVEL

In addition to deciding between one-tailed and two-tailed tests, the researcher must also select an alpha level. We have seen that the alpha level plays a crucial role in hypothesis testing. When we assign a value to alpha, we define what we mean by an "unlikely" sample outcome. If the probability of the observed sample outcome is lower than the alpha level (if the test statistic falls into the critical region), we reject the null hypothesis as untrue. Thus, the alpha level will have important consequences for our decision in step 5.

How can reasonable decisions be made with respect to the value of alpha? Recall that, in addition to defining what will be meant by *unlikely,* the alpha level is also the probability that the decision to reject the null hypothesis, if the test statistic falls into the critical region, will be incorrect. In hypothesis testing, the error of incorrectly rejecting the null hypothesis or rejecting a null hypothesis that is actually true is called **Type I error,** or **alpha error.** To minimize this type of error, use very small values for alpha.

To elaborate: when an alpha level is specified, the sampling distribution is divided into two sets of possible sample outcomes. The critical region includes all unlikely or rare sample outcomes. Outcomes in this region will cause us to reject the null hypothesis. The remainder of the area consists of all sample outcomes that are "nonrare." The lower the level of alpha, the smaller the critical region and the greater the distance between the mean of the sampling distribution and the beginnings of the critical region. Compare, for the sake of illustration, the following alpha levels and values for Z (critical) for two-tailed tests.

If Alpha Equals	The Two-Tailed Critical Region Will Begin at Z (Critical) Equal to
0.10	±1.65
0.05	±1.96
0.01	±2.58
0.001	±3.29

As alpha goes down, the critical region becomes smaller and moves farther away from the mean of the sampling distribution. The lower the alpha level, the harder it will be to reject the null hypothesis and, since a Type I error can be made only if our decision in step 5 is to reject, the lower the probability of Type I error. To minimize the probability of rejecting a null hypothesis that is in fact true, use very low alpha levels.

However, there is a complication. As the critical region decreases in size (as alpha levels decrease), the noncritical region—the area between the two Z (critical) scores in a two-tailed test—must become larger. All other things being equal, the lower the alpha level, the less likely that the sample outcome will fall into the critical region. This raises the possibility of a second type of incorrect decision, called **Type II error,** or **beta error:** failing to reject a null that is, in fact, false. The probability of Type I error decreases as alpha level decreases, but the probability of Type II error increases. Thus, the two types of error are inversely related, and it is not possible to minimize both in the same test. As the probability of one type of error decreases, the other increases, and vice versa.

It may be helpful to clarify the relationships between decision making and errors in table format. Table 8.2 lists the two decisions we can make in step 5 of the five-step model: we either reject or fail to reject the null hypothesis. The other dimension of Table 8.2 lists the two possible conditions

TABLE 8.2   DECISION MAKING AND THE NULL HYPOTHESIS

The H_0 Is Actually:	Decision	
	Reject	Fail to Reject
True	Type I or α error	OK
False	OK	Type II or β error

of the null hypothesis: it is either actually true or actually false. The table combines these possibilities into a total of four possible combinations, two of which are desirable ('OK') and two of which indicate that an error has been made.

The two desirable outcomes are rejecting null hypotheses that are actually false and failing to reject null hypotheses that are actually true. The goal of any scientific investigation is to verify true statements and reject false statements. The remaining two combinations are errors or situations that, naturally, we wish to avoid. If we reject a null hypothesis that is in fact true, we are saying that a true statement is false. Likewise, if we fail to reject a null hypothesis that is in fact false, we are saying that a false statement is true. Obviously, we would prefer to always wind up in one of the boxes labeled "OK" in Table 8.2—to always reject false statements and accept the truth when we find it. Remember, however, that hypothesis testing always carries an element of risk and that it is not possible to minimize the chances of both Type I and Type II error simultaneously.

What all of this means, finally, is that you must think of selecting an alpha level as an attempt to balance the two types of error. Higher alpha levels will minimize the probability of Type II error (saying that false statements are true), and lower alpha levels will minimize the probability of Type I error (saying that true statements are false). Normally, in social science research, we will want to minimize Type I error, and lower alpha levels (.05, .01, .001 or lower) will be used. The 0.05 level in particular has emerged as a generally recognized indicator of a significant result. However, the widespread use of the 0.05 level is simply a convention, and there is no reason that alpha cannot be set at virtually any sensible level (such as 0.04, 0.027, 0.083). The researcher has the responsibility of selecting the alpha level that seems most reasonable in terms of the goals of the research project.

8.6 THE STUDENT'S *t* DISTRIBUTION

To this point, we have considered only one type of hypothesis test. Specifically, we have focused on situations involving single sample means where the value of the population standard deviation (σ) was known. Needless to say, in most research situations, the value of σ will not be known. However, a value for σ is required in order to compute the standard error of the mean (σ/N), convert our sample outcome into a Z score, and place the Z (obtained) on the sampling distribution (step 4). How can a value for the population standard deviation reasonably be obtained?

It might seem sensible to estimate σ with s, the sample standard deviation. As we noted in Chapter 7, s is a biased estimator of σ, but the degree of bias decreases as sample size increases. For large samples (that is, samples with 100 or more cases), the sample standard deviation yields an adequate estimate of σ. Thus, for large samples, we simply substitute s for σ in the formula for Z (obtained) in step 4 and continue to use the standard normal curve to find areas under the sampling distribution.*

For smaller samples, however, when σ is unknown, an alternative distribution called the **Student's t distribution** must be used to find areas under the sampling distribution and establish the critical region. The shape of the t distribution varies as a function of sample size. The relative shapes of the t and Z distributions are depicted in Figure 8.6. For small samples, the t distribution is much flatter than the Z distribution, but, as sample size increases, the t distribution comes to resemble the Z distribution more and more until the two are essentially identical when sample size is greater than 120. As N increases, the sample standard deviation (s) becomes a more and more adequate estimator of the population standard deviation (σ), and the t distribution becomes more and more like the Z distribution.

The t distribution is summarized in Appendix B. The t table differs from the Z table in several ways. First, there is a column at the left of the table labeled df for "degrees of freedom." [†] As mentioned above, the exact shape of the t distribution and thus the exact location of the critical region for any alpha level varies as a function of sample size. Degrees of freedom, which are equal to $N - 1$ in the case of a single-sample mean, must first be computed before the critical region for any alpha can be located. Second, alpha levels are arrayed across the top of Appendix B in two rows, one row for the one-tailed tests and one for two-tailed tests. To use the table, begin by locating the selected alpha level in the appropriate row.

The third difference is that the entries in the table are the actual scores, called t **(critical),** that mark the beginnings of the critical regions and not areas under the sampling distribution. To illustrate the use of this table with

---

*Even though its effect will be minor and will decrease with sample size, we will always correct for the bias in s by using the term $N - 1$ rather than N in the computation for the standard deviation of the sampling distribution when σ is unknown.

†Degrees of freedom refer to the number of values in a distribution that are free to vary. For a sample mean, a distribution has $N - 1$ degrees of freedom. This means that, for a specific value of $\overline{X}$ and N, $N - 1$ scores are free to vary. For example, if $\overline{X} = 3$ and $N = 5$, the distribution of five scores would have $N - 1$, or four, degrees of freedom. When the value of four of the scores is known, the value of the fifth is fixed. If four scores are 1, 2, 3, and 4, the fifth must be 5 and no other value.

FIGURE 8.6 THE *t* DISTRIBUTION AND THE *Z* DISTRIBUTION

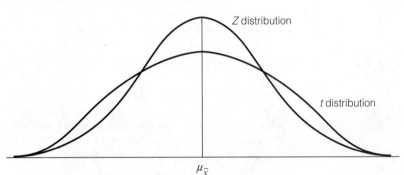

single-sample means, find the critical region for alpha equal to 0.05, two-tailed test, for $N = 30$. The degrees of freedom will be $N - 1$, or 29; reading down the proper column, you should find a value of 2.045. Thus, the critical region for this test will begin at t (critical) $= \pm 2.045$.

Take a moment to notice some additional features of the *t* distribution. First, note that the *t* (critical) we found above is larger in value than the comparable *Z* (critical), which for a two-tailed test at an alpha of 0.05 would be ± 1.96. This relationship reflects the fact that the *t* distribution is flatter than the *Z* distribution (see Figure 8.6). When you use the *t* distribution, the critical regions will begin farther away from the mean of the sampling distribution and, therefore, the null hypothesis will be harder to reject. Furthermore, the smaller the sample size (the lower the degrees of freedom), the larger the value of *t* (obtained) necessary for a rejection of the H_0.

Second, scan the column for an alpha of 0.05, two-tailed test. Note that, for one degree of freedom, the *t* (critical) is ± 12.706 and that the value of *t* (critical) decreases as degrees of freedom increase. For degrees of freedom greater than 120, the value of *t* (critical) is the same as the comparable value of *Z* (critical), or ± 1.96. As sample size increases, the *t* distribution comes to resemble the *Z* distribution more and more until, with sample sizes greater than 120, the two distributions are essentially identical.*

To demonstrate the uses of the *t* distribution in more detail, we will work through an example problem. Note that, in terms of the five-step model, the changes required by using *t* scores occur mostly in steps 3 and 4. In step 3,

---

*Appendix B abbreviates the *t* distribution by presenting a limited number of critical *t* scores for degrees of freedom between 31 and 120. If the degrees of freedom for a specific problem equal 77 and alpha equals 0.05, two-tailed, we have a choice between a *t* (critical) of ± 2.000 (df $= 60$) and a *t* (critical) of ± 1.980 (df $= 120$). In situations such as these, take the larger table value as *t* (critical). This will make rejection of the H_0 less likely and is therefore the more conservative course of action.

the sampling distribution will be the t distribution, and degrees of freedom (df) must be computed before locating the critical region as marked by t (critical). In step 4, a slightly different formula for computing the test statistic, t **(obtained),** will be used. As compared with the formula for Z (obtained), s will replace σ and $N - 1$ will replace N.

Specifically,

FORMULA 8.2
$$t\,(\text{obtained}) = \frac{\overline{X} - \mu}{s/\sqrt{N - 1}}$$

A researcher wonders if commuter students are different from the general student body in terms of academic achievement. She has gathered a random sample of 30 commuter students and has learned from the registrar that the mean grade-point average for all students is 2.50 ($\mu = 2.50$), but the standard deviation of the population (σ) has never been computed. Sample data are reported below. Is the sample from a population that has a mean of 2.50?

Student Body	Commuter Students
$\mu = 2.50\ (= \mu_{\bar{x}})$	$\overline{X} = 2.78$
$\sigma = ?$	$s = 1.23$
	$N = 30$

Step 1. Making Assumptions.

Model: Random sampling
Level of measurement is interval-ratio
Sampling distribution is normal

Step 2. Stating the Null Hypothesis.

$H_0: \mu = 2.50$
$(H_1: \mu \neq 2.50)$

You can see from the research hypothesis that the researcher has not predicted a direction for the difference. This will be a two-tailed test.

Step 3. Selecting the Sampling Distribution and Establishing the Critical Region. Since σ is unknown and sample size is small, the t distribution will be used to find the critical region. Alpha will be set at 0.01.

Sampling distribution = t distribution
$\alpha = 0.01$, two-tailed test
df $= (N - 1) = 29$
$t\,(\text{critical}) = \pm 2.756$

FIGURE 8.7 SAMPLING DISTRIBUTION SHOWING t (OBTAINED) VERSUS t (CRITICAL) ($\alpha = 0.05$, two-tailed test, df = 29)

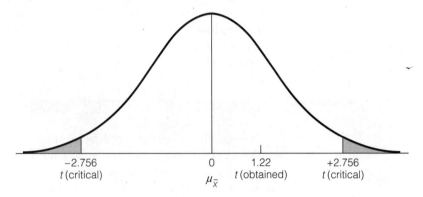

| −2.756 | | 0 | 1.22 | +2.756 |
| t (critical) | | | t (obtained) | t (critical) |

$\mu_{\overline{X}}$

Step 4. Computing the Test Statistic.

$$t \text{ (obtained)} = \frac{\overline{X} - \mu}{s/\sqrt{N - 1}}$$

$$t \text{ (obtained)} = \frac{2.78 - 2.50}{1.23/\sqrt{29}}$$

$$t \text{ (obtained)} = \frac{.28}{.23}$$

$$t \text{ (obtained)} = +1.22$$

Step 5. Making a Decision. The test statistic does not fall into the critical region. Therefore, the researcher fails to reject the H_0. The difference between the sample mean (2.78) and the population mean (2.50) is no greater than what would be expected if only random chance were operating. The test statistic and critical regions are displayed in Figure 8.7.

To summarize, when testing single-sample means we must make a choice regarding the theoretical distribution we will use to establish the critical region. The choice is straightforward. If the population standard deviation (σ) is known or sample size is large, the Z distribution (summarized in Appendix A) will be used. If σ is unknown and the sample is small, the t distribution (summarized in Appendix B) will be used. *(For practice in using the t distribution in a test of hypothesis, see problems 8.8 to 8.10 and 8.17.)*

8.7 TESTS OF HYPOTHESES FOR SINGLE-SAMPLE PROPORTIONS (LARGE SAMPLES)

In many cases, the characteristic of interest in the sample will not be measured in a way that justifies the assumption of interval-ratio level of measurement. One alternative in this situation would be to use a sample proportion (P_s) rather than a sample mean as the test statistic. As we shall see below, the overall procedures for testing single-sample proportions are the

---

Application 8.1

For a random sample of 152 felony cases tried in a local court, the average prison sentence was 27.3 months. Is this significantly different from the average prison term for felons nationally? The necessary information for conducting a test of the null hypothesis is

$$\overline{X} = 27.3 \qquad \mu = 28.7$$
$$s = 3.7$$
$$N = 152$$

The null hypothesis would be

$$H_0: \mu = 28.7$$

The null states that the sample comes from a population that has a mean of 28.7.

The test statistic, Z (obtained), would be

$$Z\text{(obtained)} = \frac{\overline{X} - \mu}{s/\sqrt{N-1}}$$

$$Z\text{(obtained)} = \frac{27.3 - 28.7}{3.7/\sqrt{152-1}}$$

$$Z\text{(obtained)} = \frac{-1.40}{3.7/\sqrt{151}}$$

$$Z\text{(obtained)} = \frac{-1.40}{0.30}$$

$$Z\text{(obtained)} = -4.67$$

If alpha were set at the 0.05 level, the critical region would begin at Z (critical) $= \pm 1.96$. With an obtained Z score of -4.67, the null would be rejected. The difference between the prison sentences of felons convicted in the local court and felons convicted nationally is statistically significant. The difference is so large that we may conclude that it did not occur by random chance. The decision to reject the null hypothesis has a 0.05 probability of being wrong.

---

same as those for testing means. The central question is still "Does the population from which the sample was drawn have a certain characteristic?" We still conduct the test based on the assumption that the null hypothesis is true, and we still evaluate the probability of the obtained sample outcome against a sampling distribution of all possible sample outcomes. Our decision at the end of the test is also the same. If the obtained test statistic falls into the critical region (is unlikely, given the assumption that the H_0 is true), we reject the H_0.

Having stressed the continuity in procedures and logic, I must hastily point out the important differences as well. These differences are best related in terms of the five-step model for hypothesis testing. In step 1, when working with sample proportions, we assume that the variable is measured at the nominal level of measurement. In step 2, the symbols used to state the null hypothesis are different even though the null is still a statement of "no difference."

In step 3, we will use only the standardized normal curve (the Z distribution) to find areas under the sampling distribution and locate the critical region. This will be appropriate as long as sample size is large. We will not consider small-sample tests of hypothesis for proportions in this text.

Application 8.2

In a random sample drawn from the most affluent neighborhood in a community, 76% of the respondents reported that they had voted Republican in the most recent presidential election. For the community as a whole, 66% of the electorate voted Republican. Was the affluent neighborhood significantly more likely to have voted Republican? The information necessary for a test of the null hypothesis, expressed in the form of proportions, is

Neighborhood	Community
$P_s = 0.76$	$P_u = 0.66$
$N = 103$	

Our null hypothesis would be

$$H_0: P_u = 0.66$$

or, the sample was taken from a population that voted 66% Republican.

The test statistic, Z (obtained), would be

$$Z \text{ (obtained)} = \frac{P_s - P_u}{\sqrt{P_u(1 - P_u)/N}}$$

$$Z \text{ (obtained)} = \frac{0.76 - 0.66}{\sqrt{(0.66)(1 - 0.66)/103}}$$

$$Z \text{ (obtained)} = \frac{0.100}{0.047}$$

$$Z \text{ (obtained)} = 2.13$$

Because we want to see if the affluent neighborhood was "significantly more likely to have voted Republican," a one-tailed test of hypothesis is called for. With alpha set at the 0.05 level, our critical region will begin at Z (critical) = +1.65. The critical region is placed in the upper tail because we are predicting that the sample comes from a population that voted Republican at a higher rate than the community.

With an obtained Z score of 2.13, the null hypothesis would be rejected. The difference between the affluent neighborhood and the community as a whole is statistically significant and in the predicted direction. Residents of the affluent neighborhood were significantly more likely to have voted Republican in the last presidential election.

In step 4, computing the test statistic, the form of the formula remains the same. That is, the test statistic, Z (obtained), equals the sample statistic minus the mean of the sampling distribution, divided by the standard deviation of the sampling distribution. However, the symbols will change because we are basing the tests on sample proportions. The formula can be stated as

FORMULA 8.3

$$Z \text{ (obtained)} = \frac{P_s - P_u}{\sqrt{P_u(1 - P_u)/N}}$$

Step 5, making a decision, is exactly the same as before. If the test statistic, Z (obtained), falls into the critical region, as marked by Z (critical), reject the H_0.

An example should clarify these procedures. A random sample of 122 households in a low-income neighborhood revealed that 53 (or a proportion of 0.43) of the households were headed by females. In the city as a whole, the proportion of female-headed households is .39. Are households in the lower-income neighborhood significantly different from the city as a whole in terms of this characteristic?

Step 1. Making Assumptions.

> Model: Random sampling
> Level of measurement is nominal
> Sampling distribution is normal in shape

Step 2. Stating the Null Hypothesis. The research question, as stated above, asks only if the sample proportion is different from the population proportion. Since no direction is predicted for the difference, a two-tailed test will be used.

$$H_0: P_u = .39$$
$$(H_1: P_u \neq .39)$$

Step 3. Selecting the Sampling Distribution and Establishing the Critical Region.

> Sampling distribution = Z distribution
> $\alpha = 0.10$, two-tailed
> $Z \,(\text{critical}) = \pm 1.65$

Step 4. Computing the Test Statistic.

$$Z\,(\text{obtained}) = \frac{P_s - P_u}{\sqrt{P_u(1 - P_u)/N}}$$

$$Z\,(\text{obtained}) = \frac{0.43 - 0.39}{\sqrt{(0.39)(0.61)/122}}$$

$$Z\,(\text{obtained}) = +0.91$$

Step 5. Making a Decision. The test statistic, Z (obtained), does not fall into the critical region. Therefore, we fail to reject the H_0. There is no statistically significant difference between the low-income community and the city as a whole in terms of the proportion of households headed by females. Figure 8.8

FIGURE 8.8   SAMPLING DISTRIBUTION SHOWING Z (OBTAINED) VERSUS Z (CRITICAL) ($\alpha = 0.10$, two-tailed test)

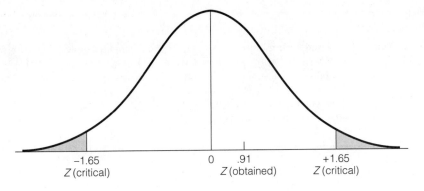

displays the sampling distribution, the critical region, and the Z (obtained). *(For practice in tests of significance using sample proportions, see problems 8.1 c, 8.11 to 8.14, 8.15 a to d, and 8.16.)*

SUMMARY

1. All the basic concepts and techniques for testing hypotheses were presented in this chapter. We saw how to test the null hypothesis of "no difference" for single sample means and proportions. In both cases, the central question is whether the population represented by the sample has a certain characteristic.

2. All tests of a hypothesis involve finding the probability of the observed sample outcome, given that the null hypothesis is true. If the outcomes have low probabilities, we reject the null hypothesis. In the usual research situation, we will wish to reject the null hypothesis and thereby support the research hypothesis.

3. The five-step model will be our framework for decision making throughout the hypothesis-testing chapters. We will always (1) make assumptions, (2) state the null hypothesis, (3) select a sampling distribution, specify alpha, and find the critical region, (4) compute a test statistic, and (5) make a decision. What we do during each step, however, will vary, depending on the specific test being conducted.

4. If we can predict a direction for the difference in stating the research hypothesis, a one-tailed test is called for. If no direction can be predicted, a two-tailed test is appropriate. There are two kinds of errors in hypothesis testing. Type I, or alpha, error is rejecting a true null; Type II, or beta, error is failing to reject a false null. The probabilities of committing these two types of error are inversely related and cannot be simultaneously minimized in the same test. By selecting an alpha level, we try to balance the probability of these two kinds of error.

5. When testing sample means, the t distribution must be used to find the critical region when the population standard deviation is unknown and sample size is small.

6. Sample proportions can also be tested for significance. Tests are conducted using the five-step model. Compared to the test for the sample mean, the major differences lie in the level-of-measurement assumption (step 1), the statement of the null (step 2), and the computation of the test statistic (step 4).

7. If you are still confused about the uses of inferential statistics described in this chapter, don't be alarmed or discouraged. A sizable volume of rather complex material has been presented and only rarely will a beginning student fully comprehend the unique logic of hypothesis testing on the first exposure. After all, it is not every day that you learn how to test a statement you don't believe (the null hypothesis) against a distribution that doesn't exist (the sampling distribution)!

SUMMARY OF FORMULAS

Single-sample means, large samples:

8.1 $\qquad Z \,(\text{obtained}) = \dfrac{\overline{X} - \mu}{\sigma/\sqrt{N}}$

Single-sample means when samples are small and population standard deviation is unknown:

8.2 $\qquad t \,(\text{obtained}) = \dfrac{\overline{X} - \mu}{\sigma/\sqrt{N-1}}$

Single-sample proportions, large samples:

8.3 $\qquad Z \,(\text{obtained}) = \dfrac{P_s - P_u}{\sqrt{P_u(1 - P_u)/N}}$

GLOSSARY

Alpha level (α). The proportion of area under the sampling distribution that contains unlikely sample outcomes, given that the null hypothesis is true. Also, the probability of Type I error.

Critical region (region of rejection). The area under the sampling distribution that, in advance of the test itself, is defined as including unlikely sample outcomes, given that the null hypothesis is true.

Five-step model. A step-by-step guideline for conducting tests of hypotheses. A framework that organizes decisions and computations for all tests of significance.

Hypothesis testing. Statistical tests that estimate the probability of sample outcomes if assumptions about the population (the null hypothesis) are true.

Null hypothesis (H_0). A statement of "no difference." In the context of single-sample tests of significance, the population from which the sample was drawn is assumed to have a certain characteristic or value.

One-tailed test. A type of hypothesis test used when (1) the direction of the difference can be predicted or (2) concern focuses on outcomes in only one tail of the sampling distribution.

Research hypothesis (H_1). A statement that contradicts the null hypothesis. In the context of single-sample tests of significance, the research hypothesis says that the population from which the sample was drawn does not have a certain characteristic or value.

Significance testing. See Hypothesis testing.

***t* (critical).** The t score that marks the beginning of the critical region of a t distribution.

Student's *t* distribution. A distribution used to find the critical region for tests of sample means when σ is unknown and sample size is small.

***t* (obtained).** The test statistic computed in step 4 of the five-step model. The sample outcome expressed as a t score.

Test statistic. The value computed in step 4 of the five-step model that converts the sample outcome into either a t score or a Z score.

Two-tailed test. A type of hypothesis test used when (1) the direction of the difference cannot be predicted or (2) concern focuses on outcomes in both tails of the sampling distribution.

Type I error (alpha error). The probability of rejecting a null hypothesis that is, in fact, true.

Type II error (beta error). The probability of failing to reject a null hypothesis that is, in fact, false.

***Z* (critical).** The Z score that marks the beginnings of the critical region on a Z distribution.

***Z* (obtained).** The test statistic computed in step 4 of the five-step model. The sample outcomes expressed as a Z score.

MULTIMEDIA RESOURCES

The Wadsworth Sociology Resource Center: Virtual Society
http://sociology.wadsworth.com/

Visit the companion web site for the sixth edition of *Statistics: A Tool for Social Research* to access a wide range of student resources. Begin by clicking on the Student Resources section of the book's web site to access the following study tools:

- Basic math review
- Statistics review

- Flash cards
- Internet links
- Additional chapter problems
- Table of random numbers
- MicroCase and SPSS examples and exercises
- "Find the text" flowcharts
- Hypothesis testing for variables measured at the ordinal level

PROBLEMS

8.1 a. For each situation, find Z (critical).

Alpha	Form	Z (Critical)
.05	One-tailed	
.10	Two-tailed	
.06	Two-tailed	
.01	One-tailed	
.02	Two-tailed	

b. For each situation, find the critical t score.

Alpha	Form	N	t (Critical)
.10	Two-tailed	31	
.02	Two-tailed	24	
.01	Two-tailed	121	
.01	One-tailed	31	
.05	One-tailed	61	

c. Compute the appropriate test statistic (Z or t) for each situation:

1. $\mu = 2.40$ $\overline{X} = 2.20$
     $\sigma = 0.75$ $N = 200$

2. $\mu = 17.1$ $\overline{X} = 16.8$
                 $s = 0.9$
                 $N = 45$

3. $\mu = 10.2$ $\overline{X} = 9.4$
                 $s = 1.7$
                 $N = 150$

4. $P_u = .57$ $P_s = 0.60$
                 $N = 117$

5. $P_u = 0.32$ $P_s = 0.30$
                 $N = 322$

8.2 ⎡SOC⎤ **a.** The student body at St. Algebra College attends an average of 3.3 parties per month. A random sample of 117 sociology majors averages 3.8 parties per month with a standard deviation of 0.53. Are sociology majors significantly different from the student body as a whole? *(HINT: the wording of the research question suggests a two-tailed test. This means that the alternative or research hypothesis in step 2 will be stated as H_1: $\mu \neq 3.3$ and that the critical region will be split between the upper and lower tails of the sampling distribution. See Table 8.1 for values of Z (critical) for various alpha levels.)*

b. What if the research question were changed to "Do sociology majors attend a significantly *greater* number of parties"? How would the test conducted in 8.2a change? *(HINT: this wording implies a one-tailed test of significance. How would the research hypothesis change? For the alpha you used in problem 8.2a, what would the value of Z (critical) be?)*

8.3 ⎡SW⎤ **a.** Nationally, social workers average 10.2 years of experience. In a random sample, 203 social workers in greater metropolitan Shinbone average only 8.7 years with a standard deviation of 0.52. Are social workers in Shinbone significantly less experienced? *(Note the wording of the research hypotheses. These situations may justify one-tailed tests of significance. If you chose a one-tailed test, what form would the research hypothesis take, and where would the critical region begin?)*

b. The same sample of social workers reports an average annual salary of $25,782 with a standard deviation of $622. Is this figure significantly higher than the national average of $24,509? *(The wording of the research hypotheses suggests a one-tailed test. What form would the research hypothesis take, and where would the critical region begin?)*

8.4 ⎡SOC⎤ Nationally, the average score on the college entrance exams (verbal test) is 453 with a standard deviation of 95. A random sample of 152 freshmen entering St. Algebra College shows a mean score of 502. Is there a significant difference?

8.5 ⎡SOC⎤ A random sample of 423 Chinese Americans has finished an average of 12.7 years of formal education with a standard deviation of 1.7. Is this significantly different from the national average of 12.2 years?

8.6 ⎡SOC⎤ A sample of 105 workers in the Overkill Division of the Machismo Toy Factory earns an average of $24,375 per year. The average salary for all workers is $24,230 with a standard deviation of $523. Are workers in the Overkill Division overpaid? Conduct both one- and two-tailed tests.

8.7 GER **a.** Nationally, the population as a whole watches 6.2 hours of TV per day. A random sample of 1017 senior citizens report watching an average of 5.9 hours per day with a standard deviation of .7. Is the difference significant?

b. The same sample of senior citizens reports that they belong to an average of 2.1 voluntary organizations and clubs with a standard deviation of .5. Nationally, the average is 1.7. Is the difference significant?

8.8 SOC A school system has assigned several hundred "chronic and severe underachievers" to an alternative educational experience. To assess the program, a random sample of 35 has been selected for comparison with all students in the system.

a. In terms of GPA, did the program work?

Systemwide GPA	Program GPA
$\mu = 2.47$	$\overline{X} = 2.55$
	$s = 0.70$
	$N = 35$

b. In terms of absenteeism (number of days missed per year), what can be said about the success of the program?

Systemwide	Program
$\mu = 6.13$	$\overline{X} = 4.78$
	$s = 1.11$
	$N = 35$

c. In terms of standardized test scores in math and reading, was the program a success?

Math Test Systemwide	Program
$\mu = 103$	$\overline{X} = 106$
	$s = 2.0$
	$N = 35$

Reading Test Systemwide	Program
$\mu = 110$	$\overline{X} = 113$
	$s = 2.0$
	$N = 35$

(*HINT: note the wording of the research questions. Is a one-tailed test justified? Is the program a success if the students in the program are no different from students systemwide?*

What if the program students were performing at lower levels? If a one-tailed test is used, what form should the research hypothesis take? Where will the critical region begin?)

8.9 SOC A random sample of 26 local sociology graduates scored an average of 458 on the GRE advanced sociology test with a standard deviation of 20. Is this significantly different from the national average ($\mu = 440$)?

8.10 PA Nationally, the per capita property tax is $130. A random sample of 36 southeastern cities average $98 with a standard deviation of $5. Is the difference significant? Summarize your conclusions in a sentence or two.

8.11 GER/CJ A survey shows that 10% of the population is victimized by property crime each year. A random sample of 527 older citizens (65 years or more of age) shows a victimization rate of 14%. Are older people more likely to be victimized? Conduct both one- and two-tailed tests of significance.

8.12 CJ A random sample of 113 convicted rapists in a state prison system completed a program designed to change their attitudes towards women, sex, and violence before being released on parole. Fifty-eight eventually became repeat sex offenders. Is this recidivism rate significantly different from the rate for all offenders (57%) in that state? Summarize your conclusions in a sentence or two. (*HINT: you must use the information given in the problem to compute a sample proportion. Remember to convert the population percentage to a proportion*).

8.13 PS In a recent statewide election, 55% of the voters rejected a proposal to institute a state lottery. In a random sample of 150 urban precincts, 49% of the voters rejected the proposal. Is the difference significant? Summarize your conclusions in a sentence or two.

8.14 CJ Statewide, the police clear by arrest 35% of the robberies and 42% of the aggravated assaults reported to them. A researcher takes a random sample of all the robberies ($N = 207$) and aggravated assaults ($N = 178$) reported to a metropolitan police department in one year and finds that 83 of the robberies and 80 of the assaults were cleared by arrest. Are the local arrest rates significantly different from the statewide rate? Write a sentence or two interpreting your decision.

8.15 [SOC/SW] A researcher has compiled a file of information on a random sample of 317 families in a city that has chronic, long-term patterns of child abuse. Below are reported some of the characteristics of the sample along with values for the city as a whole. For each trait, test the null hypothesis of "no difference" and summarize your findings.

  a. Mothers' educational level (proportion completing high school):

City	Sample
$P_u = 0.63$	$P_s = 0.61$

  b. Family size (proportion of families with four or more children):

City	Sample
$P_u = 0.21$	$P_s = 0.26$

  c. Mothers' work status (proportion of mothers with jobs outside the home):

City	Sample
$P_u = 0.51$	$P_s = 0.27$

  d. Relations with kin (proportion of families that have contact with kin at least once a week):

City	Sample
$P_u = 0.82$	$P_s = 0.43$

  e. Fathers' educational achievement (average years of formal schooling):

City	Sample
$\mu = 12.3$	$\overline{X} = 12.5$
	$s = 1.7$

  f. Fathers' occupational stability (average years in present job):

City	Sample
$\mu = 5.2$	$\overline{X} = 3.7$
	$s = 0.5$

8.16 [SW] You are the head of an agency seeking funding for a program to reduce unemployment among teenage males. Nationally, the unemployment rate for this group is 18%. A random sample of 323 teenage males in your area reveals an unemployment rate of 21.7%. Is the difference significant? Can you demonstrate a need for the program? Should you use a one-tailed test in this situation? Why? Explain the result of your test of significance as you would to a funding agency.

8.17 [PA] The city manager of Shinbone has received a complaint from the local union of firefighters to the effect that they are underpaid. Not having much time, the city manager gathers the records of a random sample of 27 firefighters and finds that their average salary is $38,073 with a standard deviation of $575. If she knows that the average salary nationally is $38,202, how can she respond to the complaint? Should she use a one-tailed test in this situation? Why? What would she say in a memo to the union that would respond to the complaint?

8.18 The following essay questions review the basic principles and concepts of inferential statistics. The order of the questions roughly follows the five-step model.

  a. Hypothesis testing or significance testing can be conducted only with a random sample. Why?

  b. Under what specific conditions can it be assumed that the sampling distribution is normal in shape?

  c. Explain the role of the sampling distribution in a test of hypothesis.

  d. The null hypothesis is an assumption about reality that makes it possible to test sample outcomes for their significance. Explain.

  e. What is the critical region? How is the size of the critical region determined?

f. Describe a research situation in which a one-tailed test of hypothesis would be appropriate.

g. Thinking about the shape of the sampling distribution, why does use of the t distribution (as opposed to the Z distribution) make it more difficult to reject the null hypothesis?

h. What exactly can be concluded in the one-sample case when the test statistic falls into the critical region?

HYPOTHESIS TESTING II
THE TWO-SAMPLE CASE

LEARNING OBJECTIVES   By the end of this chapter, you will be able to

1. Identify and cite examples of situations in which the two-sample test of hypothesis is appropriate.
2. Explain the logic of hypothesis testing as applied to the two-sample case.
3. Explain what an independent random sample is.
4. Perform a test of hypothesis for two sample means or two sample proportions following the five-step model and correctly interpret the results.
5. List and explain each of the factors (especially sample size) that affect the probability of rejecting the null hypothesis. Explain the difference between statistical significance and importance.

9.1 INTRODUCTION   In Chapter 8, we dealt with hypothesis testing in the one-sample case. In that situation, our concern was with the significance of the difference between a sample value and a population value. In this chapter, we will consider research situations where we will be concerned with the significance of the difference between two separate populations. For example, do men and women in the United States vary in their support for gun control? Obviously, we cannot ask every male and female for their opinions on this issue. Instead, we must draw random samples of both groups and use the information gathered from these samples to infer population patterns.

The central question asked in hypothesis testing in the two-sample case is "Is the difference between the samples large enough to allow us to conclude (with a known probability of error) that the populations represented by the samples are different?" Thus, if we find a large enough difference in support for gun control between random samples of men and women, we can argue that the difference between the samples did not occur by simple random chance but, rather, represents a real difference between men and women in the population.

In this chapter, we will consider tests for the significance of the difference between sample means and sample proportions. In both tests, the five-step model will serve as a framework for organizing our decision making.

Although the general flow of the hypothesis-testing process is very similar to the one followed in the one-sample case, some important differences will be identified as we proceed through the material.

9.2 HYPOTHESIS TESTING WITH SAMPLE MEANS (LARGE SAMPLES)

One important difference between the one- and two-sample situations occurs in step 1 of the five-step model. In the one-sample case, we assumed "random sampling" or that the sample was selected following the principle of EPSEM. As you recall, this means that each case in the population must have an equal chance of being selected for the sample. In the two-sample situation, the samples must be selected independently as well as randomly. This requirement is met when the selection of a case for one sample has no effect on the probability that any particular case will be included in the other sample. In our example concerning gender differences in support of gun control, this would mean that the selection of a specific male for the sample would have no effect on the probability of selecting any particular female. This new assumption will be stated as **independent random sampling** in step 1.

Drawing EPSEM samples from separate lists (for example, one for females and one for males) can satisfy the requirement of independent random sampling. It is usually more convenient, however, to draw a single EPSEM sample from a single population list and then subdivide the cases into separate groups (males and females, for example). As long as the original sample is selected randomly, any sub-samples created by the researcher will meet the assumption of independent random samples.

The second major difference in the five-step model for the two-sample case is in the form of the null hypothesis. The null is still a statement of "no difference." Now, however, instead of saying that the population from which the sample is drawn has a certain characteristic, it will say that the two populations are no different. ("There is no significant difference between men and women in their support of gun control.") If the test statistic falls in the critical region, the null hypothesis of no difference between the populations can be rejected, and the argument that the populations are different on the trait of interest will be supported.

A third important new element is the sampling distribution: the distribution of all possible sample outcomes. In Chapter 8, the sample outcome was either a mean or a proportion. Now, we are dealing with two samples (e.g., samples of men and women) and the sample outcome is the *difference* between the sample statistics. In terms of our example, the sampling distribution would include all possible differences in sample means for support of gun control between men and women. If the null hypothesis is true and men and women do *not* have different views about gun control, the difference between the population means would be zero, the mean of the sampling distribution will be zero, and the huge majority of differences between sample means would be small. The greater the differences between the sample means, the further the sample outcome (the *difference* between the two sample means) will be from the mean of the sampling distribution (zero), and the more likely that the difference reflects a real difference between the populations represented by the samples.

To illustrate the procedure for testing sample means, assume that a researcher has access to a nationally representative random sample and that the individuals in the sample have responded to a scale that measures attitudes toward gun control. The sample is divided by sex, and sample statistics are computed for males and females. Assuming that the scale yields interval-ratio-level data, a test for the significance of the difference in sample means can be conducted.

As long as sample size is large (that is, as long as the combined number of cases in the two samples exceeds 100), the sampling distribution of the differences in sample means will be normal, and the normal curve (Appendix A) can be used to establish the critical regions. The test statistic, Z (obtained), will be computed by the usual formula: sample outcome (the difference between the sample means) minus the mean of the sampling distribution, divided by the standard deviation of the sampling distribution. The formula is presented as Formula 9.1. Note that numerical subscripts are used to identify the samples and the two populations they represent. The subscript $\bar{x} - \bar{x}$ is used to indicate that we are dealing with the sampling distribution of the differences in sample means.

FORMULA 9.1
$$Z\,(\text{obtained}) = \frac{(\bar{X}_1 - \bar{X}_2) - (\mu_1 - \mu_2)}{\sigma_{\bar{x} - \bar{x}}}$$

where $(\bar{X}_1 - \bar{X}_2)$ = the difference in the sample means
$(\mu_1 - \mu_2)$ = the difference in the population means
$\sigma_{\bar{x} - \bar{x}}$ = the standard deviation of the sampling distribution
of the differences in sample means

The second term in the numerator, $(\mu_1 - \mu_2)$, reduces to zero because we assume that the null hypothesis (which will be stated as H_0: $\mu_1 = \mu_2$) is true. Recall that tests of significance are always based on the assumption that the null hypothesis is true. If the means of the two populations are equal, then the term $(\mu_1 - \mu_2)$ will be zero and can be dropped from the equation. In effect, then, the formula we will actually use to compute the test statistic in step 4 will be

FORMULA 9.2
$$Z\,(\text{obtained}) = \frac{(\bar{X}_1 - \bar{X}_2)}{\sigma_{\bar{x} - \bar{x}}}$$

For large samples, the standard deviation of the sampling distribution of the difference in sample means is defined as

FORMULA 9.3
$$\sigma_{\bar{x} - \bar{x}} = \sqrt{\frac{\sigma_1^2}{N_1} + \frac{\sigma_2^2}{N_2}}$$

Since we will rarely be in a position to know the values of the population standard deviations (σ_1 and σ_2), we must use the sample standard deviations, suitably corrected for bias, to estimate them. Formula 9.4 displays the equation used to estimate the standard deviation of the sampling distribution in this situation. This is called a **pooled estimate** since it combines information from both samples.

FORMULA 9.4

$$\sigma_{\bar{x}-\bar{x}} = \sqrt{\frac{s_1^2}{N_1 - 1} + \frac{s_2^2}{N_2 - 1}}$$

The sample outcomes for support of gun control are reported below, and a test for the significance of the difference can now be conducted.

Sample 1 Men	Sample 2 Women
$\bar{X}_1 = 6.2$	$\bar{X}_2 = 6.5$
$s_1 = 1.3$	$s_2 = 1.4$
$N_1 = 324$	$N_2 = 317$

We see from the sample statistics that men have a lower average score on the Support for Gun Control Scale and are thus less supportive of gun control. The test of hypothesis will tell us if this difference is large enough to justify the conclusion that it did not occur by random chance alone but rather reflects an actual difference between the populations of men and women on this issue.

Step 1. Making Assumptions. Note that, although we now assume that the random samples are independent, the remaining model assumptions are the same as in the one-sample case.

> Model: Independent random samples
> Level of measurement is interval-ratio
> Sampling distribution is normal

Step 2. Stating the Null Hypothesis. The null hypothesis states that the *populations* represented by the samples are not different on this variable. Since no direction for the difference has been predicted, a two-tailed test is called for, as reflected in the research hypothesis.

$$H_0: \mu_1 = \mu_2$$
$$(H_1: \mu_1 \neq \mu_2)$$

Step 3. Selecting the Sampling Distribution and Establishing the Critical Region. For large samples, the Z distribution can be used to find areas under the sampling distribution and establish the critical region. Alpha will be set at 0.05.

$$\text{Sampling distribution} = Z \text{ distribution}$$
$$\text{Alpha} = 0.05, \text{ two-tailed}$$
$$Z \text{ (critical)} = \pm 1.96$$

Step 4. Computing the Test Statistic. Since the population standard deviations are unknown, Formula 9.4 will be used to estimate the standard deviation of the sampling distribution. This value will then be substituted into Formula 9.2 and Z (obtained) will be computed.

$$\sigma_{\bar{x}-\bar{x}} = \sqrt{\frac{s_1^2}{N_1 - 1} + \frac{s_2^2}{N_2 - 1}}$$

$$\sigma_{\bar{x}-\bar{x}} = \sqrt{\frac{(1.3)^2}{324 - 1} + \frac{(1.4)^2}{317 - 1}}$$

$$\sigma_{\bar{x}-\bar{x}} = \sqrt{(0.0052) + (0.0062)}$$

$$\sigma_{\bar{x}-\bar{x}} = \sqrt{0.0114}$$

$$\sigma_{\bar{x}-\bar{x}} = 0.107$$

$$Z \text{ (obtained)} = \frac{(\bar{X}_1 - \bar{X}_2)}{\sigma_{\bar{x}-\bar{x}}}$$

$$Z \text{ (obtained)} = \frac{(6.2 - 6.5)}{0.107}$$

$$Z \text{ (obtained)} = \frac{-0.300}{0.107}$$

$$Z \text{ (obtained)} = -2.80$$

Step 5. Making a Decision. Comparing the test statistic with the critical region:

$$Z \text{ (obtained)} = -2.80$$
$$Z \text{ (critical)} = \pm 1.96$$

We see that the Z score clearly falls into the critical region. This outcome indicates that a difference as large as -0.3 ($6.2 - 6.5$) between the sample means is unlikely if the null is true. The null hypothesis of no difference can be rejected, and the notion that men and women are different in terms of their support of gun control is supported. The decision to reject the null hypothesis has only a 0.05 probability (the alpha level) of being incorrect.

Note that the value for Z (obtained) is negative, indicating that men have significantly lower scores than women for support for gun control. The sign of the test statistics reflects our arbitrary decision to label men sample 1 and women sample 2. If we had reversed the labels and called women sample 1 and men sample 2, the sign of the Z (obtained) would have been positive, but its value (2.80) would have been exactly the same, as would our decision in step 5 ("reject the null hypothesis"). *(For practice in testing the significance of the difference between sample means for large samples, see problems 9.1 to 9.6 and 9.15 d to f.)*

9.3 HYPOTHESIS TESTING WITH SAMPLE MEANS (SMALL SAMPLES)

As with single-sample means, when the population standard deviation is unknown and sample size is small (combined N's of less than 100), the Z distribution can no longer be used to find areas under the sampling distribution. Instead, we will use the t distribution to find the critical region and thus to identify unlikely sample outcomes. To utilize the t distribution for testing two sample means, we need to perform one additional calculation and make one additional assumption. The calculation is for degrees of freedom, a quantity required for proper use of the t table (Appendix B). In the two-sample case, degrees of freedom are equal to $N_1 + N_2 - 2$.

The additional assumption is a more complex matter. With small samples, to justify the assumption of a normal sampling distribution and to form a pooled estimate of the standard deviation of the sampling distribution, we must assume that the variances of the populations of interest are equal $(\sigma_1^2 = \sigma_2^2)$. The assumption of equal variance in the population can be tested by an inferential statistical technique known as the analysis of variance or ANOVA (see Chapter 10). For our purposes here, however, we will simply assume equal population variances without formal testing. This assumption can be considered justified as long as sample sizes are approximately equal.

To illustrate this procedure, assume that a researcher believes that center-city families are significantly larger than suburban families, as measured by number of children. Random samples from both areas are gathered and sample statistics computed.

Sample 1 Suburban	Sample 2 Center-city
$\overline{X}_1 = 2.37$	$\overline{X}_2 = 2.78$
$s_1 = 0.63$	$s_2 = 0.95$
$N_1 = 42$	$N_2 = 37$

The sample data reveal a difference in the predicted direction. The significance of this observed difference can be tested with the five-step model.

Step 1. Making Assumptions. Sample size is small, and the population standard deviation is unknown. Hence, we must assume equal population variances in the model.

Model: Independent random samples
Level of measurement is interval-ratio
Equal population variances $(\sigma_1^2 = \sigma_2^2)$
Sampling distribution is normal

Step 2. Stating the Null Hypothesis. Since a direction has been predicted (center-city families are larger), a one-tailed test will be used, and the research hypothesis is stated in accordance with this decision.

$$H_0: \mu_1 = \mu_2$$
$$(H_1: \mu_1 < \mu_2)$$

Step 3. Selecting the Sampling Distribution and Establishing the Critical Region. With small samples, the t distribution is used to establish the critical region. Alpha will be set at 0.05, and a one-tailed test will be used.

Sampling distribution = t distribution
Alpha = 0.05, one-tailed
Degrees of freedom = $N_1 + N_2 - 2 = 42 + 37 - 2 = 77$
t (critical) = -1.671

Note that the critical region is placed in the lower tail of the sampling distribution in accordance with the direction specified in H_1.

Step 4. Computing the Test Statistic. With small samples, a different formula (Formula 9.5) is used for the pooled estimate of the standard deviation of the sampling distribution. This value is then substituted directly into the denominator of the formula for t (obtained) given in Formula 9.6.

FORMULA 9.5
$$\sigma_{\bar{x}-\bar{x}} = \sqrt{\frac{N_1 s_1^2 + N_2 s_2^2}{N_1 + N_2 - 2}} \sqrt{\frac{N_1 + N_2}{N_1 N_2}}$$

$$\sigma_{\bar{x}-\bar{x}} = \sqrt{\frac{(42)(.63)^2 + (37)(.95)^2}{42 + 37 - 2}} \sqrt{\frac{42 + 37}{(42)(37)}}$$

$$\sigma_{\bar{x}-\bar{x}} = \sqrt{\frac{50.06}{77}} \sqrt{\frac{79}{1554}}$$

$$\sigma_{\bar{x}-\bar{x}} = (.81)(.23)$$

$$\sigma_{\bar{x}-\bar{x}} = .19$$

FORMULA 9.6
$$t\,(\text{obtained}) = \frac{(\bar{X}_1 - \bar{X}_2)}{\sigma_{\bar{x}-\bar{x}}}$$

$$t\,(\text{obtained}) = \frac{(2.37 - 2.78)}{.19}$$

$$t\,(\text{obtained}) = \frac{-.41}{.19}$$

$$t\,(\text{obtained}) = -2.16$$

Step 5. Making a Decision. Comparing the test statistic with the critical region,

$$t\,(\text{obtained}) = -2.16$$
$$t\,(\text{critical}) = -1.671$$

Application 9.1

An attitude scale measuring satisfaction with family life has been administered to a sample of married respondents. On this scale, higher scores indicate greater satisfaction. The sample has been divided into respondents with no children and respondents with at least one child, and means and standard deviations have been computed for both groups. Is there a significant difference in satisfaction with family life between these two groups? The information necessary for conducting a test of the null hypothesis ($H_0: \mu_1 = \mu_2$) is

Sample 1 (No Children)	Sample 2 (At Least One Child)
$\overline{X}_1 = 11.3$	$\overline{X}_2 = 10.8$
$s_1 = 0.6$	$s_2 = 0.5$
$N_1 = 78$	$N_2 = 93$

The estimate of the standard deviation of the sampling distribution would be

$$\sigma_{\overline{x}-\overline{x}} = \sqrt{\frac{s_1^2}{N_1 - 1} + \frac{s_2^2}{N_2 - 1}}$$

$$\sigma_{\overline{x}-\overline{x}} = \sqrt{\frac{(0.6)^2}{78 - 1} + \frac{(0.5)^2}{93 - 1}}$$

$$\sigma_{\overline{x}-\overline{x}} = \sqrt{0.008}$$

$$\sigma_{\overline{x}-\overline{x}} = 0.09$$

The test statistic, Z (obtained), would be

$$Z\,(\text{obtained}) = \frac{(\overline{X}_1 - \overline{X}_2)}{\sigma_{\overline{x}-\overline{x}}}$$

$$Z\,(\text{obtained}) = \frac{11.3 - 10.8}{0.09}$$

$$Z\,(\text{obtained}) = \frac{0.50}{0.09}$$

$$Z\,(\text{obtained}) = 5.56$$

If alpha were set at the 0.05 level, the critical region would begin at Z (critical) $= \pm 1.96$. With an obtained Z of 5.56, we would reject the null hypothesis. This test supports the conclusion that parents and childless couples are different with respect to satisfaction with family life.

---

you can see that the test statistic falls into the critical region. If the null ($\mu_1 = \mu_2$) were true, this would be a very unlikely outcome, so the null can be rejected. There is a statistically significant difference (a difference so large that it is unlikely to be due to random chance) in the sizes of center-city and suburban families. Furthermore, center-city families are significantly larger in size. The test statistic and sampling distribution are depicted in Figure 9.1. *(For practice in testing the significance of the difference between sample means for small samples, see problems 9.7 to 9.9.)*

9.4 HYPOTHESIS TESTING WITH SAMPLE PROPORTIONS (LARGE SAMPLES)

Testing for the significance of the difference between two sample proportions is analogous to testing sample means. The null hypothesis states that no difference exists between the populations from which the samples are drawn on the trait being tested. The sample proportions form the basis of the test statistic computed in step 4, which is then compared with the critical region. When sample sizes are large (combined N's of more than 100), the Z distribution may be used to find the critical region. We will not consider

FIGURE 9.1 THE SAMPLING DISTRIBUTION WITH CRITICAL REGION AND TEST STATISTIC DISPLAYED

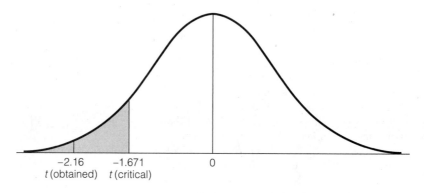

$$-2.16 \qquad -1.671 \qquad 0$$
$$t\,(\text{obtained}) \quad t\,(\text{critical})$$

tests of significance for proportions based on small samples in this text. The formula for finding Z (obtained) in step 4 is presented as Formula 9.7. The subscript $_{p-p}$ for σ reminds us that we are now dealing with the sampling distribution of the differences in sample proportions.

FORMULA 9.7
$$Z\,(\text{obtained}) = \frac{(P_{s1} - P_{s2}) - (P_{u1} - P_{u2})}{\sigma_{p-p}}$$

where $(P_{s1} - P_{s2})$ = the difference between the sample proportions
$(P_{u1} - P_{u2})$ = the difference between the population proportions
σ_{p-p} = the standard deviation of the sampling distribution of the differences in sample proportions

As was the case with sample means, the second term in the numerator is assumed to be 0 by the null hypothesis. Therefore, the formula reduces to

FORMULA 9.8
$$Z\,(\text{obtained}) = \frac{(P_{s1} - P_{s2})}{\sigma_{p-p}}$$

The value of the standard deviation of the sampling distribution of the differences in sample proportions is found by solving Formula 9.9:

FORMULA 9.9
$$\sigma_{p-p} = \sqrt{P_u(1 - P_u)}\ \sqrt{(N_1 + N_2)/N_1 N_2}$$

Solving this formula requires that a value for P_u be estimated. This is done by pooling information from both samples as specified in Formula 9.10:

FORMULA 9.10
$$P_u = \frac{N_1 P_{s1} + N_2 P_{s2}}{N_1 + N_2}$$

To find Z (obtained) in step 4, we must begin with Formula 9.10 and work back to Formula 9.8. An example will clarify these procedures.

Random samples of black and white senior citizens have been selected, and each respondent has been classified as high or low in terms of the number of memberships he or she holds in voluntary associations. Is there a statistically significant difference in the participation patterns of black and white

---

Application 9.2

Do attitudes toward sex vary by gender? The respondents in a national survey have been asked if they think that premarital sex is "always wrong" or only "sometimes wrong." The proportion of each sex that feels that premarital sex is always wrong is

Sample 1 Females	Sample 2 Males
$P_{s1} = 0.35$	$P_{s2} = 0.32$
$N_1 = 450$	$N_2 = 417$

This is all the information we will need to conduct a test of the null hypothesis (H_0: $P_{u1} = P_{u2}$). The estimate of the population proportion is

$$P_u = \frac{N_1 P_{s1} + N_2 P_{s2}}{N_1 + N_2}$$

$$P_u = \frac{(450)(0.35) + (417)(0.32)}{450 + 417}$$

$$P_u = \frac{290.94}{867}$$

$$P_u = 0.34$$

The standard deviation of the sampling distribution would be

$$\sigma_{p-p} = \sqrt{P_u(1 - P_u)} \sqrt{\frac{N_1 + N_2}{N_1 N_2}}$$

$$\sigma_{p-p} = \sqrt{(0.34)(0.66)} \sqrt{\frac{450 + 417}{(450)(417)}}$$

$$\sigma_{p-p} = \sqrt{0.2244} \sqrt{\frac{867}{187650}}$$

$$\sigma_{p-p} = \sqrt{0.2244} \sqrt{0.0046}$$

$$\sigma_{p-p} = (0.47)(0.068)$$

$$\sigma_{p-p} = 0.032$$

The test statistic, Z (obtained), is

$$Z \text{ (obtained)} = \frac{P_{s1} - P_{s2}}{\sigma_{p-p}}$$

$$Z \text{ (obtained)} = \frac{0.35 - 0.32}{0.032}$$

$$Z \text{ (obtained)} = \frac{0.030}{0.032}$$

$$Z \text{ (obtained)} = 0.94$$

If alpha is set at the 0.05 level, the critical region would begin at Z (critical) = ± 1.96. With an obtained Z score of 0.94, we would fail to reject the null hypothesis. There is no statistically significant difference between males and females on attitudes toward premarital sex.

---

elderly? The proportion of each group classified as "high" in participation and sample size for both groups is reported below.

Sample 1 Black senior citizens	Sample 2 White senior citizens
$P_{s1} = 0.34$	$P_{s2} = 0.25$
$N_1 = 83$	$N_2 = 103$

Step 1. Making Assumptions.

Model: Independent random samples
Level of measurement is nominal
Sampling distribution is normal

Step 2. Stating the Null Hypothesis. Since no direction has been predicted, this will be a two-tailed test.

$$H_0: P_{u1} = P_{u2}$$
$$(H_1: P_{u1} \neq P_{u2})$$

Step 3. Selecting the Sampling Distribution and Establishing the Critical Region. Since sample size is large, the Z distribution will be used to establish the critical region. Setting alpha at 0.05, we have

$$\text{Sampling distribution} = Z \text{ distribution}$$
$$\text{Alpha} = 0.05, \text{two-tailed}$$
$$Z \text{(critical)} = \pm 1.96$$

Step 4. Computing the Test Statistic. Remember to begin with the formula for estimating P_u (Formula 9.10), substitute the resultant value in the formula for σ_{p-p} (Formula 9.9), and then solve for Z (obtained) by Formula 9.8.

$$P_u = \frac{N_1 P_{s1} + N_2 P_{s2}}{N_1 + N_2}$$

$$P_u = \frac{(83)(.34) + (103)(.25)}{83 + 103}$$

$$P_u = .29$$

$$\sigma_{p-p} = \sqrt{P_u(1 - P_U)} \sqrt{\frac{N_1 + N_2}{N_1 N_2}}$$

$$\sigma_{p-p} = \sqrt{(.29)(.71)} \sqrt{\frac{(83 + 103)}{(83)(103)}}$$

$$\sigma_{p-p} = (.45)(.15)$$

$$\sigma_{p-p} = .07$$

$$Z \text{(obtained)} = \frac{(P_{s1} - P_{s2})}{\sigma_{p-p}}$$

$$Z \text{(obtained)} = \frac{.34 - .25}{.07}$$

$$Z \text{(obtained)} = 1.29$$

Step 5. Making a Decision. Since the test statistic, Z (obtained) = 1.29, does not fall into the critical region as marked by the Z (critical) of ± 1.96, we fail to reject the null hypothesis. The difference between the sample proportions is no greater than what would be expected if the null hypothesis were true and only random chance were operating. Black and white senior citizens are not significantly different in terms of participation patterns as measured in this test. *(For practice in testing the significance of the difference between sample proportions, see problems 9.10 to 9.14 and 9.15 a to c.)*

READING STATISTICS 6: Hypothesis Testing

Professional researchers use a vocabulary that is much terser than ours when presenting the results of tests of significance. This is partly because of space limitations in scientific journals and partly because professional researchers can assume a certain level of statistical literacy in their audiences. Thus, they omit many of the elements—such as the null hypothesis or the critical region—that we have been so careful to state.

Instead, researchers report only the sample values (for example, means or proportion), the value of the test statistic (for example, a Z or t score), the alpha level, the degrees of freedom (if applicable), and sample size. The results of the example problem in Section 9.3 might be reported in the professional literature as "the difference between the sample means of 2.37 (suburban families) and 2.78 (center-city families) was tested and found to be significant ($t = -2.16$, df = 77, $p < 0.05$)." Note that the alpha level is reported as "$p < 0.05$." This is shorthand for "the probability of a difference of this magnitude occurring by chance alone, if the null hypothesis of no difference is true, is less than 0.05" and is a good illustration of how researchers can convey a great deal of information in just a few symbols. In a similar fashion, our somewhat long-winded phrase "the test statistic falls in the critical region and, therefore, the null hypothesis is rejected" is rendered tersely and simply: "the difference . . . was . . . found to be significant."

When researchers need to report the results of many tests of significance, they will often use a summary table to report the sample information and whether the difference is significant at a certain alpha level. If you read the researcher's description and analysis of such tables, you should have little difficulty interpreting and understanding them. As a final note, these comments about how significance tests are reported in the literature apply to all of the tests of hypotheses covered in Part II of this text.

Statistics in the Professional Literature

Sociologists John Bartkowski and Bradford Wilcox use data from a sample representative of the U.S. population to study parental discipline in conservative Protestant families. They note that there are two views of such families. The more common view, among both researchers and the general public, is that these families are strongly authoritarian and stress harsh punishment, strict discipline, frequent spanking, and arbitrary assertions of power. An alternate view, less widespread, is that child rearing in conservative Protestant families stresses "positive parental emotional work." Discipline may be strict and physical punishment may be frequent, but the parents are nurturing, frequently express their love and affection for their children, and are unlikely to belittle or intimidate their child with verbal abuse.

Bartkowski and Wilcox investigated the accuracy of these views focusing, in particular, on the frequency of parental yelling. Excessive yelling, along with other severe disciplinary practices, have been linked to a variety of negative developmental outcomes for children, including antisocial behavior and poor school performance. If the more common view of conservative Protestant families mentioned above is accurate, the use of verbal intimidation should be common.

The authors measured the extent of parental yelling by averaging scores from a four-point scale in which parents ranked their frequency of yelling as never (1), seldom (2), sometimes (3), and often (4). The results of their tests comparing conservative and non-conservative Protestant families are presented in the table that follows for families with preschool children and school-age (5−18)

(continued)

READING STATISTICS 6: *(continued)*

Religious Differences in Parental Yelling Means

	Preschool Children	School Age Children
Conservative Protestant	$\overline{X} = 2.29^*$ $s = 0.84$	$\overline{X} = 2.56^{**}$ $s = 0.81$
Nonconservative Protestant	$\overline{X} = 2.40^*$ $s = 0.86$	$\overline{X} = 2.68^{**}$ $s = 0.79$
	Total $N = 1,051$	3,199

$^* p < .10$ $^{**} p < .05$

children. The authors argue that these results support the second, less popular view of conservative Protestant families.

Mean differences in yelling among parents of preschool children (column 1) are marginally significant ($p < .10$), such that parents who affiliate with conservative Protestant denominations . . . report lower rates of verbal reproof. We note more robust mean differences ($p < .05$) in reported rates of parental yelling among caregivers of school-age children . . .

The authors go on to suggest that sociologists need to rethink their view of conservative Protestant families and that contextual factors—the circumstances under which discipline is administered—may mediate the relationship between disciplinary tactics (such as spanking) and child development outcomes.

Bartkowski, J. P., and Wilcox, W. B. 2000. "Conservative Protestant Child Discipline: The Case of Parental Yelling." *Social Forces:* 79: 265–290.

9.5 THE LIMITATIONS OF HYPOTHESIS TESTING: SIGNIFICANCE VERSUS IMPORTANCE

Given that we are usually interested in rejecting the null hypothesis, we should take a moment to consider systematically the factors that affect our decision in step 5. Generally speaking, the probability of rejecting the null hypothesis is a function of four independent factors:

1. The size of the observed difference(s)
2. The alpha level
3. The use of one- or two-tailed tests
4. The size of the sample

Only the first of these four is not under the direct control of the researcher. The size of the difference (either between the sample outcome and the population value or between two sample outcomes) is a function, in part, of the testing procedures (that is, how variables are measured), but for the most part it should reflect the underlying realities we are trying to probe.

The relationship between alpha level and the probability of rejection is straightforward. The higher the alpha level, the larger the critical region, the higher the percentage of all possible sample outcomes that fall in the critical region, and the greater the probability of rejection. Thus, it is easier to reject the H_0 at the 0.05 level than at the 0.01 level, and easier still at the 0.10

level. The danger here, of course, is that higher alpha levels will lead to more frequent Type I errors, and we might find ourselves declaring small differences to be statistically significant. In similar fashion, using a one-tailed test will increase the probability of rejection (assuming that the proper direction has been predicted).

The final factor is sample size: the larger the sample, with all other factors constant, the higher the probability of rejecting H_0. This may appear to be a surprising relationship, but the reasons for it can be appreciated with a brief consideration of the formulas used to compute test statistics in step 4. In all these formulas, sample size (N) is in the "denominator of the denominator." Algebraically, this is equivalent to being in the numerator of the formula and means that the value of the test statistic is directly proportional to N and that the two will increase together. To illustrate, consider Table 9.1, which shows the value of the test statistic for single sample means from samples of various sizes. The value of the test statistic, Z (obtained), increases as N increases even though none of the other terms in the formula changes. This pattern of higher probabilities for rejecting H_0 with larger samples holds for all tests of significance.

TABLE 9.1   TEST STATISTICS FOR SINGLE-SAMPLE MEANS COMPUTED FROM SAMPLES OF VARIOUS SIZES ($\overline{X} = 80$, $\mu = 79$, $s = 5$ throughout)

Sample Size	Test Statistic, Z (Obtained)
100	1.99
200	2.82
500	4.47

On one hand, the relationship between sample size and the probability of rejecting the null should not alarm us unduly. Larger samples are, after all, better approximations of the populations they represent. Thus, decisions based on larger samples can be regarded as more trustworthy than decisions based on small samples.

On the other hand, this relationship clearly underlines what is perhaps the most significant limitation of hypothesis testing. Simply because a difference is statistically significant does not guarantee that it is important in any other sense. Particularly with very large samples (say, N's in excess of 1000), relatively small differences may be statistically significant. Even with small samples, of course, differences that are otherwise trivial or uninteresting may be statistically significant. The crucial point is that statistical significance and theoretical or practical importance can be two very different things. Statistical significance is a necessary but not sufficient condition for theoretical or practical importance. A difference that is not statistically significant is almost certainly unimportant. However, significance by itself does not guarantee importance. Even when it is clear that the research results were not produced by random chance, the researcher must still assess their importance. Do they firmly support a theory or hypothesis? Are they clearly consistent with a prediction or analysis? Do they strongly indicate a line of action in solving some problem? These are the kinds of questions a researcher must ask when assessing the importance of the results of a statistical test.

Also, we should note that researchers have access to some very powerful ways of analyzing the importance (vs. the statistical significance) of research results. These statistics, including bivariate measures of association and multivariate statistical techniques, will be introduced in Parts III and IV of this text.

9.6 INTERPRETING STATISTICS: ARE THERE SIGNIFICANT DIFFERENCES IN INCOME BETWEEN MEN AND WOMEN?

In the United States, as in many other nations around the globe, concerted efforts have been made to equalize working conditions for men and women. How successful have these efforts been? Do significant differences in the earnings of men and women persist? Is there a "gender gap" in income?

Note that we could answer these questions with absolute certainty only if we knew the earnings of every single man and woman in the United States, information that not even the census bureau has. (The Internal Revenue Service has the necessary information, but their records are confidential.) Because the data are not available for the population, we will have to investigate the relationship between gender and income by using statistics calculated on randomly selected samples of men and women. If the difference between the samples of men and women are large enough, we can infer that there is a difference in average salaries between men and women in the population. The General Social Survey or GSS (see Appendix G) is given to randomly selected samples of Americans and will be used as a basis for this test.

Before conducting the test, we need to deal with several issues. First and more importantly, the GSS measures personal income with a series of categories rather than in actual dollars. In other words, respondents were asked to choose from a list of salary ranges (e.g., $10,000 to $12,500, 12,501 to 15,000, and so forth. See the "income98" variable in Appendix G for the complete scale). Thus, income is measured at the ordinal level and we need an interval-ratio dependent variable for the ANOVA test. To deal with this problem, we can convert the variable to a form that more closely approximates interval-ratio data by substituting the midpoints for each of the salary intervals. For example, instead of working with the interval "12,501 to 15,000," we would let the midpoint (13,750) stand for the interval. While this technique makes income a more numerical variable, it also creates possible inaccuracies. Estimates to the population parameters from this modified variable should be treated only as gross approximations. However, since we are concerned only with the size of the difference in income rather than actual income itself, we should still be able to come to some conclusions.

Second, we will restrict the comparison to respondents who are employed full time. This will eliminate any differences in average incomes that are the result of different employment patterns (e.g., if members of one gender are more likely to be employed part time).

Once these adjustments have been made, we can test for the significance of the difference in income. The sample information calculated from the 1998 GSS database is

Males	Females
$\overline{X}_1 = 38{,}603$	$\overline{X}_2 = 28{,}261$
$s_1 = 24{,}786$	$s_2 = 18{,}924$
$N_1 = 724$	$N_2 = 666$

It appears that there is a large gender gap and that, on the average, males earn about $10,000 a year more than females. Is the difference in sample means significant? Could it have occurred by random chance? Since we are dealing with large samples, we use the test of significance described in Section 9.2. The null hypothesis is that, in the population, males and females have the same average salary ($\mu_1 = \mu_2$). We will skip the customary trip through the five-step model and simply report that the Z (obtained) calculated with Formula 9.2 (step 4) is 8.78, much larger than the customary Z (critical) score of ± 1.96 associated with an alpha level of 0.05. This difference in sample means is so large that we can reject the null and conclude (with a probability of error of 5%) that the population means are different: males earn significantly more than females.

Is the gender gap in income due to differences in levels of education? If females were significantly less educated than males, this would account for at least some of the difference in income. Using the 1998 General Social Survey again to get information on years of education for random samples of males and females, we have

Males	Females
$\overline{X}_1 = 13.36$	$\overline{X}_2 = 13.17$
$s_1 = 2.97$	$s_2 = 2.90$
$N_1 = 1225$	$N_2 = 1595$

We can see that both males and females average more than 13 years of schooling (one year more than a high school education) and that the average for males is higher than the average for females. The difference seems quite small (at least compared to the difference in income reported above) and, sure enough, the test statistic computed in step 4—Z (obtained) $= 1.70$ —does not justify rejection of the null hypothesis at the .05 level. Thus, males and females have essentially equal levels of schooling in the population and this factor cannot account for the differences in income.

These results suggest that females are getting a lower return in income for their education. Although they are essentially equal to males in preparation for work and careers (at least as measured by level of schooling), a significant difference in income persists. Could it be that the income differential is maintained, in part, because men and women pursue different kinds of careers and jobs? It is often argued that men tend to dominate the more lucrative, higher-prestige occupations (lawyer, doctor) while women are concentrated in jobs (elementary school teacher, nurse) that have lower levels of remuneration. Thus, it could be that a form of "occupational segregation" perpetuates the gender gap in income.

We can test this idea with data collected by the Bureau of Labor Statistics of the United States Department of Labor. Every March, the Bureau collects information on the median weekly earnings for over 300 occupations.* From these data, we can calculate which occupations are predominantly male (over half of the work force is male), and which are predominantly female. A random sample of 120 occupations was selected from the list of over 300 jobs and used to test the hypothesis that "male" and "female" occupations are significantly different in earnings. The dependent variable is the median weekly earning for each occupation. In this situation, we treat the median weekly incomes as scores (X_i) and use them to compute our sample means and standard deviations. The null hypothesis ($\mu_1 = \mu_2$) is that there is no difference between the average earnings of *the population* of predominantly male occupations and *the population* of predominantly female occupations.

The descriptive statistics for the two groups of occupations are

Predominately Male Occupations	Predominately Female Occupations
$\overline{X}_1 = 620.23$	$\overline{X}_2 = 488.09$
$s_1 = 206.59$	$s_2 = 179.70$
$N_1 = 77$	$N_2 = 43$

We can see from the sample statistics that the difference in earning is about $132 a week or almost $7000 per year. The Z (obtained) calculated from these statistics is 3.62, greater than the usual Z (critical) of ± 1.96. We can reject the null hypothesis and conclude that the difference is statistically significant. Occupations that are predominantly male have significantly higher weekly earnings than occupations that are predominantly female. These results support the idea that the gender gap in income is sustained, in part, by the fact that women are more likely to be employed in occupations with lower earnings. This difference in income will persist until the genders are more evenly spread across occupations or until salaries are equalized in some other way.

SUMMARY

1. A common research situation is to test for the significance of the difference between two populations. Sample statistics are calculated for random samples of each population, and then we test for the significance of the difference between the samples as a way of inferring differences between the specified populations.

2. When sample information is summarized in the form of sample means, and N is large, the Z distribution is used to find the critical region. When N is

---

*Current Population Survey (CPS), "Educational Attainment in the United States: March 1999," *Employment and Earnings,* August 2000.

small, the t distribution is used to establish the critical region. In the latter circumstance, we must also assume equal population variances before forming a pooled estimate of the standard deviation of the sampling distribution.

3. Differences in sample proportions may also be tested for significance. For large samples, the Z distribution is used to find the critical region.

4. In all tests of hypothesis, a number of factors affect the probability of rejecting the null: the size of the

difference, the alpha level, the use of one- versus two-tailed tests, and sample size. Statistical significance is not the same thing as theoretical or practical importance. Even after a difference is found to be statistically significant, the researcher must still demonstrate the relevance or importance of his or her findings. The statistics presented in Parts III and IV of this text will give us the tools we need to deal directly with issues beyond statistical significance.

SUMMARY OF FORMULAS

Test statistic for two sample means, large samples:

9.1 $\qquad Z \text{(obtained)} = \dfrac{(\overline{X}_1 - \overline{X}_2) - (\mu_1 - \mu_2)}{\sigma_{\overline{x} - \overline{x}}}$

Test statistic for two sample means, large samples (simplified formula):

9.2 $\qquad Z \text{(obtained)} = \dfrac{(\overline{X}_1 - \overline{X}_2)}{\sigma_{\overline{x} - \overline{x}}}$

Standard deviation of the sampling distribution of the difference in sample means, large samples:

9.3 $\qquad \sigma_{\overline{x} - \overline{x}} = \sqrt{\dfrac{\sigma_1^2}{N_1} + \dfrac{\sigma_2^2}{N_2}}$

Pooled estimate of the standard deviation of the sampling distribution of the difference in sample means, large samples:

9.4 $\qquad \sigma_{\overline{x} - \overline{x}} = \sqrt{\dfrac{s_1^2}{N_1 - 1} + \dfrac{s_2^2}{N_2 - 1}}$

Pooled estimate of the standard deviation of the sampling distribution of the difference in sample means, small samples:

9.5 $\qquad \sigma_{\overline{x} - \overline{x}} = \sqrt{\dfrac{N_1 s_1^2 + N_2 s_2^2}{N_1 + N_2 - 2}} \sqrt{\dfrac{N_1 + N_2}{N_1 N_2}}$

Test statistic for two sample means, small samples:

9.6 $\qquad t \text{(obtained)} = \dfrac{(\overline{X}_1 - \overline{X}_2)}{\sigma_{\overline{x} - \overline{x}}}$

Test statistic for two sample proportions, large samples:

9.7 $\qquad Z \text{(obtained)} = \dfrac{(P_{s1} - P_{s2}) - (P_{u1} - P_{u2})}{\sigma_{p-p}}$

Test statistic for two sample proportions, large samples (simplified formula):

9.8 $\qquad Z \text{(obtained)} = \dfrac{(P_{s1} - P_{s2})}{\sigma_{p-p}}$

Standard deviation of the sampling distribution of the difference in sample proportions, large samples:

9.9 $\qquad \sigma_{p-p} = \sqrt{P_u(1 - P_u)} \sqrt{(N_1 + N_2)/N_1 N_2}$

Pooled estimate of population proportion, large samples:

9.10 $\qquad P_u = \dfrac{N_1 P_{s1} + N_2 P_{s2}}{N_1 + N_2}$

GLOSSARY

Independent random samples. Random samples gathered in such a way that the selection of a particular case for one sample has no effect on the

probability that any other particular case will be selected for the other samples.

Pooled estimate. An estimate of the standard deviation of the sampling distribution of the difference in sample means based on the standard deviations of both samples.

σ_{p-p}. Symbol for the standard deviation of the sampling distribution of the differences in sample proportions.

$\sigma_{\bar{x}-\bar{x}}$. Symbol for the standard deviation of the sampling distribution of the differences in sample means.

MULTIMEDIA RESOURCES

The Wadsworth Sociology Resource Center: Virtual Society
http://sociology.wadsworth.com/

Visit the companion web site for the sixth edition of *Statistics: A Tool for Social Research* to access a wide range of student resources. Begin by clicking on the Student Resources section of the book's web site to access the following study tools:

- Basic math review
- Statistics review

- Flash cards
- Internet links
- Additional chapter problems
- Table of random numbers
- MicroCase and SPSS examples and exercises
- "Find the text" flowcharts
- Hypothesis testing for variables measured at the ordinal level

PROBLEMS

9.1 For each problem below, test for the significance of the difference in sample statistics using the five-step model. *(HINT: remember to solve Formula 9.4 for $\sigma_{\bar{x}-\bar{x}}$ before attempting to solve Formula 9.2. Also, in Formula 9.4, perform the mathematical operations in the proper sequence. First square each sample standard deviation, then divide by $N - 1$, add the resultant values, and then find the square root of the sum.)*

a. Sample 1   Sample 2

Sample 1	Sample 2
$\overline{X}_1 = 72.5$	$\overline{X}_2 = 76.0$
$s_1 = 14.3$	$s_2 = 10.2$
$N_1 = 136$	$N_2 = 257$

b. Sample 1   Sample 2

Sample 1	Sample 2
$\overline{X}_1 = 107$	$\overline{X}_2 = 103$
$s_1 = 14$	$s_2 = 17$
$N_1 = 175$	$N_2 = 200$

9.2 SOC Gessner and Healey administered questionnaires to samples of undergraduates. Among other things, the questionnaires contained a scale that measured attitudes toward interpersonal violence (higher scores indicate greater approval of interpersonal violence). Test the results as reported below for sexual, racial, and social-class differences.

a. Males   Females

Males	Females
$\overline{X}_1 = 2.99$	$\overline{X}_2 = 2.29$
$s_1 = 0.88$	$s_2 = 0.91$
$N_1 = 122$	$N_2 = 251$

b. Blacks   Whites

Blacks	Whites
$\overline{X}_1 = 2.76$	$\overline{X}_2 = 2.49$
$s_1 = 0.68$	$s_2 = 0.91$
$N_1 = 43$	$N_2 = 304$

c. White Collar   Blue Collar

White Collar	Blue Collar
$\overline{X}_1 = 2.46$	$\overline{X}_2 = 2.67$
$s_1 = 0.91$	$s_2 = 0.87$
$N_1 = 249$	$N_2 = 97$

d. Summarize your results in terms of the significance and the direction of the differences. Which of these three factors seems to make the biggest difference in attitudes toward interpersonal violence?

9.3 | SOC | Do athletes in different sports vary in terms of intelligence? Below are reported College Board scores of random samples of college basketball and football players. Is there a significant difference? Write a sentence or two explaining the difference.

Basketball Players	Football Players
$\overline{X}_1 = 460$	$\overline{X}_2 = 442$
$s_1 = 92$	$s_2 = 57$
$N_1 = 102$	$N_2 = 117$

What about male and female college athletes?

Male	Female
$\overline{X}_1 = 452$	$\overline{X}_2 = 480$
$s_1 = 88$	$s_2 = 75$
$N_1 = 107$	$N_2 = 105$

9.4 | PA | A number of years ago, the fire department in Shinbone, Kansas, began recruiting minority group members through an affirmative action program. In terms of efficiency ratings as compiled by their superiors, how do the affirmative action employees rate? The ratings of random samples of both groups were collected, and the results are reported below (higher ratings indicate greater efficiency).

Affirmative Action	Regular
$\overline{X}_1 = 15.2$	$\overline{X}_2 = 15.5$
$s_1 = 3.9$	$s_2 = 2.0$
$N_1 = 97$	$N_2 = 100$

Write a sentence or two of interpretation.

9.5 | SOC | Are middle-class families more likely than working-class families to maintain contact with kin? Write a paragraph summarizing the results of these tests.
  a. A sample of middle-class families reported an average of 7.3 visits per year with close kin while a sample of working-class families averaged 8.2 visits. Is the difference significant?

Middle Class	Working Class
$\overline{X}_1 = 7.3$	$\overline{X}_2 = 8.2$
$s_1 = 0.3$	$s_2 = 0.5$
$N_1 = 89$	$N_2 = 55$

  b. The middle-class families averaged 2.3 phone calls and 8.7 email messages per month with close kin. The working-class families averaged 2.7 calls and 5.7 email messages per month. Are these differences significant?

Phone Calls	
Middle Class	Working Class
$\overline{X}_1 = 2.3$	$\overline{X}_2 = 2.7$
$s_1 = 0.5$	$s_2 = 0.8$
$N_1 = 89$	$N_2 = 55$

Email Messages	
Middle Class	Working Class
$\overline{X}_1 = 8.7$	$\overline{X}_2 = 5.7$
$s_1 = 0.3$	$s_2 = 1.1$
$N_1 = 89$	$N_2 = 55$

9.6 | SOC | Are college students who live in dormitories significantly more involved in campus life than students who commute to campus? The data below report the average number of hours per week students devote to extracurricular activities. Is the difference between these randomly selected samples of commuter and residential students significant?

Residential	Commuter
$\overline{X}_1 = 12.4$	$\overline{X}_2 = 10.2$
$s_1 = 2.0$	$s_2 = 1.9$
$N_1 = 158$	$N_2 = 173$

9.7 | SOC | Are senior citizens who live in retirement communities more socially active than those who live in age-integrated communities? Write a sentence or two explaining the results of these tests. (*HINT: remember to use the proper formulas for small sample sizes.*)
  a. A random sample of senior citizens living in a retirement village reported that they had an average of 1.42 face-to-face interactions per day with their neighbors. A random sample of those living in age-integrated communities reported 1.58 interactions. Is the difference significant?

Retirement Community	Age-integrated Neighborhood
$\overline{X}_1 = 1.42$	$\overline{X}_2 = 1.58$
$s_1 = 0.10$	$s_2 = 0.78$
$N_1 = 43$	$N_2 = 37$

b. Senior citizens living in the retirement village reported that they had 7.43 telephone calls with friends and relatives each week while those in the age-integrated communities reported 5.50 calls. Is the difference significant?

Retirement Community	Age-integrated Neighborhood
$\overline{X}_1 = 7.43$	$\overline{X}_2 = 5.50$
$s_1 = 0.75$	$s_2 = 0.25$
$N_1 = 43$	$N_2 = 37$

9.8 $\boxed{\text{SW}}$ As the director of the local Boys Club, you have claimed for years that membership in your club reduces juvenile delinquency. Now, a cynical member of your funding agency has demanded proof of your claim. Fortunately, your local sociology department is on your side and springs to your aid with student assistants, computers, and hand calculators at the ready. Random samples of members and nonmembers are gathered and interviewed with respect to their involvement in delinquent activities. Each respondent is asked to enumerate the number of delinquent acts he has engaged in over the past year. The results are in and reported below (the average number of admitted acts of delinquency). What can you tell the funding agency?

Members	Nonmembers
$\overline{X}_1 = 10.3$	$\overline{X}_2 = 12.3$
$s_1 = 2.7$	$s_2 = 4.2$
$N_1 = 40$	$N_2 = 55$

9.9 $\boxed{\text{SOC}}$ Are professional basketball teams significantly more oriented toward offense these days? Do they score more points, on the average, than teams in past years? Games are randomly selected from the most recent season and from 10 years ago. Is the difference in average points per game significant?

This Season	10 Years Ago
$\overline{X}_1 = 110$	$\overline{X}_2 = 102$
$s_1 = 10$	$s_2 = 9$
$N_1 = 35$	$N_2 = 37$

9.10 For each problem, test the sample statistics for the significance of the difference. *(HINT: in test-ing proportions, remember to begin with Formula 9.10, then solve Formulas 9.9 and 9.8.)*

a.

Sample 1	Sample 2
$P_{s1} = 0.17$	$P_{s2} = 0.20$
$N_1 = 101$	$N_2 = 114$

b.

Sample 1	Sample 2
$P_{s1} = 0.62$	$P_{s2} = 0.60$
$N_1 = 532$	$N_2 = 478$

9.11 $\boxed{\text{CJ}}$ About half of the police officers in Shinbone, Kansas, have completed a special course in investigative procedures. Has the course increased their efficiency in clearing crimes by arrest? The proportions of cases cleared by arrest for samples of trained and untrained officers are reported below.

Trained	Untrained
$P_{s1} = 0.47$	$P_{s2} = 0.43$
$N_1 = 157$	$N_2 = 113$

9.12 $\boxed{\text{SW}}$ A large counseling center needs to evaluate several experimental programs. Write a paragraph summarizing the results of these tests. Did the new programs work?

a. One program is designed for divorce counseling; the key feature of the program is its counselors, who are married couples working in teams. About half of all clients have been randomly assigned to this special program and half to the regular program, and the proportion of cases that eventually ended in divorce was recorded for both. The results for random samples of couples from both programs are reported below. In terms of preventing divorce, did the new program work?

Special Program	Regular Program
$P_{s1} = 0.53$	$P_{s2} = 0.59$
$N_1 = 78$	$N_2 = 82$

b. The agency is also experimenting with peer counseling for depressed children. About half of all clients were randomly assigned to peer counseling. After the program had run for a year, a random sample of children from the new program were compared with a random sample of children who did not receive peer counseling. In terms of the percentage who were judged to be "much improved," did the new program work?

Peer Counseling	No Peer Counseling
$P_{s1} = 0.10$	$P_{s2} = 0.15$
$N_1 = 52$	$N_2 = 56$

9.13 $\boxed{\text{SOC}}$ At St. Algebra College, the sociology and psychology departments have been feuding for years about the respective quality of their programs. In an attempt to resolve the dispute, you have gathered data about the graduate school experience of random samples of both groups of majors. The results are presented below: the proportion of majors who applied to graduate schools, the proportion of majors accepted into their preferred programs, and the proportion of these who completed their programs. As measured by these data, is there a significant difference in program quality?

a. Proportion of majors who applied to graduate school:

Sociology	Psychology
$P_{s1} = 0.53$	$P_{s2} = 0.40$
$N_1 = 150$	$N_2 = 175$

b. Proportion accepted by program of first choice:

Sociology	Psychology
$P_{s1} = 0.75$	$P_{s2} = 0.85$
$N_1 = 80$	$N_2 = 70$

c. Proportion completing the programs:

Sociology	Psychology
$P_{s1} = 0.75$	$P_{s2} = 0.69$
$N_1 = 60$	$N_2 = 60$

9.14 $\boxed{\text{CJ}}$ The local police chief started a "crimeline" program some years ago and wonders if it's really working. The program publicizes unsolved violent crimes in the local media and offers cash rewards for information leading to arrests. Are "featured" crimes more likely to be cleared by arrest than other violent crimes? Results from random samples of both types of crimes are reported as follows:

Crimeline Crimes Cleared by Arrest	Noncrimeline Crimes Cleared by Arrest
$P_{s1} = 0.35$	$P_{s2} = 0.25$
$N_1 = 178$	$N_2 = 212$

9.15 $\boxed{\text{SOC}}$ Some results from the 1998 General Social Survey are reported below in terms of differences by sex. The actual questions are listed in Appendix G. Which of these differences, if any, are significant? Write a sentence or two of interpretation for each test.

a. Proportion favoring gun control (GUNLAW):

Males	Females
$P_{s1} = 0.77$	$P_{s2} = 0.89$
$N_1 = 404$	$N_2 = 513$

b. Proportion agreeing that premarital sex is "always wrong" (PREMARSX):

Males	Females
$P_{s1} = 0.21$	$P_{s2} = 0.29$
$N_1 = 389$	$N_2 = 484$

c. Proportion voting for President Clinton in 1996 (PRES92):

Males	Females
$P_{s1} = 0.46$	$P_{s2} = 0.59$
$N_1 = 372$	$N_2 = 458$

d. Average occupational prestige (PRESTG80):

Males	Females
$\overline{X}_1 = 43.78$	$\overline{X}_2 = 43.95$
$s_1 = 13.41$	$s_2 = 13.62$
$N_1 = 605$	$N_2 = 713$

e. Average rate of church attendance (ATTEND):

Males	Females
$\overline{X}_1 = 3.20$	$\overline{X}_2 = 4.02$
$s_1 = 2.67$	$s_2 = 2.76$
$N_1 = 610$	$N_2 = 755$

f. Number of children (CHILDS):

Males	Females
$\overline{X}_1 = 1.63$	$\overline{X}_2 = 1.95$
$s_1 = 1.67$	$s_2 = 1.66$
$N_1 = 620$	$N_2 = 765$

Using *SPSS for Windows* to Test the Significance of the Difference Between Two Means

SPSS DEMONSTRATION 9.1 Do Men or Women Watch More TV?

SPSS for Windows includes several tests for the significance of the difference between means. In this demonstration, we'll use the **Independent-Samples T Test,** the test we covered in Sections 9.2 and 9.3, to test for the significance of the difference between men and women in average hours of reported TV watching. If there is a statistically significant difference between the sample means for men and women, we can conclude that the populations (all U.S. adult men and women) are different on this variable.

Start *SPSS for Windows* and load the 1998 GSS database. From the main menu bar, click **Analyze,** then **Compare Means,** and then **Independent-Samples T Test.** The **Independent-Samples T Test** dialog box will open with the usual list of variables on the left. Find and move the cursor over *tvhours* (the variable label is HOURS PER DAY WATCHING TV) and click the top arrow in the middle of the window to move *tvhours* to the **Test Variable(s)** box. Next, find and highlight *sex* (the label is RESPONDENTS SEX) and click the bottom arrow in the middle of the window to move *sex* to the **Grouping Variable** box. Two question marks will appear in the **Grouping Variable** box, and the **Define Groups** button will become active. SPSS needs to know which cases go in which groups and, in the case at hand, the instructions we need to supply are straightforward. Males (indicated by a score of 1 on *sex*) go into group 1 and females (a score of 2) will go into group 2.

Click the **Define Groups** button, and the **Define Groups** window will appear. The cursor will be blinking in the box beside Group 1—SPSS is asking for the score that will determine which cases go into this group. Type a 1 in this box (for males) and then click the box next to Group 2 and type a 2 (for females). Click **Continue** to return to the **Independent-Samples T Test** window and click **OK** and the following output will be produced. (The 95% confidence interval automatically produced by this program has been deleted to conserve space.)

Group Statistics

	RESPONDENTS SEX	N	Mean	Std. Deviation	Std. Error Mean
HOURS PER DAY WATCHING TV	MALE	502	2.84	2.22	9.90E-02
	FEMALE	632	2.87	2.18	8.67E-02

Independent Samples Test

		Levene's Test for Equality of Variances		t-test for Equality of Means				
		F	Sig.	t	df	Sig. (2-tailed)	Mean Difference	Std. Error Difference
HOURS PER DAY WATCHING TV	Equal variances assumed	.010	.919	-.225	1132	.822	-2.96E-02	.13
	Equal variances not assumed			-.225	1066.134	.822	-2.96E-02	.13

In the first block of output are some descriptive statistics. There were 502 males in the sample, and they watched TV an average of 2.84 hours a day with a standard deviation of 2.22. The 632 females watched an average of 2.87 hours per day with a standard deviation of 2.18.

We will skip over the first columns of the next block of output (which reports the results of a test for equality of the population variances). The results of the test for significance are reported in this block. *SPSS for Windows* does a separate test for each assumption about the population variance (see Sections 9.2 and 9.3), but we will look only at the 'Equal variances assumed' reported in the top row. This is basically the same model used in this chapter.

SPSS for Windows reports the *t* value (−.225), the degrees of freedom (df = 1132), and the 'Sig. (2-tailed)' (.822). This last piece of information is an alpha level (or a "*p*" level—see Reading Statistics 6) except that it is the *exact* probability of getting the observed difference in sample means if only chance is operating. Thus, there is no need to look up the test statistic in a *t* table. This value is greater than .05, our usual indicator of significance. We will fail to reject the null hypothesis and conclude that the difference is not statistically significant. On the average, men and women do not have significantly different TV-viewing habits.

SPSS DEMONSTRATION 9.2   Using the COMPUTE Command to Test for Gender Differences in Attitudes About Abortion

The **Compute** command was introduced in Demonstration 4.2. To refresh your memory, I used **Compute** to create a summary scale (*abscale*) for attitudes toward abortion by adding the scores on the two constituent items (*abhlth* and *abany*). Remember that, once created, a computed variable is added to the active file and can be used like any of the variables actually recorded in the file. If you did not save the data file with *abscale* included, you can quickly recreate the variable by following the instructions in Demonstration 4.2.

Here, we will test *abscale* for the significance of the difference by gender. Our question is: Do men and women have different attitudes toward abortion? If the difference in the sample means is large enough, we can reject the null hypothesis and conclude that the populations are different. Before we conduct the test, I should point out that the abortion scale used in this test is only ordinal in level of measurement. Scales like this are often treated as interval-ratio variables, but we should still be cautious in interpreting our results.

Follow the instructions in Demonstration 9.1 for using the **Independent-Samples T TEST** command, and move *abscale* rather than *tvhours* to the **Test Variable(s)** box. If necessary, repeat the procedure for making *sex* the grouping variable. Your output will be as shown:

Group Statistics

	RESPONDENTS SEX	N	Mean	Std. Deviation	Std. Error Mean
ABINDEX	MALE	378	2.6349	.6383	3.283E-02
	FEMALE	477	2.7275	.6990	3.200E-02

Independent Samples Test

		Levene's Test for Equality of Variances		t-test for Equality of Means		
		F	Sig.	t	df	Sig. (2-tailed)
ABINDEX	Equal variances assumed	2.221	.137	-1.997	853	.046
	Equal variances not assumed			-2.018	836.010	.044

		95% Confidence Interval of the Difference	
Mean Difference	Std. Error Difference	Lower	Upper
-9.2543E-02	4.633E-02	-.1835	-1.5999E-03
-9.2543E-02	4.585E-02	-.1825	-2.5467E-03

The sample means are quite close in value (2.6349 versus 2.7275), but the test statistic ($t = -1.997$) is (barely) significant since the "Sig. (2-tailed)" value is .046, less than .05, the conventional marker of a significant result. Women are significantly more opposed to legal abortion (at least for the reasons cited in *abany* and *abhlth*) than men.

As a final point, let me direct your attention to the "Number of Cases" column. This test was based on 378 men and 477 women, for a total of 855 people. The original sample included almost 1500 people. What happened to all those cases?

Recall from Appendix F that many of the items on the GSS are given to only about two-thirds of the sample. Furthermore, as I mentioned at the end of Chapter 4, the **Compute** command is designed to drop a case from the computations if it is missing a score on any of the constituent variables. So, most of the "missing cases" were not asked these questions, and the rest did not answer one or both of the constituent abortion items. This phenomenon of diminishing sample size is a common problem in survey research that, at some point, may jeopardize the integrity of the inquiry.

Using MicroCase to Test the Significance of the Difference Between Two Means

MICROCASE DEMONSTRATION 9.1 Do Men or Women Watch More TV?

In this demonstration, we'll use the **t Test** (located on the **Statistics** menu for the student version and the **Basic Statistics** menu for the full version) procedure to test for the significance of the difference between men and women in average hours of reported TV watching. If there is a statistically significant difference between the sample means for men and women, we can conclude that the populations (all U.S. adult men and women) are different on this variable.

Start MicroCase and load the 1998 GSS database. Click **t Test,** and the **t Test** dialog box will open with the usual list of variables on the left. Find and move the cursor over *tvhours* in the list (variable #42) and click the top arrow to move *tvhours* to the **Dependent Variable** box. Next, find and highlight *sex* (variable #10) and click the arrow to move *sex* to the **Independent Variable** box. Remember that the *t* test requires an independent variable with exactly two categories (e.g., male and female). If you attempt to use an independent variable with more than two categories, MicroCase will issue an error message and refuse to proceed.

When you click **OK,** the first thing you will see is a graph called a "box and whiskers" plot. Our independent variable (*sex*) is arrayed along the horizontal axis and the dependent variable (*tvhours*) along the vertical axis. Each vertical line of dots represents the scores of the males and females ranging from zero hours at the bottom to 21 hours at the top of the graph. The means for each group are noted by vertical lines, as is the overall mean (the mean for the entire sample). In this case the two sample means and the overall mean are nearly the same value, and the line appears to be parallel to the horizontal axis. If there had been larger differences between men and women, the line would be at an angle, a quick visual indicator of a substantial difference between groups.

Note the vertical boxes for each group. These extend one standard deviation above and below the mean of each subgroup and provide a visual indication of the relative dispersion in the groups. In this case, the boxes are nearly identical, underscoring the impression that there is little difference in television viewing between men and women. The dots extending above and below the two boxes are the "whiskers," another visual indication of relative dispersion.

To conduct the *t* test, click on the button next to **Means** in the **Statistics** box at the left of the screen, and the following output will be produced:

```
Means, Standard Deviations and Number of Cases of Dependent Var: TVHOURS
by Categories of Independent Var: SEX
Difference of means across groups is NOT statistically significant (Prob. = 0.822)
            N    Mean   Std.Dev.
MALE      502   2.841     2.219
FEMALE    632   2.870     2.180
```

There were 502 males in the sample, and they watched TV an average of 2.841 hours a day with a standard deviation of 2.219. The 632 females watched an average of 2.870 hours per day with a standard deviation of 2.180. MicroCase does not report the value of the test statistic. Instead, it reports the *exact* probability (Prob. = 0.822) of getting the observed difference in sample means if only chance is operating. This value is greater than .05, our usual indicator of significance, and MicroCase reminds us that the "difference of means across groups is NOT statistically significant." We will fail to reject the null hypothesis and conclude that men and women do not have significantly different TV-viewing habits.

MICROCASE DEMONSTRATION 9.2 Using the SUM RECODE Command to Test for Gender Differences in Attitudes About Abortion

The **Sum Recode** command was introduced in Demonstration 4.2. To refresh your memory, I used **Sum Recode** to create a summary scale (*abscale*) for attitudes toward abortion by adding the scores on the two constituent items (*abhlth* and *abany*). Remember that, once created, a computed variable is added to the active file and can be used like any of the variables actually recorded in the file. If you did not save the data file with *abscale* included, you can quickly recreate the variable by following the instructions in Demonstration 4.2.

Here, we will test *abscale* for the significance of the difference by gender. Our question is: Do men and women have different attitudes toward abortion? If the difference in the sample means is large enough, we can reject the null hypothesis and conclude that the populations are different. Before we conduct the test, I should point out that the abortion scale used in this test is only ordinal in level of measurement. Scales like this are often treated as interval-ratio variables, but we should still be cautious in interpreting our results.

Follow the instructions in MicroCase Demonstration 9.1 for using the **t TEST** command, and move *abscale* (variable #51) rather than *tvhours* to the **Dependent Variable** box. Your independent variable, once again, is *sex*. Click the button next to **Means** in the **Statistics** box at the left of the screen, and the following output will be produced

```
Means, Standard Deviations and Number of Cases of Dependent Var: abscale
by Categories of Independent Var: SEX
Difference of means across groups is statistically significant (Prob. = 0.046)
            N    Mean   Std.Dev.
MALE      378   2.635     0.638
FEMALE    477   2.727     0.699
```

The sample means are quite close in value (2.64 versus 2.73), but the difference is (barely) significant (Prob. = .046), less than .05, the conventional marker of a significant result. Women are significantly more opposed to legal abortion (at least for the reasons cited in *abany* and *abhlth*) than men.

As a final point, let me direct your attention to the "Number of Cases" column. This test was based on 378 men and 477 women, for a total of 855 people. The original sample included almost 1500 people. What happened to all those cases?

Recall from Appendix F that many of the items on the GSS are given to only about two-thirds of the sample. Furthermore, as I mentioned at the end of Chapter 4, the **Sum Recode**command is designed to drop a case from the computations if it is missing a score on any of the constituent variables. So, most of the "missing cases" were not asked these questions, and the rest did not answer one or both of the constituent abortion items. This phenomenon of diminishing sample size is a common problem in survey research and, at some point, may jeopardize the integrity of the inquiry.

Exercises

9.1 Are men significantly different from women in occupational prestige, education, or income? Using Demonstration 9.1 as a guide, substitute *prestg80, educ,* and *income91* for *tvhours*. Write a sentence or two summarizing the results of this test. (See Reading Statistics 6 for some ideas on writing up results.)

9.2 Using Demonstration 9.2 as a guide, construct a summary scale for attitudes about traditional gender roles using *fehelp* and *fefam* in place of *abhlth* and *abany*. Test for the significance of the difference using *sex* as the independent variable. Write a sentence or two summarizing the results of this test.

9.3 What other independent variables might explain differences in opinions about abortion? Select several more independent variables besides *sex* and conduct additional *t* tests with *abscale* as the dependent variable. Remember that the *t* test requires the independent variable to have *only* two categories so your choices in the 1998 GSS data set are quite limited. One option in this situation is to collapse the scores of variables with more than two categories, using the **Recode** command (SPSS) or the **Collapse Categories** command (MicroCase). Use Demonstration 2.3 as a guide and collapse your selected independent variables at their median scores or some other logical point. (SPSS: Don't forget to choose the "recode into different variable" option). Write a sentence or two summarizing the results of these tests.

10

HYPOTHESIS TESTING III
THE ANALYSIS OF VARIANCE

LEARNING OBJECTIVES

By the end of this chapter, you will be able to

1. Identify and cite examples of situations in which ANOVA is appropriate.
2. Explain the logic of hypothesis testing as applied to ANOVA.
3. Perform the ANOVA test, using the five-step model as a guide, and correctly interpret the results.
4. Define and explain the concepts of population variance, total sum of squares, the sum of squares between, and the sum of squares within, mean square estimates, and post hoc tests.
5. Explain the difference between the statistical significance and the importance of relationships between variables.

10.1 INTRODUCTION

In this chapter, we will examine a very flexible and widely used test of significance called the **analysis of variance** (often abbreviated as **ANOVA**). This test can be used in a number of situations where previously discussed tests are less than optimum or entirely inappropriate. ANOVA is designed to be used with interval-ratio-level dependent variables and is a powerful tool for analyzing the most sophisticated and precise measurements you are likely to encounter.

It is perhaps easiest to think of ANOVA as an extension of the t test for the significance of the difference between two sample means, which was presented in Chapter 9. The t test can be used only in situations in which our independent variable has exactly two categories (e.g., Protestants and Catholics). The analysis of variance, on the other hand, is appropriate for independent variables with more than two categories (e.g., Protestants, Catholics, Jews, people with no religious affiliation, and so forth).

To illustrate, suppose we were interested in examining the social basis of support for capital punishment. Why does support for the death penalty vary from person to person? Could there be a relationship between religion (the independent variable) and support for capital punishment (the dependent variable)? Opinion about the death penalty has an obvious moral dimension and should reflect a person's religious background.

Suppose that we administered a scale that measures support for capital punishment at the interval-ratio level to a randomly selected sample that includes Protestants, Catholics, Jews, and people with no religious affiliation ("Nones"). We will have four categories of subjects, and we will want to see if a particular attitude or opinion varies significantly by the category (religion) into which a person is classified. We will also want to answer other questions such as, Which religion shows the least or most support for capital punishment? Are Protestants significantly more supportive than Catholics or Jews? How do people with no religious affiliation compare to people in the other categories? The analysis of variance provides a very useful statistical context in which the questions can be addressed.

10.2 THE LOGIC OF THE ANALYSIS OF VARIANCE

For ANOVA, the null hypothesis is that the populations from which the samples are drawn are equal on the characteristic of interest. As applied to our problem, the null hypothesis could be phrased as "People from different religious denominations do not vary in their support for the death penalty" or, symbolically as $\mu_1 = \mu_2 = \mu_3 = \cdots = \mu_k$. (Note that this is an extended version of the null hypothesis for the two-sample t test). As usual, the researcher will normally be interested in rejecting the null and, in this case, showing that support is related to religion.

If the null hypothesis of "no difference" in the populations is true, then any means calculated from randomly selected samples should be roughly equal in value. The average score for the Protestant sample should be about the same as the average score for the Catholics and the Jews, and so forth. Note that the averages are unlikely to be exactly the same value even if the null hypothesis really is true, since we will always encounter some error or chance fluctuations in the measurement process. We are *not* asking: "are there differences between the samples or categories of the independent variable (or, in our example, the religions)?" Rather, we are asking: "are the differences between the samples large enough to reject the null hypothesis and justify the conclusion that the populations represented by the samples are different?"

Now, consider what kinds of outcomes we might encounter if we actually administered a "Support of Capital Punishment Scale" and organized the scores by religion. Of the infinite variety of possibilities, let's focus on two extreme outcomes as exemplified by Tables 10.1 and 10.2. In the first set of hypothetical results (Table 10.1), we see that the means and standard deviations of the groups are quite similar. The average scores are about the same for every religious group, and all four groups exhibit about the same dispersion. These results would be quite consistent with the null hypothesis of

TABLE 10.1   SUPPORT FOR CAPITAL PUNISHMENT BY RELIGION (fictitious data)

	Protestant	Catholic	Jew	None
Mean =	10.3	11.0	10.1	9.9
Standard Deviation =	2.4	1.9	2.2	1.7

TABLE 10.2   SUPPORT FOR CAPITAL PUNISHMENT BY RELIGION (fictitious data)

	Protestant	Catholic	Jew	None
Mean =	14.7	11.3	5.7	8.3
Standard Deviation =	.5	1.3	.6	.8

no difference. Neither the average score nor the dispersion of the scores changes in any important way by religion.

Now consider another set of fictitious results as displayed in Table 10.2. Here we see substantial differences in average score from category to category, with Jews showing the lowest support and Protestants showing the highest. Also, the standard deviations are low and similar from category to category, indicating that there is not much variation within the religions. Table 10.2 shows marked differences *between* religions combined with homogeneity *within* religions. In other words, there are marked differences from religion to religion but little difference within each religion. These results would contradict the null hypothesis and support the notion that support for the death penalty does vary by religion.

In principle, ANOVA proceeds by making the kinds of comparisons outlined above. The test compares the amount of variation between categories (for example, from Protestants to Catholics to Jews to "Nones") with the amount of variation within categories (among Protestants, among Catholics, and so forth). The greater the differences between categories, relative to the differences within categories, the more likely that the null hypothesis of "no difference" is false and can be rejected. If support for capital punishment truly varies by religion, then the sample mean for each religion should be quite different from the others and dispersion within the categories should be relatively low.

10.3 THE COMPUTATION OF ANOVA

Even though we have been thinking of ANOVA as a test for the significance of the difference between sample means, the computational routine actually involves developing two separate estimates of the population variance, σ^2 (hence the name 'analysis of variance'). Recall from Chapter 4 that the variance and standard deviation both measure dispersion and that the variance is simply the standard deviation squared. One estimate of the population variance is based on the amount of variation within each of the categories of the independent variable and the other is based on the amount of variation between categories.

Before constructing these estimates, we need to introduce some new concepts and statistics. The first new concept is the total variation of the scores, which is measured by a quantity called the **total sum of squares, or SST**

FORMULA 10.1
$$SST = \Sigma(X_i - \overline{X})^2$$

To find this quantity, we would take each score, subtract the mean, square the difference, and then add to get the total of the squared differences. If this formula seems vaguely familiar, it's because the same expression appears in the numerator of the formula for the sample variance and the standard deviation (see Chapter 4). This redundancy is not surprising, considering that the SST is also a way of measuring dispersion.

To construct the two separate estimates of the population variance, the total variation (SST) is divided into two components. One of these reflects the pattern of variation within the categories and is called the **sum of squares within (SSW).** The other component is based on the variation between categories and is called the **sum of squares between (SSB).** SSW and SSB are components of SST, as reflected in Formula 10.2:

FORMULA 10.2
$$SST = SSB + SSW$$

The sum of squares within is defined as

FORMULA 10.3
$$SSW = \Sigma(X_i - \overline{X}_k)^2$$

where SSW = the sum of the squares within the categories
$\overline{X}_k$ = the mean of a category

This formula directs us to take each score, subtract the mean of the category from the score, square the result, and then sum the squared differences. We will do this for each category separately and then sum the squared differences for all categories to find SSW.

The between sum of squares (SSB) reflects the variation between the samples, with the category means serving as summary statistics for each category. Each category mean will be treated as a "case" for purposes of this estimate, and the formula for the sum of squares between (SSB) is

FORMULA 10.4
$$SSB = \Sigma N_k(\overline{X}_k - \overline{X})^2$$

where SSB = the sum of squares between the categories
N_k = the number of cases in a category
X_k = the mean of a category

To find SSB, subtract the overall mean of all scores from each category mean, square the difference, multiply by the number of cases in the category, and add the results across all the categories.

Let's pause for a second to remember what we are after here. If the null hypothesis is true, then there should not be much variation from category to category, relative to the variation within categories, and the two estimates to the population variance based on SSW and SSB should be roughly equal. The larger the difference between the two estimates, the more likely we will be to reject the null hypothesis. If the category means are about the same value, the differences will not be significant. The larger the differences between category means and the more homogeneous the categories, the more likely that the differences are statistically significant.

The next step in the computational routine is to construct the estimates of the population variance. To do this, we will divide each sum of squares by its respective degrees of freedom. To find the degrees of freedom associated with SSW, subtract the number of categories (k) from the number of cases (N). The degrees of freedom associated with SSB are the number of categories minus one. In summary,

FORMULA 10.5
$$\text{dfw} = N - k$$

where dfw = degrees of freedom associated with SSW
N = number of cases
k = number of categories

FORMULA 10.6
$$\text{dfb} = k - 1$$

where dfb = degrees of freedom associated with SSB
k = number of categories

The actual estimates of the population variance, called the **mean square estimates,** are calculated by dividing each sum of squares by its respective degrees of freedom:

FORMULA 10.7
$$\text{Mean square within} = \frac{\text{SSW}}{\text{dfw}}$$

FORMULA 10.8
$$\text{Mean square within} = \frac{\text{SSB}}{\text{dfb}}$$

The actual test statistic, which we will calculate in step 4 of the five-step model, is called the **F ratio,** and its value is determined by the following formula:

FORMULA 10.9
$$F = \frac{\text{Mean square between}}{\text{Mean square within}}$$

As you can see, the value of the F ratio will be a function of the amount of variation between categories to the amount of variation within the categories. The greater the variation between the categories relative to the variation within, the higher the value of the F ratio and the more likely we will reject the null hypothesis.

10.4 A COMPUTATIONAL SHORTCUT

The computational routine for ANOVA, as summarized in the previous section, requires a number of separate steps and different formulas and will surely seem complicated the first time you see it. As is almost always the case, if you proceed systematically from formula to formula (in the correct order, of course), you will see that the computations aren't nearly as formidable as they appear at first glance. Unfortunately, they will still be lengthy and time-consuming. So, at the risk of stretching your patience, let me introduce a way to save some time and computational effort. This will require the introduction of even more formulas, but the eventual savings in time will be worth the effort.

The formulas presented to this point have been definitional formulas and, as we have seen in previous chapters, these are often inconvenient for computation. The inconvenience is particularly a problem for SST and SSW and, fortunately, a computational formula is available for SST:

FORMULA 10.10

$$\text{SST} = \Sigma \ X^2 - N\overline{X}^2$$

To solve this formula, first square the value of the overall mean, multiply the result by the total number of cases in the sample, and subtract that sum from the sum of the squared scores. In finding the latter quantity, $\Sigma \ X^2$, be sure that you use all of the scores, square each one, and then sum the squared scores.

Once we have the values of SST and SSB, we can find SSW by manipulating Formula 10.2 and doing some simple subtraction:

FORMULA 10.11

$$\text{SSW} = \text{SST} - \text{SSB}$$

The computational routine for ANOVA can be summarized as follows:

1. Find SST by Formula 10.10.
2. Find SSB by Formula 10.4.
3. Find SSW by subtraction (Formula 10.11).
4. Calculate the degrees of freedom (Formulas 10.5 and 10.6).
5. Construct the two mean square estimates to the population variance by dividing SSB and SSW by their respective degrees of freedom (Formulas 10.7 and 10.8).
6. Find the obtained F ratio by dividing the between estimate by the within estimate (Formula 10.9).

These computations and the actual test of significance will be illustrated in the next sections.

10.5 A COMPUTATIONAL EXAMPLE

Assume that we have administered our "Support for Capital Punishment Scale" to a sample of 16 individuals who are equally divided into the four religions. (Obviously, this sample is much too small for any serious research and is intended solely for purposes of illustration.) All scores are reported

TABLE 10.3 SUPPORT FOR CAPITAL PUNISHMENT BY RELIGION FOR 16 SUBJECTS (fictitious data)

Protestant		Catholic		Jew		None	
X	X^2	X	X^2	X	X^2	X	X^2
8	64	12	144	12	144	15	225
12	144	20	400	13	169	16	256
13	169	25	625	18	324	23	529
17	289	27	729	21	441	28	784
50	666	84	1898	64	1078	82	1794

$\overline{X}_k =$ 12.5   21.0   16.0   20.5

$$\overline{X}_k = 280/16 = 17.5$$

in Table 10.3 along with the squared scores, the category means, and the overall mean.

To organize our computations, we'll follow the routine summarized at the end of Section 10.4. We begin by finding SST by Formula 10.10:

$$SST = \Sigma\, X^2 - N\overline{X}^2$$
$$SST = (666 + 1898 + 1078 + 1794) - (16)(17.5)^2$$
$$SST = 5436 - (16)(306.25)$$
$$SST = 5436 - 4900$$
$$SST = 536$$

The sum of squares between is found by Formula 10.4:

$$SSB = \Sigma\, N_k(\overline{X}_k - \overline{X})^2$$
$$SSB = (4)(12.5 - 17.5)^2 + (4)(21.0 - 17.5)^2$$
$$\qquad + (4)(16.0 - 17.5)^2 + (4)(20.5 - 17.5)^2$$
$$SSB = 4(-5)^2 + 4(3.5)^2 + 4(-1.5)^2 + 4(3)^2$$
$$SSB = 100 + 49 + 9 + 36$$
$$SSB = 194$$

Now SSW can be found by subtraction (Formula 10.11):

$$SSW = SST - SSB$$
$$SSW = 536 - 194$$
$$SSW = 342$$

To find the degrees of freedom for the two sums of squares, we use Formulas 10.5 and 10.6:

$$dfw = N - k = 16 - 4 = 12$$
$$dfb = k - 1 = 4 - 1 = 3$$

Finally, we are ready to construct the mean square estimates to the population variance. For the estimate based on SSW, we use Formula 10.7:

$$\text{Mean square within} = \frac{SSW}{dfw} = \frac{342}{12} = 28.5$$

For the between estimate, we use Formula 10.8:

$$\text{Mean square between} = \frac{\text{SSB}}{\text{dfb}} = \frac{194}{3} = 64.67$$

The test statistic, or F ratio, is found by Formula 10.9:

$$F(\text{obtained}) = \frac{\text{Mean square between}}{\text{Mean square within}}$$

$$F(\text{obtained}) = \frac{64.67}{28.5}$$

$$F(\text{obtained}) = 2.27$$

This statistic must still be evaluated for its significance. *(For practice in computing these quantities and solving these formulas see problem 10.1.)*

10.6 A TEST OF SIGNIFICANCE FOR ANOVA

In this section, we will see how to test an F ratio for significance and also take a look at some of the assumptions underlying the ANOVA test. As usual, we will follow the five-step model as a convenient way of organizing the decision-making process.

Step 1. **Making Assumptions.**

> Model: Independent random samples
> Level of measurement is interval-ratio
> Populations are normally distributed
> Population variances are equal

The model assumptions are quite stringent and underscore the fact that this technique should be used only with dependent variables that have been carefully and precisely measured. As long as sample sizes are equal (or nearly so), ANOVA can tolerate some violation of the model assumptions, but—in situations where you are uncertain or have samples of very different size—it is probably advisable to use an alternative test. (Chi square in Chapter 11 is one option.)

Step 2. **Stating the Null Hypothesis.** For ANOVA, the null hypothesis always states that the means of the populations from which the samples were drawn are equal. For our example problem, we are concerned with four different populations or categories, so our null hypothesis would be

$$H_0: \mu_1 = \mu_2 = \mu_3 = \mu_4$$

where μ_1 represents the mean for Protestants, μ_2, the mean for Catholics, and so forth.

The alternative hypothesis states simply that at least one of the population means is different. The wording here is important. If we reject the null,

Application 10.1

An experiment in teaching introductory biology was recently conducted at a large university. One section was taught by the traditional lecture-lab method, a second was taught by an all-lab/demonstration approach with no lectures, and a third was taught entirely by a series of videotaped lectures and demonstrations that the students were free to view at any time and as often as they wanted. Students were randomly assigned to each of the three sections and, at the end of the semester, random samples of final exam scores were collected from each section. Is there a significant difference in student performance by teaching method?

Final Exam Scores by Teaching Method

	Lecture		Demonstration		Videotape	
	X	X^2	X	X^2	X	X^2
	55	3,025	56	3,136	50	2,500
	57	3,249	60	3,600	52	2,704
	60	3,600	62	3,844	60	3,600
	63	3,969	67	4,489	61	3,721
	72	5,184	70	4,900	63	3,969
	73	5,329	71	5,041	69	4,761
	79	6,241	82	6,724	71	5,041
	85	7,225	88	7,744	80	6,400
	92	8,464	95	9,025	82	6,724
$\Sigma X =$	636		651		588	
$\Sigma X^2 =$		46,286		48,503		39,420
$\overline{X}_k =$	70.67		72.33		65.33	

$$\overline{X} = 1,875/27 = 69.44$$

We can see by inspection that the "Videotape" group had the lowest average score and that the "Demonstration" group had the highest average score. The ANOVA test will tell us if these differences are large enough to justify the conclusion that they did not occur by chance alone. Following the computational routine established at the end of Section 10.4

$$SST = \Sigma X^2 - N\overline{X}^2$$
$$SST = (46286 + 48503 + 39420) - (27)(69.44)^2$$
$$SST = 134209 - 130191.67$$
$$SST = 4017.33$$

$$SSB = \Sigma N_k(\overline{X}_k - \overline{X})^2$$
$$SSB = 9(70.67 - 69.44)^2 + 9(72.33 - 69.44)^2$$
$$\quad + 9(65.33 - 69.44)^2$$
$$SSB = 13.62 + 75.17 + 152.03$$
$$SSB = 240.82$$

$$SSW = SST - SSB$$
$$SSW = 4017.33 - 240.82$$
$$SSW = 3776.51$$

$$dfw = N - k = 27 - 3 = 24$$
$$dfb = k - 1 = 3 - 1 = 2$$

$$\text{Mean square within} = \frac{SSW}{dfw} = \frac{3776.51}{24} = 157.36$$

$$\text{Mean square between} = \frac{SSB}{dft} = \frac{240.82}{2} = 120.41$$

$$F = \frac{\text{Mean square between}}{\text{Mean square within}}$$

$$F = \frac{120.41}{157.36}$$

$$F = 0.77$$

If we set alpha equal to .05, the critical region for 2 and 24 degrees of freedom would begin at 3.40, and our obtained F score is less than 1.00. We would clearly fail to reject the null hypothesis ("the population means are equal") and would conclude that the observed differences among the category means were the results of random chance. Student performance in this course does not vary significantly by teaching method.

ANOVA does not identify which of the means are significantly different. In the final section of the chapter, we will briefly consider some advanced tests that can help us identify which pairs of means are significantly different.

(H_1: At least one of the population means is different.)

Step 3. Selecting the Sampling Distribution and Establishing the Critical Region. The sampling distribution for ANOVA is the F distribution, which is summarized in Appendix D. Note that there are separate tables for alphas of .05 and .01, respectively. As with the t table, the value of the critical F score will vary by degrees of freedom. For ANOVA, there are two separate degrees of freedom, one for each estimate of the population variance. The numbers across the top of the table are the degrees of freedom associated with the between estimate (dfb), and the numbers down the side of the table are those associated with the within estimate (dfw). In our example, dfb is ($k - 1$), or 3, and dfw is ($N - k$), or 12 (see Formulas 10.5 and 10.6). So, if we set alpha at .05, our critical F score will be 3.49.

Summarizing these considerations:

$$\text{Sampling distribution} = F \text{ distribution}$$
$$\text{Alpha} = .05$$
$$\text{Degrees of freedom (within)} = (N - k) = 12$$
$$\text{Degrees of freedom (between)} = (k - 1) = 3$$
$$F(\text{critical}) = 3.49$$

Taking a moment to inspect the two F tables, you will notice that all the values are greater than 1.00. This is because ANOVA is a one-tailed test and we are concerned only with outcomes in which there is more variance between categories than within categories. F values of less than 1.00 would indicate that the between estimate was lower in value than the within estimate and, since we would always fail to reject the null in such cases, we simply ignore this class of outcomes.

Step 4. Computing the Test Statistic. This was done in the previous section, where we found an obtained F ratio of 2.27.

Step 5. Making a Decision. Compare the test statistic with the critical value:

$$F(\text{obtained}) = 2.27$$
$$F(\text{critical}) = 3.49$$

Since the test statistic does not fall into the critical region, our decision would be to fail to reject the null. Support for capital punishment does not differ significantly by religion, and the variation we observed in the sample means is unimportant.

10.7 AN ADDITIONAL EXAMPLE FOR COMPUTING AND TESTING THE ANALYSIS OF VARIANCE

In this section, we will work through an additional example of the computation and interpretation of the ANOVA test. We will first review matters of computation, find the obtained F ratio, and then test the statistic for its significance. In the computational section, we will follow the step-by-step guidelines presented at the end of Section 10.4.

A researcher has been asked to evaluate the efficiency with which each of three social service agencies is administering a particular program. One area of concern is the speed of the agencies in processing paperwork and determining the eligibility of potential clients. The researcher has gathered information on the number of days required for processing a random sample of 10 cases in each agency. Is there a significant difference? The data are reported in Table 10.4, which also includes some additional information we will need to complete our calculations.

To find SST by the computational formula (10.10)

$$SST = \Sigma X^2 - N\overline{X}^2$$
$$SST = (524 + 1816 + 2462) - 30(11.67)^2$$
$$SST = 4802 - 30(136.19)$$
$$SST = 4802 - 4085.70$$
$$SST = 716.30$$

To find SSB by Formula 10.4

$$SSB = \Sigma N_k(\overline{X}_k - \overline{X})^2$$
$$SSB = (10)(7 - 11.67)^2 + 10(13 - 11.67)^2 + 10(15 - 11.67)^2$$
$$SSB = 10(-4.67)^2 + 10(1.33)^2 + 10(3.33)^2$$
$$SSB = 10(21.81) + 10(1.77) + 10(11.09)^2$$
$$SSB = 218.09 + 17.69 + 110.89$$
$$SSB = 346.7$$

TABLE 10.4 NUMBER OF DAYS REQUIRED TO PROCESS CASES FOR THREE AGENCIES (fictitious data)

Client	Agency A		Agency B		Agency C	
	X	X^2	X	X^2	X	X^2
1	5	25	12	144	9	81
2	7	49	10	100	8	64
3	8	64	19	361	12	144
4	10	100	20	400	15	225
5	4	16	12	144	20	400
6	9	81	11	121	21	441
7	6	36	13	169	20	400
8	9	81	14	196	19	361
9	6	36	10	100	15	225
10	6	36	9	81	11	121
$\Sigma X =$	70		130		150	
$\Sigma X^2 =$		524		1816		2462
$\overline{X}_k =$	7.0		13.0		15.0	

$$\overline{X} = 350/30 = 11.67$$

Now we can find SSW by Formula 10.11

$$SSW = SST - SSB$$
$$SSW = 716.30 - 346.7$$
$$SSW = 369.6$$

The degrees of freedom are found by Formulas 10.5 and 10.6

$$dfw = N - k = 30 - 3 = 27$$
$$dfb = k - 1 = 3 - 1 = 2$$

The estimates to the population variances are found by Formulas 10.7 and 10.8

$$\text{Mean square within} = \frac{SSW}{dfw} = \frac{369.60}{27} = 13.69$$

$$\text{Mean square between} = \frac{SSB}{dfb} = \frac{346.7}{2} = 173.35$$

The F ratio (Formula 10.9) is

$$F(\text{obtained}) = \frac{\text{Mean square between}}{\text{Mean square within}}$$

$$F(\text{obtained}) = \frac{173.35}{13.69}$$

$$F(\text{obtained}) = 12.66$$

And we can now test this value for its significance.

Step 1. **Making Assumptions.**

> Model: Independent random samples
> Level of measurement is interval-ratio
> Populations are normally distributed
> Population variances are equal

The researcher will always be in a position to judge the adequacy of the first two assumptions in the model. The second two assumptions are more problematical, but remember that ANOVA will tolerate some deviation from its assumptions as long as sample sizes are roughly equal.

Step 2. **Stating the Null Hypothesis.**

$$H_0: \mu_1 = \mu_2 = \mu_3$$

(H_1: At least one of the population means is different.)

Step 3. **Selecting the Sampling Distribution and Establishing the Critical Region.**

> Sampling distribution $= F$ distribution
> Alpha $= .05$
> Degrees of freedom (within) $= (N - k) = (30 - 3) = 27$
> Degrees of freedom (between) $= (k - 1) = (3 - 1) = 2$
> $F(\text{critical}) = 3.35$

Step 4. Computing the Test Statistic. We found an obtained F ratio of 12.66.

Step 5. Making a Decision. Compare the test statistic with the critical value:

$$F(\text{critical}) = 3.35$$
$$F(\text{obtained}) = 12.66$$

The test statistic is in the critical region, and we would reject the null of no difference. The differences between the three agencies are very unlikely to have occurred by chance alone. The agencies are significantly different in the speed with which they process paperwork and determine eligibility. *(For practice in conducting the ANOVA test, see problems 10.2 to 10.8. Begin with the lower-numbered problems since they have smaller data sets, fewer categories, and, therefore, the simplest calculations.)*

10.8 THE LIMITATIONS OF THE TEST

ANOVA is appropriate whenever you want to test differences between the means of an interval-ratio level variable across three or more categories of an independent variable. This application is called **one-way analysis of variance,** since it involves the effect of a single variable (for example, religion) on another (for example, support for capital punishment). This is the simplest application of ANOVA, and you should be aware that the technique has numerous more advanced and complex forms. For example, you may encounter research projects in which the effects of two separate variables (for example, religion and gender) on some third variable were observed.

One important limitation of ANOVA is that it requires interval-ratio measurement of the dependent variable and roughly equal numbers of cases in each of the categories of the independent variable. The former condition may be difficult to meet with complete confidence for many variables of interest to the social sciences. The latter condition may create problems when the research hypothesis calls for comparisons between groups that are, by their nature, unequal in numbers (for example, white versus black Americans) and may call for some unusual sampling schemes in the data-gathering phase of a research project. Neither of these limitations should be particularly crippling, since ANOVA can tolerate some deviation from its model assumptions, but you should be aware of these limitations in planning your own research as well as in judging the adequacy of research conducted by others. ·

A second limitation of ANOVA actually applies to all forms of significance testing and was introduced in Section 9.6. These tests are designed to detect nonrandom differences: differences so large that they are very unlikely to be produced by random chance alone. The problem is that differences that are statistically significant are not necessarily important in any other sense. The last two parts of this text will provide statistical techniques that can assess the importance of results directly.

A final limitation of ANOVA relates to the research hypothesis. As you

READING STATISTICS 7: Investigating Perceptions of Disabled Workers

The employment picture for disabled workers is grim: only about 29% are employed, compared with 79% of the general population. For adults with severe disabilities, the employment rate drops as low as 10%. What role do the perceptions of employers play in this situation? To find out, researchers Bricout and Bently mailed three applications for the job of administrative assistant to the human resource managers at 1000 randomly selected employers throughout the United States. The three hypothetical applicants had similar backgrounds and qualifications except one was identified as having a physical disability (a brain injury which necessitated the use of a wheelchair) and one was identified as having a psychiatric disability (schizophrenia).

The 246 managers that responded to the survey were asked to rate the employability of the three candidates using a standard rating scale (the "Employment Characteristic Scale"), which measures attributes such as professionalism and competence. The researchers expected that the disabled applicants would be perceived as less

qualified in spite of the essential equivalence of all three applicants. Table 1 shows that these expectations were supported.

The mean for the nondisabled candidate was the highest, and the means for the two disabled candidates are quite similar. The researchers conducted a post hoc analysis of the results and verified that ". . . there was a significant difference between the applicants with no disability and those with a physical disability, as well as between the applicants with no disability and those with a psychiatric disability. The employability ratings of the two disability conditions are essentially indistinguishable." The authors discuss the implications of these findings for the job prospects of the disabled—whom they call the "most financially disadvantaged minority group"—and some strategies that social workers, the disabled, and others concerned with this problem might take to ameliorate this condition.

John C. Bricout and Kia J. Bentley. 2000. "Disability Status and Perceptions of Employability by Employers." *Social Work Research*, 24: 87–94.

TABLE 1   RESULTS OF ONE-WAY ANOVA OF EMPLOYABILITY RATINGS

	No Disability	Physical Disability	Psychiatric Disability	ANOVA	
				F ratio	Alpha
N of raters	74	88	86	17.37	$p<.000$
Mean Score	139.59	125.95	126.15		
Std. Dev.	17.04	17.54	15.04		

recall, when the null hypothesis is rejected, the alternative hypothesis is supported. The limitation is that the alternative hypothesis is not specific: it simply asserts that at least one of the population means is different from the others. Obviously, we would like to know which differences are significant. We can sometimes make this determination by simple inspection. In our problem involving social service agencies, for example, it is pretty clear from Table 10.4 that Agency A is the source of most of the differences. This informal, "eyeball" method can be misleading, however, and you should exercise caution in making conclusions about which means are significantly different. In the next section, we provide an extended example of the interpretation of ANOVA and introduce a technique (called **post hoc** or "after the fact" analysis) that permits us to reliably identify significant differences between the sample means. The computational routines for post hoc analysis are beyond the scope of this text, but the tests are commonly available in computerized statistical packages such as SPSS.

10.9 INTERPRETING STATISTICS: DOES SEXUAL ACTIVITY VARY BY MARITAL STATUS?

Does a respondent's marital status make a difference in his or her sexual activity? Do married individuals have sex more often? Do unmarried individuals report more sexual partners than married individuals? We will use the data from the 1998 General Social Survey to answer these questions scientifically. The respondents to this survey are a representative sample (see Chapter 6) of the population of the United States.

This analysis involves three variables: marital status, the number of sexual partners in the past year, and the number of times the respondent had sex during the last 12 months. The first of these is a nominal-level variable with five categories, and it will serve as our independent variable. The other two variables have been measured at the ordinal level rather than the interval-ratio level required by ANOVA. For example, when asked for the number of times they had sex over the past year, the respondents were asked to select broad categories such as "2 or 3 times a month" (see the listing in Appendix G for "partners5" and "sexfreq" for the complete coding scheme). We encountered a similar problem in Section 9.6 when we assessed the gender gap in income. Then, we solved the problem by substituting midpoints for each of the intervals of the variable, and we will do the same here. For example, on number of sex partners, we will substitute 7.5 for the category "5 to 10" and, using the most conservative estimate, a value of 100 will be substituted for the "More than 100 partners" category. The variable measuring frequency of sexual activity has been transformed in a similar way: a value of 12 has been substituted for "about once a month," 52 for "about once a week," and so forth.

The summary statistics for the two sexual activity variables are presented in Table 10.5. As you can see, the sample averaged slightly more than one

TABLE 10.5   DESCRIPTIVE STATISTICS FOR SEXUAL ACTIVITY VARIABLES

	Number of Sex Partners During Last 12 months	Number of Times Respondent Had Sex During Last 12 Months
Mean =	1.23	55.30
Standard Deviation =	3.41	67.62
N =	2403	2320

TABLE 10.6 NUMBER OF SEXUAL PARTNERS BY MARITAL STATUS

	Marital Status					ANOVA	
	Married	Widowed	Divorced	Sepa-rated	Never Married	F ratio	Alpha
Mean =	1.08	0.26	1.31	1.37	1.83	9.84	<0.05
Standard Deviation =	2.94	0.53	3.38	1.41	4.76		
N =	1,157	224	365	82	575		

sex partner (1.23) and had sex at a rate of a little more than once a week over the past year (55.30). Note that for both variables, the standard deviation is high relative to the mean. This is caused by extreme positive skews: most cases are grouped in the low range of the variable (e.g., they had only one or two sexual partners), but a few respondents have very high scores.

The means and standard deviations for number of partners for each marital status are reported in Table 10.6. The F ratio for these differences is 9.84, which is significant at less than the .05 level. We would reject the null hypothesis of "no difference" and conclude that marital status does make a significant difference in the number of sex partners. Comparing the category means, the results are also what one would expect. Married individuals averaged essentially one partner; divorced and separated individuals had a slightly higher numbers of sex partners; individuals who were never married had the highest average number; and widowed respondents had the lowest.

Table 10.7 reports the means and standard deviations for the number of times the respondent reported having sex in the last twelve months by marital status. Married individuals, on the average, have sex more often than any other group (63.8 times a year). Never married and separated individuals both reported having sex about 59 times a year. Divorced respondents report lower rates of sexual activity, and widowed respondents have the lowest rate. The F ratio of 27.690 is significant at less than .05, so we can conclude that rate of sexual activity is significantly affected by marital status.

TABLE 10.7 FREQUENCY OF SEX BY MARITAL STATUS

	Marital Status					ANOVA	
	Married	Widowed	Divorced	Sepa-rated	Never Married	F ratio	Alpha
Mean =	63.83	12.47	48.23	58.79	58.58	27.960	<0.05
Standard Deviation =	62.66	39.72	69.29	78.80	75.69		
N =	1,102	209	364	81	564		

Tables 10.6 and 10.7 indicate significant relationships, but which differences in the tables are most important and contribute the most to the significant F ratios? We can see by inspection that the widowed group is dramatically different from the others, but what other differences might be important? A post hoc analysis of the differences in sample means can provide objective answers to these questions.

We will not present the computational routines for post hoc tests, but the interpretation of the results of these tests is straightforward. The tests essentially compare the means of all possible pairs of categories (i.e., married with divorced, widowed with separated, and so forth) and tell us exactly which combinations of means contribute the most to a significant F ratio. Comparing sample means in this way increases the probability of making an alpha error (falsely rejecting a true null hypothesis—see Chapter 8) and, to correct for this, post hoc tests use more stringent criteria to identify significant differences.

Table 10.8 presents the results of post hoc tests for both measures of sexual activity. The categories are listed at the left-hand side of the table and again across the top. The entries in the table are the differences in means for each possible combination of categories. For example, the average number of partners was 1.0795 for married respondents and 0.2634 for widowed respondents. The difference between the two category means is 1.0795 − 0.2634 or 0.8161, and this value is reported at the intersection of the two

TABLE 10.8 A POST HOC TEST FOR DIFFERENCES IN SEXUAL ACTIVITY BY MARITAL STATUS

Marital Status	Group Mean	Mean Differences in Number of Partners				
		Married	Widowed	Divorced	Separated	Never Married
Married	1.0795		0.8161*	−0.2287	−0.2924	−0.7553*
Widowed	0.2634			−1.0448*	−1.1086*	−1.5714*
Divorced	1.3082				−0.0638	−0.5266*
Separated	1.3720					−0.4628
Never Married	1.8348					

Marital Status	Group Mean	Mean Difference in the Number of Times One Has Sex				
		Married	Widowed	Divorced	Separated	Never Married
Married	63.83		51.35*	15.60*	5.04	5.25
Widowed	12.47			−35.76*	−46.32*	−46.11*
Divorced	48.23				−10.56	−10.35*
Separated	58.79					0.21
Never Married	58.58					

*Mean difference is Significant at the 0.05 level, Modified Least Significant Differences Test

categories. An asterisk (*) next to the value means that the difference is significant at the .05 level.

The differences for number of sexual partners are reported in the top portion of the table. The mean for the widowed group is significantly different from every other group, but there are two other significant differences in the table: the never married group has significantly more partners than both the married group and the divorced group. The differences for frequency of sexual activity are reported in the bottom portion of Table 10.8. Once again, the widowed group is significantly different from every other group. However, the test also finds some less obvious differences: divorced respondents have sex significantly less often than either married or never married respondents. That the widowed group was the "most different" was obvious from Tables 10.6 and 10.7, but these additional significant differences might well have escaped our attention had a post hoc test not been conducted.

In conclusion, the analysis of variance test shows that marital status makes a significant difference in levels of sexual activity for adult Americans. Married individuals have sex significantly more often with significantly fewer partners. People who "never married" have significantly more sex partners than married or divorced individuals and widowed people rank the lowest on both measures of sexual activity.

SUMMARY

1. One-way analysis of variance is a powerful test of significance that is commonly used when comparisons across more than two categories or samples are of interest. It is perhaps easiest to conceptualize ANOVA as an extension of the test for the difference in sample means.

2. ANOVA compares the amount of variation within the categories to the amount of variation between categories. If the null of no difference is false, there should be relatively great variation between categories and relatively little variation within categories. The greater the differences from category to category relative to the differences within the categories, the more likely we will be able to reject the null.

3. The computational routine for even simple applications of ANOVA can quickly become quite complex. The basic process is to construct separate estimates to the population variance based on the variation within the categories and the variation between the categories. The test statistic is the F ratio, which is based on a comparison of these two estimates. The

basic computational routine is summarized at the end of Section 10.4, and this is probably an appropriate time to mention the widespread availability of statistical packages such as SPSS, the purpose of which is to perform complex calculations such as these accurately and quickly. If you haven't yet learned how to use such programs, ANOVA may provide you with the necessary incentive.

4. The ANOVA test can be organized into the familiar five-step model for testing the significance of sample outcomes. Although the model assumptions (step 1) require high-quality data, the test can tolerate some deviation as long as sample sizes are roughly equal. The null takes the familiar form of stating that there is no difference of any importance among the population values, while the alternative hypothesis asserts that at least one population mean is different. The sampling distribution is the F distribution, and the test is always one-tailed. The decision to reject or to fail to reject the null is based on a comparison of the obtained F ratio with the critical F ratio as determined for a given alpha level and degrees of freedom. The decision to reject the null indicates only that one or more of the population means is different from the others. We can often determine which sample mean(s) account for

the difference by inspecting the sample data, but this informal method should be used with caution, and post hoc tests are more reliable indicators of significant differences.

SUMMARY OF FORMULAS

Total sum of squares:

10.1 $\quad$ $SST = \Sigma(X_i - \overline{X})^2$

The two components of the total sum of squares:

10.2 $\quad$ $SST = SSB + SSW$

Sum of squares within:

10.3 $\quad$ $SSW = \Sigma(X_i - \overline{X}_k)^2$

Sum of squares between:

10.4 $\quad$ $SSB = \Sigma\, N_k(\overline{X}_k - \overline{X})^2$

Degrees of freedom for SSW:

10.5 $\quad$ $dfw = N - k$

Degrees of freedom for SSB:

10.6 $\quad$ $dfb = k - 1$

Mean square within:

10.7 $\quad$ Mean square within $= \dfrac{SSW}{dfw}$

Mean square between:

10.8 $\quad$ Mean square between $= \dfrac{SSB}{dfb}$

F ratio:

10.9 $\quad$ $F = \dfrac{\text{Mean square between}}{\text{Mean square within}}$

Computational formula for SST:

10.10 $\quad$ $SST = \Sigma\, X^2 - N\overline{X}^2$

Finding SSW by subtraction:

10.11 $\quad$ $SSW = SST - SSB$

GLOSSARY

Analysis of variance. A test of significance appropriate for situations in which we are concerned with the differences among more than two sample means.

ANOVA. See Analysis of variance.

F ratio. The test statistic computed in step 4 of the ANOVA test.

Mean square estimate. An estimate of the variance calculated by dividing the sum of squares within (SSW) or the sum of squares between (SSB) by the proper degrees of freedom.

One-way analysis of variance. Applications of

ANOVA in which the effect of a single independent variable on a dependent variable is observed.

Post hoc test. A technique for determining which pairs of means are significantly different.

Sum of squares between (SSB). The sum of the squared deviations of the sample means from the overall mean, weighted by sample size.

Sum of squares within (SSW). The sum of the squared deviations of scores from the category means.

Total sum of squares (SST). The sum of the squared deviations of the scores from the overall mean.

MULTIMEDIA RESOURCES

The Wadsworth Sociology Resource Center: Virtual Society
http://sociology.wadsworth.com/

Visit the companion web site for the sixth edition of *Statistics: A Tool for Social Research* to access a wide range of student resources. Begin by clicking on the Student Resources section of the book's web site to access the following study tools:

- Basic math review
- Statistics review

- Flash cards
- Internet links
- Additional chapter problems
- Table of random numbers
- MicroCase and SPSS examples and exercises
- "Find the text" flowcharts
- Hypothesis testing for variables measured at the ordinal level

PROBLEMS

10.1 Conduct the ANOVA test for each set of scores below. *(HINT: follow the computational shortcut outlined in Section 10.4 and keep track of all sums and means by constructing computational tables like Table 10.3 or 10.4.)*

a.

	Category	
A	B	C
5	10	12
7	12	16
8	14	18
9	15	20

b.

	Category	
A	B	C
1	2	3
10	12	10
9	2	7
20	3	14
8	1	1

c.

	Category		
A	B	C	D
13	45	23	10
15	40	78	20
10	47	80	25
11	50	34	27
10	45	30	20

10.2 SOC What type of person is most involved in the neighborhood and community? Who is more likely to volunteer for organizations such as PTA, scouts, or Little League? A random sample of

15 people have been asked for their number of memberships in community voluntary organizations and some other information. Which differences are significant?

a. Membership by education:

Less than High School	High School	College
0	1	0
1	3	3
2	3	4
3	4	4
4	5	4

b. Membership by length of residence in present community:

Less than 2 Years	2–5 Years	More than 5 Years
0	0	1
1	2	3
3	3	3
4	4	4
4	5	4

c. Membership by extent of television watching:

Little or None	Moderate	High
0	3	4
0	3	4
1	3	4
1	3	4
2	4	5

d. Membership by number of children

None	One Child	More than One Child
0	2	0
1	3	3
1	4	4
3	4	4
3	4	5

10.3 SOC In a local community, a random sample of 18 couples has been assessed on a scale that measures the extent to which power and decision making are shared (lower scores) or monopolized by one party (higher scores) and on marital happiness (lower scores indicate lower levels of unhappiness). The couples were also classified by type of relationship: traditional (only the husband works outside the home), dual-career (both parties work), and cohabitational (parties living together but not legally married, regardless of work patterns). Does decision making or happiness vary significantly by type of relationship?

a. Decision Making

Traditional	Dual-career	Cohabitational
7	8	2
8	5	1
2	4	3
5	4	4
7	5	1
6	5	2

b. Happiness

Traditional	Dual-career	Cohabitational
10	12	12
14	12	14
20	12	15
22	14	17
23	15	18
24	20	22

10.4 CJ Two separate crime-reduction programs have been implemented in the city of Shinbone. One involves a neighborhood watch program with citizens actively involved in crime prevention. The second involves officers patrolling the neighborhoods on foot rather than in patrol cars. In terms of the percentage reduction in crimes reported to the police over a one-year period, were the programs successful? The results are for random samples of 18 neighborhoods drawn from the entire city.

Neighborhood Watch	Foot Patrol	No Program
−10	−21	+3
−2	−15	−10
+1	−8	+14
+2	−10	+8
+7	−5	+5
+10	−1	−2

10.5 SOC Are sexually active teenagers any better informed about AIDS and other potential health problems related to sex than teenagers who are sexually inactive? A 15-item test of general knowledge about sex and health was administered to random samples of teens who are sexually inactive, teens who are sexually active but with only a single partner ("going steady"), and teens who are sexually active with more than one partner. Is there any significant difference in the test scores?

Inactive	Active— One Partner	Active— More than One Partner
10	11	12
12	11	12
8	6	10
10	5	4
8	15	3
5	10	15

10.6 PS Does the rate of voter turnout vary significantly by the type of election? A random sample of voting precincts displays the following pattern of voter turnout by election type. Assess the results for significance.

Local Only	State	National
33	35	42
78	56	40
32	35	52
28	40	66
10	45	78
12	42	62
61	65	57
28	62	75
29	25	72
45	47	51
44	52	69
41	55	59

10.7 GER Do older citizens lose interest in politics and current affairs? A brief quiz on recent headline stories was administered to random samples of respondents from each of four different age groups. Is there a significant difference? The data below represent numbers of correct responses.

High School (15–18)	Young Adult (21–30)	Middle-aged (40–55)	Retired (65+)
0	0	2	5
1	0	3	6
1	2	3	6
2	2	4	6
2	4	4	7
2	4	5	7
3	4	6	8
5	6	7	10
5	7	7	10
7	7	8	10
7	7	8	10
9	10	10	10

10.8 SOC A small random sample of respondents has been selected from the General Social Survey database. Each respondent has been classified as either a city dweller, a suburbanite, or a rural dweller. Are there statistically significant differences by place of residence for any of the variables listed below?

a. Occupational prestige (PRESTG80)

Urban	Suburban	Rural
32	40	30
45	48	40
42	50	40
47	55	45
48	55	45
50	60	50
51	65	52
55	70	55
60	75	55
65	75	60

b. Number of children (CHILDS)

Urban	Suburban	Rural
1	0	1
1	1	4
0	0	2
2	0	3
1	2	3
0	2	2
2	3	5
2	2	0
1	2	4
0	1	6

c. Family income (INCOME98)

Urban	Suburban	Rural
5	6	5
7	8	5
8	11	11
11	12	10
8	12	9
9	11	6
8	11	10
3	9	7
9	10	9
10	12	8

d. Church attendance (ATTEND)

Urban	Suburban	Rural
0	0	1
7	0	5
0	2	4
4	5	4
5	8	0
8	5	4
7	8	8
5	7	8
7	2	8
4	6	5

e. Hours of TV watching per day (TVHOURS)

Urban	Suburban	Rural
5	5	3
3	7	7
12	10	5
2	2	0
0	3	1
2	0	8
3	1	5
4	3	10
5	4	3
9	1	1

Using *SPSS for Windows* to Conduct Analysis of Variance

SPSS DEMONSTRATION 10.1 Does Political Conservatism Increase with Age?

SPSS provides several different ways of conducting the analysis of variance test. The procedure summarized below is the most accessible of these, but it still incorporates options and capabilities that we have not covered in this chapter. If you wish to explore these possibilities, please consult the SPSS manual or use the online Help facility.

In designing examples to demonstrate the ANOVA procedure, my choices are constrained by the scarcity of interval-ratio variables in the GSS data set. Some variables which "should be" interval-ratio are actually measured at the ordinal level (e.g., income), while others are unsuitable dependent variables on logical grounds (e.g., age). To have something to discuss, I have taken some liberties with respect to level-of-measurement criteria in several of the examples that follow (a practice that is, in fact, common in social science research).

Let's begin by exploring the idea that political ideology is linked to age in U.S. society. People are often said to become more conservative about a wide range of issues as they age, and this might result in greater political conservatism in older age groups. For a measure of ideology, we will use *polviews.* Higher scores on this variable indicate higher levels of conservatism. We need to collapse *age* into a few categories to fit the ANOVA design. To find reasonable cutting points for age, I ran the **Frequencies** command to find the ages that divided the sample into three groups of roughly equal size. Use these scores to recode *age*

Interval	% of Sample
18–34	32.3%
35–50	34.4%
51–89	33.4%

To use the ANOVA procedure, click **Analyze, Compare Means,** and then **One-way Anova.** The **One-way Anova** window appears. Find *polviews* (the label for this variable is 'THINK OF SELF AS LIBERAL OR CONSERVATIVE') in the variable list on the left and click the arrow to move the variable name into the **Dependent List** box. Note that you can request more than one dependent variable at a time. Next, find the name of the recoded *age* variable (*ager?*) and click the arrow to move the variable name into the **Factor** box.

Click **Options** and then click the box next to **Descriptive** in the **Statistics** box to request means and standard deviations along with the analysis of variance. Click **Continue** and then click **OK,** and the following output will be produced

Descriptives: THINK OF SELF AS LIBERAL OR CONSERVATIVE

	N	Mean	Std. Deviation	Std. Error	95% Confidence Interval for Mean		Minimum	Maximum
					Lower Bound	Upper Bound		
1.00	422	3.91	1.36	6.62E-02	3.78	4.04	1	7
2.00	455	4.04	1.40	6.58E-02	3.91	4.16	1	7
3.00	440	4.28	1.36	6.48E-02	4.15	4.40	1	7
Total	1317	4.08	1.38	3.81E-02	4.00	4.15	1	7

ANOVA: THINK OF SELF AS LIBERAL OR CONSERVATIVE

	Sum of Squares	df	Mean Square	F	Sig.
Between Groups	29.489	2	14.744	7.790	.000
Within Groups	2486.918	1314	1.893		
Total	2516.407	1316			

The output box labeled ANOVA includes the various degrees of freedom, all of the sums of squares, the Mean Square estimates, the *F* ratio (7.790), and, at the far right, the exact probability ("Sig.") of getting these results if the null hypothesis is true. This is reported as .000, which is lower than our usual alpha level of .05. The differences in *polviews* for the various age groups are statistically significant.

The report also displays some summary statistics, with the three age groups identified as 1 (this is the youngest group), 2, and 3 (the oldest group). An inspection of the means shows that the average score for the entire sample was 4.08. The oldest age group is the most conservative (remember that higher scores indicate greater conservatism). These results indicate that conservatism is significantly related to age and that older people are more conservative.

SPSS DEMONSTRATION 10.2
Does Political Ideology Vary by Social Class?

Let's continue our analysis of *polviews* to see if there are any significant differences in political ideology by social class. The GSS includes several variables that might be used as indicators of class, including *degree, income, class,* and *prestg80.* We should probably investigate all of these, but — to conserve space — we confine our attention to *class* (the label for this variable is SUBJECTIVE CLASS IDENTIFICATION). Follow the instructions for **One-way ANOVA** in Demonstration 10.1 and specify *polviews* again as the dependent variable and *class* as the factor, or independent variable. The output will look like this

Descriptives: THINK OF SELF AS LIBERAL OR CONSERVATIVE

	N	Mean	Std. Deviation	Std. Error	95% Confidence Interval for Mean Lower Bound	95% Confidence Interval for Mean Upper Bound	Minimum	Maximum
LOWER CLASS	58	3.90	1.25	.16	3.57	4.23	1	7
WORKING CLASS	590	3.99	1.34	5.50E-02	3.89	4.10	1	7
MIDDLE CLASS	621	4.12	1.42	5.70E-02	4.00	4.23	1	7
UPPER CLASS	43	4.91	1.25	.19	4.52	5.29	2	6
Total	1312	4.08	1.38	3.81E-02	4.00	4.15	1	7

ANOVA: THINK OF SELF AS LIBERAL OR CONSERVATIVE

	Sum of Squares	df	Mean Square	F	Sig.
Between Groups	36.426	3	12.142	6.462	.000
Within Groups	2457.644	1308	1.879		
Total	2494.070	1311			

The F Ratio (6.462) and Sig. (.000) reported at the far right of the ANOVA output box show that the difference between groups is statistically significant at the .05 level. Inspect the group means and you will see that lower-class respondents were most liberal (had the lowest average score on *polviews*) and that the level of conservatism increased as score on *class* increased. These results indicate that, at the .05 level, political conservatism does vary significantly by social class in the population.

SPSS DEMONSTRATION 10.3 Another Test for Differences in Attitudes on Abortion

In Demonstration 9.2, we found that *sex* did have a (barely) statistically significant impact on *abscale*. In this demonstration, we will investigate the impact of religious affiliation (*relig*). Click **Analyze, Compare Means,** and **One way Anova** and name *abscale* as the dependent variable and *relig* as the factor. If necessary, see Demonstration 9.2 for instructions on computing *abscale*. The output will look like this

Descriptives: ABSCALE

	N	Mean	Std. Deviation	Std. Error	95% Confidence Interval for Mean		Minimum	Maximum
					Lower Bound	Upper Bound		
Protestant	448	2.7701	.6678	3.155E-02	2.7081	2.8321	2.00	4.00
Catholic	227	2.7665	.7059	4.685E-02	2.6742	2.8588	2.00	4.00
Jewish	14	2.2143	.4258	.1138	1.9684	2.4601	2.00	3.00
None	118	2.3220	.5044	4.643E-02	2.2301	2.4140	2.00	4.00
OTHER (SPECIFY)	40	2.5500	.6385	.1010	2.3458	2.7542	2.00	4.00
Total	847	2.6871	.6738	2.315E-02	2.6417	2.7326	2.00	4.00

ANOVA: ABSCALE

	Sum of Squares	df	Mean Square	F	Sig.
Between Groups	24.125	4	.031	14.108	.000
Within Groups	359.965	842	.428		
Total	384.090	846			

The F Ratio (14.108) and Sig. (.000) indicate a significant difference at the .05 level. Protestant and Catholic respondents were the most opposed and Jews were the most supportive of the right to a legal abortion.

Post Hoc Analysis. To conduct a post hoc analysis to determine which differences are significant, you must select one of the many available tests from the **One Way Anova** window. To do this, rerun the **One Way Anova** command for *relig* and *abscale,* click the **Post Hoc** button on the **One Way Anova** window, and click the box next to **LSD** (the "least significant difference" test). Then click **Continue** and **OK** and you will get new ANOVA output with the post hoc test included. The output is lengthy and will not be reproduced here, but the essential information is contained in the column labeled "Mean Difference." This column lists the difference between the mean scores of each category of the independent variable. For example, the top line compares Protestants with Catholics and reports a difference in mean scores of .003. (The value is reported in scientific notation as 3.569E-03. To translate to actual values, move the decimal point three places to the left). The next line compares Protestants with Jews (a difference of .5558), and differences that are significant at the .05 level are marked with an asterisk (*). Overall, the table shows that most of the significance in this relationship is accounted for by differences between Protestants and Catholics, on one hand, and Jews and "Nones" on the other.

Using MicroCase to Conduct Analysis of Variance

MICROCASE DEMONSTRATION 10.1
Does Political Conservatism Increase with Age?

In designing examples to demonstrate the ANOVA procedure, my choices are constrained by the scarcity of interval-ratio variables in the GSS data set. Some variables which "should be" interval-ratio are actually measured at the ordinal level (e.g., income), while others are unsuitable dependent variables on logical grounds (e.g., age). To have something to discuss, I have taken some liberties with respect to level-of-measurement criteria in several of the examples that follow (a practice that is, in fact, common in social science research).

Let's begin by exploring the idea that political ideology is linked to age in U.S. society. People are often said to become more conservative about a wide range of issues as they age, and this might result in greater political conservatism in older age groups. For a measure of ideology, we will use *polviews*. Higher scores on this variable indicate higher levels of conservatism. We need to collapse *age* into a few categories to fit the ANOVA design. To find reasonable cutting points for age, I ran the **Univariate** command to find the ages that divided the sample into three groups of roughly equal size. Use these scores and the **Collapse Variables** command to recode *age:*

Interval	% of Sample
18–34	32.3%
35–50	34.4%
51–89	33.4%

To use the ANOVA procedure, click **Anova** from the **Statistics** menu. The **Anova** window appears. Find *polviews* (variable #16) in the variable list on the left and click the arrow to move the variable name into the **Dependent Variable** box. Next, find the recoded *age* variable (probably variable #50) and click the arrow to move the variable name into the **Independent Variable** box. Click **OK,** and the first screen will present a "box and whiskers" graph (see Microcase Demonstration 9.1). From the **Statistics** box on the left of the screen, click **Means,** and the following output will be produced

```
Means, Standard Deviations and Number of Cases of Dependent Var: POLVIEWS
by Categories of Independent Var: AGER
Difference of means across groups is statistically significant (Prob. = 0.000)

    N    Mean   Std.Dev.
1  422  3.912    1.361
2  455  4.035    1.404
3  440  4.275    1.360
```

This screen shows the number of cases (N), means, and standard deviations (Std.Dev.) for each category of the independent variable. The three age groups are identified as 1 (this is the youngest group), 2, and 3 (the oldest group). The oldest age group is the most conservative (remember that higher scores indicate greater conservatism).

Click **Anova** in the **Statistics** box, and the following output will appear

```
Analysis Of Variance
Dependent Variable: POLVIEWS
Independent Variable: AGER
N: 1317 Missing: 70

ETA Square = 0.012

TEST FOR NON-LINEARITY:

R Square = 0.011 F = 0.538 Prob. = 0.463
        Source    Sum of Squares    DF   Mean Square     F    Prob.
        Between         29.489        2      14.744     7.790  0.000
        Within        2486.918     1314       1.893
        TOTAL         2516.407     1316
```

This output includes the various degrees of freedom, all of the sums of squares, the Mean Squares estimates, the *F* ratio (7.790), and, at the far right, the exact probability of getting these results if the null is true. This is reported as .000, which is lower than our usual alpha level of .05. The differences in *polviews* for the various age groups are statistically significant.

MICROCASE DEMONSTRATION 10.2
Does Political Ideology Vary by Social Class?

Let's continue our analysis of *polviews* and see if there are any significant differences in political ideology by social class. The GSS includes several variables that might be used as indicators of class, including *degree, income, class,* and *prestg80.* We should probably investigate all of these, but—to conserve space— we confine our attention to *class* (variable #30). Follow the instructions for **Anova** in Demonstration 10.1 and specify *polviews* again as the dependent variable and *class* as the independent variable. The output for **Means** and **Anova** will look like this

```
Means, Standard Deviations and Number of Cases of Dependent Var: POLVIEWS
by Categories of Independent Var: CLASS
Difference of means across groups is statistically significant (Prob. = 0.000)

              N    Mean   Std.Dev.
LOWER CLAS    58   3.897    1.252
WORKING CL   590   3.995    1.336
MIDDLE CLA   621   4.116    1.421
UPPER CLAS    43   4.907    1.250

Analysis Of Variance
Dependent Variable: POLVIEWS
Independent Variable: CLASS
N: 1312 Missing: 75

ETA Square = 0.015

TEST FOR NON-LINEARITY:
R Square = 0.008 F = 4.100 Prob. = 0.017

        Source   Sum of Squares   DF   Mean Square     F    Prob.
        Between          36.426    3       12.142    6.462  0.000
        Within         2457.644 1308        1.879
        TOTAL          2494.070 1311
```

The *F* Ratio (6.462) and Sig. (.000) reported at the far right of the ANOVA output box show that the difference between groups is statistically significant at the .05 level. Inspect the group means and you will see that lower-class respondents were most liberal (had the lowest average score on *polviews*) and that the level of conservatism increased as score on *class* increased. These results indicate that, at the .05 level, political conservatism does vary significantly by social class in the population.

MICROCASE DEMONSTRATION 10.3 Another Test for Differences in Attitudes on Abortion

In Demonstration 9.2, we found that *sex* did have a (barely) statistically significant impact on *abscale*. In this demonstration, we will investigate the impact of religious affiliation (*relig*). Click **Anova** and name *abscale* as the dependent variable and *relig* (variable #22) as the independent variable. If necessary, see Demonstration 9.2 for instructions on computing *abscale*. The output will look like this

```
Means, Standard Deviations and Number of Cases of Dependent Var: abscale
by Categories of Independent Var: RELIG
Difference of means across groups is statistically significant (Prob. = 0.000)

            N   Mean  Std.Dev.
Protestant  448  2.770    0.668
Catholic    227  2.767    0.706
Jewish       14  2.214    0.426
None        118  2.322    0.504
OTHER (SPE   40  2.550    0.639

Analysis Of Variance
Dependent Variable: abscale
Independent Variable: RELIG
N: 847 Missing: 540

ETA Square = 0.063

TEST FOR NON-LINEARITY:
R Square = 0.043 F = 5.823 Prob. = 0.001

        Source  Sum of Squares  DF  Mean Square      F   Prob.
        Between         24.125   4        6.031  14.108  0.000
        Within         359.965 842        0.428
        TOTAL          384.090 846
```

The *F* Ratio (14.108) and Sig. (.000) indicate a significant difference at the .05 level. Protestant and Catholic respondents were the most opposed and Jews were the most supportive of the right to a legal abortion.

Exercises

10.1 As a follow-up on Demonstration 10.2, test *income91, prestg80,* and *papres-80* as independent variables (factors) against *polviews.* Do these measures of social class display the same type of relationship with *polviews* as *class?* First, recode each of these independents into three categories. Run **Frequencies** (SPSS) or **Univariate** (MicroCase) for each and find cutting points that divide the sample into three groups of roughly equal size.

10.2 What other variable might have a significant relationship with *abscale?* Pick three potential independent variables (factors) and test their relationships with *abscale.* Some possible independent variables would be *race, attend,* or any of the measures of social class. If necessary, recode the independent variables into a few (three or four) categories.

HYPOTHESIS TESTING IV
CHI SQUARE

LEARNING OBJECTIVES

By the end of this chapter, you will be able to

1. Identify and cite examples of situations in which the chi square test is appropriate.
2. Explain the structure of a bivariate table and the concept of independence as applied to expected and observed frequencies in a bivariate table.
3. Explain the logic of hypothesis testing as applied to a bivariate table.
4. Perform the chi square test using the five-step model and correctly interpret the results.
5. Explain the limitations of the chi square test and, especially, the difference between statistical significance and importance.

11.1 INTRODUCTION

The **chi square (χ^2) test** has probably been the most frequently used test of hypothesis in the social sciences, a popularity that is due largely to the fact that the model assumptions in step 1 of the five-step model are easy to satisfy. This test assumes only that the variables are measured at the nominal level (the lowest level of measurement) and, because it is a **nonparametric** or "distribution-free" test,* it requires no assumption at all about the shape of the population or sampling distribution. Compare these minimal assumptions with those necessary for a test of significance between two sample proportions, the test chi square resembles most closely. As you recall from Section 9.4, the test for proportions assumes that the sampling distribution is normal in shape, an assumption that may be questionable for smaller samples.

Why is it an advantage to have "weak assumptions"? The decision to reject the null hypothesis (step 5) is not specific: it means only that one of the assumptions—either the model assumptions or the null hypothesis—underlying the test is wrong. Usually, of course, we single out the null hypothesis for rejection. The more certain we are of the model, the greater our confi-

---

*See the web site for this text for other nonparametric tests of significance.

dence that the null hypothesis is the faulty assumption. A "weak" or easily satisfied model means that our decision to reject the null hypothesis can be made with even greater certainty.

Chi square has also been popular for its flexibility. Not only can it be used with variables at any level of measurement, it can be used with variables that have many categories or scores. For example, in Chapter 9, we tested the significance of the difference in the proportions of black and white citizens who were "highly participatory" in voluntary associations. What if the researcher wished to expand the test to include Americans of Hispanic and Asian descent? The two-sample test would no longer be applicable, but chi square handles the more complex variable easily.

11.2 BIVARIATE TABLES

Chi square is computed from **bivariate tables,** so called because they display the scores of cases on two different variables at the same time. Bivariate tables are used to ascertain if there is a significant relationship between the variables and for other purposes that we will investigate in later chapters. In fact, these tables are very commonly used in research, and a detailed examination of them is in order.

First of all, bivariate tables have (of course) two dimensions. The horizontal (across) dimension is referred to in terms as **rows,** and the vertical dimension (up and down) is referred to in terms of **columns.** Each column or row represents a score on a variable, and the intersections of the row and columns **(cells)** represent the various combined scores on both variables.

Let's use an example to clarify. Suppose a researcher is interested in the relationship between racial group membership and participation in voluntary groups, community-service organizations, and so forth. Is there a difference in the level of involvement in volunteer groups between the races? We have two variables here (race and membership) and, for the sake of simplicity, assume that both are simple dichotomies. That is, people have been classified as either black or white and as either high or low in their level of involvement in voluntary associations.

By convention, the independent variable (the variable that is taken to be the cause) is placed in the columns and the dependent variable in the rows. In the example at hand, race would be the causal variable, and each column will represent a score on this variable. Each row, on the other hand, will represent a score on level of membership (high or low). Table 11.1 displays the outline of the bivariate table for a sample of 100. Note that the table has a descriptive title that includes the names of the variables with the dependent variable listed first. Clear, concise titles should be included in all tables, graphs, and charts.

To finish Table 11.1, we would classify people in terms of both their race and their level of participation and keep a count of how often each combination of scores occurred. Each variable has two scores, so there are four possible combinations of scores, and each is represented by a specific cell.

TABLE 11.1 RATES OF PARTICIPATION IN VOLUNTARY ASSOCIATIONS BY RACIAL GROUP FOR 100 SENIOR CITIZENS

	Racial Group		
Participation Rates	Black	White	
High			50
Low			50
	50	50	100

Blacks with high levels of participation would be counted in the upper-left-hand cell, whites with low levels of participation would be counted in the lower-right-hand cell, and so forth. When we had finished sorting cases into cells, each cell would display the number of times each combination of scores occurred. Tables also generally include the row and column subtotals or **marginals.** The total number of cases in the table (N) is reported at the intersection of the row and column marginals.

Finally, note how the bivariate table could be expanded to accommodate more complex variables. If we wished to include more groups in the test (e.g., Asian Americans or Hispanic Americans) we would simply add additional columns to the table. More elaborate dependent variables could also be easily accommodated. If we had measured participation rates with three categories (e.g., high, moderate, and low) rather than two, we would simply add an additional row to the table.

11.3 THE LOGIC OF CHI SQUARE

The chi square test has several different uses. One application, called the goodness-of-fit test, is covered on the web site for this text. In this chapter, we will deal with an application called the chi square test for independence. We have encountered the term *independence* in connection with the requirements for the two-sample case and for the ANOVA test. In those situations, we noted that independent random samples are gathered such that the selection of a particular case for one sample has no effect on the probability that any particular case will be selected for the other sample (see Section 9.2).

In the context of chi square, the concept of **independence** takes on a slightly different meaning. Two variables are independent if the classification of a case into a particular category of one variable has no effect on the probability that the case will fall into any particular category of the second variable. For example, race and participation in voluntary association would be independent of each other if the classification of a case as black or white has no effect on the classification of the case as high or low on participation. In other words, the variables would be independent if level of participation and race were completely unrelated to each other.

TABLE 11.2 THE CELL FREQUENCIES THAT WOULD BE EXPECTED IF RATES OF PARTICIPATION AND RACIAL GROUP WERE INDEPENDENT

	Racial Group		
Participation Rates	Black	White	
High	25	25	50
Low	25	25	50
	50	50	100

Consider Table 11.1 again. If these two variables are independent, then the cell frequencies will be determined solely by random chance. For example, we would find that, just as an honest coin will show heads about 50% of the time when flipped, about half of the black respondents will rank high on participation and half will rank low. If the variables are independent, the same pattern would hold for the 50 white respondents and, therefore, each of the four cells should have about 25 cases in it, as illustrated in Table 11.2. This pattern of cell frequencies indicates that the racial classification of the subjects has no effect on the probability that they would be either high or low in participation. The probability of being classified as high or low would be 0.5 for both blacks and whites, and the variables would therefore be independent.

In the case of chi square, the null hypothesis is that the variables are independent. Under the assumption that the null hypothesis is true, the cell frequencies we would expect to find if only random chance were operating are computed. These frequencies, called **expected frequencies** (symbolized f_e), are then compared, cell by cell, with the frequencies actually observed in the table ("**observed frequencies,**" symbolized f_o). If the null hypothesis is true and the variables are independent, then there should be little difference between the expected and observed frequencies. If the null is false, however, there should be large differences between the two. The greater the differences between expected (f_e) and observed (f_o) frequencies, the less likely that the variables are, in fact, independent and the more likely that we will be able to reject the null hypothesis.

11.4 THE COMPUTATION OF CHI SQUARE

As is the case with all tests of hypothesis, with chi square we compute a test statistic, χ^2 **(obtained),** from the sample data and then place that value on the sampling distribution of all possible sample outcomes. Specifically, the χ^2 (obtained) will be compared with the value of χ^2 **(critical)** that will be determined by consulting a chi square table (Appendix C) for a particular alpha level and degrees of freedom. Prior to conducting the formal test of hypothesis, let us take a moment to consider the calculation of chi square:

FORMULA 11.1
$$\chi^2 \text{ (obtained)} = \sum \frac{(f_o - f_e)^2}{f_e}$$

where f_o = the cell frequencies observed in the bivariate table
f_e = the cell frequencies that would be expected if the variables were independent

Solving this formula requires that an expected frequency be computed for each cell in the table. In Table 11.2, the marginals are the same value for all rows and columns, and the expected frequencies are obvious by intuition. They will be the same value ($f_e = 25$) for all four cells. In the more usual case, the expected frequencies will not be obvious, marginals will be unequal, and we must use Formula 11.2 to find the expected frequency for each cell:

FORMULA 11.2
$$f_e = \frac{\text{Row marginal} \times \text{Column marginal}}{N}$$

That is, the expected frequency for any cell is equal to the total number of cases in the row (the row marginal) times the total number of cases in the column (the column marginal) divided by the total number of cases in the table (N).

Once the expected frequencies have been calculated, Formula 11.1 directs us to subtract the expected frequency from the observed frequency for each cell, square this difference, divide by the expected frequency for that cell, and then sum the resultant values for all cells.

An example using Table 11.3 should clarify these procedures. A random sample of 100 social work majors have been classified in terms of whether the Council on Social Work Education has accredited their undergraduate programs (the column variable) and whether they were hired in social work positions within three months of graduation (the row variable).

Beginning with the upper-left-hand cell (graduates of accredited programs who are working as social workers), the expected frequency for this cell, using Formula 11.2, is (40)(55)/100, or 22. For the other cell in this row (graduates of nonaccredited programs who are working as social workers), the expected frequency is (40)(45)/100, or 18. For the two cells in the bottom

TABLE 11.3 EMPLOYMENT OF 100 SOCIAL WORK MAJORS BY ACCREDITATION STATUS OF UNDERGRADUATE PROGRAM

Employment Status	Accreditation Status		Totals
	Accredited	Not Accredited	
Working as a social worker	30	10	40
Not working as a social worker	25	35	60
Totals	55	45	100

TABLE 11.4   EXPECTED FREQUENCIES FOR TABLE 11.3

Employment Status	Accreditation Status		Totals
	Accredited	Not Accredited	
Working as a social worker	22	18	40
Not working as a social worker	33	27	60
Totals	55	45	100

row, the expected frequencies are (60)(55)/100, or 33, and (60)(45)/100, or 27, respectively. The expected frequencies for all four cells are displayed in Table 11.4.

Note that the row and column marginals as well as the total number of cases in Table 11.4 are exactly the same as those in Table 11.3. The row and column marginals for the expected frequencies must *always* equal those of the observed frequencies. This relationship provides a convenient way of checking your arithmetic to this point.

The value for chi square for these data can now be found by making the appropriate substitutions into Formula 11.1.

$$\chi^2 \text{ (obtained)} = \sum \frac{(f_o - f_e)^2}{f_e}$$

$$\chi^2 \text{ (obtained)} = \frac{(30 - 22)^2}{22} + \frac{(10 - 18)^2}{18} + \frac{(25 - 33)^2}{33} + \frac{(35 - 27)^2}{27}$$

$$\chi^2 \text{ (obtained)} = \frac{64}{22} + \frac{64}{18} + \frac{64}{33} + \frac{64}{27}$$

$$\chi^2 \text{ (obtained)} = 2.91 + 3.56 + 1.94 + 2.37$$

$$\chi^2 \text{ (obtained)} = 10.78$$

This sample value for chi square must still be tested for its significance. *(For practice in computing chi square, see problem 11.1.)*

11.5 THE CHI SQUARE TEST FOR INDEPENDENCE

As always, the five-step model for significance testing will provide the framework for organizing our decision making. The data presented in Table 11.3 will serve as our example.

Step 1. **Making Assumptions.**

Model: Independent random samples
Level of measurement is nominal

Step 2. **Stating the Null Hypothesis.** As stated previously, the null hypothesis in the case of chi square states that the two variables are independent. If the null is true, the differences between the observed and expected frequencies should be small. As usual, the research hypothesis directly contradicts the null. Thus, if we reject H_0, the research hypothesis will be supported.

H_0: The two variables are independent
(H_1: The two variables are dependent)

Step 3. **Selecting the Sampling Distribution and Establishing the Critical Region.** The sampling distribution of sample chi squares, unlike the Z and t distributions, is positively skewed, with higher values of sample chi squares in the upper tail of the distribution (to the right). Thus, with the chi square test, the critical region is established in the upper tail of the sampling distribution.

Values for χ^2 (critical) are given in Appendix C. This table is similar to the t table, with alpha levels arrayed across the top and degrees of freedom down the side. A major difference, however, is that degrees of freedom (df) for chi square are found by the following formula:

FORMULA 11.3
$$df = (r - 1)(c - 1)$$

where df = degrees of freedom
$(r - 1)$ = number of rows minus 1
$(c - 1)$ = number of columns minus 1

A table with two rows and two columns (a 2 × 2 table) has one degree of freedom regardless of the number of cases in the sample.* A table with two rows and three columns would have $(2 - 1)(3 - 1)$, or two degrees of freedom. Since our sample problem involves a 2 × 2 table (df = 1), if we set alpha at 0.05, the critical chi square score would be 3.841. Summarizing these decisions, we have

Sampling distribution = χ^2 distribution
Alpha = 0.05
Degrees of freedom = 1
χ^2 (critical) = 3.841

The sampling distribution of χ^2 is depicted in Figure 11.1.

---

*Degrees of freedom are the number of values in a distribution that are free to vary for any particular statistic. A 2 × 2 table has one degree of freedom because, for a given set of marginals, once one cell frequency is determined, all other cell frequencies are fixed (that is, they are no longer free to vary). In Table 11.3, for example, if any cell frequency is known, all others are determined. If the upper-left-hand cell is known to be 30, the remaining cell in that row must be 10, since there are 40 cases total in the row and 40 − 30 = 10. Once the frequencies of the cells in the top row are established, cell frequencies for the bottom row are determined by subtraction from the column marginals. Incidentally, this relationship can be used to good advantage when computing expected frequencies. For example, in a 2 × 2 table, only one expected frequency needs to be computed. The f_e's for all other cells can then be found by subtraction.

FIGURE 11.1   THE CHI SQUARE DISTRIBUTION WITH $\alpha = 0.05$, df = 1

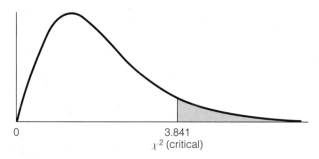

0   3.841
                                 χ^2 (critical)

Step 4. Computing the Test Statistic. The mechanics of these computations were introduced in Section 11.4. As you recall, we had

$$\chi^2 \text{ (obtained)} = \sum \frac{(f_o - f_e)^2}{f_e}$$

$$\chi^2 \text{ (obtained)} = 10.78$$

Step 5. Making a Decision. Comparing the test statistic with the critical region,

$$\chi^2 \text{ (obtained)} = 10.78$$
$$\chi^2 \text{ (critical)} = 3.841$$

we see that the test statistic falls into the critical region and, therefore, we reject the null hypothesis of independence. The pattern of cell frequencies observed in Table 11.3 is unlikely to have occurred by chance alone. The variables are dependent. Specifically, based on these sample data, the probability of securing employment in the field of social work is dependent on the accreditation status of the program. *(For practice in conducting and interpreting the chi square test for independence, see problems 11.2 to 11.15.)*

Note that while chi square tells us that the variables are dependent, it does not tell us the actual relationship between variables. Are graduates of accredited programs more or less likely to be working as social workers? To answer this question, we can compare the two types of graduates by performing some additional calculations. From Table 11.3, we can see that 30 of the 55 graduates of accredited programs are working as social workers, a percentage of 54.55%. (This percentage is found by dividing 30 by 55 and multiplying by 100. See Chapter 2.) In comparison, only 10 of the 45 graduates of nonaccredited programs, or 22.22%, are working as social workers. The relationship is significant (according to chi square), and graduates of accredited programs are more likely to find jobs as social workers (according to the percentages). We will explore the use of percentages for bivariate tables more extensively when we discuss bivariate association in Chapters 12–14.

11.6 THE CHI SQUARE TEST: AN EXAMPLE

To this point, we have confined our attention to 2×2 tables. For purposes of illustration, we will work through the computational routines and decision-making process for a larger table. As you will see, larger tables require more computations (because they have more cells); but, in all other essentials, they are dealt with in the same way as the 2×2 table.

A researcher is concerned with the possible effects of marital status on the academic progress of college students. Do married students, with their extra burden of family responsibilities, suffer academically as compared to unmarried students? Is academic performance dependent on marital status? A random sample of 453 students is gathered, and each student is classified as either married or unmarried, and—using grade-point average (GPA) as a measure—as a good, average, or poor student. Results are presented in Table 11.5.

For the top-left-hand cell (married students with good GPA's) the expected frequency would be (160)(175)/453, or 61.8. For the other cell in this row, expected frequency is (160)(278)/453, or 98.2. In similar fashion, all expected frequencies are computed (being very careful to use the correct row and column marginals) and displayed in Table 11.6.

The next step is to solve the formula for χ^2 (obtained), being very careful to be certain that we are using the proper f_o's and f_e's for each cell.

TABLE 11.5 GRADE-POINT AVERAGE (GPA) BY MARITAL STATUS FOR 453 COLLEGE STUDENTS

	Marital Status		
GPA	Married	Not Married	Totals
Good	70	90	160
Average	60	110	170
Poor	45	78	123
Totals	175	278	453

TABLE 11.6 EXPECTED FREQUENCIES FOR TABLE 11.5

	Marital Status		
GPA	Married	Not Married	Totals
Good	61.8	98.2	160
Average	65.7	104.3	170
Poor	47.5	75.5	123
Totals	175	278	453

$$\chi^2 \text{ (obtained)} = \sum \frac{(f_o - f_e)^2}{f_e}$$

$$\chi^2 \text{ (obtained)} = \frac{(70 - 61.8)^2}{61.8} + \frac{(90 - 98.2)^2}{98.2} + \frac{(60 - 65.7)^2}{65.7} + \frac{(110 - 104.3)^2}{104.3}$$

$$+ \frac{(45 - 47.5)^2}{47.5} + \frac{(78 - 75.5)^2}{75.5}$$

$$\chi^2 \text{ (obtained)} = \frac{(8.2)^2}{61.8} + \frac{(-8.2)^2}{98.2} + \frac{(-5.7)^2}{65.7} + \frac{(5.7)}{104.3}$$

$$+ \frac{(-2.5)^2}{47.5} + \frac{(2.5)^2}{75.5}$$

$$\chi^2 \text{ (obtained)} = 1.09 + 0.69 + 0.49 + 0.31 + 0.13 + 0.08$$

$$\chi^2 \text{ (obtained)} = 2.79$$

This value for the test statistic can now be tested for its significance.

Step 1. Making Assumptions.

> Model: Independent random samples
> Level of measurement is nominal

Step 2. Stating the Null Hypothesis.

> H_0: The two variables are independent.
> (H_1: The two variables are dependent.)

Step 3. Selecting the Sampling Distribution and Establishing the Critical Region.

> Sampling distribution = χ^2 distribution
> Alpha = 0.05
> Degrees of freedom = $(r - 1)(c - 1) = (3 - 1)(2 - 1) = 2$
> χ^2 (critical) = 5.991

Step 4. Computing the Test Statistic.

$$\chi^2 \text{ (obtained)} = \sum \frac{(f_o - f_e)^2}{f_e}$$

$$\chi^2 \text{ (obtained)} = 2.79$$

Step 5. Making a Decision. The test statistic, χ^2 (obtained) = 2.79, does not fall into the critical region, which, for alpha = 0.05, df = 2, begins at χ^2 (critical) of 5.991. Therefore, we fail to reject the null. The observed frequencies are not significantly different from the frequencies we would expect to find if the variables were independent and only random chance were operating.

Application 11.1

Do members of different groups have different levels of "narrow-mindedness"? A random sample of 47 white and black Americans have been rated as high or low on a scale that measures intolerance of viewpoints or belief systems different from their own. The results are

| | Group | | |
Intolerance	White	Black	Totals
High	15	5	20
Low	10	17	27
Totals	25	22	47

The frequencies we would expect to find if the null hypothesis (H_0: the variables are independent) were true are

| | Group | | |
Intolerance	White	Black	Totals
High	10.64	9.36	20.00
Low	14.36	12.64	27.00
Totals	25.00	22.00	47.00

Expected frequencies are found on a cell-by-cell basis by the formula

$$f_e = \frac{(\text{Row marginal})(\text{Column marginal})}{N}$$

for "White-High": $f_e = \dfrac{(20)(25)}{47} = 10.64$

for "Black-High": $f_e = \dfrac{(20)(22)}{47} = 9.36$

for "White-Low": $f_e = \dfrac{(27)(25)}{47} = 14.36$

for "Black-Low": $f_e = \dfrac{(27)(22)}{47} = 12.64$

The value of χ^2 (obtained) would be

$$\chi^2 \text{ (obtained)} = \sum \frac{(f_o - f_e)^2}{f_e}$$

$$\chi^2 \text{ (obtained)} = \frac{(15 - 10.64)^2}{10.64} + \frac{(5 - 9.36)^2}{9.36}$$
$$+ \frac{(10 - 14.36)^2}{14.36} + \frac{(17 - 12.64)^2}{12.64}$$

$$\chi^2 \text{ (obtained)} = \frac{19.01}{10.64} + \frac{19.01}{9.36} + \frac{19.01}{14.36} + \frac{19.01}{12.64}$$

$$\chi^2 \text{ (obtained)} = 1.79 + 2.03 + 1.32 + 1.50$$

$$\chi^2 \text{ (obtained)} = 6.64$$

If alpha is set at the 0.05 level, the critical region, with 1 degree of freedom, would begin at χ^2 (critical) = 3.841. With an obtained χ^2 of 6.64, we would reject the null hypothesis of independence. For this sample there is a statistically significant relationship between group membership and intolerance.

Based on these sample results, we can conclude that the academic performance of college students is not dependent on their marital status.

11.7 THE LIMITATIONS OF THE CHI SQUARE TEST

Like any other test, chi square has limits, and you should be aware of several potential difficulties. First, even though chi square is very flexible and handles many different types of variables, it becomes difficult to interpret when the variables have many categories. For example, two variables with 5 categories each would generate a 5 × 5 table with 25 cells—far too many combinations of scores to be easily absorbed or understood. As a very rough rule of

In Section 9.6, we tested the significance of the difference in income by gender for the United States population. We found a substantial, statistically significant gender gap in income and some evidence that suggested that the gap was due, in part, to occupational segregation—that is, the tendency of women to congregate in occupations that paid less than those dominated by males. In this section, we'll look at a research project that examines gender differences within a specific occupation (lawyer) and finds a similar pattern of gender segregation.

Sociologists Hull and Nelson interviewed a randomly selected sample of 788 lawyers practicing in the Chicago area about their career history. The law is a highly stratified profession. Large, corporate law firms occupy the top rank in the hierarchy, and lawyers in these positions enjoy the highest prestige and earnings. Other practice settings—corporate counsels, solo and small-firm lawyers, and government and public-interest law—generate lower earnings and less prestige. Table 1 shows the distribution of men and women across these settings at the start of their careers and in their current positions.

Gender is significantly related to both entry-level position and career outcome. Note that the relationship is more significant for "current position," suggesting that the career patterns of males and females grow more divergent with time. The authors summarize these patterns as follows:

> . . . women are somewhat underrepresented at the start of their careers in small and medium firms and heavily overrepresented in government, public-interest law, and legal education. These differences by gender grow over the course of the lawyers' careers. . . . women move out of both large and small firms and become more overrepresented in nonfirm settings, especially internal counsel jobs. (p. 239)

What forces might account for this pattern? The authors test a number of different explanations and conclude that the evidence strongly points toward a combination of gender discrimination and the career choices made by women in response to discrimination and other factors. Women drift away from the most demanding and time-consuming (and lucrative) positions toward positions that allow them to balance work and family responsibilities, a tendency that is reinforced by gender inequality and systematic discrimination within the profession (Hull and Nelson, p. 237).

Table and excerpts reprinted from *Social Forces* 79(2000). "Assimilation, Choice, or Constraint? Testing Theories of Gender Differences in the Careers of Lawyers." Kathleen Hull and Robert Nelson. Copyright © The University of North Carolina Press.

TABLE 1 FIRST POSITION AND CURRENT POSITION OF CHICAGO LAWYERS BY GENDER. Frequency and (Percentages*)[1]

	First Position		Current Position	
	Male	Female	Male	Female
Large firm (100 or more lawyers)	101(24%)	55(26%)	99(24%)	44(20%)
Solo practice or small-to-medium-size firm	206(49%)	79(37%)	221(52%)	61(28%)
Government, public interest law, legal education	55(13%)	52(24%)	36(9%)	55(25%)
Internal counsel or nonlegal setting	56(13%)	30(14%)	62(15%)	54(25%)
Retired or unemployed	—	—	3(1%)	4(2%)
Totals	418(99%)	216(101%)	421(101%)	218(100%)
	$\chi^2 = 15.30$ ($p < .01$)		$\chi^2 = 57.96$ ($p < .001$)	

[1]Adapted from Table 1, p. 240, Hull and Nelson, 2000.
*Percentages may not total to 100% because of rounding error.

thumb, the chi square test is easiest to interpret and understand when both variables have 4 or fewer scores.

Two limitations of the test are related to sample size. When sample size is small, it can no longer be assumed that the sampling distribution of all possible sample outcomes is accurately described by the chi square distribution. For chi square, a small sample is defined as one where a high percentage of the cells have expected frequencies (f_e) of 5 or less. Various rules of thumb have been developed to help the researcher decide what constitutes a "high percentage of cells." Probably the safest course is to take corrective action whenever *any* of the cells have expected frequencies of 5 or less.

In the case of 2 × 2 tables, the value of χ^2 (obtained) can be adjusted by applying Yates' correction for continuity, the formula for which is

FORMULA 11.4

$$\chi_c^2 = \sum \frac{(|f_o - f_e| - .5)^2}{f_e}$$

where χ_c^2 = corrected chi square
$|f_o - f_e|$ = the absolute value of the difference between the observed and expected frequency for each cell

The correction factor is applied by reducing the absolute value* of the term ($f_o - f_e$) by .5 before squaring the difference and dividing by the expected frequency for the cell.

For tables larger than 2 × 2, there is no correction formula for computing χ^2 (obtained) for small samples. It may be possible to combine some of the categories of the variables and thereby increase cell sizes. Obviously, however, this course of action should be taken only when it is sensible to do so. In other words, distinctions that have clear theoretical justifications should not be erased merely to conform to the requirements of a statistical test. When you feel that categories cannot be combined to build up cell frequencies, and the percentage of cells with expected frequencies of 5 or less is small, it is probably justifiable to continue with the uncorrected chi square test as long as the results are regarded with a suitable amount of caution.

A second potential problem related to sample size occurs with large samples. I pointed out in Chapter 9 that all tests of hypothesis are sensitive to sample size. That is, the probability of rejecting the null hypothesis increases as the number of cases increases, regardless of any other factor. It turns out that chi square is especially sensitive to sample size and that larger samples may lead to the decision to reject the null when the actual relationship is trivial. In fact, chi square is more responsive to changes in sample size than other test statistics, since the value of χ^2 (obtained) will increase at the same rate as sample size. That is, if sample size is doubled, the value of χ^2 (obtained) will be doubled. *(For an illustration of this principle, see problem 11.14.)*

---

*Absolute values ignore plus or minus signs.

My major purpose in stressing the relationship between sample size and the value of chi square is really to point out, once again, the distinction between statistical significance and theoretical importance. On one hand, tests of significance play a crucial role in research. As long as we are working with random samples, we must know if our research results could have been produced by mere random chance.

On the other hand, like any other statistical technique, tests of hypothesis are limited in the range of questions they can answer. Specifically, these tests will tell us whether our results are statistically significant or not. They will not necessarily tell us if the results are important in any other sense. To deal more directly with questions of importance, we must use an additional set of statistical techniques called measures of association. These techniques will be the subject of Part III of this text.

11.8 INTERPRETING STATISTICS: CAPITAL PUNISHMENT AND VIOLENT CRIME RATES

Over the past several decades, local, state, and federal officials have adopted, with near invariance, a "get tough" approach to crime in the United States. Among other consequences, this has resulted in an increased use of the death penalty, a punishment that is often justified in terms of deterrence, or the power of extreme punishment to lower the rate of violent crime. Is it true that the death penalty deters violent crime? Is the rate of murder or rape dependent on capital punishment? We can use the chi square test to explore this relationship.*

The first step in the analysis is to take a random sample of 25 cases from a list of the 50 states and the District of Columbia for the year 1998 (the latest year for which data are available) and to get the homicide and rape rates for each case. Crime rates are interval-ratio level variables with wide ranges and many possible scores, a form of data that does not lend itself to bivariate tables and chi square.† Crime rates will have to be collapsed into a few broad categories to make them suitable for this test. One way to do this is to categorize each case as "high" if it is above the national average for the crime in question and "low" if it is below that average. The independent variable is the presence or absence of the death penalty in the jurisdiction. Tables 11.7 and 11.8 display the bivariate relationships. If the death penalty acts as a deterrent, we should find a statistically significant relationship between these variables.

For both tables, the test statistic does not fall into the critical region, which—at the .05 level for a 2 × 2 table—begins at 3.841. We would fail

---

*Our data on crime comes from the FBI (http://www.ojp.usdoj.gov/bjs/datast.htm), and information on which states use the death penalty is from the Bureau of Justice Statistics of the U.S. Department of Justice (Correctional Populations in the United States, 1997, NCJ 177613. "Prisoners executed under civil authority by region and jurisdiction, 1930–1997," Washington DC: U.S. Department of Justice, 2000.)

†An alternative test that would take advantage of the numerical quality of the crime rates would be the two sample tests discussed in Chapter 9.

TABLE 11.7 HOMICIDE RATE BY PRESENCE OF THE DEATH PENALTY

| Homicide Rate | Death Penalty | | Totals |
	No	Yes	
Low	4	6	10
High	3	12	15
Totals	7	18	25
	χ^2 (obtained) = 1.19		

TABLE 11.8 RAPE RATE BY PRESENCE OF THE DEATH PENALTY

| Rape Rate | Death Penalty | | Totals |
	No	Yes	
Low	3	11	14
High	4	7	11
Totals	7	18	25
	χ^2 (obtained) = 0.68		

to reject the null hypothesis and conclude that there is no statistically significant relationship between the death penalty and violent crime rate in the population (all 50 states and the District of Columbia). The variables are independent of each other, an outcome that does not support the argument that the death penalty is a deterrent to violent crime.

Additional analysis of the pattern of cell frequencies in the tables underscores the weakness of the argument for deterrence. For example, in Table 11.7, the majority (12 out of 18 or 66.7%) of the cases with the death penalty had high homicide rates. The patterns in Table 11.8 are more consistent with the deterrence argument (e.g., the majority of the cases without the death penalty had high rates of rape), but the differences are very slight.

Do these results refute the deterrence argument once and for all? Even the most ardent opponent of the death penalty must admit that these tests are partial at best. Among other limitations, the tests reported in Tables 11.7 and 11.8 have some statistical problems. For example, recall that the value of obtained chi square (and the probability of rejecting the null hypothesis) is a direct function of sample size (see Section 11.7 and especially problem 11.14). Would we have found a significant relationship with a larger sample? Also, the test is limited to a single year: What relationships would we see if the test included a broader time span?

We can address both of these issues, at least partially. With a small population of only 51 cases, sample size can never be very large, but we can maximize N by including all cases—the entire population—and conducting the test for several different years. In fact, it is not unusual in the professional literature to see tests of significance conducted on entire populations in order to eliminate the possibility that relationships between variables reflect the operation of random factors other than those associated with selecting an EPSEM sample.

Table 11.9 displays relationships between death penalty and homicide rates for each of three different years for all 50 states and the District of Columbia. Cases are rated as low if they were below the national average homicide rate and high if they were above that standard for the year in question. At the .05 level, none of the chi squares is significant. More importantly, the death penalty states were more likely to have high homicide rates, the oppo-

TABLE 11.9 HOMICIDE RATE BY PRESENCE OF THE DEATH PENALTY FOR THREE DIFFERENT YEARS FOR ALL STATES AND THE DISTRICT OF COLUMBIA

A. 1986

Homicide Rate	Death Penalty		Totals
	No	Yes	
Low	10	22	32
High	3	16	19
Totals	13	38	51

χ^2 (obtained) = 1.50

B. 1991

Homicide Rate	Death Penalty		Totals
	No	Yes	
Low	12	21	33
High	2	16	18
Totals	14	37	51

χ^2 (obtained) = 3.73

C. 1998

Homicide Rate	Death Penalty		Totals
	No	Yes	
Low	10	20	30
High	3	18	21
Totals	13	38	51

χ^2 (obtained) = 2.360

TABLE 11.10 RAPE RATE BY PRESENCE OF THE DEATH PENALTY FOR THREE DIFFERENT YEARS FOR ALL STATES AND THE DISTRICT OF COLUMBIA

A. 1986

Rape Rate	Death Penalty		Totals
	No	Yes	
Low	10	21	31
High	3	17	20
Totals	13	38	51

χ^2 (obtained) = 1.91

B. 1991

Rape Rate	Death Penalty		Totals
	No	Yes	
Low	11	18	29
High	3	19	22
Totals	14	37	51

χ^2 (obtained) = 3.71

C. 1998

Rape Rate	Death Penalty		Totals
	No	Yes	
Low	8	20	28
High	5	18	23
Totals	13	38	51

χ^2 (obtained) = 0.31

site of the pattern which would be expected from the deterrence argument. In 1986, for example, 42% (16 out of 38) of the death-penalty states had high homicide rates. By comparison, only 23% (3 out of 13) of the non-death-penalty states had high homicide rates.

Following the same procedures, Table 11.10 displays the relationships for the rate of rape, and once again we see that none of the obtained chi squares exceeds the threshold for significance at the .05 level. The 1991 results are the closest to significance, but the pattern in the table is not consistent with the deterrence argument because the majority of the death-penalty states (19 out of 37, or 51%) had high rates of rape (vs. only 21% of the non-death-penalty states).

The results of these chi-square analyses are not consistent with the assertion that capital punishment acts as a deterrent for violent crime. For both the sample and the populations, the states' murder and rape rates are not significantly affected by the death penalty.

SUMMARY

1. The chi square test for independence is appropriate for situations in which the variables of interest have been organized into table format. The null hypothesis is that the variables are independent or that the classification of a case into a particular category on one variable has no effect on the probability that the case will be classified into any particular category of the second variable.

2. Since chi square is nonparametric and requires only nominally measured variables, its model assumptions are easily satisfied. Furthermore, since it is computed from bivariate tables, in which the number of rows and columns can be easily expanded, the chi square test can be used in many situations in which other tests are inapplicable.

3. In the chi square test, we first find the frequencies that would appear in the cells if the variables were independent (f_e) and then compare those frequencies, cell by cell, with the frequencies actually observed in the cells (f_o). If the null is true, expected and observed frequencies should be quite close in value. The greater the difference between the observed and expected frequencies, the greater the possibility of rejecting the null.

4. The chi square test has several important limitations. It is often difficult to interpret when tables have many (more than 4 or 5) dimensions. Also, as sample size (N) decreases, the chi square test becomes less trustworthy, and corrective action may be required. Finally, with very large samples, we may declare relatively trivial relationships to be statistically significant. As is the case with all tests of hypothesis, statistical significance is not the same thing as "importance" in any other sense. As a general rule, statistical significance is a necessary but not sufficient condition for theoretical or practical importance.

SUMMARY OF FORMULAS

Chi square (obtained):

11.1 $$\chi^2 \text{ (obtained)} = \sum \frac{(f_o - f_e)^2}{f_e}$$

Expected frequencies:

11.2 $$f_e = \frac{(\text{Row marginal})(\text{Column marginal})}{N}$$

Degrees of freedom, bivariate tables:

11.3 $$df = (r - 1)(c - 1)$$

Yates' correction for continuity:

11.4 $$\chi_c^2 = \sum \frac{(|f_o - f_e| - .5)^2}{f_e}$$

GLOSSARY

Bivariate table. A table that displays the joint frequency distributions of two variables.

Cells. The cross-classification categories of the variables in a bivariate table.

χ^2 (critical). The score on the sampling distribution of all possible sample chi squares that marks the beginning of the critical region.

χ^2 (obtained). The test statistic as computed from sample results.

Chi square test. A nonparametric test of hypothesis for variables that have been organized into a bivariate table.

Column. The vertical dimension of a bivariate table. By convention, each column represents a score on the independent variable.

Expected frequency (f_e). The cell frequencies that would be expected in a bivariate table if the variables were independent.

Independence. The null hypothesis in the chi square test. Two variables are independent if, for all cases, the classification of a case on one variable has no effect on the probability that the case will be classified in any particular category of the second variable.

Marginals. The row and column subtotals in a bivariate table.

Nonparametric. A "distribution-free" test. These tests do not assume a normal sampling distribution.

Observed frequency (f_o). The cell frequencies actually observed in a bivariate table.

Row. The horizontal dimension of a bivariate table, conventionally representing a score on the dependent variable.

MULTIMEDIA RESOURCES

The Wadsworth Sociology Resource Center: Virtual Society
http://sociology.wadsworth.com/

Visit the companion web site for the sixth edition of *Statistics: A Tool for Social Research* to access a wide range of student resources. Begin by clicking on the Student Resources section of the book's web site to access the following study tools:

- Basic math review
- Statistics review

- Flash cards
- Internet links
- Additional chapter problems
- Table of random numbers
- MicroCase and SPSS examples and exercises
- "Find the text" flowcharts
- Hypothesis testing for variables measured at the ordinal level

PROBLEMS

11.1 For each table below, calculate the obtained chi square. *(HINT: calculate the expected frequencies for each cell with Formula 11.2. Double-check to make sure you are using the correct row and column marginals for each cell. It may be helpful to record the expected frequencies in table format as well—see Tables 11.4 and 11.3, 11.6 and 11.5. Next, use Formula 11.1 to calculate chi square. For each cell, subtract expected frequency from obtained frequency, square the result, and divide by expected frequency. Sum the results to determine the value of chi square. Double-check to make sure that you are using the correct values for each cell.)*

a.

20	25	45
25	20	45
45	45	90

b.

10	15	25
20	30	50
30	45	75

c.

25	15	40
30	30	60
55	45	100

d.

20	45	65
15	20	35
35	65	100

11.2 SOC A sample of 25 cities have been classified as high or low on their homicide rates and on the number of handguns sold within the city limits. Is there a relationship between these two variables? Explain your results in a sentence or two.

Volume of Gun Sales	Homicide Rate		Totals
	Low	High	
High	8	5	13
Low	4	8	12
Totals	12	13	25

11.3 SW A local politician is concerned that a program for the homeless in her city is discriminating against blacks and other minorities. The data below were taken from a random sample of black and white homeless people. Is there a statistically significant relationship between race and whether or not the person has received services from the program?

| Received | Race | | |
Services?	Black	White	Totals
Yes	6	7	13
No	4	9	13
Totals	10	16	26

11.4 PS Many analysts have noted a "gender gap" in elections for the U.S. presidency with women more likely to vote for the Democratic candidate. A sample of university faculty has been asked about their political party preference. Do their responses indicate a significant relationship between gender and party preference?

| Party | Gender | | |
Preference	Male	Female	Totals
Democratic	10	15	25
Republican	15	10	25
Totals	25	25	50

11.5 PA Is there a relationship between salary levels and unionization for public employees? The data below represent this relationship for fire departments in a random sample of 100 cities of roughly the same size. Salary data have been dichotomized at the median. Summarize your findings.

Salary	Union	Nonunion	Totals
High	21	29	50
Low	14	36	50
Totals	35	65	100

11.6 SOC A program of pet therapy has been running at a local nursing home. Are the participants in the program more alert and responsive than nonparticipants? The results, drawn from a random sample of residents, are reported below.

Alertness	Participants	Non-participants	Totals
High	23	15	38
Low	11	18	29
Totals	34	33	67

11.7 SOC The state Department of Education has rated a sample of local school systems for compliance with state-mandated guidelines for quality. Is the quality of a school system significantly related to the affluence of the community as measured by per capita income?

| | Per Capita Income | | |
Quality	Low	High	Totals
Low	16	8	24
High	9	17	26
Totals	25	25	50

11.8 CJ A local judge has been allowing some individuals convicted of "driving under the influence" to work in a hospital emergency room as an alternative to fines, suspensions, and other penalties. A random sample of offenders has been drawn. Do participants in this program have lower rates of recidivism for this offense?

Recidivist?	Participants	Non-participants	Totals
Yes	60	123	183
No	55	108	163
Totals	115	231	346

11.9 SOC Is there a relationship between length of marriage and satisfaction with marriage? The necessary information has been collected from a random sample of 100 respondents drawn from a local community. Write a sentence or two explaining your decision.

| | Length of Marriage (in years) | | | |
Satisfaction	Less than 5	5–10	More than 10	Totals
Low	10	20	20	50
High	20	20	10	50
Totals	30	40	30	100

11.10 PS Is there a relationship between political ideology and class standing? Are upperclass students significantly different from underclass students on this variable? The table below reports the relationship between these two variables for a random sample of 267 college students.

| Political | Class Standing | | |
Ideology	Underclass	Upperclass	Totals
Liberal	43	40	83
Moderate	50	50	100
Conservative	40	44	84
Totals	133	134	267

11.11 SOC At a large urban college, about half of the students live off campus in various arrangements, and the other half live in dormitories on campus. Is academic performance dependent on living arrangements? The results based on a random sample of 300 students are presented below.

GPA	Off Campus with Roommates	Off Campus with Parent	On Campus	Totals
Low	22	20	48	90
Moderate	36	40	54	130
High	32	10	38	80
Totals	90	70	140	300

11.12 SOC An urban sociologist has built up a database describing a sample of the neighborhoods in her city and has developed a scale by which each area can be rated for the "quality of life" (this includes measures of pollution, noise, open space, services available, and so on). She has also asked samples of residents of these areas about their level of satisfaction with their neighborhoods. Is there significant agreement between the sociologist's objective ratings of quality and the respondents' self-reports of satisfaction?

	Quality of Life			
Satisfaction	Low	Moderate	High	Totals
Low	21	15	6	42
Moderate	12	25	21	58
High	8	17	32	57
Totals	41	57	59	157

11.13 SOC Does support for the legalization of marijuana vary by region of the country? The table displays the relationship between the two variables for a random sample of 1020 adult citizens. Is the relationship significant?

Legalize Marijuana?	Region				
	North	Midwest	South	West	Totals
Yes	60	65	42	78	245
No	245	200	180	150	775
Totals	305	265	222	228	1020

11.14 SOC A researcher is concerned with the relationship between attitudes toward violence and violent behavior. If attitudes "cause" behavior (a very debatable proposition), then people who have positive attitudes toward violence should have high rates of violent behavior. A pretest was conducted on 70 respondents and, among other things, the respondents were asked, "Have you been involved in a violent incident of any kind over the past six months?" The researcher established the following relationship:

	Attitude Toward Violence		
Involvement	Favorable	Unfavorable	Totals
Yes	16	19	35
No	14	21	35
Totals	30	40	70

The chi square calculated on these data is .23, which is not significant at the .05 level (confirm this conclusion with your own calculations). Undeterred by this result, the researcher proceeded with the project and gathered a random sample of 7000. In terms of percentage distributions, the results for the full sample were exactly the same as for the pretest:

	Attitude Toward Violence		
Involvement	Favorable	Unfavorable	Totals
Yes	1600	1900	3500
No	1400	2100	3500
Totals	3000	4000	7000

However, the chi square obtained is a very healthy 23.4 (confirm with your own calculations). Why is the full-sample chi square significant when the pretest was not? What happened? Do you think that the second result is important?

11.15 SOC Some results from the General Social Survey are presented below. The survey is administered every year to a nationally representative sample. See Appendix G for the complete questions and other details about the survey. For each table, conduct the chi square test of significance. Write a sentence or two of interpretation for each test.

a. Support for the legal right to an abortion "for any reason" by age:

Support for Abortion:	Age			
	Younger than 30	30–49	50 and Older	Totals
Yes	154	360	213	727
No	179	441	429	1049
Totals	333	801	642	1776

b. Support for the death penalty by age:

Support for Death Penalty:	Age			
	Younger than 30	30–49	50 and Older	Totals
Favor	361	867	675	1903
Oppose	144	297	252	693
Totals	505	1164	927	2596

c. Fear of walking alone at night by age:

Fear:	Age			
	Younger than 30	30–49	50 and Older	Totals
Yes	147	325	300	772
No	202	507	368	1077
Totals	349	832	668	1849

d. Support for legalizing marijuana by age:

Legalize?:	Age			
	Younger than 30	30–49	50 and Older	Totals
Should	128	254	142	524
Should not	224	534	504	1262
Totals	352	788	646	1786

e. Support for suicide when a person has an incurable disease by age:

Suicide?:	Age			
	Younger than 30	30–49	50 and Older	Totals
Yes	225	537	367	1129
No	107	270	266	643
Totals	332	807	633	1772

SPSS for Windows

Using *SPSS for Windows* to Conduct the Chi Square Test

SPSS DEMONSTRATION 11.1 Do Childcare Practices Vary by Social Class?

The **Crosstabs** procedure in SPSS produces bivariate tables and a wide variety of statistics. This procedure is very commonly used in social science research at all levels, and you will see many references to **Crosstabs** in chapters to come. I will introduce the command here, and we will return to it often in later sessions.

Do people in different social classes raise their children differently? Which social class would be most likely to approve of spanking as a disciplinary technique? We can answer these questions, at least for the 1998 GSS sample, by constructing a bivariate table to display the relationship between *class* and *spanking*. We will also request a chi square test for the table.

Begin by clicking **Analyze,** then click **Descriptive Statistics** and **Crosstabs.** The **Crosstabs** dialog box will appear with the variables listed in a box on the left. Highlight *spanking* (the label for this variable is FAVOR SPANKING TO DISCIPLINE CHILD) and click the arrow to move the variable name into the **Rows** box, and then highlight *class* (the label for this variable is SUBJECTIVE CLASS IDENTIFICATION) and move it into the **Columns** box. Click the **Statistics** button at the bottom of the window and click the box next to Chi-square. Click **Continue** and **OK,** and the following output will be produced:

FAVOR SPANKING TO DISCIPLINE CHILD SUBJECTIVE CLASS IDENTIFICATION
Crosstabulation Count

| | | SUBJECTIVE CLASS IDENTIFICATION | | | | |
		LOWER CLASS	WORKING CLASS	MIDDLE CLASS	UPPER CLASS	Total
FAVOR SPANKING TO DISCIPLINE CHILD	STRONGLY AGREE	13	110	100	11	234
	AGREE	20	204	179	11	414
	DISAGREE	8	65	84	11	168
	STRONGLY DISAGREE	1	29	32	2	64
Total		42	408	395	35	880

Chi-Square Tests

	Value	df	Asymp. Sig. (2-sided)
Pearson Chi-Square	11.486	9	.244
Likelihood Ratio	11.811	9	.224
Linear-by-Linear Association	3.068	1	.080
N of Valid Cases	880		

The output displays the number of cases in each cell and the row and column totals. The results of the chi square test are reported in the output block that follows the table. The value of chi square (obtained) is 11.486, the degrees of freedom are 9, and the exact significance of the chi square is .244. This is well above the standard indicator of a significant result (alpha = .05), so we may conclude that there is no statistically significant relationship between *class* and *spanking*. Support for spanking is independent of social class.

SSPS DEMONSTRATION 11.2 Who Watches X-Rated Movies?

Is there a relationship between social class and the viewing of X-rated movies? Run the **Crosstabs** procedure again with *xmovie* (the label for this variable is SEEN X-RATED MOVIE IN LAST YEAR) as the row variable (in place of *spanking*) and class as the column variable. Don't forget to request chi square. The output will be as shown:

SEEN X-RATED MOVIE IN LAST YEAR SUBJECTIVE CLASS IDENTIFICATION
Crosstabulation Count

| | | SUBJECTIVE CLASS IDENTIFICATION | | | | |
		LOWER CLASS	WORKING CLASS	MIDDLE CLASS	UPPER CLASS	Total
SEEN X-RATED MOVIE IN LAST YEAR	YES	13	110	97	4	224
	NO	33	307	333	26	699
Total		46	417	430	30	923

Chi-Square Tests

	Value	df	Asymp. Sig. (2-sided)
Pearson Chi-Square	4.046	3	.257
Likelihood Ratio	4.295	3	.231
Linear-by-Linear Association	3.527	1	.060
N of Valid Cases	923		

There is no significant relationship between the variables. The exact probability of getting this pattern of cell frequencies by random chance alone is .257, well above the standard alpha level of .05. What other variables might be significantly related to *xmovie?* See exercise 11.2 for an opportunity to investigate further.

SPSS DEMONSTRATION 11.3 Is Ignorance Bliss?

We can test the accuracy of the ancient folk wisdom that "ignorance is bliss" with the 1998 GSS. Are more-educated people less happy? To test the relationship, we will use *degree* as our measure of education and, to measure the dependent variable, we will use an item from the GSS (*happy*), which asked respondents to rate their overall level of happiness. As originally coded, *degree* has five categories (see Appendix G), but we will simplify the analysis and distinguish only between people with a high school education or less and people with at least some education beyond high school. When you recode *degree,* remember to choose "Recode into Different Variable" and give the recoded variable a new name (say, *rdeg*). The recode instruction in the **Old → New** box of the **Recode into Different Variable** window should look like this:

0 thru 1 → 1
2 thru 4 → 2

Follow the instructions in Demonstration 11.1 for the **Crosstabs** command. Specify *happy* as the row variable (the label for this variable is GENERAL HAPPINESS) and recoded *degree* as the column variable. Don't forget to request chi square. The output should look like this:

GENERAL HAPPINESS RDEG Crosstabulation Count

		RDEG		Total
		1.00	2.00	
GENERAL HAPPINESS	VERY HAPPY	276	174	450
	PRETTY HAPPY	525	245	770
	NOT TOO HAPPY	121	29	150
Total		922	448	1370

Chi-Square Tests

	Value	df	Asymp. Sig. (2-sided)
Pearson Chi-Square	19.730	2	.000
Likelihood Ratio	20.714	2	.000
Linear-by-Linear Association	18.691	1	.000
N of Valid Cases	1370		

The table displays the observed frequencies for the cells, along with row and column marginals and total number of cases. The value of chi square (19.730), the degrees of freedom (2), and the exact probability that this pattern of cell frequencies occurred by chance (.000) are reported in the output block below the table. The reported significance is less than .000, so we would reject the null hypothesis of independence. There is a relationship between level of education and degree of happiness.

Remember that chi square tells us only that the overall relationship between the variables is significant. To assess the specific idea that "ignorance is bliss," we need to do some further analysis. If the idea is true, a higher percentage of the less educated (category 1 of *rdeg*) should be happy. Unfortunately for the old adage, the bivariate table shows the reverse pattern. While 29.94% (276 out of 922) of the less-educated respondents are "very happy," 38.84% (174 out of 448) for the more-educated place themselves in this category.

MicroCase

Using Microcase to Conduct the Chi Square Test

MICROCASE DEMONSTRATION 11.1
Do Childcare Practices Vary by Social Class?

The **Cross-tabulation** procedure in MicroCase produces bivariate tables and a wide variety of statistics. This procedure is very commonly used in social science research at all levels, and you will see many references to **Cross-tabulation** in chapters to come. I will introduce the command here, and we will return to it often in later sessions.

Do people in different social classes raise their children differently? Which social class would be most likely to approve of spanking as a disciplinary technique? We can answer these questions, at least for the 1998 GSS sample, by constructing a bivariate table to display the relationship between *class* and *spanking*. We will also request a chi square test for the table.

Begin by clicking **Cross-tabulation** from the **Statistics** menu. The **Cross-tabulation** dialog box will appear with the variables listed in the window to the left. Highlight *spanking* (variable #38) and click the arrow to move the variable name into the **Row Variable** box, and then highlight *class* (variable #30) and move it into the **Column Variable** box. Click **OK,** and the following table will be produced:

SPANKING by CLASS

	LOWER	WORKING	MIDDLE	UPPER	MISSING	TOTAL
STRONGLY A	13	110	100	11	2	234
AGREE	20	204	179	11	1	414
DISAGREE	8	65	84	11	0	168
STRONGLY D	1	29	32	2	1	64
Missing	25	220	242	12	4	503
TOTAL	42	408	395	35	8	880

The output displays the number of cases in each cell and the row and column totals. Note that MicroCase reports missing cases in the table as well as actual scores. The values in the column and row labeled "missing" are ignored in all computations.

Next, click **Summary** in the **Statistics** box on the left of the screen.

```
Chi-Square: 11.486 (DF = 9; Prob. = 0.244)
```

The value of chi square (obtained) is 11.486, the degrees of freedom are 9, and the exact significance of this chi square is .244. This is well above the standard indicator of a significant result (alpha = .05), so we may conclude that there is no statistically significant relationship between *class* and *spanking*. Support for spanking is independent of social class.

MICROCASE DEMONSTRATION 11.2 Who Watches X-Rated Movies?

Is there a relationship between social class and the viewing of X-rated movies? Run the **Crosstabs** procedure again with *xmovie* as the row variable (in place of *spanking*) and class as the column variable. The following table and results for the chi square test will be displayed

XMOVIE by CLASS

	LOWER	WORKING	MIDDLE	UPPER	Missing	TOTAL
YES	13	110	97	4	1	224
NO	33	307	333	26	5	699
Missing	21	211	207	17	2	458
TOTAL	46	417	430	30	8	923

Chi-Square: 4.046 (DF = 3; Prob. = 0.257)

There is no significant relationship between the variables. The exact probability of getting this pattern of cell frequencies by random chance alone is .257, well above the standard alpha level of .05. What other variables might be significantly related to *xmovie?* See exercise 11.2 for an opportunity to investigate further.

MICROCASE DEMONSTRATION 11.3 Is Ignorance Bliss?

We can test the accuracy of the ancient folk wisdom that "ignorance is bliss" with the 1998 GSS. Are more-educated people less happy? To test the relationship, we will use *degree* (variable #9) as our measure of education and, to measure the dependent variable, we will use an item from the GSS (*happy,* variable #26), which asked respondents to rate their overall level of happiness. As originally coded, *degree* has five categories (see Appendix G), but we will simplify the analysis and distinguish only between people with a high school education or less and people with at least some education beyond high school. Use the **Collapse Variable** command to recode *degree* and remember to give the recoded variable a new name (say, *rdeg*). Collapse the scores of 0 and 1 together and then combine the scores of 2 through 4. Follow the instructions in Demonstration 11.1 for the **Cross-tabulation** command. Specify *happy* as the row variable and recoded *degree* (probably variable #51) as the column variable. The output should look like this:

HAPPY by RDEG

	HS or less	HS +	Missing	TOTAL
VERY HAPPY	276	174	2	450
PRETTY HAP	525	245	1	770
NOT TOO HA	121	29	1	150
Missing	7	6	0	13
TOTAL	922	448	4	1370
Chi-Square: 19.730 (DF = 2; Prob. = 0.000)				

The table displays the observed frequencies for the cells, along with row and column marginals and total number of cases. The value of chi square (19.730), the degrees of freedom (2), and the exact probability that this pattern of cell frequencies occurred by chance (.000) are available by clicking **Summary** in the **Statistics** box on the left of the screen. The reported significance is less than .000, so we would reject the null hypothesis of independence. There is a relationship between level of education and degree of happiness.

Remember that chi square tells us only that the overall relationship between the variables is significant. To assess the specific idea that "ignorance is bliss," we need to do some further analysis. If the idea is true, a higher percentage of the less educated ("HS or less" on *rdeg*) should be happy. Unfortunately for the old adage, the bivariate table shows the reverse pattern. While 29.94% (276 out of 922) of the less educated respondents are "very happy," 38.84% (174 out of 448) for the more educated place themselves in this category.

Exercises

11.1 For a follow-up on Demonstration 11.1, pick two variables that measure social class. Recode these variables so that they have only two or three categories. Run the **Crosstabs** (SPSS) or **Cross-Tabulation** (MicroCase) procedure with *spanking* as the row variable and your new measures of social class as the column variables. How do these relationships compare with the table generated in Demonstration 11.1? Are these tables consistent with the idea that approval of spanking does not vary by social class?

11.2 Find three more variables that might be related to *xmovie* (*sex* and *relig* seem like possibilities). Use the **Crosstabs** (SPSS) or **Cross-Tabulation** (MicroCase) procedure, with chi square, to see if any of your variables have a significant relationship with *xmovie*. Which of the variables had the most significant relationship?

11.3 Since we already recoded *degree* for Demonstration 11.3, let's see if education has any significant relationship with *grass, gunlaw,* or *cappun*. Run the **Crosstabs** (SPSS) or **Cross-Tabulation** (MicroCase) procedure with these three variables in the **Rows** box and recoded *degree* in the **Columns** box. Write a paragraph summarizing the results of these three tests. Which relationships are significant? At what levels?

1. Conduct the appropriate test of significance for each research situation. Problems are stated in no particular order and include research situations from each chapter in Part II.

   a. Is there a gender gap in use of the Internet? Random samples of men and women have been questioned about the average number of minutes they spend each week on the Internet for any purpose. Is the difference significant?

Women	Men
$\bar{X}_1 = 55$	$\bar{X}_2 = 60$
$s_1 = 2.5$	$s_2 = 2.0$
$N_1 = 520$	$N_2 = 515$

   b. For high school students, is there a relationship between social class and involvement in activities such as clubs and sports? Data have been gathered for a random sample of students. Is the relationship significant?

	Class		
Involvement	Middle	Lower	Totals
High	11	19	30
Moderate	19	21	40
Low	8	12	20
Totals	38	52	90

   c. The General Social Survey asks respondents how many total sex partners they have had over their lifetimes. For a subsample of 23 respondents, does the number vary significantly by educational level?

Less than High School	High School	At Least Some College
1	1	2
1	2	1
2	3	5
1	8	9
9	10	11
2	5	2
9	4	1
	3	1

   d. A random sample of the U.S. population was asked how many times they had moved since they were 18 years of age. The results are presented below. On the average, how many times do adult Americans move?

   $$\bar{X} = 3.5$$
   $$s = 0.4$$
   $$N = 1450$$

e. On the average, school districts in a state receive budget support from the state government of $623 per student. A random sample of 107 rural schools reports that they received an average of $605 per pupil. Is the difference significant?

$$\mu = 623 \quad \overline{X} = 605$$
$$s = \ \ 74$$
$$N = 107$$

2. Below are a number of research questions that can be answered by the techniques presented in Chapters 7–11. For each question, select the most appropriate test, compute the necessary statistics, and state your conclusions.

In order to complete some problems, you must first calculate the sample statistics (for example, means or proportions) that are used in the test of hypothesis. Use alpha = 0.05 throughout. The questions are presented in random order. There is at least one problem for each chapter, but I have not included a research situation for every statistical procedure covered in the chapters.

In selecting tests and procedures, you need to consider the question, the number of samples or categories being compared, and the level of measurement of the variables. The database for this problem is based on the General Social Survey (GSS). The actual questions asked and the complete response codes are presented in Appendix G. Abbreviated codes are listed below. The scores of some variables have been simplified or collapsed for this exercise. These problems are based on a small sample, and you may have to violate some assumptions about sample size in order to complete this exercise.

a. Is there a statistically significant difference in average hours of TV watching by income level? By race?

b. Is there a statistically significant relationship between age and happiness?

c. Estimate the average number of hours spent watching TV for the entire population.

d. If Americans currently average 2.3 children, is this sample representative of the population?

e. Are the educational levels of Catholics and Protestants significantly different?

f. Does average hours of TV watching vary by level of happiness?

g. Based on the sample data, estimate the proportion of Black Americans in the population.

SURVEY ITEMS

1. How many children have you ever had? (CHILDS) Scores are actual numbers.
2. Respondent's educational level (DEGREE)
   0. Less than HS
   1. HS
   2. At least some college
3. Race (RACE)
   1. White
   2. Black
4. Age (AGE)
   1. Younger than 35
   2. 35 and older

5. Number of hours of TV watched per day. (TVHOURS) Values are actual number of hours.
6. What is your religious preference? (RELIG)
    1. Protestant
    2. Catholic
7. Respondent's income (INCOME91)
    1. 24999 or less
    2. 25000 or more
8. Respondent's overall level of happiness (HAPPY)
    1. Very happy
    2. Pretty happy
    3. Not too happy

Case	No. of Children	Educational Level	Race	Age	TV Hours	Religious Preference	Income	Happiness
1	3	1	1	1	3	1	1	2
2	2	0	1	1	1	1	2	3
3	4	2	1	2	3	1	1	1
4	0	2	1	1	2	1	1	1
5	5	1	1	1	2	1	2	2
6	1	1	1	2	3	1	2	1
7	9	0	1	1	6	1	1	1
8	6	1	1	2	4	1	1	2
9	4	2	1	1	2	2	2	2
10	2	1	1	1	1	1	2	3
11	2	0	1	2	4	1	1	3
12	4	1	2	1	5	2	1	2
13	0	1	1	2	2	2	1	2
14	2	1	1	2	2	1	1	2
15	3	1	2	2	4	1	1	1
16	2	0	1	2	2	1	2	1
17	2	1	1	2	2	1	2	3
18	0	2	1	2	2	1	2	1
19	3	0	1	2	5	2	1	1
20	2	1	2	1	10	1	1	3
21	2	1	1	2	4	1	2	1
22	1	0	1	2	5	1	2	1
23	0	2	1	1	2	2	1	1
24	0	1	1	2	0	2	2	1
25	2	2	1	1	1	2	2	1

12 BIVARIATE ASSOCIATION
INTRODUCTION AND BASIC CONCEPTS

LEARNING OBJECTIVES

By the end of this chapter, you will be able to

1. Explain how we can use measures of association to describe and analyze the importance of a relationship (vs. its significance).
2. Define "association" in the context of bivariate tables and in terms of changing conditional distributions.
3. List and explain the three characteristics of a bivariate relationship (existence, strength, and pattern or direction).
4. Investigate a bivariate association by properly calculating percentages for a bivariate table and interpreting the results.

12.1 STATISTICAL SIGNIFICANCE AND THEORETICAL IMPORTANCE

As we have seen over the past several chapters, tests of statistical significance are extremely important in social science research. As long as social scientists must work with random samples rather than populations, these tests are indispensable for dealing with the possibility that our research results are the products of mere random chance. However, tests of significance are often merely the first step in the analysis of research results. These tests do have limitations, and statistical significance is not necessarily the same thing as relevance or importance. Furthermore, all tests of significance are affected by sample size: tests performed on large samples may result in decisions to reject the null hypothesis when, in fact, the observed differences are quite minor.

Beginning with this chapter, we will be working with **measures of association.** Whereas tests of significance detect nonrandom relationships, measures of association provide information about the strength and direction of relationships. Furthermore, measures of association can help us reach some important scientific goals. First, they can help us trace causal relationships among variables. The theories that guide scientific research are almost always stated in cause-and-effect terms (for example: "variable X causes variable *Y*"). Recall our discussion of the contact hypothesis in Chapter 1. In that theory, the causal (or independent) variable was equal status contacts between groups, and the effect (or dependent) variable was level of individual prejudice. The theory asserts that involvement in equal-status-contact situations *causes* prejudice to decline. Measures of association are our most important and powerful statistical tools for documenting, measuring, and analyzing causal relationships.

As useful as they are, measures of association do have their limitations. Most importantly, these statistics cannot *prove* that two variables are causally related. Even if there is a strong statistical association between two variables, we cannot necessarily conclude that one variable is a cause of the other. We will explore causation in more detail in Part IV, but for now you should keep in mind that causation and association are two different things. We can use a statistical association between variables as evidence for a causal relationship, but association by itself is not proof of a causal relationship.

Prediction is a second important use for measures of association. If two variables are associated, we can predict the score of a case on one variable from the score of that case on the other variable. For example, if equal status contacts and prejudice are associated, we can predict that people who have experienced many such contacts will be less prejudiced than those who have had few or no contacts. Note that prediction and causation can be two separate matters. If variables are associated, we can predict from one to the other even if the variables are not causally related.

This chapter will introduce the concept of **association** between variables in the context of bivariate tables and will stress the use of percentages to analyze associations between variables. In the chapters that follow, we will concentrate on the logic, calculation, and interpretation of the various measures of association. Finally, in Part IV, we will extend some of these ideas to the multivariate (more than two variables) case.

12.2 ASSOCIATION BETWEEN VARIABLES AND THE BIVARIATE TABLE

Most generally, two variables are said to be associated if the distribution of one of them changes under the various categories or scores of the other. For example, suppose that an industrial sociologist was concerned with the relationship between job satisfaction and productivity for assembly-line workers. If these two variables are associated, then scores on productivity will change under the different conditions of satisfaction. Highly satisfied workers will have different scores on productivity than workers who are low on satisfaction, and levels of productivity will vary by levels of satisfaction.

This relationship will become clearer with the use of bivariate tables. As you recall (see Chapter 11), bivariate tables are devices for displaying the scores of cases on two different variables. By convention, the **independent** or **X variable** (that is, the variable taken as causal) is arrayed in the columns,

TABLE 12.1   PRODUCTIVITY BY JOB SATISFACTION (frequencies)

Productivity (Y)	Job Satisfaction (X)			Totals
	Low	Moderate	High	
Low	30	21	7	58
Moderate	20	25	18	63
High	10	15	27	52
Totals	60	61	52	173

and the **dependent** or **Y variable** in the rows.* That is, each column of the table (the vertical dimension) represents a score or category of the independent variable (X), and each row (the horizontal dimension) represents a score or category of the dependent variable (Y).

Table 12.1 displays a relationship between productivity and job satisfaction for a fictitious sample of 173 factory workers. We focus on the columns to detect the presence of an association between variables displayed in table format. Each column shows the pattern of scores on the dependent variable for each score on the independent variable. For example, the left-hand column indicates that 30 of the 60 workers who were low on job satisfaction were low on productivity, 20 were moderately productive, and 10 were high on productivity. The middle column shows that 21 of the 61 moderately satisfied workers were low on productivity, 25 were moderately productive, and 15 were high on productivity. What is the pattern in the right-hand column for the 52 highly satisfied workers?

By inspecting the table from column to column we can observe the effects of the independent variable on the dependent variable (provided, of course, that the table is constructed with the independent variable in the columns). These "within-column" frequency distributions are called the **conditional distributions of Y,** since they display the distribution of scores on the dependent variable for each condition (or score) of the independent variable.

Table 12.1 indicates that productivity and satisfaction are associated: the distribution of scores on Y (productivity) changes across the various conditions of X (satisfaction). For example, half of the workers who were low on satisfaction were also low on productivity. On the other hand, over half of the workers who were high on satisfaction were high on productivity.

Although it is intended to be a test of significance, the chi square statistic provides another way to detect the existence of an association between two variables that have been organized into table format. Any nonzero value

---

* In the material that follows, we will often, for the sake of brevity, refer to the independent variable as X and the dependent variable as Y.

for the test statistic computed in step 4 indicates that the variables are associated. For example, the obtained chi square for Table 12.1 is 24.2, a value that affirms our previous conclusion, based on the conditional distributions of Y, that an association of some sort exists between job satisfaction and productivity.

Often, the researcher will have already conducted a chi square test before considering matters of association. In such cases, an extended inspection of the conditional distributions of Y will not be necessary to ascertain whether or not the two variables are associated. If the obtained chi square is zero, the two variables are independent and not associated. Any value other than zero indicates some association between the variables. Remember, however, that statistical significance and association are two different things. It is perfectly possible for two variables to be associated (as indicated by a non-zero chi square) but still independent (if we fail to reject the null hypothesis).

In this section, we have defined, in a general way, the concept of association between two variables. We have also shown two different ways to detect the presence of an association. In the next section, we will extend the analysis beyond questions of the mere presence or absence of an association and, in a systematic way, show how additional very useful information about the relationship between two variables can be developed.

12.3 THREE CHARACTERISTICS OF BIVARIATE ASSOCIATIONS

Bivariate associations possess three different characteristics, each of which must be analyzed for a full investigation of the relationship. Investigating these characteristics may be thought of as a process of finding answers to three questions:

1. Does an association exist?
2. If an association does exist, how strong is it?
3. What is the pattern and/or the direction of the association?

We will consider each of these questions separately.

Does an Association Exist?   We have already discussed the general definition of association, and we have seen that we can detect an association by observing the conditional distributions of Y in a table or by using chi square. In Table 12.1, we know that the two variables are associated to some extent because the conditional distributions of productivity (Y) are different across the various categories of satisfaction (X) and because the chi square statistic is a nonzero value.

Comparisons from column to column in Table 12.1 are relatively easy to make because the column totals are roughly equal. This will not always be the case, and it is usually helpful to compute percentages to control for varying column totals. Calculating percentages also helps to make the pattern of association more visible.

TABLE 12.2   PRODUCTIVITY BY JOB SATISFACTION (percentages)

Productivity (Y)	Job Satisfaction (X)		
	Low	Moderate	High
Low	50.00%	34.43%	13.46%
Moderate	33.33%	40.98%	34.62%
High	16.67%	24.59%	51.92%
Totals	100%	100%	100%
	(60)	(61)	(52)

The general rule for calculating percentages with bivariate tables is to compute them in the direction of the independent variable and then compare in the opposite direction. When the independent variable has been arrayed in the columns, compute the percentages within each column separately. Table 12.2 presents column percentages calculated from the data in Table 12.1. Note that this table omits the row marginals and reports the column marginals in parentheses. Besides controlling for any differences in column totals, tables in percentage form are usually easier to read because changes in the conditional distributions of Y are easier to detect.

In Table 12.2, we can readily see that the largest cell changes position from column to column. For workers who are low on satisfaction, the single largest cell is in the top row (low on productivity). For the middle column (moderate on satisfaction), the largest cell is in the middle row (moderate on productivity), and, for the right-hand column (high on satisfaction), it is in the bottom row (high on productivity). Even a cursory glance at the conditional distributions of Y in Table 12.2 reinforces our conclusion that an association does exist between these two variables.

If two variables are not associated, then the conditional distributions of Y will not change across the columns. The distribution of Y would be the same for each condition of X. Table 12.3 illustrates a "perfect nonassociation" between hair color and productivity. Table 12.3 is only one of many patterns that indicate "no association." The important point is that the conditional distributions of Y are the same. Levels of productivity do not change

TABLE 12.3   PRODUCTIVITY BY HAIR COLOR (an illustration of no association)

Productivity (Y)	Hair Color (X)		
	Brown	Blonde	Red
Low	33.33%	33.33%	33.33%
Moderate	33.33%	33.33%	33.33%
High	33.33%	33.33%	33.33%
	100%	100%	100%

at all for the various hair colors and, therefore, no association exists between these variables. Also, the obtained chi square computed from this table would have a value of zero, again indicating no association.

How Strong Is the Association?   Once we establish the existence of the association, we need to develop some idea of how strong the association is. This is essentially a matter of determining the amount of change in the conditional distributions of *Y*. At one extreme, of course, there is the case of "no association" where the conditional distributions of *Y* do not change at all (see Table 12.3). At the other extreme is a perfect association, the strongest possible relationship. In general, a perfect association exists between two variables if each value of the dependent variable is associated with one and only one value of the independent variable.* In a bivariate table, all cases in each column would be located in a single cell and there would be no variation in *Y* for a given value of *X* (see Table 12.4).

A perfect relationship would be taken as very strong evidence of a causal relationship between the variables, at least for the sample at hand. In fact, the results presented in Table 12.4 would indicate that, for this sample, hair color is the sole cause of productivity. Also, in the case of a perfect relationship, predictions from one variable to the other could be made without error. If we knew that a particular worker had brown hair, for example, we could precisely predict that that worker would be highly productive.

Of course, the huge majority of relationships will fall somewhere between the two extremes of no association and perfect association. We need to develop some way of describing these intermediate relationships consistently and meaningfully. For example, Tables 12.1 and 12.2 show that there is an association between productivity and job satisfaction. How could this relationship be described in terms of strength? How close is the relationship to perfect? How far away from no association?

Although you might be able to describe this association impressionistically, the task would certainly be easier if you had a set of objective indices to consistently describe the strength of the association (for example, as strong, moderate, or weak). As we shall see in Chapters 13–15, these objective indices of the strength of a relationship are supplied by the various measures of association. Virtually all of these statistics are designed so that they have a lower limit of 0.00 and an upper limit of 1.00 (±1.00 for ordinal and interval-ratio measures of association). A measure that equals 0.00 indicates no association between the variables (the conditional distributions

---

* Each measure of association that will be introduced in the following chapters incorporates its own definition of a "perfect association," and these definitions vary somewhat, depending on the specific logic and mathematics of the statistic. That is, for different measures computed from the same table, some measures will possibly indicate perfect relationships when others will not. We will note these variations in the mathematical definitions of a perfect association at the appropriate times.

TABLE 12.4 PRODUCTIVITY BY HAIR COLOR (an illustration of perfect association)

Productivity (Y)	Hair Color (X)		
	Brown	Blonde	Red
Low	0%	0%	100%
Moderate	0%	100%	0%
High	100%	0%	0%
	100%	100%	100%

of Y do not vary), and a measure of 1.00 (±1.00 in the case of ordinal and interval-ratio measures) indicates a perfect relationship. The exact meaning of values between 0.00 and 1.00 varies from measure to measure, but, for all measures, the closer the value is to 1.00, the stronger the relationship (the greater the change in the conditional distributions of Y).

Application 12.1

Why are many Americans attracted to movies that emphasize graphic displays of violence? One idea is that "slash" movie fans feel threatened by violence in their daily lives and use these movies as a means of coping with their fears. In the safety of the theater, violence can be vicariously experienced, and feelings and fears can be expressed privately. Also, highly violent movies almost always, as a necessary plot element, provide a role model of one character who does deal with violence successfully (usually, of course, with more violence).

Is fear of violence associated with frequent attendance at high-violence movies? The following table reports the joint frequency distributions of "fear" and "attendance" in percentages for a sample of 600.

Attendance	Fear		
	Low	Moderate	High
Rare	50%	20%	30%
Occasional	30	60	30
Frequent	20	20	40
	100%	100%	100%
	(200)	(200)	(200)

The conditional distributions of attendance (Y) do change across the values of fear (X), so these variables are associated. The clustering of cases in the diagonal from upper left to lower right suggests a substantial relationship in the predicted direction. People who are low on fear are infrequent attenders; people who are high on fear are frequent attenders. These results do suggest an important relationship between fear and attendance. Notice that the results, as presented here, pose an interesting causal problem. The table supports the idea that fearful and threatened people attend violent movies as a coping mechanism (X causes Y). However, the table is also consistent with the reverse causal argument: attendance at violent movies increases fears for one's personal safety (Y causes X). The results support both causal arguments, and I remind you that association is not the same thing as causation.

READING STATISTICS 9: Bivariate Tables

The conventions for constructing and interpreting bivariate tables presented in this text are commonly but not universally followed in the professional literature. Tables will usually be constructed with the independent variable in the columns, the dependent variable in the rows, and percentages calculated in the columns. However, you should be careful to check the format of every table you attempt to read to see if these conventions have been observed. If the table is presented with the independent variable in the rows, for example, you will have to reorient your analysis (or redraw the table) to account for this. Above all, you should convince yourself that the percentages have been calculated in the correct direction. Even skilled professionals occasionally calculate percentages incorrectly and misinterpret the data.

Once you have assured yourself that the table is properly presented, you can apply the analytical techniques developed in this chapter. By comparing the conditional distributions of the dependent variable, you can ascertain for yourself if the variables are associated and check the strength and pattern of the association. You may then compare your conclusions with those of the researchers. As an aid in the interpretation of bivariate tables, researchers will usually compute and report other statistics in addition to percentages. We'll talk about these statistics in Reading Statistics 11 in Chapter 14.

Statistics in the Professional Literature

Researchers Michelle Kelly and Beth Parsons used a survey to measure the extent of sexual harassment in a university located in the Southeastern United States. They wanted to explore how recent federal legislation, national sex scandals, and changing university policy had affected the experiences of women on one particular campus. They included all female employees (faculty, staff, and administrators) and undergraduate and graduate students in their sample of 765 respon-

dents. The undergraduate sample consisted of students enrolled in psychology courses, and the graduate students were randomly selected from a list of all graduate students at the university. Sexual harassment was defined as having three dimensions: gender harassment (e.g., sexual insults or slurs, hostile and degrading remarks or attitudes), unwanted sexual attention, and sexual coercion (extortion of sex in return for job or academic benefits).

Kelly and Parsons were particularly interested in measuring the overall extent of the problem, the differences in type of sexual harassment for different groups, and the differences in type of response by different groups. They reported some results in the body of the report. For example, they found that 30% of the staff, 22% of the faculty, and 20% of the students reported being sexually harassed. To provide some perspective on these figures, they note that other studies estimate the prevalence of sexual harassment for students from 20% to 40% and at about 50% for female faculty.

They reported other results in bivariate table format. For example, Table 1 shows that employees and students experience different forms of sexual harassment.

The authors report in the body of the paper that the table is based on 158 incidents reported by employees and 87 incidents reported by students.

TABLE 1 TYPE OF SEXUAL HARASSMENT BY POSITION OF RESPONDENT

Type of Sexual Harassment	Position	
	Employee	Student
Gender harassment	62%	43%
Unwanted sexual attention	30%	41%
Sexual coercion	8%	16%
	100%	100%

Employees are more likely to experience gender harassment, and students report more incidents of unwanted attention and sexual coercion. What other patterns can you detect? A chi square test conducted on the table shows that the relationship is significant ($\chi^2 = 149$, df = 2, $p < .001$), but this test may not be very meaningful since the sample includes *all* employees and a nonrandom sample of undergraduates. Remember that only the graduate student respondents were selected randomly.

Kelly and Parsons also found that the women in their sample generally responded to sexual harassment in nonassertive (64%) rather than direct (36%) ways. That is, the women tended to ignore the behavior or simply tell a friend rather than file a formal complaint. Also, they found that the response of the victim was very dependent on the status of the perpetrator, as shown in Table 2. Incidents were twice as likely to be ignored if the perpetrator was a student, whereas more formal, direct action (telling a supervisor, filing a complaint) was more likely when the perpetrator was an employee.

Among other points, Kelly and Parsons conclude that their data show a continuing reluctance of women to respond formally to incidents of sexual harassment:

TABLE 2   RESPONSE TO SEXUAL HARASSMENT BY THE POSITION OF THE PERPETRATOR

Response	Position of Perpetrator	
	Employee	Student
Ignored it	15%	31%
Told a friend	20%	22%
Told a coworker	18%	8%
Told person to stop	17%	21%
Told a supervisor	14%	7%
Filed a complaint	4%	1%
Other	12%	10%

Despite changes in public awareness, an expressed stance on sexual harassment by many universities, and a specific policy at the university surveyed, women continue to prefer non-formal and non-assertive responses to sexual harassment.

The authors suggest that simply having a policy may not be enough to deter harassers or entice victims to come forward.

Quoted material from "Sexual Harassment in the 1990s: A University-Wide Survey of Female Faculty, Administrators, Staff, and Students." Michelle Kelly and Beth Parsons. *The Journal of Higher Education.* Vol 71, #5 (September/October 2000).

You may be tempted to regard the chi square test as an indication of the strength of a relationship. Remember that significance and association are two separate matters and chi square, by itself, is not a measure of association. While a nonzero value indicates that there is an association, the magnitude of chi square bears no particular relationship to the strength of the association. That is, chi square might be quite large in value while the actual association is weak. Chapter 13 will introduce some ways to transform chi square into other statistics that do measure the strength of the association between two variables. *(For practice in computing percentages and judging the existence and strength of an association, see any of the problems at the end of this chapter.)*

What Is the Pattern and/or the Direction of the Association?   Investigating the pattern of the association requires that we ascertain which values or categories of one variable are associated with which values or categories

TABLE 12.5 LIBRARY USE BY EDUCATION (an illustration of a positive relationship)

Library Use	Education		
	Low	Moderate	High
Low	60%	20%	10%
Moderate	30	60	30
High	10	20	60
	100%	100%	100%

TABLE 12.6 AMOUNT OF TELEVISION VIEWING BY EDUCATION (an illustration of a negative relationship)

Television Viewing	Education		
	Low	Moderate	High
Low	10%	20%	60%
Moderate	30	60	30
High	60	20	10
	100%	100%	100%

of the other. We have already remarked on the pattern of the relationship between productivity and satisfaction. Table 12.2 indicates that low scores on satisfaction are associated with low scores on productivity, moderate satisfaction with moderate productivity, and high satisfaction with high productivity.

When both variables are at least ordinal in level of measurement, the pattern of association may also have a direction to it.* The direction of the association can be either positive or negative. An association is positive if the variables vary in the same direction. That is, in a **positive association,** high scores on one variable are associated with high scores on the other variable, and low scores on one variable are associated with low scores on the other. In a positive association, as one variable increases in value, the other also increases; and as one variable decreases, the other also decreases. Table 12.5 displays, with fictitious data, a positive relationship between education and use of public libraries. As education increases, library use also increases. The association between job satisfaction and productivity, as displayed in Tables 12.1 and 12.2, is also a positive association.

In a **negative association,** the variables vary in opposite directions. High scores on one variable are associated with low scores on the other, and increases in one variable are accompanied by decreases in the other. Table 12.6 displays a negative relationship, again with fictitious data, between education and television viewership. The amount of television viewing decreases as education increases.

Measures of association for ordinal and interval-ratio variables are designed so that they will take on positive values for positive associations and negative values for negative associations. Thus, a measure of association preceded by a plus sign indicates a positive relationship between the two variables, with the value +1.00 indicating a perfect positive relationship. A negative sign indicates a negative relationship, with −1.00 indicating a perfect

---

* Variables measured at the nominal level have no numerical order to them (by definition). Therefore, associations including nominal-level variables, while they may have a pattern, cannot be said to have a direction.

negative relationship. *(For practice in determining the pattern of an association, see any of the end-of-chapter problems. For practice in determining
the direction of a relationship, see problems 12.1, 12.6, 12.7, 12.8, 12.9,
12.10, and 12.12.)*

SUMMARY

1. Analyzing the association between variables provides information that is complementary to tests of
significance. The latter are designed to detect nonrandom relationships, whereas measures of association are designed to quantify the importance or
strength of a relationship.
2. Relationships between variables have three characteristics: the existence of an association, the strength
of the association, and the direction or pattern of
the association. These three characteristics can be
investigated by calculating percentages for a bivariate table in the direction of the independent variable
(vertically) and then comparing in the opposite direction (horizontally).
3. Tables 12.1 and 12.2 can be analyzed in terms of
these three characteristics. Clearly, a relationship
does exist between job satisfaction and productivity, since the conditional distributions of the dependent variable (productivity) are different for the
three different conditions of the independent variable (job satisfaction). Even though we cannot
quantify the strength of the relationship yet, we
can see that the association is substantial in that the
change in Y (productivity) across the three categories of X (satisfaction) is marked. Furthermore, the
relationship is positive in direction. Productivity increases as job satisfaction rises, and workers who

report high job satisfaction tend also to be high on
productivity. Workers with little job satisfaction tend
to be low on productivity.
4. Given the nature and strength of the relationship, it
could be predicted with fair accuracy that highly satisfied workers tend to be highly productive ("happy
workers are busy workers"). These results might be
taken as evidence of a causal relationship between
these two variables, but they cannot, by themselves,
prove that a causal relationship exists: association
is not the same thing as causation. In fact, although
we have presumed that job satisfaction is the independent variable, we could have argued the reverse causal sequence ("busy workers are happy
workers"). The results presented in Tables 12.1 and
12.2 are consistent with both causal arguments.

   Thus, the analysis of the association between
variables may be seen as a very important process
by which evidence for (or against) a causal relationship is systematically gathered. Ultimate proof
for causal relationships depends more on logical,
theoretical, and methodological grounds (actually
proving causation is a rather difficult task). As we
shall see in Part IV, some of the multivariate techniques are quite useful for probing possible causal
relationships, and we will return to some of these
concerns at that point.

GLOSSARY

Association. The relationship between two (or more)
variables. Two variables are said to be associated if
the distribution of one variable changes for the various categories or scores of the other variable.

Conditional distribution of Y. The distribution of
scores on the dependent variable for a specific
score or category of the independent variable when
the variables have been organized into table format.

Dependent variable. In a bivariate relationship, the
variable that is taken as the effect.

Independent variable. In a bivariate relationship,
the variable that is taken as the cause.

Measures of association. Statistics that quantify the
strength of the association between variables.

Negative association. A bivariate relationship where
the variables vary in opposite directions. As one

variable increases, the other decreases, and high scores on one variable are associated with low scores on the other.

Positive association. A bivariate relationship where the variables vary in the same direction. As one variable increases, the other also increases, and high scores on one variable are associated with high scores on the other.

X. Symbol used for any independent variable.

Y. Symbol used for any dependent variable.

MULTIMEDIA RESOURCES

The Wadsworth Sociology Resource Center: Virtual Society
http://sociology.wadsworth.com/

Visit the companion web site for the sixth edition of *Statistics: A Tool for Social Research* to access a wide range of student resources. Begin by clicking on the Student Resources section of the book's web site to access the following study tools:

- Basic math review
- Statistics review

- Flash cards
- Internet links
- Additional chapter problems
- Table of random numbers
- MicroCase and SPSS examples and exercises
- "Find the text" flowcharts
- Hypothesis testing for variables measured at the ordinal level

PROBLEMS

12.1 PA Various supervisors in the city government of Shinbone, Kansas, have been rated on the extent to which they practice authoritarian styles of leadership and decision making. The efficiency of each department has also been rated, and the results are summarized below. Calculate percentages for the table so that it shows the effect of leadership style on efficiency. Is there an association between these two variables? What are the strength and direction of the relationship?

| | Authoritarianism | |
Efficiency	Low	High
Low	10	12
High	17	5
Totals	27	17

12.2 SOC The administration of a local college campus has proposed an increase in the mandatory student fee in order to finance an upgrading of the intercollegiate football program. A sample of the faculty has completed a survey on the issues.

Is there any association between support for raising fees and the gender, discipline, or tenured status of the faculty? Describe the strength and direction of the association.

a. Support for raising fees by gender:

Support	Males	Females
For	12	8
Against	15	12
Totals	27	20

b. Support for raising fees by discipline:

Support	Liberal Arts	Science & Business
For	6	13
Against	14	14
Totals	20	27

c. Support for raising fees by tenured status:

Support	Tenured	Nontenured
For	15	4
Against	18	10
Totals	33	14

12.3 PS How consistent are people in their voting habits? Do people vote for the same party from election to election? Below are the results of a poll in which people were asked if they had voted Democrat or Republican in each of the last two presidential elections. Assess the strength of this relationship.

| | 1996 Election | |
2000 Election	Democrat	Republican
Democrat	117	23
Republican	17	178
Totals	134	201

12.4 SOC A needs assessment survey has been distributed in a large retirement community. Residents were asked to check off the services or programs they thought should be added. Is there any association between gender and the perception of a need for more social occasions? Write a few sentences describing the relationship in terms of pattern and strength of the association.

More Parties?	Male	Female
Yes	321	426
No	175	251
Totals	496	677

12.5 For problems 11.13 to 11.15 in Chapter 11, calculate percentages in the direction of the independent variable. Is there an association between the variables? Characterize each association in terms of the strength and direction of the relationship.

12.6 SW As the state director of mental health programs, you note that some local mental health facilities have very high rates of staff turnover. You believe that part of this problem is a result of the fact that some of the local directors have very little training in administration and poorly developed leadership skills. Before implementing a program to address this problem, you collect some data to make sure that your beliefs are supported by the facts. Is there a relationship between staff turnover and the administrative experience of the directors? Describe the rela-

tionship in terms of pattern and strength of the association.

| | Director Experienced? | |
Turnover	No	Yes
Low	4	9
Moderate	9	8
High	15	5
Totals	28	22

12.7 CJ About half the neighborhoods in a large city have instituted programs to increase citizen involvement in crime prevention. Do these areas experience less crime? Write a few sentences describing the relationship in terms of pattern and strength of the association.

| Crime Rate | Program | |
	No	Yes
Low	29	15
Moderate	33	27
High	52	45
Totals	114	87

12.8 PS Are people who think of themselves as political liberals really more liberal on social issues? The table below summarizes the relationship between political ideology and the respondent's position on the legalization of marijuana. Are the variables associated? Write a few sentences describing the relationship in terms of pattern and strength of the association.

Should Marijuana Be Legal?	Liberal	Moderates	Conservatives
Yes	132	78	52
No	101	87	109
Totals	233	165	161

12.9 SOC A researcher hypothesizes that physical attractiveness is related to academic achievement (that is, the more attractive the student, the higher the grade). One hundred students are rated on physical attractiveness and asked for their grade-point averages. The bivariate table that follows summarizes the data. Calculate percentages for the table in the proper direction and compare the conditional distributions. Is there an association between these variables? What is your impression of the strength of the association? What is the direction of the association? Is the hypothesis supported? Why or why not?

	Attractiveness		
GPA	Low	Moderate	High
Low	7	8	15
Moderate	10	10	16
High	8	12	14
Totals	25	30	45

12.10 SOC The latest fad to sweep college campuses is streaking to panty raids while swallowing live goldfish. A researcher is interested in how closely the spread of this bizarre behavior is linked to the amount of coverage and publicity provided by local campus newspapers. For a sample of 25 universities, the researcher has rated the amount of press coverage (as extensive, moderate, or no coverage) and how much the student body was involved in this new fad. The data for each campus are reported below. Organize the data into a properly labeled table in percentage form. Does the table indicate an association between press coverage and fad behavior?

Campus	Amount of Press Coverage	Extent of Student Involvement
1	Extensive	Extensive
2	Extensive	Some
3	Moderate	Some
4	Moderate	Some
5	Moderate	Extensive
6	Extensive	Some
7	Extensive	Extensive
8	Moderate	Some
9	None	Some
10	Moderate	None
11	None	None
12	Extensive	Extensive
13	None	Some
14	Extensive	Extensive
15	Moderate	None
16	Moderate	Some
17	Moderate	Extensive
18	Moderate	None
19	None	None
20	Extensive	Some

Campus	Amount of Press Coverage	Extent of Student Involvement
21	None	Extensive
22	Moderate	None
23	None	None
24	Extensive	Extensive
25	Moderate	Extensive

12.11 In any social science journal, find an article that includes a bivariate table. Inspect the table and the related text carefully and answer the following questions:

a. Identify the variables in the table. What values (categories) does each possess? What is the level of measurement for each variable?

b. Is the table in percentage form? In what direction are the percentages calculated? Are comparisons made between columns or rows?

c. Is one of the variables identified by the author as independent? Are the percentages in the direction of the independent variable?

d. How is the relationship characterized by the author in terms of the strength of the association? In terms of the direction (if any) of the association?

e. Find the measure of association (if any) calculated for the table. What is the numerical value of the measure? What is the sign (if any) of the measure?

12.12 Let's see if political ideology is associated with a variety of different issues and concerns. The tables that follow are taken from the 1998 General Social Survey. Political ideology (POLVIEWS) has been collapsed into three categories and is used as the "column," or independent, variable. If liberals are generally progressive and conservatives are generally traditional (with moderates in between), what relationships would you expect to find between political ideology and these issues?

a. Support for the legal right to an abortion

b. The death penalty

c. The legal right to commit suicide for people with incurable disease

d. Sex education in schools

e. Support for traditional gender roles

For each table, compute percentages and describe the pattern and strength of the relationship. Are the patterns consistent with your predictions? Overall, is political ideology an important factor in shaping people's views and opinions? See Appendix G for the exact wording of each item.*

a. Support for the legal right to an abortion for any reason (ABANY) by political ideology:

Right to Abortion?	Political Ideology		
	Liberal	Moderate	Conservative
Yes	309	234	154
No	211	360	419
Totals	520	594	573

b. Support for capital punishment (CAPPUN) by political ideology:

Capital Punishment	Political Ideology		
	Liberal	Moderate	Conservative
Favor	440	693	693
Oppose	265	214	186
Totals	705	907	879

c. Approve of suicide for people with incurable disease (SUICIDE1) by political ideology:

Right to Suicide?	Political Ideology		
	Liberal	Moderate	Conservative
Yes	381	394	319
No	120	229	261
Totals	501	623	580

d. Support for sex education in public schools (SEXEDUC) by political ideology:

Sex Education?	Political Ideology		
	Liberal	Moderate	Conservative
Favor	481	572	476
Oppose	28	63	123
Totals	509	635	599

e. Support for traditional gender roles (FEHOME) by political ideology:

Women should take care of running their homes and leave running the country to men	Political Ideology		
	Liberal	Moderate	Conservative
Agree	59	90	108
Disagree	454	548	484
Totals	513	638	592

SSPS for Windows

Using *SPSS for Windows* to Analyze Bivariate Association

SPSS DEMONSTRATION 12.1 Do Religious Beliefs Vary by Social Class?

In Demonstrations 11.1 and 11.2, we conducted tests using social class as the independent variable and found no significant relationships with either attitude about spanking or watching X-rated movies. In this demonstration, we'll continue to assess the importance of social class as an independent variable, this time focusing on its relationship with belief in an afterlife.

Click **Analyze, Descriptive Statistics,** and **Crosstabs** and name *postlife* as the row variable and *class* as the column variable. Click the **Cells** button and request column percentages by clicking the box next to **Column** in the **Percentages** box. Also, request chi square by clicking the **Statistics** button. Click **Continue** and **OK,** and the following output will be produced:

---

*Sample size (*N*) varies from table to table because not all respondents were asked every question.

BELIEF IN LIFE AFTER DEATH SUBJECTIVE CLASS IDENTIFICATION Crosstabulation

			SUBJECTIVE CLASS IDENTIFICATION				
			LOWER CLASS	WORKING CLASS	MIDDLE CLASS	UPPER CLASS	Total
BELIEF IN LIFE AFTER DEATH	YES	Count	30	370	374	25	799
		% within SUBJECTIVE CLASS IDENTIFICATION	65.2%	81.7%	82.4%	73.5%	81.0%
	NO	Count	16	83	80	9	188
		% within SUBJECTIVE CLASS IDENTIFICATION	34.8%	18.3%	17.6%	26.5%	19.0%
Total		Count	46	453	454	34	987
		% within SUBJECTIVE CLASS IDENTIFICATION	100.0%	100.0%	100.0%	100.0%	100.0%

Chi-Square Tests

	Value	df	Asymp. Sig. (2-sided)
Pearson Chi-Square	9.355	3	.025
Likelihood Ratio	8.184	3	.042
Linear-by-Linear Association	1.197	1	.274
N of Valid Cases	987		

There is an association between the variables (the conditional distribution of *Y* or *postlife* changes from column to column), and—judging by the amount of change —the relationship appears to be moderate in strength. What makes this relationship "moderate"? Basically, the conditional distributions change substantially in some instances but not in all.

Observe the top row of the table ("yes"), and you will see a change of about 15% from column 1 (lower class) to column 2 (working class). However, the percentage of cases in this row is about the same for the working and middle classes and then drops by about 10% for upper-class respondents. In a strong relationship, we would expect to find more pronounced changes between every column in the table.

As for the direction of the relationship, it is generally positive: as class increases, belief in an afterlife (% "yes") increases. Remember, however, that the increase in percentage across the table is not constant since there is a decrease in the percentage that believe in an afterlife between the middle and upper classes. Finally, the chi square of 9.355 is significant at less than .05. These variables are dependent on one another.

SPSS DEMONSTRATION 12.2 Does Support for Gun Control Vary by Political Ideology?

Gun control is a highly politicized issue in our society and, given the way in which this issue is usually debated, we would expect liberals to support gun control, conservatives to oppose it, and moderates to be intermediate. In the 1998 GSS, support for gun control is measured by the variable *gunlaw,* and political ideology is measured by *polviews*. The latter has too many categories (7) to be used in a bivariate table, so we must first recode the variable. Click **Transform** from the main menu and remember to choose "Recode into Different Variable." In this demonstration, I named the recoded version of the variable *pol*. The original scores for this variable are given in Appendix G, and the recoding instructions that should appear in the **Old New** box of the **Recode into Different Variable** window should be

$$1 \text{ thru } 3 \rightarrow 1$$
$$4 \rightarrow 2$$
$$5 \text{ thru } 7 \rightarrow 3$$

This scheme groups all liberals (a score of 1 on *pol*) and conservatives (a score of 3 on *pol*) together and changes the score associated with "moderates" (a score of 2 on *pol*) so that they remain intermediate between the other two scores.

The **Crosstabs** procedure is accessed by following the steps explained in Demonstration 12.1. Click **Analyze, Descriptive Statistics,** and **Crosstabs.** Specify *gunlaw* as the row variable, or the dependent variable (the label for this variable is FAVOR OR OPPOSE GUN PERMITS), and *pol* (recoded *polviews*) as the column, or independent, variable. On the **Crosstabs** window, click **Cells** and, under **Percentages,** click the radio button next to **Columns.** With the dependent variable in the rows and the independent variable in the columns and with percentages calculated within columns, we will be able to read the table by following the rules developed in this chapter. Also, click the **Statistics** button and request chi square. The bivariate table for *gunlaw* and *pol* is shown below.

FAVOR OR OPPOSE GUN PERMITS POL Crosstabulation

			POL			
			1.00	2.00	3.00	Total
FAVOR OR OPPOSE GUN PERMITS	FAVOR	Count	235	247	239	721
		% within POL	87.0%	83.2%	78.9%	82.9%
	OPPOSE	Count	35	50	64	149
		% within POL	13.0%	16.8%	21.1%	17.1%
Total		Count	270	297	303	870
		% within POL	100.0%	100.0%	100.0%	100.0%

Chi-Square Tests

	Value	df	Asymp. Sig. (2-sided)
Pearson Chi-Square	6.724	2	.035
Likelihood Ratio	6.780	2	.034
Linear-by-Linear Association	6.710	1	.010
N of Valid Cases	870		

In each cell of the table, the count is reported first. For example, there are 235 cases in the upper-left-hand cell. These are liberals (1 on *pol*) who favor gun control. Below the count is the percentage of all cases in that column (all liberals) that are in favor of gun control. Reading across the columns, 87.0% of the liberals favor gun control, as compared with 83.2% of the moderates and 78.9% of the conservatives. These results are similar to what we expected and suggest that the variables are associated. This conclusion is reinforced by the fact that chi square is significant at less than .05 ("Significance" is reported as .035, to be exact). However, although an association exists, the great majority of cases in each column are in favor of gun control and there is not much variation in support from column to column. The relationship between support for gun control and political ideology is, at best, moderate in strength.

SPSS DEMONSTRATION 12.3 The Direction of Relationships: Do Attitudes Toward Sex Vary by Age?

Let's take a look at a relationship between two ordinal-level variables so that you can develop some experience in describing the direction as well as the strength of bivariate relationships. Both of our variables have to be at least ordinal in level of measurement, so let's start with *age* as an independent variable and **Recode** it into three categories, as in Demonstration 10.1. The recoding instructions were

$$18 \text{ thru } 34 \rightarrow 1$$
$$35 \text{ thru } 50 \rightarrow 2$$
$$51 \text{ thru } 89 \rightarrow 3$$

I used these particular cutting points to divide the sample into three categories of roughly equal size.

Without looking at the data, I'm willing to bet that *age* will have a negative relationship with approval of premarital sex (*premarsx*). Run the **Crosstabs** procedure (see Demonstration 12.1) with *premarsx* as the row variable and *ager* (recoded *age*) as the column variable. Click the **Cells** button and make sure that the radio button next to **Columns** under **Percentages** is checked. If you wish, click the **Statistics** button and request a chi square test. The bivariate table for these two variables is shown.

Crosstabs
SEX BEFORE MARRIAGE AGER Crosstabulation

			AGER			
			1.00	2.00	3.00	Total
SEX BEFORE MARRIAGE	ALWAYS WRONG	Count	54	64	104	222
		% within AGER	18.8%	20.6%	38.0%	25.5%
	ALMOST ALWAYS WRONG	Count	17	25	31	73
		% within AGER	5.9%	8.1%	11.3%	8.4%
	SOMETIMES WRONG	Count	68	56	59	183
		% within AGER	23.7%	18.1%	21.5%	21.0%
	NOT WRONG AT ALL	Count	148	165	80	393
		% within AGER	51.6%	53.2%	29.2%	45.1%
Total		Count	287	310	274	871
		% within AGER	100.0%	100.0%	100.0%	100.0%

Chi-Square Tests

	Value	df	Asymp. Sig. (2-sided)
Pearson Chi-Square	54.350	6	.000
Likelihood Ratio	54.686	6	.000
Linear-by-Linear Association	39.239	1	.000
N of Valid Cases	871		

Is there a relationship between these two variables? Do the conditional distributions change? Inspect the table column by column, and I think you will agree that there is a relationship. Is this a positive or negative relationship? Remember that in a positive association, high scores on one variable will be associated with high scores on the other and low scores will be associated with low. In a negative relationship, high scores on one variable are associated with low scores on the other.

Now go back and look at the table again. Find the single largest cell in each column and see if you can detect the pattern. The lowest score (youngest age group) on *ager* is in the left-hand column. For this age group, the most common score (51.6% of this age group) on *premarsx* is 4 (not wrong at all). (To view the numerical codes associated with each response on a variable, see Appendix G or click **Utilities** and **Variables.**) In other words, a slight majority of the youngest respondents supports the most permissive position on premarital sex. This score is also the most common response for the middle group on age (35 to 50 years old), but the older respondents are much less likely to endorse this view. Less than 30% of the 51 to 89 age group felt than premarital sex was "not wrong at all." So, we could say that permissive attitudes tend to *decline* as age *increases*— the relationship is *negative* in direction.

Now look at the top row of the table. This is the least permissive position on premarital sex, and the percentage of cases in this row increases as age increases. So, we could say that opposition to premarital sex *increases* as age *increases*— the variables change in the *same* direction, so the relationship is *positive*.

Which characterization of the relationship is correct? Both: the relationship is negative if you think of *premarsx* as measuring approval and positive if you think of the variable as measuring opposition to premarital sex. The point, of course, is to call your attention to the fact that the direction of a relationship can be ambiguous when we are dealing with ordinal-level variables like *premarsx*. In a positive association the scores always vary in the same direction and, in a negative relationship, they always vary in opposite directions. However, because the coding for ordinal-level variables is arbitrary, higher scores don't necessarily mean that the quantity being measured is increasing, and lower scores don't always mean that the quantity is decreasing. Always pay careful attention to the coding scheme for the variable and exercise caution when analyzing the direction of relationships that involve ordinal-level variables.

MicroCase

Using MicroCase to Analyze Bivariate Association

MICROCASE DEMONSTRATION 12.1
Do Religious Beliefs Vary by Social Class?

In Demonstrations 11.1 and 11.2, we conducted tests using social class as the independent variable and found no significant relationships with either attitude about spanking or watching X-rated movies. In this demonstration, we'll continue to assess the importance of social class as an independent variable, this time focusing on its relationship with belief in an afterlife.

Click **Cross-tabulation** from the **Statistics** menu. The **Cross-tabulation** dialog box will appear with the variables listed in the widow to the left. Highlight *post-*

life (variable # 24) and click the arrow to move the variable name into the **Row Variable** box, and then highlight *class* (variable #30) and move it into the **Column Variable** box. Click **OK** to get the bivariate table. Click **Column %** in the **Tables** box in the upper-left-hand corner, and the table will look like this:

```
POSTLIFE by CLASS
           LOWER   WORKING  MIDDLE   UPPER   Missing   TOTAL
YES           30       370     374      25         4     799
           65.2%     81.7%   82.4%   73.5%             81.0%
NO            16        83      80       9         1     188
           34.8%     18.3%   17.6%   26.5%             19.0%
Missing      21       175     183      13         3     395
TOTAL        46       453     454      34         8     987
          100.0%    100.0%  100.0%  100.0%
```

There is an association between the variables (the conditional distribution of *Y* or *postlife* changes from column to column), and—judging by the amount of change—the relationship appears to be moderate in strength. What makes this relationship "moderate"? Basically, the conditional distributions change substantially in some instances but not in all.

Observe the top row of the table ("yes"), and you will see a change of about 15% from column 1 (lower class) to column 2 (working class). However, the percentage of cases in this row stays about the same for the working and middle classes and then drops by about 10% for upper-class respondents. In a strong relationship, we would expect to find much more pronounced changes across the columns.

As for the direction of the relationship, it is generally positive: as class increases, belief in an afterlife (% "yes") increases. Remember, however, that the increase in percentage across the table is not constant since there is a decrease in the percentage "yes" between the middle and upper classes. Finally, the chi square of 9.355 is significant at less than .05. (You can get chi square by clicking **Summary** in the **Statistics** box). These variables are dependent on one another.

MICROCASE DEMONSTRATION 12.2 Does Support for Gun Control Vary by Political Ideology?

Gun control is a highly politicized issue in our society and, given the way in which this issue is usually debated, we would expect liberals to support gun control, conservatives to oppose it, and moderates to be intermediate. In the 1998 GSS, support for gun control is measured by the variable *gunlaw,* and political ideology is measured by *polviews.* The latter has too many categories (7) to be used in a bivariate table, so we must first recode the variable. Use the **Collapse Variable** command to recode *polviews.* We want to group all liberals and all conservatives into single categories and keep the moderates intermediate between the other two categories. The original scores for this variable are given in Appendix G. Collapse scores 1–3 (liberals) into one category, 4 (moderates) into the second category, and scores of 5–7 (conservatives) into a third category. I named the new (collapsed) variable *pol* and labeled each of the three categories.

The **Cross-tabulation** procedure is accessed by following the steps explained in Demonstration 12.1. Specify *gunlaw* (variable #20) as the row variable, or the dependent variable, and *pol* (recoded *polviews*, probably variable #51) as the column, or independent, variable. Once the table is displayed, click **Column %** in the **Tables** box in the upper-left-hand corner, and the table will look like this:

GUNLAW by POL	liberals	moderates	conserv's	Missing	TOTAL
FAVOR	235	247	239	44	721
	87.0%	83.2%	78.9%		82.9%
OPPOSE	35	50	64	3	149
	13.0%	16.8%	21.1%		17.1%
Missing	119	189	141	21	470
TOTAL	270	297	303	68	870
	100.0%	100.0%	100.0%		

In each cell of the table, the count is reported first. For example, there are 235 cases in the upper-left-hand cell. These are liberals who favor gun control. Below the count is the percentage of all cases in that column (all liberals) that are in favor of gun control. Reading across the columns, 87.0% of the liberals favor gun control, as compared with 83.2% of the moderates and 78.9% of the conservatives. These results are similar to what we expected and suggest that the variables are associated. This conclusion is reinforced by the fact that chi square (click **Summary** in the **Statistics** box) is significant at less than .05 ("Prob" or alpha is reported as 0.035, to be exact). However, although an association exists, the great majority of cases in each column are in favor of gun control, and there is not much variation in support from column to column. The relationship between support for gun control and political ideology is, at best, moderate in strength.

MICROCASE DEMONSTRATION 12.3 The Direction of Relationships: Do Attitudes Toward Sex Vary by Age?

Let's take a look at a relationship between two ordinal-level variables so that you can develop some experience in describing the direction as well as the strength of bivariate relationships. Both of our variables have to be at least ordinal in level of measurement, so let's start with *age* (variable # 6) as an independent variable and use the **Collapse Categories** command to reduce it to three categories, as in Demonstration 10.1. The recoding scheme was to group ages 18 through 34 into category 1, 35 through 50 into category 2, and ages 51 through 89 into category 3. I used these particular cutting points to divide the sample into three categories of roughly equal size.

Without looking at the data, I'm willing to bet that *age* will have a negative relationship with approval of premarital sex (*premarsx*). Run the **Cross-tabulation** procedure (see Demonstration 12.1) with *premarsx* (variable # 35) as the row variable and *ager* (recoded *age*) as the column variable. Once the table is displayed, click **Column %** in the **Tables** box in the upper-left-hand corner, and the table will look like this:

```
PREMARSX by AGER
              18-34     35-50     51-89   Missing   TOTAL
ALWAYS WRO      54        64        104        0      222
              18.8%     20.6%     38.0%             25.5%
ALMST ALWA      17        25         31        0       73
               5.9%      8.1%     11.3%              8.4%
SOMETIMES       68        56         59        0      183
              23.7%     18.1%     21.5%             21.0%
NOT WRONG      148       165         80        2      393
              51.6%     53.2%     29.2%             45.1%
Missing        160       166        188        0      514
TOTAL          287       310        274        2      871
             100.0%    100.0%    100.0%
```

Is there a relationship between these two variables? Do the conditional distributions change? Inspect the table column by column, and I think you will agree that there is a relationship. Is this a positive or negative relationship? Remember that in a positive association, high scores on one variable will be associated with high scores on the other and low scores will be associated with low. In a negative relationship, high scores on one variable are associated with low scores on the other.

Now go back and look at the table again. Find the single largest cell in each column and see if you can detect the pattern. The lowest score (or youngest age group) on *ager* is in the left-hand column. For this age group, the most common score (51.6% of this age group) on *premarsx* is 4 (not wrong at all). (To view the numerical codes associated with each response on a variable, see Appendix G). In other words, a slight majority of the youngest respondents supports the most permissive position on premarital sex. This score is also the most common response for the middle group on age (35 to 50 years old), but older respondents are much less likely to endorse this view. Less than 30% of the 51 to 89 age group felt than premarital sex was "not wrong at all." So, we could say that permissive attitudes tend to *decline* as age *increases*—the relationship is *negative* in direction.

Now look at the top row of the table. This is the least permissive position on premarital sex, and the percentage of cases in this row increases as age increases. So, we could say that opposition to premarital sex *increases* as age *increases*—the variables change in the *same* direction, so the relationship is *positive.*

Which characterization of the relationship is correct? Both: the relationship is negative if you think of *premarsx* as measuring approval and positive if you think of the variable as measuring opposition to premarital sex. The point, of course, is to call your attention to the fact that the direction of a relationship can be ambiguous when we are dealing with ordinal-level variables like *premarsx*. In a positive association the scores always vary in the same direction and, in a negative relationship, they always vary in opposite directions. However, because the coding for ordinal-level variables is arbitrary, higher scores don't necessarily mean that the quantity being measured is increasing, and lower scores don't always mean that the quantity is decreasing. Always pay careful attention to the coding scheme for the variable and exercise caution when analyzing the direction of relationships that involve ordinal-level variables.

Exercises

12.1 As long as *polviews* has already been recoded, pick a few more "social is-sues" (such as *cappun* or *busing*) and, with *polviews* as the independent variable, see if the patterns of association conform to expectations. Are "lib-erals" really more liberal on these issues? For each table, be sure to request column percentage in the cells and the chi square test. For each table, write a sentence or two of interpretation.

12.2 See if you can assess the strongest determinant of attitudes on an issue such as *cappun, grass, gunlaw,* or *busing.* You already have results for *polviews* as an independent variable. Run the same dependent or row variables against *sex, relig, class,* and *race* and compare the strength and direction or pattern of the relationships with each other and with *polviews*. Which independent variable has the strongest effect on the "social issue" you choose to analyze?

13

ASSOCIATION BETWEEN VARIABLES MEASURED AT THE NOMINAL LEVEL

LEARNING OBJECTIVES   By the end of this chapter, you will be able to

1. Calculate and interpret phi, Cramer's V, and lambda.
2. Explain the logic of proportional reduction in error in terms of lambda.
3. Use any of the three measures of association to analyze and describe a bivariate relationship in terms of the three questions introduced in Chapter 12.

13.1 INTRODUCTION   Measures of association are descriptive statistics that summarize the overall strength of the association between two variables. Because they represent that relationship in a single number, these statistics are more efficient methods of expressing an association than conditional distributions and column percentages. As with any summarizing technique, however, a certain amount of information (detail and nuance) about the relationship will be lost if the researcher considers only the measure of association. Always make it a habit to inspect the patterns of cell frequencies or percentages in the table along with the summary measure of association in order to maximize the amount of information you have about the relationship. You should do this regardless of the level of measurement of the data or the specific measure that has been calculated.

As we shall see, there are many measures of association. In this text, these statistics have been organized according to the level of measurement for which they are most appropriate. In this chapter, we will consider measures appropriate for nominally measured variables. You will note that several of the research situations used as examples involve variables measured at different levels (for example, one nominal-level variable and one ordinal-level variable). The general procedure in the situation of "mixed levels" is to be conservative and select measures of association appropriate for the lower of the two levels of measurement.

13.2 CHI SQUARE-BASED MEASURES OF ASSOCIATION

Over the years, statisticians have relied heavily on measures of association based on the value of chi square. When the value of chi square is already known, these measures are easy to calculate. To illustrate, let us reconsider Table 11.3, which displayed, with fictitious data, a relationship between accreditation and employment for social work majors. For the sake of convenience, this table is reproduced here as Table 13.1.

We saw in Chapter 11 that this relationship is statistically significant ($\chi^2 = 10.78$, which is significant at $\alpha = .05$). The question now concerns the strength of the association. A brief glance at Table 13.1 shows that the conditional distribution of employment status does change, so the variables are associated. To emphasize this point, it is always helpful to calculate the percentages, as in Table 13.2.

So far, we know that the relationship between these two variables is statistically significant and that there is an association of some kind between accreditation and employment. To assess the strength of the association, we will compute a **phi (ϕ).** This statistic is a frequently used chi square–based measure of association appropriate for 2 × 2 tables (that is, tables with two rows and two columns). The formula for phi is

FORMULA 13.1

$$\phi = \sqrt{\frac{\chi^2}{N}}$$

One of the attractions of phi is that it is easy to calculate. Simply divide the value of the obtained chi square by N and take the square root of the result.

TABLE 13.1   EMPLOYMENT STATUS OF SOCIAL WORK MAJORS BY ACCREDITATION STATUS OF UNDERGRADUATE PROGRAM (fictitious data)

| | Accreditation Status | | |
	Accredited	Not Accredited	Totals
Employment			
Employed as a social worker	30	10	40
Not employed as a social worker	25	35	60
Totals	55	45	100

TABLE 13.2   EMPLOYMENT STATUS BY ACCREDITATION STATUS (percentages)

| | Accreditation Status | |
Employment	Accredited	Not Accredited
Employed as a social worker	54.55%	22.22%
Not employed as a social worker	45.45%	77.78%
	100.00%	100.00%

For the data displayed in Table 13.1, the chi square was 10.78. Therefore, phi is

$$\phi = \sqrt{\frac{\chi^2}{N}}$$

$$\phi = \sqrt{\frac{10.78}{100}}$$

$$\phi = 0.33$$

For a 2 × 2 table, phi ranges in value from 0 (no association) to 1.00 (perfect association). The phi of .33 for Table 13.1 indicates a moderate relationship between accreditation and employment status. The relationship between these two variables is statistically significant and moderately strong. As for the pattern of the association, Table 13.2 shows that graduates of accredited programs were more often employed as social workers.

For tables larger than 2 × 2 (specifically, for tables with more than two columns and more than two rows), the upper limit of phi can exceed 1.00. This makes phi difficult to interpret, and a more general form of the statistic called **Cramer's V** must be used for larger tables. The formula for Cramer's V is

FORMULA 13.2
$$V = \sqrt{\frac{\chi^2}{(N)(\text{Minimum of } r - 1, \, c - 1)}}$$

where (Minimum of $r - 1$, $c - 1$) = the minimum value of $r - 1$ (number of rows minus 1) or $c - 1$ (number of columns minus 1)

To calculate V, find the square root of chi square divided by the quantity N multiplied by the number of rows minus 1 or the number of columns minus 1, whichever is the lower (or "minimum") value. Cramer's V has an upper limit of 1.00 for any size table and will be the same value as phi if the table has either two rows or two columns. Like phi, Cramer's V can be interpreted as an index that measures the strength of the association between two variables.

To illustrate the computation of V, suppose you had gathered the data displayed in Table 13.3, which shows the relationship between membership in student organizations and academic achievement for a sample of college

TABLE 13.3   ACADEMIC ACHIEVEMENT BY CLUB MEMBERSHIP

Academic Achievement	Membership			Totals
	Fraternity or Sorority	Other Organization	No Memberships	
Low	4	4	17	25
Moderate	15	6	4	25
High	4	16	5	25
Totals	23	26	26	75

TABLE 13.4   ACADEMIC ACHIEVEMENT BY CLUB MEMBERSHIP (percentages)

Academic Achievement	Membership		
	Fraternity or Sorority	Other Organization	No Memberships
Low	17.4	15.4	65.4
Moderate	65.2	23.1	15.4
High	17.4	61.5	19.2
	100.0	100.0	100.0

students. The obtained chi square for this table is 31.5, a value that is significant at the .05 level. Cramer's V is

$$V = \sqrt{\frac{\chi^2}{(N)(\text{Minimum of } r - 1, c - 1)}}$$

$$V = \sqrt{\frac{31.5}{(75)(2)}}$$

$$V = \sqrt{\frac{31.5}{150}}$$

$$V = .\sqrt{.21}$$

$$V = .46$$

Since Table 13.3 has the same number of rows and columns, we may use either $(r - 1)$ or $(c - 1)$ in the denominator. In either case, the value of the denominator is N multiplied by $(3 - 1)$, or 2. The computed value of V of .46 suggests a moderate to strong association between club membership and academic achievement.

Calculating percentages as in Table 13.4 will help identify the pattern of the relationship. Fraternity and sorority members tend to be moderate, members of other organizations tend to be high, and nonmembers tend to be low in academic achievement.

A limitation of phi and Cramer's V, as you may have noticed, is the absence of a direct or meaningful interpretation for values between the extremes of 0.00 and 1.00. Both measures are indices of the strength of the association; for example, a value of .75 indicates a stronger relationship than a value of .50. The problem is that the values between 0.00 and 1.00 cannot be interpreted as anything other than an index of the relative strength of the association. On the other hand, these statistics are easy to calculate (once the value of chi square has been obtained) and are commonly used indicators of the importance of an association.* *(For practice in computing phi and*

---

*Two other chi square–based measures of association, T^2 and C (the contingency coefficient), are sometimes reported in the literature. Both of these measures have serious limitations. T^2 has an upper limit of 1.00 only for tables with an equal number of rows and columns, and the upper limit of C varies, depending on the dimensions of the table. These characteristics make these measures more difficult to interpret and thus less useful than phi or Cramer's V.

READING STATISTICS 10: The Importance of Percentages

In this chapter and the next, our primary concern is with measures of association for bivariate tables. These statistics are extremely useful for summarizing the strength and—as we will see in Chapter 14—the direction of relationships. Nonetheless, the first step in analyzing bivariate tables should always be to apply the techniques introduced in Chapter 12. The column percentages and conditional distributions will give you more detail about the relationship than measures of association. As useful as they are, the latter should be regarded as summary statements rather than analysis-in-depth.

Of course, even statistics as humble as percentages can be miscalculated and misunderstood. Sociologists Wilbert Leonard and John Phillips illustrate the importance of calculating percentages in the correct direction in a research note concerning the phenomenon of racial stacking in professional sports. This is the practice of reserving certain positions for white players—usually the positions that require leadership skills or decision making (e.g., the quarterback in football). Racial stacking has been widely documented but sometimes misinterpreted because of incorrectly calculated percentages.

Leonard and Phillips illustrate their point with the tables reproduced below. Note that the first table has the independent variable (race) in the columns—consistent with the convention followed in this text—but incorrectly computes the percentages within the rows. The second table correctly computes the percentages in the direction of the independent variable.

The authors point out that the first table does not provide information about the proportion of players from each ethnic group in the various positions. For example, according to Table A:

> . . . Utility infielder (player capable of "filling in" at most infield positions . . .) appears to be a "white" job. One might be tempted to construct a hypothesis to explain the absence of minority players from this unglamorous but important role. In fact, Table [B] shows that Black, White, and Latin American players are about equally likely to be utility infielders. The differences in Table [A] say more about the proportions of White, Black, and Latin American players in the major leagues than they do about whether

TABLE A RACE BY POSITION IN MAJOR LEAGUE BASEBALL (IN PERCENTAGES)*

| | Race | | | | |
Position	White	Black	Latin American	N	%
Catcher	87	1	12	98	100
1st base	69	19	11	36	99
2nd base	59	15	26	34	100
3rd base	76	12	12	33	100
Shortstop	42	8	50	36	100
Outfield	34	49	17	194	100
Utility	54	21	25	111	100

*Some percentages do not total to 100 because of rounding error.

TABLE B RACE BY POSITION IN MAJOR LEAGUE BASEBALL (IN PERCENTAGES)*

| | Race | | |
Position	White	Black	Latin American
Catcher	29	1	11
1st base	8	5	4
2nd base	7	4	8
3rd base	8	3	4
Shortstop	5	2	17
Outfield	22	69	31
Utility	20	17	26
Total =	99	101	101
	(296)	(138)	(108)

*Percentages do not total to 100 because of rounding error. Data are from R. E. Lapchick and J. R. Benedict. 1993. *Racial Report Card*. Boston: Center for the Study of Sport in Society.

one racial/ethnic group is better or more competent to fill the job of utility infielder.

Table A shows the percentage of players at each position who are from each of the groups. White players appear to dominate certain positions in Table A only because they outnumber players from the other groups.

Table B shows the percentage of players in each group who are in each position. By computing percentages within the columns, the differences in group size are controlled, and we can see the effect of race (the independent variable)

on position (the dependent variable). The groups are about evenly distributed across most positions, but—consistent with a pattern of racial stacking—whites dominate the position of catcher (a primary decision maker on the field), and blacks are overrepresented among outfielders (the position most distant from decision-making responsibilities).

Table and excerpts reprinted by permission from W. M. Leonard II and J. Phillips, 1997, "The Cause and Effect Rule for Percentaging Tables: An Overdue Statistical Correction for 'Stacking' Studies,"; *Sociology of Sport Journal*, 14(5):283–289.

Cramer's V, see any of the problems at the end of this chapter or at the end of Chapter 12. To minimize computations, however, use problems from the end of Chapter 11 for which you already have a value for chi square. Otherwise, use the problems based on 2 × 2 tables. Remember that for tables that have either two rows or two columns, phi and Cramer's V will have the same value.)

13.3 PROPORTIONAL REDUCTION IN ERROR (PRE)

In recent years, a group of measures based on a logic known as **proportional reduction in error (PRE)** have been developed to complement the older chi square–based measures of association. For nominal-level variables, the logic of PRE involves first attempting to guess or predict the category into which each case will fall on the dependent variable (Y) while ignoring the independent variable (X). Since we would, in effect, be predicting blindly in this case, we would make many errors (that is, incorrectly predict the category of many cases on the dependent variable).

The second step would be to predict again the category of each case on the dependent variable, this time taking the independent variable into account. If the two variables are associated, the additional information supplied by the independent variable should reduce our errors of prediction (that is, we should misclassify fewer cases). The stronger the association between the variables, the greater the reduction in errors. In the case of a perfect association, we would make no errors at all when predicting score on Y from score on X. When there is no association between the variables, on the other hand, knowledge of the independent variable will not improve the accuracy of our predictions. We would make just as many errors of prediction with knowledge of the independent variable as we did without knowledge of the independent variable.

An illustration should make these principles clearer. Suppose you were placed in the rather unusual position of having to predict whether each of the next 100 people you meet will be shorter or taller than 5 feet 9 inches in height under the condition that you would have no knowledge of these people at all. With absolutely no information about these people, your predictions will be wrong quite often (you will frequently misclassify a tall person as short and vice versa).

Now assume that you must go through this ordeal twice; but, on the second round, just prior to your prediction, you are supplied with information about the sex of the person whose height you must predict. Since height is associated with sex and females are, on the average, shorter than males, the optimal strategy would be to predict that all females are short and all males are tall. Of course, you will still make errors on this second round; but, given an association between the variables, the number of errors on the second round will be less than the number of errors on the first. That is, by taking information from an associated variable into account, you will have reduced your errors of prediction. How can these unusual thoughts be translated into a useful statistic?

13.4 A PRE MEASURE FOR NOMINAL-LEVEL VARIABLES: LAMBDA

One hundred individuals have been categorized by gender and height, and the data are displayed in Table 13.5. It is clear, even without percentages, that the two variables are associated. To measure the strength of this association, a PRE measure called **lambda** (symbolized by the Greek letter λ) will be calculated. Following the logic introduced in the previous section, we must find two quantities. First, the number of prediction errors made while ignoring the independent variable (gender) must be found. Then, we will find the number of prediction errors made while taking gender into account. These two sums will then be compared to derive the statistic.

First, the information given by the independent variable (gender) can be ignored, in effect, by working only with the row marginals. Two different predictions can be made by using these marginals. We can predict either that all subjects are tall or that all subjects are short.* For the first prediction (all subjects are tall), 48 errors will be made. That is, for this prediction, all 100 cases would be placed in the first row. Since only 52 of the cases actually belong in this row, this prediction would result in $(100 - 52)$, or 48, errors. If we had predicted that all subjects were short, on the other hand, we would have made 52 errors $(100 - 48 = 52)$. We will take the lesser of these two numbers and refer to this quantity as E_1 for the number of errors made while ignoring the independent variable. So, $E_1 = 48$.

The second step in the computation of lambda is again to predict score on Y (height), this time taking X (gender) into account. Follow the same procedure as in the first step, but this time move from column to column. Since each column is a category of X, we thus take X into account in making our predictions. For the left-hand column (males), we predict that all 50 cases

TABLE 13.5 HEIGHT BY GENDER

Height	Gender		
	Male	Female	Totals
Tall	44	8	52
Short	6	42	48
Totals	50	50	100

---

*Other predictions are, of course, possible, but these are the only two permitted by lambda.

will be tall and make six errors ($50 - 44 = 6$). For the second column (females), our prediction is that all females are short, and eight errors will be made. By moving from column to column, we have taken X into account and have made a total of 14 errors of prediction, a quantity we will label $\boldsymbol{E_2}$ ($E_2 = 6 + 8 = 14$).

The logic of lambda is that, if the variables are associated, fewer errors will be made under the second procedure than under the first. Clearly, gender and height are associated, since we made fewer errors of prediction while taking gender into account ($E_2 = 14$) than while ignoring gender ($E_1 = 48$). Our errors have been reduced from 48 to only 14. To find the proportional reduction in error, use Formula 13.3:

FORMULA 13.3

$$\lambda = \frac{E_1 - E_2}{E_1}$$

For the sample problem, the value of lambda would be

$$\lambda = \frac{E_1 - E_2}{E_1}$$

$$\lambda = \frac{48 - 14}{48}$$

$$\lambda = \frac{34}{48}$$

$$\lambda = .71$$

The value of lambda ranges from 0.00 to 1.00. A lambda of 0.00 would indicate that the information supplied by the independent variable does not improve our ability to predict the dependent and, therefore, that there is no association between the variables. A lambda of 1.00 would mean that we could predict Y without error from X.

Part of what has made PRE measures popular is that values between 0.00 and 1.00, such as the lambda of .71 calculated above, have a direct and meaningful interpretation. These values can be seen as indices of the extent to which X helps us to predict (or, more loosely, understand) Y. When multiplied by 100, the value of the lambda indicates the percentage reduction in error and, therefore, the strength of the association. Thus, the lambda above would be interpreted by concluding that knowledge of gender improves our ability to predict height by 71%. Or, we are 71% better off knowing gender when attempting to predict height than we are not knowing gender.

13.5 THE COMPUTATION OF LAMBDA

Let us work through another example in order to state the computational routine for lambda in more general terms. Suppose a researcher was concerned with the relationship between religious denomination and attitude toward capital punishment and had collected the data presented in Table 13.6 from a sample of 130 respondents.

TABLE 13.6 ATTITUDE TOWARD CAPITAL PUNISHMENT BY RELIGIOUS DENOMINATION (fictitious data)

	Religion				
Attitude	Catholic	Protestant	Other	None	Totals
Favors	10	9	5	14	38
Neutral	14	12	10	6	42
Opposed	11	4	25	10	50
Totals	35	25	40	30	130

Step 1. To find E_1, the number of errors made while ignoring X (religion, in this case), subtract the largest row total from N. For Table 13.6, E_1 will be

$$E_1 = N - \text{(Largest row total)}$$
$$E_1 = 130 - 50$$
$$E_1 = 80$$

Thus, we will misclassify 80 cases on attitude toward capital punishment while ignoring religion.

Step 2. Next, E_2—the number of errors made when taking the independent variable into account—must be found. For each column, subtract the largest cell frequency from the column total and then add the subtotals together. For the data presented in Table 13.6:

$$\begin{aligned} \text{For Catholics: } 35 - 14 &= 21 \\ \text{For Protestants: } 25 - 12 &= 13 \\ \text{For "Others": } 40 - 25 &= 15 \\ \text{For "None": } 30 - 14 &= 16 \\ E_2 &= 65 \end{aligned}$$

A total of 65 errors are made when predicting attitude on capital punishment while taking religion into account.

Step 3. In step 1, 80 errors of prediction were made as compared to 65 errors in step 2. Since the number of errors has been reduced, the variables are associated. To find the proportional reduction in error, the values for E_1 and E_2 can be directly substituted into Formula 13.3:

$$\lambda = \frac{E_1 - E_2}{E_1}$$

$$\lambda = \frac{80 - 65}{80}$$

$$\lambda = \frac{15}{80}$$

$$\lambda = .19$$

Application 13.1

A random sample of students at a large urban university have been classified as either "traditional" (18–23 years of age and unmarried) or "nontraditional" (24 or older or married). Subjects have also been classified as "vocational," if their primary motivation for college attendance is career or job oriented, or "academic," if their motivation is to pursue knowledge for its own sake. Are these two variables associated?

	Type		
Motivation	Traditional	Non-traditional	Totals
Vocational	25	60	85
Academic	75	15	90
Totals	100	75	175

A lambda will be computed to measure the association between these two variables.

$$E_1 = 175 - 90 = 85$$

$$E_2 = (100 - 75) + (75 - 60)$$

$$= 25 + 15 = 40$$

$$\lambda = \frac{E_1 - E_2}{E_1}$$

$$\lambda = \frac{85 - 40}{85} = \frac{45}{85} = 0.53$$

A lambda of 0.53 indicates that we would make 53% fewer errors in predicting motivation from student type, as opposed to predicting motivation while ignoring student type. The association is moderate to strong, and, by inspection of the table, we can see that traditional students are more likely to have academic motivations (75%), and nontraditional types are more likely to be vocational in motivation (80%)

When attempting to predict attitude toward capital punishment, we would make 19% fewer errors by taking religion into account. Knowledge of a respondent's religious denomination does improve the accuracy of our predictions (the variables are associated) but not by a very large margin. The relatively low value of this lambda indicates that factors other than religion are associated with the dependent variable.

13.6 THE LIMITATIONS OF LAMBDA

As a measure of association, lambda has two characteristics that should be stressed. First, lambda is asymmetric. This means that the value of the statistic will vary, depending on which variable is taken as independent. For example, in Table 13.6, the value of lambda would be .14 if attitude toward capital punishment had been taken as the independent variable (verify this with your own computation). Thus, you should exercise some caution in the designation of an independent variable. If you consistently follow the convention of arraying the independent variable in the columns and compute lambda as outlined above, the asymmetry of the statistic should not be confusing.

Second, when one of the row totals is much larger than the others, lambda can be misleading. It can be 0.00 even when other measures of association are greater than 0.00 and the conditional distributions for the table indicate that there is an association between the variables. This anomaly is a function of the way lambda is calculated and suggests that great caution should be exercised in the interpretation of lambda when the row marginals are very unequal. In fact, in the case of very unequal row marginals, a chi square–based measure of association would be the preferred measure of association *(For practice in computing lambda, see any of the problems at the end of this chapter, Chapter 12, or Chapter 11. As with phi and Cramer's V, it's probably a good idea to start with small samples and 2 × 2 tables.)*

SUMMARY

1. Three measures of association—phi, Cramer's *V*, and lambda—were introduced. Each is used to summarize the overall strength of the association between two variables that have been organized into a bivariate table.

2. Phi and Cramer's *V* are chi square-based measures of association and have the advantage of being easy to compute (once the value of chi square is found). Phi is used for 2 × 2 tables; Cramer's *V* can be used for any size table. Both indicate the strength of the relationship, but values between 0.00 and 1.00 have no direct interpretation.

3. Lambda is a PRE-based measure and provides a more direct interpretation for values between the extremes of 0.00 and 1.00. Lambda indicates the improvement in predicting the dependent variable with knowledge of the independent, compared to predicting the dependent without knowledge of the independent. Because of the meaningfulness of values between the extremes, lambda is often preferred over the more traditional chi square–based measures (except when row totals are very unequal).

SUMMARY OF FORMULAS

Phi	13.1	$\phi = \sqrt{\dfrac{\chi^2}{N}}$
Cramer's *V*	13.2	$V = \sqrt{\dfrac{\chi^2}{(N)(\text{Minimum of } r - 1, c - 1)}}$
Lambda	13.3	$\lambda = \dfrac{E_1 - E_2}{E_1}$

GLOSSARY

Cramer's V. A chi square–based measure of association. Appropriate for nominally measured variables that have been organized into a bivariate table of any number of rows and columns.

E_1. For lambda, the number of errors of prediction made when predicting which category of the dependent variable cases will fall into while ignoring the independent variable.

E_2. For lambda, the number of errors of prediction made when predicting which category of the dependent variable cases will fall into while taking account of the independent variable.

Lambda (λ). A measure of association appropriate for nominally measured variables that have been organized into a bivariate table. Lambda is based on the logic of proportional reduction in error (PRE).

Phi (ϕ). A chi square–based measure of association. Appropriate for nominally measured variables that have been organized into a 2 × 2 bivariate table.

Proportional reduction in error (PRE). The logic that underlies the definition and computation of

lambda. The statistic compares the number of errors made when predicting the dependent variable while ignoring the independent variable (E_1) with the number of errors made while taking the independent variable into account (E_2).

MULTIMEDIA RESOURCES

The Wadsworth Sociology Resource Center: Virtual Society
http://sociology.wadsworth.com/

Visit the companion web site for the sixth edition of *Statistics: A Tool for Social Research* to access a wide range of student resources. Begin by clicking on the Student Resources section of the book's web site to access the following study tools:

- Basic math review
- Statistics review

- Flash cards
- Internet links
- Additional chapter problems
- Table of random numbers
- MicroCase and SPSS examples and exercises
- "Find the text" flowcharts
- Hypothesis testing for variables measured at the ordinal level

PROBLEMS

13.1 SOC Who is most likely to be victimized by crime? A small sample of city residents has been asked if they were the victims of burglary or robbery over the past year. The tables below report relationships between several variables and victimization. Compute phi and lambda for each table. Which relationship is strongest?

a. Victimization by sex of respondent:

	Sex		
Victimized?	Male	Female	Totals
Yes	10	12	22
No	15	18	33
Totals	25	30	55

b. Victimization by age of respondent:

	Age		
Victimized?	21 or Younger	22 or Older	Totals
Yes	12	10	22
No	15	18	33
Totals	27	28	55

c. Victimization by race of respondent:

	Race		
Victimized?	Black	White	Totals
Yes	9	13	22
No	6	27	33
Totals	15	40	55

13.2 Compute a phi and a lambda for problems 12.1 to 12.4. Compare the value of the measure of association with your impressions of the strength of the relationships based solely on the percentages you calculated in Chapter 12.

13.3 SOC There is concern that suicides are motivated, in part, by imitation. Especially among young people, it may be that "epidemics" of self-destructive behaviors follow publication of suicides in local media. A number of cities have been classified by rate of suicide and by whether or not they experienced a publicized suicide within the past year. Is there an association between these two variables? Summarize your conclusions in a sentence or two.

Suicide Rate	Publicized Suicide?		
	Yes	No	Totals
Low	15	20	35
High	15	10	25
Totals	30	30	60

13.4 SW The director of a shelter for battered women has noticed that many of the women who are referred to the shelter eventually return to their violent husbands even when there is every indication that the husband will continue the pattern of abuse. The director suspects that the women who return to their husbands do so because they have no place else to go—for example, no close relatives in the area with whom the women could reside. Do the data below support the director's suspicion? (Data are from the case files of former clients.)

Return to Husband?	Relatives Nearby	No Relatives Nearby	Totals
Yes	10	23	33
No	50	17	67
Totals	60	40	100

13.5 PA Traditionally, bus ridership in your town has been confined to lower-income and blue-collar patrons. As head of transportation planning for the city, you believe that ridership from white-collar, middle-income neighborhoods can be increased if bus routes linking these neighborhoods to the downtown area (where most people work) are increased. A survey is conducted, and the results are displayed below. Is willingness to ride the bus related to job location? What is the pattern of the relationship (if any)?

Potential Ridership	Job Location		
	Downtown	Other	Totals
Would use bus	55	20	75
Would not use bus	15	21	36
Totals	70	41	111

13.6 GER A survey of senior citizens who live in either a housing development specifically designed for retirees or an age-integrated neighborhood has been conducted. Is type of living arrangement related to sense of social isolation?

Sense of Isolation	Living Arrangement		
	Housing Development	Integrated Neighborhood	Totals
Low	80	30	110
High	20	120	140
Totals	100	150	250

13.7 SOC Is there an association between the gender of college instructors and the teaching effectiveness ratings they receive from students? Write a few sentences summarizing your findings.

Teaching Effectiveness	Gender		
	Female	Male	Totals
High	115	241	356
Low	54	113	167
Totals	169	354	523

13.8 SOC A researcher has conducted a survey on sexual attitudes for a sample of 317 teenagers. The respondents were asked whether they considered premarital sex to be "always wrong" or "OK under certain circumstances." The tables below summarize the relationship between responses to this item and several other variables. For each table, assess the strength and pattern of the relationship and write a paragraph interpreting these results.

a. Attitudes toward premarital sex by gender:

Premarital Sex	Sex		
	Male	Female	Totals
Always wrong	90	105	195
Not always wrong	65	57	122
Totals	155	162	317

b. Attitudes toward premarital sex by courtship status:

Premarital Sex	Ever "Gone Steady"?		
	No	Yes	Totals
Always wrong	148	47	195
Not always wrong	42	80	122
Totals	190	127	317

c. Attitudes toward premarital sex by social class:

| | Social Class | | |
Premarital Sex	Blue Collar	White Collar	Totals
Always wrong	72	123	195
Not always wrong	47	75	122
Totals	119	198	317

13.9 SOC St. Algebra College has a problem with attrition. A sizable number of nongraduating students do not return to classes each semester. Is attrition importantly related to race, status, or age? For each table below, how strong is the association (if any) between each of the independent variables and attrition? Does the relationship have a pattern? (Calculating percentages for the table will help to answer the second question.) Write a paragraph summarizing the results presented in these three tables.

a. Attrition by race for 532 students enrolled in fall semester:

| | Race | | |
Attrition	White	Black	Totals
Returned spring semester	280	100	380
Did not return	105	47	152
Totals	385	147	532

b. Attrition by status for 532 students enrolled in fall semester:

| | Status | | |
Attrition	Part-time	Full-time	Totals
Returned spring semester	42	353	395
Did not return	87	50	137
Totals	129	403	532

c. Attrition by age for 532 students enrolled in fall semester:

| | Age | | |
Attrition	18–24	25 or More	Totals
Returned spring semester	322	73	395
Did not return	57	80	137
Totals	379	153	532

13.10 SOC/CJ A sociologist is researching public attitudes toward crime and has asked a sample of residents of his city if they think that the crime rate in their neighborhoods is rising. Is there a relationship between sex and perception of the crime rate? Between race and perception of the crime rate? What is the pattern of the relationship? Write a paragraph summarizing the information presented in these tables.

a. Perception of crime rate by sex:

| | Sex | | |
Crime Rate Is	Male	Female	Totals
Rising	200	225	425
Stable	175	150	325
Falling	125	125	250
Totals	500	500	1000

b. Perception of crime rate by race:

| | Race | | |
Crime Rate Is	White	Black	Totals
Rising	300	150	450
Stable	230	85	315
Falling	170	65	235
Totals	700	300	1000

13.11 PS You are running for mayor of Shinbone, Kansas, and realize that, if you are to win, you must win the support of blue-collar voters. (You already have strong support in the white-collar neighborhoods.) You have a very limited advertising budget and wonder how best to reach your intended audience. An aide has found the data below, which show the relationship between social class and "main source of news" for a sample of Shinboneites. Will this information help you make a decision?

| | Social Class | | |
Main Source of News	Blue Collar	White Collar	Totals
Television	140	200	340
Radio	25	40	65
Newspapers	85	100	185
Totals	250	340	590

13.12 Problem 12.12 analyzed some bivariate relationships taken from the 1998 General Social Survey data set. Political ideology was used as the independent variable for five different dependent variables and, using only percentages, you were asked to characterize the relationships in terms of strength and direction. Now, with the aid of measures of association, these characterizations should be easier to develop. Compute a Cramer's *V* and a lambda for each table in problem 12.12. Compare the measures of association with your characterizations based on the percentages.

Below are the same five dependent variables cross-tabulated against sex as an independent variable. Compare the strength of these relationships with those for political ideology. Which independent variable has the stronger associations?

a. Support for the legal right to an abortion (ABANY) by sex:

Right to Abortion?	Sex		Totals
	Male	Female	
Yes	310	418	728
No	432	618	1050
Totals	742	1036	1778

b. Support for capital punishment (CAPPUN) by sex:

Capital Punishment?	Sex		Totals
	Male	Female	
Favor	908	998	1906
Oppose	246	447	693
Totals	1154	1445	2559

c. Approve of suicide for people with incurable disease (SUICIDE1) by sex:

Right to Suicide?	Sex		Totals
	Male	Female	
Yes	524	608	1132
No	246	398	644
Totals	770	1006	1776

d. Support for sex education in public schools (SEXEDUC) by sex:

Sex Education?	Sex		Totals
	Male	Female	
Favor	685	900	1585
Oppose	102	134	236
Totals	787	1034	1821

e. Support for traditional gender roles (FEHOME) by sex:

Women Should Take Care of Running Their Homes and Leave Running the Country to Men	Sex		Totals
	Male	Female	
Agree	116	164	280
Disagree	669	865	1534
Totals	785	1029	1814

SPSS for Windows

Using *SPSS for Windows* to Produce Nominal-Level Measures of Association

SPSS DEMONSTRATION 13.1 Does Support for Gun Control Vary by Political Ideology? Another Look

In Demonstration 12.2, we used **Crosstabs** to look for an association between *gunlaw* and *polviews*. We saw that these variables were associated and that liberals were the most supportive of gun control and conservatives were the least supportive. The sample was fairly homogeneous on this issue, and the majority of people from each of the three political ideologies were in favor of gun control. The differences in conditional distributions were not great, and the relationship seemed to be, at best, moderate in strength. In this demonstration, we will reexamine the relationship and have SPSS compute some measures of association. The commands should repeat those in Demonstration 12.2 (remember to recode *polviews* into three categories). Click **Analyze,** then **Descriptive Statistics,** and then **Crosstabs.** Move *gunlaw* into the Row(s) box and recoded *polviews* into the Column(s) box. Click the **Cells** button and request column percentages. Click the **Statistics** button and request chi square, phi, Cramer's *V,* and lambda. Click **Continue** and **OK,** and the following output will be produced:

FAVOR OR OPPOSE GUN PERMITS POL Crosstabulation

			POL			Total
			1.00	2.00	3.00	
FAVOR OR OPPOSE GUN PERMITS	FAVOR	Count	235	247	239	721
		% within POL	87.0%	83.2%	78.9%	82.9%
	OPPOSE	Count	35	50	64	149
		% within POL	13.0%	16.8%	21.1%	17.1%
Total		Count	270	297	303	870
		% within POL	100.0%	100.0%	100.0%	100.0%

Chi-Square Tests

	Value	df	Asymp. Sig. (2-sided)
Pearson Chi-Square	6.724	2	.035
Likelihood Ratio	6.780	2	.034
Linear-by-Linear Association	6.710	1	.010
N of Valid Cases	870		

Directional Measures

			Value	Asymp. Std. Error	Approx. T	Approx. Sig.
Nominal by Nominal	Lambda	Symmetric	.011	.031	.363	.717
		FAVOR OR OPPOSE GUN PERMITS Dependent	.000	.000		
		POL Dependent	.014	.039	.363	.717
	Goodman and Kruskal tau	FAVOR OR OPPOSE GUN PERMITS Dependent	.008	.006		.035
		POL Dependent	.004	.003		.036

Symmetric Measures

		Value	Approx. Sig.
Nominal by Nominal	Phi	.088	.035
	Cramer's V	.088	.035
N of Valid Cases		870	

The measures of association are reported below the output for the chi square tests. Three values for lambda are reported in the Directional Measures output block. Remember that lambda is asymmetric and will change value depending on which variable is taken as dependent. In this case, *gunlaw* (FAVOR OR OPPOSE GUN PERMITS) is the dependent variable, so lambda is .000, a value that indicates no relationship between the variables. We saw previously, however, that the conditional distributions in the table do change, indicating that there *is* a relationship. The problem here is that the row totals are very unequal, and lambda is misleading and should be disregarded (see Section 13.6).

Phi and Cramer's *V* are reported in the Symmetric Measures output block. The statistics are identical in value (.088), as they will be whenever the table has either two rows or two columns. These measures and the small changes in the conditional distributions indicate, as we suspected, a weak association between the variables. Political ideology does not seem to be an especially important cause of support for gun control.

SSPS DEMONSTRATION 13.2 What Variables Affect Support for Gun Control?

Let's see if we can do any better in "explaining" support of gun control with some other commonly used independent variables. As you are no doubt aware, attitudes are often associated with various measures of social location. In keeping with our focus on the nominal level of measurement, let's investigate relationships between *gunlaw* and *sex, marital* (marital status), *race,* and religious preference, or *relig.* Incorporating a number of potential independent variables, although quite common in social science research, generates a large volume of output, and we will abbreviate the actual output from *SPSS* in this demonstration.

	Percent in Favor	Significance of Chi Square	Phi* or V	Lambda
SEX				
Male	77.0			
Female	88.5	.000	.15	.00
RELIGION				
Protestant	81.0			
Catholic	87.1			
Jew	93.3			
None	85.9			
Other	81.0	.172	.08	.00
RACE				
White	81.6			
Black	88.5			
Other	93.8	.010	.10	.00

	Percent in Favor	Significance of Chi Square	Phi* or V	Lambda
MARITAL				
Married	83.6			
Widowed	89.1			
Divorced	82.7			
Separated	82.4			
Never married	81.4	.574	.06	.00

*Phi may be displayed as a minus value. If so, disregard the sign.

Basically, these results indicate that *gunlaw* is not particularly related to any of these demographic variables. The strongest relationship is with *sex,* with males less supportive than females. Since the rows are very unequal, lambda should be disregarded.

MicroCase

Using MicroCase to Produce Nominal-Level Measures of Association

MICROCASE DEMONSTRATION 13.1 Does Support for Gun Control Vary by Political Ideology? Another Look

In Demonstration 12.2, we used **Cross-tabulation** to look for an association be-tween *gunlaw* and *polviews.* We saw that these variables were associated and that liberals were the most supportive of gun control and conservatives were the least supportive. The sample was fairly homogeneous on this issue, and the ma-jority of people from each of the three political ideologies were in favor of gun con-trol. The differences in conditional distributions were not great, and the relation-ship seemed to be, at best, "moderate" in strength. In this demonstration, we will reexamine the relationship and have MicroCase compute some measures of as-sociation. The commands should repeat those in Demonstration 12.2 (remember to recode *polviews* into three categories). Click **Cross-tabulation** from the **Statis-tics** menu. The **Cross-tabulation** dialog box will appear with the variables listed in the window to the left. Highlight *gunlaw* (variable #20) and click the arrow to move the variable name into the **Row Variable** box, and then highlight recoded *polviews* (probably variable #51) and move it into the **Column Variable** box. Click **OK** to get the bivariate table. Click **Column %** in the **Tables** box in the upper-left-hand corner, and the table will look like this:

```
GUNLAW by POL
         liberals  moderates  conserv's  Missing   TOTAL
FAVOR         235        247        239       44     721
            87.0%      83.2%      78.9%            82.9%
OPPOSE         35         50         64        3     149
            13.0%      16.8%      21.1%            17.1%
Missing       119        189        141       21     470
TOTAL         270        297        303       68     870
           100.0%     100.0%     100.0%
```

Click **Summary** in the Statistics box at the left of the screen, and the following statistics will be computed:

```
Row Variable: GUNLAW
Column Variable: POL

Nominal Statistics:

Chi-Square: 6.724 (DF = 2; Prob. = 0.035)
V:      0.088  C:       0.088
Lambda: 0.014  Lambda: 0.000  Lambda: 0.011
(DV=51) (DV=20)

Ordinal Statistics:

Gamma:  0.191  Tau-b:  0.083  Tau-c:  0.072
s.error 0.073  s.error 0.032  s.error 0.028

Dyx:    0.054  Dxy:    0.127
s.error 0.021  s.error 0.049
Prob. = 0.009
```

Statistics are organized by level of measurement. Under "Nominal Statistics" we see the value for chi square and a Cramer's *V* of 0.088. This indicates, as we suspected, a weak association between the variables. (MicroCase does not report a separate value for phi). There are three different values reported for lambda. Remember that lambda is asymmetric and will change value, depending on which variable is taken as dependent. In this case, *gunlaw* (variable #20) is the dependent variable (or DV), so lambda is .000, a value that indicates no relationship between the variables. We saw previously, however, that the conditional distributions in the table do change, indicating that there *is* a (weak) relationship. The problem here is that the row totals are very unequal, and lambda is misleading and should be disregarded (see Section 13.6).

In conclusion, there is a relationship between these variables, but the small changes in the conditional distributions and the low values for the measures of association indicate that the relationship is weak. Political ideology does not seem to be an especially important cause of support for gun control.

MICROCASE DEMONSTRATION 13.2
What Variables Affect Support for Gun Control?

Let's see if we can do any better in "explaining" support of gun control with some other commonly used independent variables. As you are no doubt aware, attitudes are often associated with various measures of social location. In keeping with our focus on the nominal level of measurement, let's investigate relationships between *gunlaw* and *marital* (marital status, variable #3), *sex* (variable #10), *race* (variable #11), and religious preference, or *relig* (variable #22). Incorporating a number of potential independent variables, although quite common in social science research, generates a large volume of output, and we will abbreviate the actual output from MicroCase in this demonstration.

	Percent in Favor	Significance of Chi Square	Phi or *V*	Lambda
SEX				
Male	77.0			
Female	88.5	.000	.15	.00
RELIGION				
Protestant	81.0			
Catholic	87.1			
Jew	93.3			
None	85.9			
Other	81.0	.172	.08	.00
RACE				
White	81.6			
Black	88.5			
Other	93.8	.010	.10	.00
MARITAL				
Married	83.6			
Widowed	89.1			
Divorced	82.7			
Separated	82.4			
Never married	81.4	.574	.06	.00

Basically, these results indicate that *gunlaw* is not particularly related to any of these demographic variables. The strongest relationship is with *sex*, with males less supportive than females. Since the rows are very unequal, lambda should be disregarded.

Exercises

13.1 See if you can find a variable with a strong association with *gunlaw*. If necessary, use the **Recode** command (SPSS) or the **Collapse Variables** command (MicroCase) to reduce the number of categories in the column variable. As a suggestion, try *degree* and *income91* as independent variables.

13.2 Run some of the tables you did for Exercises 12.1 and 12.2 again with a request for phi, Cramer's *V*, and lambda. Do the measures of association help you interpret the relationships?

14

ASSOCIATION BETWEEN VARIABLES MEASURED AT THE ORDINAL LEVEL

LEARNING OBJECTIVES   By the end of this chapter, you will be able to

1. Calculate and interpret gamma and Spearman's rho.
2. Explain the logic of proportional reduction in error in terms of gamma.
3. Use gamma and Spearman's rho to analyze and describe a bivariate relationship in terms of the three questions introduced in Chapter 12.
4. Test gamma and Spearman's rho for significance.

14.1 INTRODUCTION

There are two general types of ordinal-level variables. Some have many possible scores and look, at least at first glance, like interval-ratio level variables. We will call these "continuous ordinal variables." An attitude scale that incorporated many different items and, therefore, had many possible values would produce this type of variable.

The second type, which we will call a "collapsed ordinal variable," has only a few (no more than five or six) values or scores and can be created either by collecting data in collapsed form or by collapsing a continuous ordinal scale. For example, we would produce collapsed ordinal variables by measuring social class as upper, middle, or lower or by reducing the scores on an attitude scale into just a few categories (such as high, moderate, and low).

A number of measures of association have been invented for use with collapsed ordinal-level variables. Rather than attempt a comprehensive coverage of all of these statistics, we will concentrate on **gamma (_G_).** Other measures suitable for collapsed ordinal-level data (Somer's _d_ and Kendall's tau-_b_) are covered at the web site for this text. For "continuous" ordinal variables, a statistic called **Spearman's rho (_r_s_)** is commonly used, and we will cover this measure of association toward the end of this chapter.

This chapter will expand your understanding of how bivariate associations can be described and analyzed, but it is important to remember that we are still trying to answer the three questions raised in Chapter 12: Are the variables associated? How strong is the association? What is the direction of the association?

14.2 PROPORTIONAL REDUCTION IN ERROR (PRE)

For nominal-level variables, the logic of PRE was based on two prediction rules. First, we predicted Y while ignoring X and, second, we predicted Y while taking X into account. Lambda measured the proportional reduction in the errors of prediction made with the second rule as compared to the first. For variables measured at the ordinal level, we have two similar prediction rules, and gamma, like lambda, measures the proportional reduction in error gained by predicting one variable while taking the other into account. The major difference lies in the way predictions are made. In the case of gamma, we predict the order of pairs of cases. That is, we predict whether one case will have a higher or lower score than the other on the variable in question. The first prediction rule is to predict the order of a pair of cases on one variable while ignoring their order on the other. The second rule is to predict the order on one variable while taking order on the other variable into account.

As an illustration, assume that a researcher is concerned about the causes of "burnout" (that is, demoralization and loss of commitment) among elementary school teachers and wonders if levels of burnout might be related to the number of years a teacher has been employed in the profession. Another way to state the problem would be to ask if teachers who rank higher on years of service would also rank higher on burnout. If we knew that teacher A had more years of service than teacher B, would we be able to predict accurately that teacher A is also more "burned out" than teacher B? That is, would knowledge of the order of this pair of cases on one variable help us predict their order on the other?

If the two variables are associated, we will reduce our errors when our predictions about one of the variables are based on knowledge of the other. Furthermore, the stronger the association, the fewer the errors we will make. When there is no association between the variables, gamma will be 0.00, and knowledge of the order of a pair of cases on one variable will not improve our ability to predict their order on the other. A gamma of ±1.00 denotes a perfect relationship: the order of all pairs of cases on one variable would be predictable without error from their order on the other variable

Since the scores of ordinal-level variables vary from "high" to "low," relationships between ordinal-level variables have a direction: they can be either positive or negative. In a positive relationship, cases tend to be ranked in the same order on both variables. For example, if Case A is ranked above Case B on one variable, it would also be ranked above Case B on the second variable. The relationship suggested above between years of service and burnout would be a positive relationship. In a negative relationship, the order of the cases would be reversed between the two variables. If Case A ranked above Case B on one variable, it would tend to rank below Case B on the second variable. If there is a negative relationship between prejudice and education and Case A was more educated than Case B (or, ranked above Case B on education), then Case A would be less prejudiced (or, would rank below Case B on prejudice).

14.3 THE COMPUTATION OF GAMMA

Table 14.1 summarizes the relationship between "length of service" and "burnout" for a fictitious sample of 100 teachers. To compute gamma, two sums are needed. First, we must find the number of pairs of cases that are ranked the same on both variables (we will label this N_s) and then the number of pairs of cases ranked differently on the variables (N_d). We find these sums by working with the cell frequencies.

To find the number of pairs of cases ranked the same (N_s), begin with the cell containing the cases that were ranked the lowest on both variables. In Table 14.1, this would be the upper-left-hand cell. (*NOTE: not all tables are constructed with values increasing from left to right across the columns and from top to bottom across the rows. When using other tables, always be certain that you have located the proper cell.*) The 20 cases in the upper-left-hand cell all rank low on both burnout and length of service, and we will refer to these cases as "low-lows," or "LLs."

Now, form a pair of cases by selecting one case from this cell and one from any other cell—for example, the middle cell in the table. All 15 cases in this cell are moderate on both variables and, following our practice above, can be labeled moderate-moderates or MMs. Any pair of cases formed between these two cells will be ranked the same on both variables. That is, all LLs are lower than all MMs on both variables (on *X*, low is less than moderate, and on *Y*, low is less than moderate). The total number of pairs of cases is given by multiplying the cell frequencies. So, the contribution of these two cells to the total N_s is (20)(15), or 300.

Gamma ignores all pairs of cases that are tied on either variable. For example, any pair of cases formed between the LLs and any other cell in the top row (low on burnout) or the left-hand column (low on length of service) will be tied on one variable. Also, any pair of cases formed within any

TABLE 14.1 BURNOUT BY LENGTH OF SERVICE (fictitious data)

| Burnout | Length of Service | | | |
	Low	Moderate	High	Totals
Low	20	6	4	30
Moderate	10	15	5	30
High	8	11	21	40
Totals	38	32	30	100

cell will be tied on both X and Y. Gamma ignores all pairs of cases formed within the same row, column, or cell. Practically, this means that, in computing N_s, we will work with only the pairs of cases that can be formed between each cell and the cells below and to the right of it.

In summary: To find the total number of pairs of cases ranked the same on both variables (N_s), multiply the frequency in each cell by the total of all frequencies below and to the right of that cell. Repeat this procedure for each cell and add the resultant products. The total of these products is N_s. This procedure is displayed below for each cell in Table 14.1. Note that none of the cells in the bottom row or the right-hand column can contribute to N_s because they have no cells below and to the right of them. Figure 14.1 shows the direction of multiplication for each of the four cells that, in a 3×3 table, can contribute to N_s. Computing N_s for Table 14.1, we find that a total of 1831 pairs of cases are ranked the same on both variables.

	Contribution to N_s
For LLs, $20(15 + 5 + 11 + 21)$ =	1040
For MLs, $6(21 + 5)$ =	156
For HLs, $4(0)$ =	0
For LMs, $10(11 + 21)$ =	320
For MMs, $15(21)$ =	315
For HMs, $5(0)$ =	0
For LHs, $8(0)$ =	0
For MHs, $11(0)$ =	0
For HHs, $21(0)$ =	0
	$N_s = 1831$

FIGURE 14.1   COMPUTING N_s IN A 3×3 TABLE

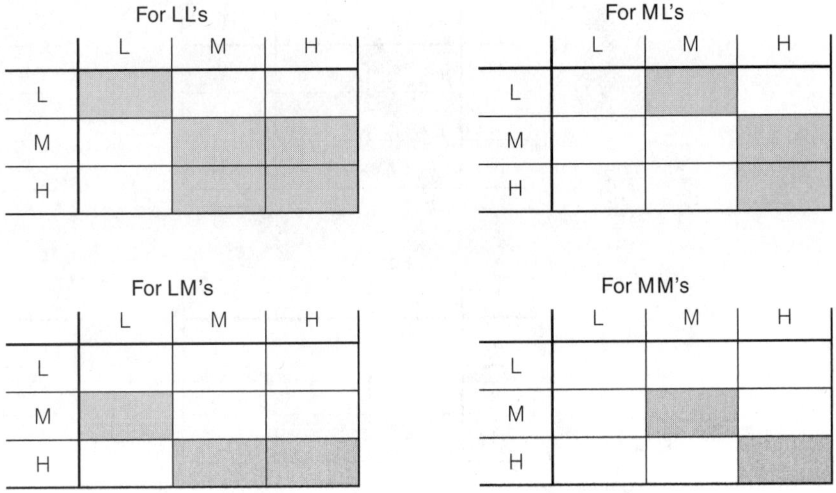

Our next step is to find the number of pairs of cases ranked differently (N_d) on both variables. To find the total number of pairs of cases ranked in different order on the variables, multiply the frequency in each cell by the total of all frequencies below and to the left of that cell. Note that the pattern for computing N_d is the reverse of the pattern for N_s. This time, we begin with the upper-right-hand cell (high-lows, or HLs) and multiply the number of cases in the cell by the total frequency of cases below and to the left. The four cases in the upper-right-hand cell are low on Y and high on X; and, if a pair is formed with any case from this cell and any cell below and to the left, the cases will be ranked differently on the two variables. For example, if a pair is formed between any HL case and any case from the middle cell (moderate-moderates, or MMs), the HL case would be less than the MM case on Y ("low" is less than "moderate") but more than the MM case on X ("high" is greater than "moderate"). The computation of N_d is detailed below and shown graphically in Figure 14.2. In the computations, we have omitted cells that cannot contribute to N_d because they have no cells below and to the left of them.

	Contribution to N_d
For HLs, $4(10 + 15 + 8 + 11) =$	176
For MLs, $6(10 + 8)$ $=$	108
For HMs, $5(8 + 11)$ $=$	95
For MMs, $15(8)$ $=$	120
	$N_d = 499$

FIGURE 14.2   COMPUTING N_d IN A 3 × 3 TABLE

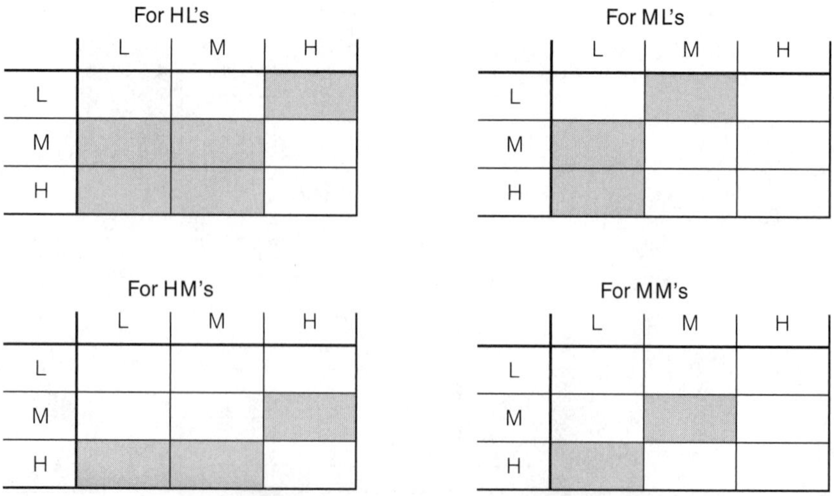

Application 14.1

For local political elections, do voters turn out at higher rates when the candidates for office spend more money on media advertising? Each of 177 localities has been rated as high or low on voter turnout. Also, the total advertising budgets for all candidates in each locality have been ascertained and classified as high, moderate, or low. Are these variables associated?

Voter	Expenditure			
Turnout	Low	Moderate	High	Totals
Low	35	32	17	84
High	23	27	43	93
Totals	58	59	60	177

Because both variables are ordinal in level of measurement, we will compute a gamma to summarize the strength and direction of the association. The number of pairs of cases ranked in the same order on both variables (N_s) would be

$N_s = 35(27 + 43) + 32(43) = 2450 + 1376 = 3826$

The number of pairs of cases ranked in different order on both variables (N_d) would be

$N_d = 17(27 + 23) + 32(23) = 850 + 736 = 1586$

Gamma is

$$G = \frac{N_s - N_d}{N_s + N_d}$$

$$G = \frac{3826 - 1586}{3826 + 1586}$$

$$G = \frac{2240}{5412}$$

$$G = 0.41$$

A gamma of 0.41 means that, when predicting the order of pairs of cases on voter turnout, we would make 41% fewer errors by taking candidates' advertising expenditures into account, as opposed to ignoring the latter variable. There is a moderate, positive association between these two variables.

Table 14.1 has 499 pairs of cases ranked in different order and 1831 pairs of cases ranked in the same order. The formula for computing gamma is

FORMULA 14.1
$$G = \frac{N_s - N_d}{N_s + N_d}$$

where N_s = the number of pairs of cases ranked the same as both variables
N_d = the number of pairs of cases ranked differently on the two variables

For Table 14.1, the value of gamma would be

$$G = \frac{1831 - 499}{1831 + 499}$$

$$G = \frac{1332}{2330}$$

$$G = .57$$

A gamma of .57 indicates that we would make 57% fewer errors if we predicted the order of pairs of cases on one variable from the order of pairs of cases on the other—as opposed to predicting order while ignoring the

READING STATISTICS 11: Bivariate Tables and Associated Statistics

The statistics associated with any bivariate table will usually be reported directly below the table itself. This information would include the name of the measure of association, its value, and (if relevant) its sign. Also, if the research involves a random sample, the results of the chi square test or a test for the significance of the measure itself (see Section 14.7) will be reported. So the information might look something like this:

$$\lambda = 0.47$$
$$\chi^2 = 13.23 \ (2 \ df, \ p \ < \ 0.05)$$

Note that the alpha level is reported in the "$p <$" format, which we discussed in Reading Statistics 5.

Besides simply reporting the value of these statistics, the researcher will also interpret them in the text of the article. The preceding statistics might be characterized by the statement: "The association between the variables is statistically significant and moderately strong." Again we see that researchers will avoid our rather wordy (but more careful) style of stating results. Where we might say "a lambda of 0.47 indicates that we will reduce our errors of prediction by 47% when predicting the dependent variable from the independent variable, as opposed to predicting the dependent while ignoring the independent," the researcher will simply report the value of lambda and characterize that value in a word or two (e.g., "The association is moderately strong"). The researcher assumes that his or her audience is statistically literate and can make the more detailed interpretations for themselves.

Statistics in the Professional Literature

In recent decades, the United States has become increasingly diverse in terms of language, ethnicity, and race. Is this level of diversity and pluralism desirable, or should all the groups that make up the American mosaic be encouraged to blend in, assimilate, and adopt the English language and Anglo-American culture and customs? Should we celebrate our differences or emphasize our similarities?

Sociologist Dennis Downey analyzed sources of support for pluralism and assimilation in the U.S. using data from a special section of the 1994 General Social Survey. Table 1 presents some of his results and reports relationships between several demographic variables and two different dependent variables. The first dependent variable measured *attitudinal content.* Lower scores are associated with support for pluralism or diversity, and higher scores indicate support for assimilation or unity. The second dependent variable measured *attitudinal extremity.* Low scores on this variable indicate a more moderate position, and higher scores indicate a more extreme position regardless of content. In other words, people who scored high on this scale supported their position, whether they were pluralists or assimilationists, with great intensity. Those who scored low felt less keenly about the issue or had more ambiguous or temperate attitudes.

Table 1 presents mean scores for each subcategory of the demographic variables and a statistic called Somer's *d* that shows the strength and direction of the relationship with each of the dependent variables. Somer's *d* is an ordinal measure of association for variables that have been organized into a bivariate table. It is very similar in computation and interpretation to gamma and, essentially, can be interpreted the same way. The major difference with gamma is that Somer's *d* includes pairs of cases tied on the dependent variable and typically will be lower in value than gamma. Somer's *d* is discussed at the web site for this text.

Looking at Table 1, race has the strongest relationship with both dependent variables. The differences in mean scores and the direction of the Somer's *d* show that blacks are generally more pluralistic and more intense in their attitudes. Interestingly, however, Downey finds that Black Americans are actually very diverse on this issue and support *both* pluralism and assimilation more extremely in higher proportions than whites. Of the other relationships, Downey says

> age tends to be associated with more intense attitudes . . . younger cohorts are more supportive [of pluralism] but also more indif-

ferent. . . . Neither [education nor income] exhibits a significant association with attitudinal content, but both are significantly associated with attitudinal extremity. . . . The poor tend to have more extreme attitudes [as do the less educated].

TABLE 1   INFLUENCE OF DEMOGRAPHIC VARIABLES ON ATTITUDINAL DISTRIBUTION

	Attitudinal Content		Attitudinal Extremity	
	Mean	Association	Mean	Association
Race†		.089*		−.188***
White	2.09		1.94	
Black	1.96		2.24	
Income		.001		−.075**
< 20 k	2.07		2.08	
20–50 k	2.06		1.94	
> 50 k	2.08		1.91	
Education		.018		−.179***
HS or less	2.05		2.11	
More than HS	2.08		1.86	
Age		.074**		.112***
18–32	1.96		1.90	
33–50	2.07		1.91	
50+	2.14		2.13	

*$p < .05$ **$p < .01$ ***$p < .001$
†Note that race is only nominal and Somer's *d* assumes ordinal level of measurement. It is common to use dichotomous nominal-level variables (often called "dummy" variables) in this way in certain statistical procedures such as regression analysis (see Chapters 15 and 17). Also, the other independent variables are ordinal in level of measurement, and using a nominal measure only for race would make it difficult to compare the different relationships.
Table and excerpts reprinted from *Social Problems Forces* 42(1) 2000. "Situating Social Attitudes Toward Cultural Pluralism: Between Culture Wars and Contemporary Racism" by Dennis Downey.

other variable. Length of service is associated with degree of burnout, and the relationship is positive. Knowing the respective rankings of two teachers on length of service (Case A is higher on length of service than Case B) will help us predict their ranking on burnout (we would predict that Case A will also be higher than Case B on burnout).

To use the computational routine for gamma presented above, you must arrange the table in the manner of Table 14.1, with the column variable increasing in value as you move from left to right and the row variable increasing from top to bottom. Be careful to construct your tables according to this format; if you are working with data already in table format, you may have to rearrange the table or rethink the direction of patterns. Gamma is a symmetrical measure of association; that is, the value of gamma will be the same regardless of which variable is taken as independent. *(To practice computing and interpreting gamma, see problems 14.1 to 14.10 and 14.15. Begin with some of the smaller, 2 × 2 tables until you are comfortable with these procedures.)*

14.4 DETERMINING THE DIRECTION OF RELATIONSHIPS

Nominal measures of association, like phi and lambda, measure only the strength of a bivariate association. Ordinal measures of association, like gamma, are more sophisticated and add information about the overall direction of the relationship (positive or negative). In one way, it is easy to determine direction: if the sign of the statistic is a plus, the direction is positive, and a minus sign indicates a negative relationship. Often, however, direction is confusing when working with ordinal-level variables, and it will be helpful if we focus on the matter specifically. We'll discuss positive relationships first and then relationships in the negative direction.

In a positive relationship, variables change in the same direction. That is, as scores on one variable increase (or decrease), scores on the other variable also increase (or decrease). Cases tend to have scores in the same range on both variables (i.e., low scores go with low scores, moderate with moderate, and so forth). Table 14.2 illustrates the general shape of a positive relationship. In a positive relationship, cases tend to fall along a diagonal from upper left to lower right (assuming, of course, that tables have been constructed with the column variable increasing from left to right and the row variable from top to bottom).

Table 14.3 presents an example of a positive relationship with actual data. The sample consists of 186 pre-industrial societies from around the globe. Each has been rated in terms of their degree of stratification and the type of political institution they have. In highly stratified societies, there is a great deal of difference in wealth and power between the top-ranked people and those at the bottom. In the societies that are the lowest in stratification,

TABLE 14.2 A GENERALIZED POSITIVE RELATIONSHIP

	Variable X		
Variable Y	Low	Moderate	High
Low	X		
Moderate		X	
High			X

TABLE 14.3 STATE STRUCTURE BY DEGREE OF STRATIFICATION*

	Degree of Stratification		
State	Low	Medium	High
Stateless	77	5	0
Semi-state	28	15	4
State	12	19	26
	117	39	30

*Data are from the Human Relation Area File, standard cross-cultural sample

there are virtually no differences between people in terms of wealth and power. Note the direction of the independent variable: inequality increases across the table from left to right.

The dependent variable is organized so that the political institution becomes more developed and comprehensive as we move down the table from the top to the bottom. Stateless societies have virtually no political institution, and decisions are made by consensus. State societies have fully developed governments, and "semi-state" societies are intermediate between the other two types.

In this table, most cases fall in the diagonal from upper left to lower right and, even without percentages, it is clear that societies with little inequality tend to be stateless and that the political institution becomes more elaborate as inequality increases. The great majority of the least stratified societies had no political institution and none of the highly stratified societies were stateless. The gamma for this table is 0.86, indicating a very strong, positive relationship between these two variables.

Negative relationships are the opposite of positive relationships. Low scores on one variable are associated with high scores on the other and high scores with low scores. This pattern means that the cases will tend to fall along a diagonal from lower left to upper right (at least for all tables in this text). Table 14.4 illustrates a generalized negative relationship. The cases with higher scores on Variable X tend to have lower scores on Variable Y and score on Y decreases as score on X increases.

Table 14.5 presents an example of a negative relationship with data taken from the 1998 General Social Survey, a survey administered to a representative sample of U.S. citizens. The independent variable is church attendance, and the dependent variable is approval of cohabitation ("Is it alright for a couple to live together without intending to get married?"). Note that rates of attendance increase from left to right, and approval of cohabitation increases from top to bottom of the table.

TABLE 14.4   A GENERALIZED NEGATIVE RELATIONSHIP

	Variable X		
Variable Y	Low	Moderate	High
Low			X
Moderate		X	
High	X		

TABLE 14.5   APPROVAL OF COHABITATION* BY CHURCH ATTENDANCE

	Attendance		
Approval	Never	Monthly or Yearly	Weekly
Low	41	132	253
Moderate	41	149	65
High	154	309	70
	236	590	388

*The labels for this variable have been changed to clarify the example. The original response categories were "approve," "neither approve nor disapprove," and "disapprove."

Once again, the pattern in the table is obvious without the aid of percentages. The great majority of people who were low on attendance were high on approval of cohabitation, and most people who were high on attendance were low on approval. As attendance increases, approval of cohabitation tends to decrease. The gamma for this table is −.56, indicating a moderately strong, negative relationship between attendance and approval of this living arrangement.

You should be aware of an additional complication. The coding for ordinal-level variables is arbitrary, and a higher score may mean "more" or "less" of the variable being measured. For example, if we measured social class as upper, middle, and lower, we could assign scores to the categories in either of two ways:

A	B
(1) Upper	(3) Upper
(2) Middle	(2) Middle
(3) Lower	(1) Lower

While coding scheme "B" might seem preferable (because higher scores go with higher class position), *both* schemes are perfectly legitimate and the direction of a relationship will change, depending on which scheme is selected. Using scheme "B," we would find positive relationships between social class and education: as education increased, so would class. Using scheme "A," however, the same relationship would be negative: as education increased, class would decrease because, in this scheme, higher scores on class are associated with lower actual class standing.

Unfortunately, this source of confusion cannot be avoided when working with ordinal-level variables. Coding schemes will always be arbitrary for these variables and you need to exercise additional caution when interpreting the direction of ordinal-level variables.

14.5 INTERPRETING ASSOCIATION WITH BIVARIATE TABLES: WHAT ARE THE SOURCES OF CIVIC ENGAGEMENT IN U.S. SOCIETY?

To this point, we have discussed the association between two variables using tables, column percentages, and both nominal and ordinal measures of association. In this section, we will summarize this material by using these techniques to address an issue in U.S. social life about which many people are concerned. Specifically, social scientists, politicians, ministers, newspaper editorialists, and others have complained that Americans are "dropping out" of community life and minimizing their involvement with other people.* They worry that the bonds of friendship and neighborliness are withering away and that Americans are deserting the groups and associations (like PTAs, Little League, and Girl Scouts) that kept our communities functioning

---

*See, for example, Putnam, Robert D. 2000. *Bowling Alone*. New York: Simon & Schuster.

in the past. This is a complex, multifaceted issue, and we cannot hope to resolve it in these few paragraphs. We can, however, use data from the 1998 General Social Survey to examine the sources of support for community participation. What kind of person is most heavily engaged in the civic life of the community?

To measure civic engagement (the dependent variable), we can group all respondents who volunteered for any community groups (political parties, churches, charities, etc.) in one category and nonvolunteers in another. This simple scale will not be a perfectly accurate measurement of involvement (e.g., there are many ways of participating in community life other than volunteering), but it should provide some insight into the issue. Researchers typically have to settle for partial or incomplete measurement of their variables.

What should we use as independent variables? In other words, what factors might have causal relationships with community participation and cause different levels of civic engagement? In this section, we investigate three common arguments. First, could it be that married people are more stable, more committed to maintaining community life, and more likely to volunteer? Second, it is sometimes argued that volunteerism depends on people who are not fully engaged in the work force (e.g., housewives or 'stay-at-home' moms). We'll use work status (full-time vs. part-time) as an independent variable for this argument. Finally, television has often been blamed for the decline in community participation: people choose to amuse themselves with the fictional lives of televised characters rather than get involved with the real life around them. Which of these arguments has support? How strong are the relationships with civic participation and what is the pattern or direction of the relationships?

Take a moment to consider the level of measurement of each of these variables. The dependent variable is ordinal since the categories can be distinguished in terms of "more or less." Television viewing, measured as low, medium, or high for this exercise, is also an ordinal level variable. Marital status, on the other hand, is clearly nominal, and work status might be considered to be either nominal (if the categories are just different) or ordinal (if "full time" is regarded as working *more than* not "full time"). Ideally, we should use gamma for ordinal variables and lambda or Cramer's V for the nominal variables, but this creates problems of comparability: if we used different measures, how could we tell which relationship was strongest? We can resolve this problem in two ways. First, we can use column percentages to make comparisons from table to table. Second, we can ignore level of measurement and compute both ordinal and nominal measures of association for each table.

Table 14.6 shows that the relationship between marital status and volunteering is significant ($p < .05$) but not particularly strong. Married people are more likely to volunteer but the difference in column percentages is relatively small (7%), and the measures of association are low in value. These

TABLE 14.6   VOLUNTEERING BY MARITAL STATUS

Volunteer?	Marital Status	
	Married	Not Married
No	37.7%	44.7%
Yes	62.3%	55.3%
Totals	100%	100%
	(575)	(640)

$\chi^2 = 6.03$ (df = 1, $p = .01$)
Phi = .07   Lambda = .00
Gamma = $-.14$

TABLE 14.7   VOLUNTEERING BY WORK STATUS

	Working Full Time?	
Volunteer?	Yes	No
No	39.1%	44.3%
Yes	60.9%	55.7%
Totals	100% (685)	100% (530)

$\chi^2 = 3.35$ (df = 1, $p = .07$)
Phi = .05   Lambda = .00
Gamma = −.107

TABLE 14.8   VOLUNTEERING BY TV WATCHING

	TV Watching per Day		
Volunteer?	Low (0–1 hours)	Medium (2–3 hours)	High (41 hours)
No	36.2%	38.9%	49.6%
Yes	63.8%	61.1%	50.4%
Totals	100% (301)	100% (532)	100% (315)

$\chi^2 = 14.72$ (df = 1, $p = .000$) $V = .11$ Lambda = .00
Gamma = −.18

results offer only weak support for the idea that married people are more likely to volunteer.

The relationship between volunteering and work status (Table 14.7) is not quite significant at the 0.05 level, and the pattern of the relationship is the opposite of the prediction: full-time workers are more likely to volunteer (61%) than people not working full time (56%). On the other hand, the change in column percentages is small and the measures of association all indicate a weak relationship between these variables.

Table 14.8 shows the relationship between volunteering and extent of television watching. The relationship is significant, and changes in the column percentages are greater than in the previous tables. Light TV viewers are more likely to volunteer (64%) than heavy viewers (50%). As indicated by gamma, this is a negative relationship: as TV viewing increases, volunteering decreases. The relationship is not very strong, however, and this means that factors other than TV viewing are important causes of volunteering. (Note also that these results cannot show which variable is cause and which is effect. Do heavy viewers avoid opportunities to volunteer, or do people who choose not to volunteer become heavy viewers? Remember that correlation is not the same thing as causation.)

In summary, we can say that married people and full-time workers are more likely to be active in their communities (but not by much). It would seem reasonable to eliminate work status as an important causal variable and to conclude that marital status, at least by itself, seems to be an unimportant cause of civic participation. The relationship between television viewing and volunteerism seems promising and worthy of additional investigation. The multivariate techniques discussed in Chapter 16 might be a logical next step in the process of analyzing this relationship.

14.6 SPEARMAN'S RHO (r_s)   To this point, we have considered ordinal variables that have a limited number of categories (possible values) and are presented in tables. However,

TABLE 14.9   THE SCORES OF 10 SUBJECTS ON INVOLVEMENT IN
JOGGING AND A MEASURE OF SELF-ESTEEM

Joggers	Involvement in Jogging (X)	Self-esteem (Y)
Wendy	18	15
Debbie	17	18
Phyllis	15	12
Frank	12	16
Evelyn	10	6
Tricia	9	10
Christy	8	8
Patsy	8	7
Marsha	5	5
Siegfried	1	2

many ordinal-level variables have a broad range of scores and many distinct values. Such data may be collapsed into a few broad categories (such as high, moderate, low), organized into a bivariate table, and analyzed with gamma or the other measures of association for collapsed ordinal variables presented on the web site for this text. Collapsing scores in this manner may be beneficial and desirable in many instances, but some important distinctions between cases may be obscured or lost as a consequence.

For example, suppose a researcher wished to test the claim that jogging is beneficial not only physically but also psychologically. Do joggers have an enhanced sense of self-esteem? To deal with this issue, 10 joggers are measured on two scales, the first measuring involvement in jogging and the other measuring self-esteem. Scores are reported in Table 14.9.

These data could be collapsed and a bivariate table produced. We could, for example, dichotomize both variables to create only two values (high and low) for both variables. Although collapsing scores in this way is certainly legitimate and often necessary,* two difficulties with this practice must be noted. First, the scores seem continuous, and there are no obvious or natural division points in the distribution that would allow us to distinguish, in a nonarbitrary fashion, between high scores and low ones. Second, and more important, grouping these cases into broader categories will cause us to lose information. That is, if both Wendy and Debbie are placed in the category "high" on involvement, we would lose sight of the fact that they had different scores on this variable. If differences like this are important and meaningful, then we should opt for a measure of association that permits the retention of as much detail and precision in the scores as possible.

---

*For example, collapsing scores may be advisable when the researcher is not sure that fine distinctions between scores are meaningful.

TABLE 14.10 COMPUTING SPEARMAN'S RHO

	Involvement (X)	Rank	Self-Image (Y)	Rank	D	D²
Wendy	18	1	15	3	−2	4
Debbie	17	2	18	1	1	1
Phyllis	15	3	12	4	−1	1
Frank	12	4	16	2	2	4
Evelyn	10	5	6	8	−3	9
Tricia	9	6	10	5	1	1
Christy	8	7.5	8	6	1.5	2.25
Patsy	8	7.5	7	7	.5	.25
Marsha	5	9	5	9	0	0
Siegfried	1	10	2	10	0	0
					$\Sigma D = 0$	$\Sigma D^2 = 22.5$

Spearman's rho (r_s) is a measure of association for ordinal-level variables that have a broad range of many different scores and few ties between cases on either variable. Scores on ordinal-level variables cannot, of course, be manipulated mathematically except for judgments of "greater than" or "less than." To compute Spearman's rho, cases are first ranked from high to low on each variable and then the ranks (not the scores) are manipulated to produce the final measure. Table 14.10 displays the original scores and the rankings of the cases on both variables.

To rank the cases, first find the highest score on each variable and assign it rank 1. Wendy has the high score on X (18) and is thus ranked number 1. Debbie, on the other hand, is highest on Y and is ranked first on that variable. All other cases are then ranked in descending order of scores. If any cases have the same score on a variable, assign them the average of the ranks they would have used up had they not been tied. Christy and Patsy have identical scores of 8 on involvement. Had they not been tied, they would have used up ranks 7 and 8. The average of these two ranks is 7.5, and this average of used ranks is assigned to all tied cases. (For example, had Marsha also had a score of 8, three ranks—7, 8, and 9—would have been used, and all three tied cases would have been ranked eighth.)

The formula for Spearman's rho is

FORMULA 14.2

$$r_s = 1 - \frac{6\Sigma D^2}{N(N^2 - 1)}$$

where D^2 = the sum of the differences in ranks, the quantity squared

To compute ΣD^2, the rank of each case on Y is subtracted from its rank on X (D is the difference between rank on Y and rank on X). A column has been provided in Table 14.10 so that these differences may be recorded on a case-by-case basis. Note that the sum of this column (ΣD) is 0. That is,

---

Application 14.2

Five cities have been rated on an index that measures the quality of life. Also, the percentage of the population that has moved into each city over the past year has been determined. Have cities with higher quality-of-life scores attracted more new residents? The table below summarizes the scores, ranks, and differences in ranks for each of the five cities.

City	Quality of Life	Rank	Percentage of New Residents	Rank	D	D^2
A	30	1	17	1	0	0
B	25	2	14	3	−1	1
C	20	3	15	2	1	1
D	10	4	3	5	−1	1
E	2	5	5	4	1	1
					0	4

Spearman's rho for these variables is

$$r_s = 1 - \frac{6\Sigma\, D^2}{N(N^2 - 1)}$$

$$r_s = 1 - \frac{(6)(4)}{5(25 - 1)}$$

$$r_s = 1 - \left(\frac{24}{120}\right)$$

$$r_s = 1 - 0.20$$

$$r_s = 0.80$$

These variables have a strong, positive association. The higher the quality-of-life score, the greater the percentage of new residents. The value of r_s^2 is 0.64 ($0.80^2 = 0.64$), which indicates that we will make 64% fewer errors when predicting rank on one variable from rank on the other, as opposed to ignoring rank on the other variable.

---

the negative differences in rank are equal to the positive differences, as will always be the case, and you should find the total of this column as a check on your computations to this point. If the $\Sigma\, D$ is not equal to 0, you have made a mistake either in ranking the cases or in subtracting the differences.

In the column headed D^2, each difference is squared to eliminate negative signs. The sum of this column is $\Sigma\, D^2$, and this quantity is entered directly into the formula. For our sample problem:

$$r_s = 1 - \frac{6\Sigma\, D^2}{N(N^2 - 1)}$$

$$r_s = 1 - \frac{6(22.5)}{10(100 - 1)}$$

$$r_s = 1 - \left(\frac{135}{990}\right)$$

$$r_s = 1 - .14$$

$$r_s = .86$$

Spearman's rho is an index of the strength of association between the variables; it ranges from 0 (no association) to ±1.00 (perfect association). A perfect positive association ($r_s = +1.00$) would exist if there were no disagreements in ranks between the two variables (if cases were ranked in exactly the same order on both variables). A perfect negative relationship ($r_s = -1.00$) would exist if the ranks were in perfect disagreement (if the case ranked highest on one variable were lowest on the other, and so forth). A Spearman's rho of .86 indicates a strong, positive relationship between these two variables. The respondents who were highly involved in jogging also ranked high on self-image. These results are supportive of claims regarding the psychological benefits of jogging.

Spearman's rho is an index of the relative strength of a relationship, and values between 0 and ± 1.00 have no direct interpretation. However, if the value of rho is squared, a PRE interpretation is possible. Rho squared (r_s^2) represents the proportional reduction in errors of prediction when predicting rank on one variable from rank on the other variable, as compared to predicting rank while ignoring the other variable. In the example above, r_s was .86 and r_s^2 would be .74. Thus, our errors of prediction would be reduced by 74% if, when predicting the rank of a subject on self-image, the rank of the subject on involvement in jogging were taken into account. *(For practice in computing and interpreting Spearman's rho, see problems 14.11 to 14.14. Problem 14.11 has the fewest number of cases and is probably a good choice for a first attempt at these procedures.)*

14.7 TESTING THE NULL HYPOTHESIS OF "NO ASSOCIATION" WITH GAMMA AND SPEARMAN'S RHO

Whenever a researcher is working with EPSEM, or random, samples, he or she will need to ascertain if the sample findings can be generalized to the population. In Part II of this text, we considered various ways that information taken from samples—for example, the difference between two sample means—could be generalized to the populations from which the samples were drawn. A test of the null hypothesis, regardless of the form or specific test used, asks essentially if the patterns (or differences, or relationships) that have been observed in the samples can be assumed to exist in the population. Measures of association can also be tested for significance. When data have been collected from a random sample, we will not only need to measure the existence, strength, and direction of the association, but we will also want to know if we can assume that the variables are related in the population.

For nominal-level variables, the statistical significance of a relationship is usually judged by the chi square test. Chi square tests could also be conducted on tables displaying the relationship between ordinal-level variables. However, chi square tests deal with the probability that the observed cell frequencies occurred by chance alone and is therefore not a direct test of the significance of the measure of association (gamma or Spearman's rho) itself.

When testing gamma and Spearman's rho for statistical significance, the null hypothesis will state that there is no association between the variables in the population and that, therefore, the population value for the measure is 0. Population values will be denoted by the Greek letters gamma (γ) and rho (ρ_s). For both measures, the test procedures will be organized around the familiar five-step model (see Chapter 8).

To illustrate the test of significance for gamma, we will use Table 14.1, where gamma was .57.

Step 1. **Making Assumptions.** When sample size is greater than 10, the sampling distribution of all possible sample gammas can be assumed to be normal in shape.

$$\text{Model:}\quad \begin{array}{l}\text{Random sampling}\\ \text{Level of measurement is ordinal}\\ \text{Sampling distribution is normal}\end{array}$$

Step 2. **Stating the Null Hypothesis.**

$$H_0\!: \gamma = 0.0$$
$$(H_1\!: \gamma \neq 0.0)$$

Step 3. **Selecting the Sampling Distribution and Establishing the Critical Region.** For samples of 10 or more, the Z distribution (Appendix A) can be used to find areas under the sampling distribution:

$$\text{Sampling distribution} = Z \text{ distribution}$$
$$\text{Alpha} = .05$$
$$Z \text{ (critical)} = \pm 1.96$$

Step 4. **Computing the Test Statistic.**

$$Z\,(\text{obtained}) = G\sqrt{\frac{N_s + N_d}{N(1 - G^2)}}$$

$$Z\,(\text{obtained}) = .57\sqrt{\frac{1831 + 499}{100(1 - .33)}}$$

$$Z\,(\text{obtained}) = .57\sqrt{\frac{2330}{100(.67)}}$$

$$Z\,(\text{obtained}) = .57\sqrt{34.78}$$

$$Z\,(\text{obtained}) = (.57)(5.90)$$

$$Z\,(\text{obtained}) = 3.36$$

Step 5. **Making a Decision.** Comparing the Z (obtained) with the Z (critical):

$$Z\,(\text{obtained}) = 3.36$$
$$Z\,(\text{obtained}) = \pm 1.96$$

We see that the null hypothesis can be rejected. The sample gamma is unlikely to have occurred by chance alone, and we may conclude that these variables are related in the population from which the sample was drawn. *(For practice in conducting and interpreting the test of significance for gamma, see problems 14.2, 14.4, 14.7, 14.10, and 14.15.)*

Spearman's rho can be tested in similar fashion. The null hypothesis is again a statement that the population value (ρ_s) is actually 0 and, therefore, that the value of the sample Spearman's rho (r_s) is the result of mere random chance. When the number of cases in the sample is 10 or more, the sampling distribution of Spearman's rho approximates the *t* distribution, and we will use this distribution to conduct the test. To illustrate, the Spearman's rho computed in Section 14.6 will be used.

Step 1. Making Assumptions.

> Model: Random sampling
> Level of measurement is ordinal
> Sampling distribution is normal

Step 2. Stating the Null Hypothesis.

$$H_0: \rho_s = 0.0$$
$$(H_1: \rho_s \neq 0.0)$$

Step 3. Selecting the Sampling Distribution and Establishing the Critical Region.

> Sampling distribution = *t* distribution
> Alpha = .05
> Degrees of freedom = $N - 2 = 8$
> t (critical) = ±2.306

Step 4. Computing the Test Statistic.

$$t\,(\text{obtained}) = r_s\sqrt{\frac{N-2}{1-r_s^2}}$$

$$t\,(\text{obtained}) = .86\sqrt{\frac{8}{1-.74}}$$

$$t\,(\text{obtained}) = .86\sqrt{\frac{8}{.26}}$$

$$t\,(\text{obtained}) = .86\sqrt{30.77}$$

$$t\,(\text{obtained}) = (.86)(5.55)$$

$$t\,(\text{obtained}) = 4.77$$

Step 5. Making a Decision. Comparing the test statistic with the critical region:

$$Z\,(\text{obtained}) = 4.77$$
$$Z\,(\text{obtained}) = ±2.306$$

We see that the null hypothesis can be rejected. We may conclude, with a .05 chance of making an error, that the variables are related in the population from which the samples were drawn. *(For practice in conducting and interpreting the test of significance for Spearman's rho, see problems 14.11 to 14.14)*

SUMMARY

1. A measure of association for variables with collapsed ordinal scales (gamma) was covered along with a measure (Spearman's rho) appropriate for "continuous" ordinal variables. Both measures summarize the overall strength and direction of the association between the variables.

2. Gamma is a PRE-based measure that shows the improvement in our ability to predict the order of pairs of cases on one variable from the order of pairs of cases on the other variable, as opposed to ignoring the order of the pairs of cases on the other variable.

3. Spearman's rho is computed from the ranks of the scores of the cases on two "continuous" ordinal variables and, when squared, can be interpreted by the logic of PRE.

4. Both gamma and Spearman's rho should be tested for their statistical significance when computed for a random sample drawn from a defined population. The null hypothesis is that the variables are not related in the population, and the test can be organized by using the familiar five-step model.

SUMMARY OF FORMULAS

Gamma	14.1	$G = \dfrac{N_s - N_d}{N_s + N_d}$
Spearman's rho	14.2	$r_s = 1 - \dfrac{6\Sigma\, D^2}{N(N^2 - 1)}$

GLOSSARY

Gamma (G). A measure of association appropriate for variables measured with "collapsed" ordinal scales that have been organized into table format; G is the symbol for any sample gamma, γ is the symbol for any population gamma.

N_s. The number of pairs of cases ranked in the same order on two variables.

N_d. The number of pairs of cases ranked in different order on two variables.

Spearman's rho (r_s). A measure of association appropriate for ordinally measured variables that are "continuous" in form; r_s is the symbol for any sample Spearman's rho; ρ_s is the symbol for any population Spearman's rho.

MULTIMEDIA RESOURCES

The Wadsworth Sociology Resource Center: Virtual Society
http://sociology.wadsworth.com/

Visit the companion web site for the sixth edition of *Statistics: A Tool for Social Research* to access a wide range of student resources. Begin by clicking on the Student Resources section of the book's web site to access the following study tools:

- Basic math review
- Statistics review

- Flash cards
- Internet links
- Additional chapter problems
- Table of random numbers
- MicroCase and SPSS examples and exercises
- "Find the text" flowcharts
- Hypothesis testing for variables measured at the ordinal level

PROBLEMS

For problems 14.1 to 14.10 and 14.15 calculate percentages for the bivariate tables as described in Chapter 12. Use the percentages to help analyze the strength and direction of the association.

14.1 SOC A small sample of non-English-speaking immigrants to the United States has been interviewed about their level of assimilation. Is the pattern of adjustment affected by length of residence in the United States? For each table compute gamma and summarize the relationship in terms of strength and direction. *(HINT: in 2 × 2 tables, only two cells can contribute to N_s or N_d. To compute N_s, multiply the number of cases in the upper-left-hand cell by the number of cases in the lower-right-hand cell. For N_d, multiply the number of cases in the upper-right-hand cell by the number of cases in the lower-left-hand cell.)*

a. Facility in English:

English Facility	Length of Residence		Totals
	Less than Five Years (Low)	More than Five Years (High)	
Low	20	10	30
High	5	15	20
Totals	25	25	50

b. Total family income:

Income	Length of Residence		Totals
	Less than Five Years (Low)	More than Five Years (High)	
Below national average (1)	18	8	26
Above national average (2)	7	17	24
Totals	25	25	50

c. Extent of contact with country of origin:

Contact	Length of Residence		Totals
	Less than Five Years (Low)	More than Five Years (High)	
Rare (1)	5	20	25
Frequent (2)	20	5	25
Totals	25	25	50

14.2 Compute gamma for the tables presented in problems 11.2, 11.6, 11.7, 11.9, 11.10, and 11.12. Since these tables are based on random samples, test the gammas you computed for significance.

14.3 Compute gamma for the table presented in problem 12.1. If you computed a nominal measure of association for this table in problem 13.2, compare the measures of association. Are they similar in value? Do they characterize the strength for the association in the same way? What information about the relationship does gamma provide that is not available from nominal measures of association?

14.4 CJ A random sample of 150 cities has been classified as small, medium, or large by population and as high or low on crime rate. Is there a relationship between city size and crime rate?

Crime Rate	Size			Totals
	Small	Medium	Large	
Low	21	17	8	46
High	29	33	42	104
Totals	50	50	50	150

a. Describe the strength and direction of the relationship.

b. Is the relationship significant?

14.5 SOC Some research has shown that families vary by how they socialize their children to sports, games, and other leisure-time activities. In middle-class families, such activities are carefully monitored by parents and are, in general, dominated by adults (for example, Little League baseball). In working-class families, children more often organize and initiate such activities themselves, and parents are much less involved (for example, sandlot or playground baseball games). Are the data below consistent with these findings? Summarize your conclusions in a few sentences.

As a Child, Did You Play Mostly Organized or Sandlot Sports?	Social Class Background	
	White-collar	Blue-collar
Organized	155	123
Sandlot	101	138
Totals	256	261

14.6 PA Do municipal employees become more or less efficient as a function of longevity? Are the "old guard" better at their jobs than the younger, less experienced workers? A city manager has gathered the yearly efficiency ratings for all municipal employees along with data on longevity. Does the table below suggest any relationship? Summarize your conclusions in a few sentences.

Efficiency	Longevity		Totals
	Low	High	
Low	54	25	79
Moderate	149	254	403
High	257	387	644
Totals	460	666	1126

14.7 PA All applicants for municipal jobs in Shinbone, Kansas, are given an aptitude test, but the test has never been evaluated to see if test scores are in any way related to job performance. The following table reports aptitude test scores and job performance ratings for a random sample of 75 city employees.

Efficiency Ratings	Test Scores			Totals
	Low	Moderate	High	
Low	11	6	7	24
Moderate	9	10	9	28
High	5	9	9	23
Totals	25	25	25	75

a. Are these two variables associated? Describe the strength and direction of the relationship in a sentence or two.

b. Is gamma statistically significant?

c. Should the aptitude test continue to be administered? Why or why not?

14.8 SW A sample of children has been observed and rated for symptoms of depression. Their parents have been rated for authoritarianism. Is there any relationship between these variables? Write a few sentences stating your conclusions.

Symptoms of Depression	Authoritarianism			Totals
	Low	Moderate	High	
Few	7	8	9	24
Some	15	10	18	43
Many	8	12	3	23
Totals	30	30	30	90

14.9 SOC Are prejudice and level of education related? State your conclusion in a few sentences.

Prejudice	Level of Education				Totals
	Elementary School	High School	Some College	College Graduate	
Low	48	50	61	42	201
High	45	43	33	27	148
Totals	93	93	94	69	349

14.10 SOC In a recent survey, a random sample of respondents was asked to indicate how happy they were with their situations in life. Are their responses related to income level?

Happiness	Income			Totals
	Low	Moderate	High	
Not happy	101	82	36	219
Pretty happy	40	227	100	367
Very happy	216	198	203	617
Totals	357	507	339	1203

a. Describe the strength and direction of the relationship.

b. Is the relationship significant?

14.11 SOC A random sample of eleven neighborhoods in Shinbone, Kansas, have been rated by an urban sociologist on a "quality-of-life" scale (which includes measures of affluence, availability of medical care, and recreational facilities) and a social cohesion scale. The results are presented below in scores. Higher scores indicate higher "quality of life" and greater social cohesion.

Neighborhood	Quality of Life	Social Cohesion
Happy Acres	17	8.8
North End	40	3.9
Brentwood	47	4.0
Rolling Meadows	90	3.1
Midtown	35	7.5
Bliss Park	52	3.5
Church View	23	6.3
Hidden Valley	67	1.7
College Park	65	9.2
Beaconsdale	63	3.0
Riverview	100	5.3

a. Are the two variables associated? What is the strength and direction of the association? Summarize the relationship in a sentence or two. (*HINT: don't forget to square the value of Spearman's rho for a PRE interpretation.*)

b. Conduct a test of significance for this relationship. Summarize your findings.

14.12 SW Several years ago, a job-training program began, and a team of social workers screened the candidates for suitability for employment. Now the screening process is being evaluated, and the actual work performance of a sample of hired candidates has been rated. Did the screening process work? Is there a relationship between the original scores and performance evaluation on the job?

Case	Original Score	Performance Evaluation
A	17	78
B	17	85
C	15	82
D	13	92
E	13	75

Case	Original Score	Performance Evaluation
F	13	72
G	11	70
H	10	75
I	10	92
J	10	70
K	9	32
L	8	55
M	7	21
N	5	45
O	2	25

14.13 SOC Below are the scores of a sample of 15 nations on a measure of ethnic diversity and a measure of civil strife (such as riots, demonstrations, and assassinations). Are these variables related? Do ethnically diverse nations experience more strife?

Nation	Ethnic Diversity	Strife
A	93	126
B	85	125
C	82	478
D	80	502
E	75	100
F	60	7
G	45	78
H	43	75
I	42	70
J	39	126
K	29	7
L	25	52
M	15	48
N	5	48
O	0	33

14.14 SOC Twenty ethnic, racial, or national groups were rated by a random sample of white and black students on a Social Distance Scale. Lower scores represent less social distance and less prejudice. How similar are these rankings? Is the relationship statistically significant?

Group	Average Social Distance Scale Scores	
	White Students	Black Students
1. White Americans	1.2	2.6
2. English	1.4	2.9
3. Canadians	1.5	3.6
4. Irish	1.6	3.6
5. Germans	1.8	3.9
6. Italians	1.9	3.3

	Average Social Distance Scale Scores	
Group	White Students	Black Students
7. Norwegians	2.0	3.8
8. American Indians	2.1	2.7
9. Spanish	2.2	3.0
10. Jews	2.3	3.3
11. Poles	2.4	4.2
12. Black Americans	2.4	1.3
13. Japanese	2.8	3.5
14. Mexicans	2.9	3.4
15. Koreans	3.4	3.7
16. Russians	3.7	5.1
17. Arabs	3.9	3.9
18. Vietnamese	3.9	4.1
19. Turks	4.2	4.4
20. Iranians	5.3	5.4

14.15 SOC In problems 12.12 and 13.12, we looked at the relationships between five dependent variables and, respectively, political ideology and sex. In this exercise, we'll use income (IN-COME91) as an independent variable and assess its relationship with this set of variables. Income has been collapsed into three categories to make it suitable for tabular analysis. For each table, calculate percentages and gamma. Describe the strength, direction, and statistical significance of each relationship in a few sentences. *Be careful in interpreting direction.* Data are from the 1998 General Social Survey, which is summarized in Appendix G.

a. Support for the legal right to an abortion (ABANY) by income:

Right to Abortion?	Less than $24,900	$24,900 to $50,000	More than $50,000	Totals
Yes	220	218	226	664
No	366	299	250	915
Totals	586	517	476	1579

b. Support for capital punishment (CAPPUN) by income:

Capital Punishment?	Less than $24,900	$24,900 to $50,000	More than $50,000	Totals
Favor	567	574	552	1693
Oppose	270	183	160	613
Totals	837	757	712	2306

c. Approval of suicide for people with an incurable disease (SUICIDE1) by income:

Right to Suicide?	Less than $24,900	$24,900 to $50,000	More than $50,000	Totals
Approve	343	341	338	1022
Oppose	227	194	147	568
Totals	570	535	485	1590

d. Support for sex education in public schools (SEXEDUC) by income:

Sex Education?	Less than $24,900	$24,900 to $50,000	More than $50,000	Totals
For	492	478	451	1421
Against	85	68	53	206
Totals	577	546	504	1627

e. Support for traditional gender roles (FEHOME) by income:

Women Should Take Care of Running Their Homes and Leave Running the Country to Men	Less than $24,900	$24,900 to $50,000	More than $50,000	Totals
Agree	130	71	39	240
Disagree	448	479	461	1388
Totals	578	550	500	1628

Using *SPSS for Windows* to Produce Ordinal-Level Measures of Association

SPSS DEMONSTRATION 14.1
Do Sexual Attitudes Vary by Age? Another Look

In Demonstration 12.3, we used percentages to look at the relationships between *premarsx* (attitude toward premarital sex) and recoded *age*. Let's reexamine this relationship and see if gamma can add any new information. We will use the **Cross-tabs** program with *ager* (recoded age) as the independent (column) variable. If you no longer have access to the recoded version of this variable, follow the directions in Demonstration 12.3. The dependent variable *premarsx* will go in the rows. On the **Crosstabs** dialog window, click the **Statistics** button and request gamma. Don't forget to click the **Cells** button and request column percentages. Click **OK,** and the output should look like this:

SEX BEFORE MARRIAGE AGER Crosstabulation

			AGER			
			1.00	2.00	3.00	Total
SEX BEFORE MARRIAGE	ALWAYS WRONG	Count % within AGER	54 18.8%	64 20.6%	104 38.0%	222 25.5%
	ALMOST ALWAYS WRONG	Count % within AGER	17 5.9%	25 8.1%	31 11.3%	73 8.4%
	SOMETIMES WRONG	Count % within AGER	68 23.7%	56 18.1%	59 21.5%	183 21.0%
	NOT WRONG AT ALL	Count % within AGER	148 51.6%	165 53.2%	80 29.2%	393 45.1%
Total		Count % within AGER	287 100.0%	310 100.0%	274 100.0%	871 100.0%

Symmetric Measures

		Value	Asymp. Std. Error	Approx. T	Approx. Sig.
Ordinal by Ordinal	Gamma	-.266	.041	-6.300	.000
N of Valid Cases		871			

A gamma of −.27 indicates a moderate negative relationship. Older respondents (a score of 3 on *ager*) were most opposed to premarital sex (they had the highest percentage who felt it was "always wrong"), and younger respondents were least opposed (they had the highest percentage who felt it was "not wrong at all"). As age increases, support of premarital sex (the percentage who say it's "not wrong at all") decreases.

SPSS DEMONSTRATION 14.2
A Note about the Direction of Ordinal Relationships

Let's pause once again to consider the direction of relationships between ordinal-level variables. Direction for ordinal variables is a surprisingly tricky matter, mostly because of the arbitrary nature of coding at the ordinal level. We usually think of *higher* scores as indicating *more* of the quantity being measured and, for every interval-ratio-level variable I can think of, this pattern will be true. For ordinal variables, however, higher scores may indicate *less* of the quantity because the codes are arbitrary. Depending on how you code the values, a high score on a scale measuring, say, prejudice might indicate great prejudice or its complete absence. Looking at the table for *premarsx* and *ager* in Demonstration 14.1, remember that a negative gamma means that *high* values on one variable are associated with *low* values on the other. This means, for example, that those who scored a 1 (the lowest possible value) on *ager* tended to score a 4 (the highest possible value) on *premarsx*. You may think of a 4 on *premarsx* as representing "high support for premarital sex" or "low opposition to premarital sex." A negative gamma always means that the scores of the variables are inversely related, but this does not necessarily mean that the underlying relationship is truly negative. Always inspect tables carefully to make sure that you are interpreting the direction of the relationship properly.

SPSS DEMONSTRATION 14.3 Interpreting the Direction
of Relationships: Are Prestigious Jobs More Satisfying?

What's the relationship between prestige of occupation and job satisfaction? Some would argue that high-prestige jobs are rewarding in a variety of ways (besides simply income) and that people in such jobs would express greater satisfaction with their work. Others might argue an opposing point of view: because people in lower-prestige jobs have fewer responsibilities and less pressure, they will be more satisfied. In an attempt to resolve the debate, we'll run a **Crosstabs** on *prestg80* and *satjob*. I first found the median on *prestg80* with the **Frequencies** command and recoded the variable into a dichotomy. I also recoded *satjob* into two approximately equal categories. Don't forget to recode "into different variable" and give the recoded variables new names (I used *rprest* and *rsat*). The recoding instructions for *prestg80* are

$$0 \text{ thru } 42 \rightarrow 1$$
$$43 \text{ thru } 86 \rightarrow 2$$

For *satjob*, I left the score of 1 alone and collapsed scores 2–4:

$$1 = 1$$
$$2 \text{ thru } 4 \rightarrow 2$$

Use **Crosstabs** once again with recoded *prestg80* as the independent (column) variable and recoded *satjob* as the dependent variable. Request gamma, chi square, and column percentages. The output should look like this:

RPREST RSAT Crosstabulation

			RPREST		Total
			1.00	2.00	
RSAT	1.00	Count	229	304	533
		% within RPREST	41.6%	54.8%	48.2%
	2.00	Count	322	251	573
		% within RPREST	58.4%	45.2%	51.8%
Total		Count	551	555	1106
		% within RPREST	100.0%	100.0%	100.0%

Chi-Square Tests

	Value	df	Asymp. Sig. (2-sided)	Exact Sig. (2-sided)	Exact Sig. (1-sided)
Pearson Chi-Square	19.337	1	.000		
Continuity Correction	18.811	1	.000		
Likelihood Ratio	19.394	1	.000		
Fisher's Exact Test				.000	.000
Linear-by-Linear Association	19.319	1	.000		
N of Valid Cases	1106				

Symmetric Measures

		Value	Asymp. Std. Error	Approx. T	Approx. Sig.
Ordinal by Ordinal	Gamma	-.260	.057	-4.436	.000
N of Valid Cases		1106			

This relationship is statistically significant (the significance of the chi square is less than .05) and, with a gamma of $-.26$, moderate in strength and negative in direction. As prestige increases, job satisfaction decreases. Right? Wrong. Look at the codes for *satjob*. A *higher* score indicates a *lower* level of satisfaction. So, "high" (or a score of 2) on *prestg80* is associated with "high" (or a score of 1) on *satjob*. In spite of the negative sign for gamma, this is a *positive* relationship, and job satisfaction increases with prestige.

MicroCase

Using MicroCase to Produce Ordinal-Level Measures of Association

MICROCASE DEMONSTRATION 14.1
Do Sexual Attitudes Vary by Age? Another Look

In Demonstration 12.3, we used percentages to look at the relationships between *premarsx* (attitude toward premarital sex) and recoded *age*. Let's reexamine this relationship and see if gamma can add any new information. We will use the **Cross-tabulation** program with *ager* (recoded age) as the independent (column) variable. If you no longer have access to the recoded version of this variable, follow the directions in Demonstration 12.3. The dependent variable *premarsx* will go in the rows. Click **OK** and then get column percents and summary statistics. To conserve space, I have deleted all statistics but gamma. The output should look like this:

```
PREMARSX by AGER
               18-34    35-50    51-89    Missing   TOTAL
ALWAYS WRO     59       59       104      0         222
               18.8%    20.8%    38.0%              25.5%
ALMST ALWA     18       24       31       0         73
               5.8%     8.5%     11.3%              8.4%
SOMETIMES      73       51       59       0         183
               23.3%    18.0%    21.5%              21.0%
NOT WRONG      163      150      80       2         393
               52.1%    52.8%    29.2%              45.1%
Missing        177      149      188      0         514
TOTAL          313      284      274      2         871
               100.0%   100.0%   100.0%

Gamma:  -0.270
```

A gamma of −.27 indicates a moderate negative relationship. Older respondents (a score of 3 on *ager*) were most opposed to premarital sex (they had the highest percentage who felt it was "always wrong"), and younger respondents were least opposed (they had the highest percentage who felt it was "not wrong at all"). As age increases, support of premarital sex (the percentage who say it's "not wrong at all") decreases.

MICROCASE DEMONSTRATION 14.2
A Note about the Direction of Ordinal Relationships

Let's pause to once again consider the direction of relationships between ordinal-level variables. Direction for ordinal variables is a surprisingly tricky matter, mostly because of the arbitrary nature of coding at the ordinal level. We usually think of *higher* scores as indicating *more* of the quantity being measured and, for every interval-ratio level variable I can think of, this pattern will be true. For ordinal variables, however, higher scores may indicate *less* of the quantity because the codes are arbitrary. Depending on how you code the values, a high score on a scale measuring, say, prejudice might indicate great prejudice or its complete absence. Looking at the table for *premarsx* and *ager* in Demonstration 14.1, remember that a negative gamma means that *high* values on one variable are associated with *low* values on the other. This means, for example, that those who scored a 1 (the lowest possible value) on *ager* tended to score a 4 (the highest possible value) on *premarsx*. You may think of a 4 on *premarsx* as representing "high support for premarital sex" or "low opposition to premarital sex." A negative gamma always means that the scores of the variables are inversely related, but this does not necessarily mean that the underlying relationship is truly negative. Always inspect tables carefully to make sure that you are interpreting the direction of the relationship properly.

MICROCASE DEMONSTRATION 14.3 Interpreting the Direction of Relationships: Are Prestigious Jobs More Satisfying?

What's the relationship between prestige of occupation and job satisfaction? Some would argue that high-prestige jobs are rewarding in a variety of ways (besides simply income) and that people in such jobs would express greater satisfaction with their work. Others might argue an opposing point of view: because people in lower-prestige jobs have fewer responsibilities and less pressure, they will be more satisfied. In an attempt to resolve the debate, we'll run a **Cross-tabulation** on *prestg80* and *satjob*. I first found the median on *prestg80* with the **Univariate** command and recoded the variable into a dichotomy. I also recoded *satjob* into two approximately equal categories. Don't forget to give the recoded variables new names (I used *rprest* and *rsat*). The recoding instructions for *prestg80* are

$$0 \text{ thru } 42 \rightarrow 1$$
$$43 \text{ thru } 86 \rightarrow 2$$

For *satjob,* I left the score of 1 alone and collapsed scores 2–4:

$$1 = 1$$
$$2 \text{ thru } 4 \rightarrow 2$$

Use **Cross-tabulation** once again with recoded *prestg80* as the independent (column) variable and recoded *satjob* as the dependent variable. Get gamma, chi square, and column percentages. The output should look like this:

RSAT by RPREST				
	0–42	43–86	Missing	TOTAL
1	217	304	12	521
	41.4%	54.8%		48.3%
2–4	307	251	15	558
	58.6%	45.2%		51.7%
Missing	121	118	42	281
TOTAL	524	555	69	1079
	100.0%	100.0%		

Chi-Square: 19.273 (DF = 1; Prob. = 0.000)
Gamma: −0.263

This relationship is statistically significant (the significance of the chi square is less than .05) and, with a gamma of −.26, moderate in strength and negative in direction. As prestige increases, job satisfaction decreases. Right? Wrong. Look at the codes for *satjob*. A *higher* score indicates a *lower* level of satisfaction. So, "high" (or a score of 2) on *prestg80* is associated with "high" (or a score of 1) on *satjob*. In spite of the negative sign for gamma, this is a *positive* relationship, and job satisfaction increases with prestige.

Exercises

14.1 Follow up on Demonstration 14.1 with some new ordinal-level independent variables. What other factors might affect attitudes towards premarital sex? Among other possibilities, consider *degree* or *class* as independent variables. Do sexual attitudes vary by education or social class? Other possibilities include *income91, prestg80,* or *educ* (these would have to be recoded), *polviews,* or *attend.*

14.2 Following the procedure in Demonstration 14.3, use other measures of social class besides prestige (*class* or *degree*) as independent variables in the relationship with recoded *satjob.* Summarize the strength and direction of the relationships in a few sentences. Are these relationships stronger or weaker than those with recoded *prestg80?* Be careful in interpreting the direction of the relationships.

14.3 Examine the relationship between recoded *polviews* and recoded *age.* Does conservatism increase with age? Interpret the strength and direction of the relationship.

ASSOCIATION BETWEEN VARIABLES MEASURED AT THE INTERVAL-RATIO LEVEL

LEARNING OBJECTIVES

By the end of this chapter, you will be able to

1. Interpret a scattergram.
2. Calculate and interpret slope (b), Y intercept (a), and Pearson's r and r^2.
3. Find and explain the least-squares regression line and use it to predict values of Y.
4. Explain the concepts of total, explained, and unexplained variance.
5. Use regression and correlation techniques to analyze and describe a bivariate relationship in terms of the three questions introduced in Chapter 12.
6. Test Pearson's r for significance.

15.1 INTRODUCTION

This chapter presents a set of statistical techniques for analyzing the association or correlation between variables measured at the interval-ratio level.* As we shall see, these techniques are rather different in their logic and computation from those covered in Chapters 13 and 14. Let me stress at the outset, therefore, that we are still asking the same three questions: Is there a relationship between the variables? How strong is the relationship? What is the direction of the relationship? You might become preoccupied with some of the technical details and computational routines in this chapter, so remind yourself occasionally that our ultimate goals are unchanged: we are trying to understand bivariate relationships, explore possible causal ties between variables, and improve our ability to predict scores.

15.2 SCATTERGRAMS

As we have seen over the past several chapters, properly percentaged tables provide important information about bivariate associations between nominal- and ordinal-level variables. In addition to measures of association like

---

*The term *correlation* is commonly used instead of *association* when discussing the relationship between interval-ratio variables. We will use the two terms interchangeably.

phi or gamma, the conditional distributions and patterns of cell frequency almost always provide useful information and a better understanding of the relationship between variables.

By the same token, the usual first step in analyzing a relationship between interval-ratio variables is to construct and examine a **scattergram.** Like bivariate tables, these graphs allow us to quickly identify several important features of the relationship. An example will illustrate the construction and use of scattergrams. Suppose a researcher is interested in analyzing how dual-wage-earner families (that is, families where both husband and wife have jobs outside the home) cope with housework. Specifically, the researcher wonders if the number of children in the family is related to the amount of time the husband contributes to housekeeping chores. The relevant data for a sample of 12 dual-wage-earner families are displayed in Table 15.1.

A scattergram, like a bivariate table, has two dimensions. The scores of the independent (X) variable are arrayed along the horizontal axis, and the scores of the dependent (Y) variable along the vertical axis. Each dot on the scattergram represents a case in the sample and is located at a point determined by the scores of the case. Figure 15.1 shows a scattergram displaying the relationship between "number of children" and "husband's housework" for the sample of 12 families presented in Table 15.1. Family A has a score of 1 on the X variable (number of children) and 1 on the Y variable (husband's housework) and is represented by the dot above the score of 1 on the X axis and directly to the right of the score of 1 on the Y axis. All 12 cases are similarly represented by dots on Figure 15.1. Also note that, as always, the scattergram is clearly titled and both axes are labeled.

The overall pattern of the dots or cases summarizes the nature of the relationship between the two variables. The clarity of the pattern can be

TABLE 15.1   NUMBER OF CHILDREN AND HUSBAND'S CONTRIBUTION TO HOUSEWORK (fictitious data)

Family	Number of Children	Hours per Week Husband Spends on Housework
A	1	1
B	1	2
C	1	3
D	1	5
E	2	3
F	2	1
G	3	5
H	3	0
I	4	6
J	4	3
K	5	7
L	5	4

FIGURE 15.1   HUSBAND'S HOUSEWORK BY NUMBER OF CHILDREN

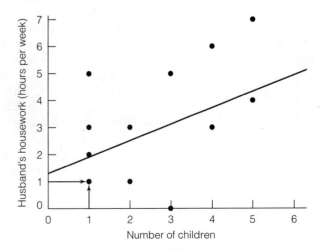

enhanced by drawing a straight line through the cluster of dots such that the line touches every dot or comes as close to doing so as possible. In Section 15.3, a precise technique for fitting this line to the pattern of the dots will be explained. For now, an "eyeball" approximation will suffice. This summarizing line is called the **regression line** and has already been added to the scattergram.

Even a crudely drawn scattergram with a freehand regression line can be used for a variety of purposes. Scattergrams provide at least impressionistic information about the existence, strength, and direction of the relationship and can also be used to check the relationship for linearity (that is, how well the pattern of dots can be approximated with a straight line). Finally, the scattergram can be used to predict the score of a case on one variable from the score of that case on the other variable. We will briefly examine each of these uses.

To ascertain the existence of a relationship, we can return to the basic definition of an association stated in Chapter 12. Two variables are associated if the distributions of Y (the dependent variable) change for the various conditions of X (the independent variable). In Figure 15.1, the conditions or scores of X are arrayed along the horizontal axis (number of children). The dots above each score on X are the conditional distributions of Y. That is, the dots represent scores on Y for each value of X. Figure 15.1 shows that there is a relationship between these variables because these conditional distributions of Y (the dots above each score on X) change as X changes. The existence of an association is further reinforced by the fact that the regression line lies at an angle to the X axis. If these two variables had not been associated, the conditional distributions of Y would not have changed, and the regression line would have been parallel to the horizontal axis.

The strength of the bivariate association can be judged by observing the spread of the dots around the regression line. In a perfect association, all dots would lie on the regression line. The more the dots are clustered around the regression line, the stronger the association.

The direction of the relationship can be detected by observing the angle of the regression line. Figure 15.1 shows a positive relationship: As X (number of children) increases, husband's housework (Y) also increases. Husbands in families with more children tend to do more housework. If the relationship had been negative, the regression line would have sloped in the opposite direction to indicate that high scores on one variable were associated with low scores on the other.

To summarize these points about the existence, strength, and direction of the relationship, Figure 15.2 shows a perfect positive and a perfect negative relationship and a "zero relationship," or "nonrelationship" between two variables.

One key assumption underlying the statistical techniques to be introduced later in this chapter is that the two variables have an essentially **lin-**

FIGURE 15.2   POSITIVE, NEGATIVE, AND ZERO RELATIONSHIPS

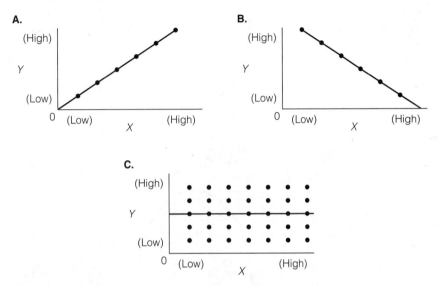

ear relationship. In other words, the observation points or dots in the scattergram must form a pattern that can be approximated with a straight line. Significant departures from linearity would require the use of statistical techniques beyond the scope of this text. Examples of some common curvilinear relationships are presented in Figure 15.3. If the scattergram shows that the variables have a nonlinear relationship, the techniques described in this chapter should be used with great caution or not at all. Checking for the linearity of the relationship is perhaps the most important reason for constructing at least a crude, hand-drawn scattergram before proceeding with the statistical analysis. If the relationship is nonlinear, you might need to treat the variables as if they were ordinal rather than interval-ratio in level of measurement. *(For practice in constructing and interpreting scattergrams, see problems 15.1 to 15.4.)*

15.3 REGRESSION AND PREDICTION

A final use of the scattergram is to predict scores of cases on one variable from their score on the other. To illustrate, suppose that, based on the relationship between number of children and husband's housework displayed in Figure 15.1, we wish to predict the number of hours of housework a husband with a family of six children would do each week. The sample has no families with six children, but if we extend the axes and regression line in Figure 15.1 to incorporate this score, a prediction is possible. Figure 15.4 reproduces the scattergram and illustrates how the prediction would be made.

The predicted score on Y—which is symbolized as Y' to distinguish predictions of Y from actual Y scores—is found by first locating the relevant

FIGURE 15.3 SOME NONLINEAR RELATIONSHIPS

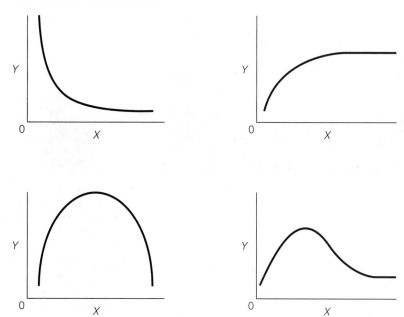

FIGURE 15.4 PREDICTING HUSBAND'S HOUSEWORK

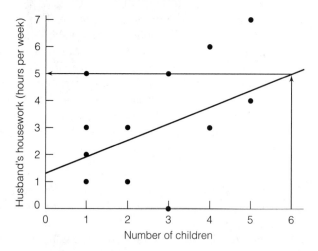

score on X ($X = 6$ in this case) and then drawing a straight line from that point to the regression line. From the regression line, another straight line parallel to the X axis is drawn across to the Y axis. The predicted Y score (Y') is found at the point where the line crosses the Y axis. In our example, we would predict that, in a dual-wage-earner family with six children, the husband would devote about five hours per week to housework.

Of course, this prediction technique is crude, and the value of Y' can change, depending on how accurately the freehand regression line is drawn. One way to eliminate this source of error would be to find the straight line that most accurately summarizes the pattern of the observation points and so best describes the relationship between the two variables. Is there such a "best-fitting" straight line? If there is, how is it defined?

Recall that our criterion for the freehand regression line was that it touch all the dots or come as close to doing so as possible. Also, recall that the dots above each value of X can be thought of as conditional distributions of Y, the dependent variable. Within each conditional distribution of Y, the mean is the point around which the variation of the scores is at a minimum. In Chapter 3, we noted that the mean of any distribution of scores is the point around which the variation of the scores, as measured by squared deviations, is minimized:

$$\Sigma(X_i - \overline{X})^2 = \text{minimum}$$

Thus, if the regression line is drawn so that it touches each **conditional mean of Y**, it would be the straight line that comes as close as possible to all the scores.

Conditional means are found by summing all Y values for each value of X and then dividing by the number of cases. For example, four families had one child ($X = 1$), and the husbands of these four families devoted 1, 2, 3, and 5 hours per week to housework. Thus, for $X = 1$, $Y = 1, 2, 3$, and 5, and the conditional mean of Y for $X = 1$ is 2.75 (11/4 = 2.75). Husbands in families with one child worked an average of 2.75 hours per week doing housekeeping chores. Conditional means of Y are computed in the same way for each value of X displayed in Table 15.2 and plotted in Figure 15.5.

TABLE 15.2   CONDITIONAL MEANS OF Y
(husband's housework) FOR VARIOUS VALUES OF X
(number of children)

Number of Children (X)	Husband's Housework (Y)	Conditional Means of Y
1	1,2,3,5	2.75
2	3,1	2.00
3	5,0	2.50
4	6,3	4.50
5	7,4	5.50

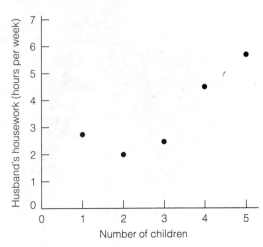

FIGURE 15.5   CONDITIONAL MEANS OF Y

Let us quickly remind ourselves of the reason for these calculations. We are seeking the single best-fitting regression line for summarizing the relationship between X and Y, and we have seen that a line drawn through the conditional means of Y will minimize the spread of the observation points. It will come as close to all the scores as possible and will therefore be the single best-fitting regression line.

Now, a line drawn through the points on Figure 15.5 (the conditional means of Y) will be the best-fitting line we are seeking, but you can see from the scattergram that the line will not be straight. In fact, only rarely (when there is a perfect relationship between X and Y) will conditional means fall in a perfectly straight line. Since we still must meet the condition of linearity, let us revise our criterion and define the regression line as the unique straight line that touches all conditional means of Y or comes as close to doing so as possible. Formula 15.1 defines the "least-squares" regression line, or the single straight regression line that best fits the pattern of the data points.

FORMULA 15.1
$$Y = a + bX$$

where Y = score on the dependent variable
a = the Y intercept, or the point where the regression line crosses the Y axis
b = the slope of the regression line, or the amount of change produced in Y by a unit change in X
X = score on the independent variable

The formula introduces two new concepts. First, the **Y intercept (a)** is the point at which the regression line crosses the vertical, or Y, axis. Second, the **slope (b)** of the least-squares regression line is the amount of change produced in the dependent variable (Y) by a unit change in the independent variable (X). Think of the slope of the regression line as a measure of the effect of the X variable on the Y variable. If the variables have a strong association, then changes in the value of X will be accompanied by substantial changes in the value of Y, and the slope (b) will have a high value. The weaker the effect of X on Y (the weaker the association between the variables), the lower the value of the slope (b). If the two variables are unrelated, the least-squares regression line would be parallel to the X axis, and b would be 0.00 (the line would have no slope).

With the least-squares formula (Formula 15.1), we can predict values of Y in a much less arbitrary and impressionistic way than through mere eyeballing. This will be so, remember, because the least-squares regression line as defined by Formula 15.1 is the single straight line that best fits the data because it comes as close as possible to all of the conditional means of Y. Before seeing how predictions of Y can be made, however, we must first calculate a and b. *(For practice in using the regression line to predict scores on Y from scores on X, see problems 15.1 to 15.3 and 15.5.)*

15.4 THE COMPUTATION OF *a* AND *b*

Since the value of b is needed to solve for a, we will begin with the computation of the slope of the least-squares regression line. The definitional formula for the slope is

FORMULA 15.2

$$b = \frac{\Sigma(X - \overline{X})(Y - \overline{Y})}{\Sigma(X - \overline{X})^2}$$

The numerator of this formula is called the *covariation* of X and Y. It is a measure of how X and Y vary together, and its value will reflect both the direction and strength of the relationship. It is tedious to find the slope with this formula, and the following computational formula, which can be derived from Formula 15.2, should be used instead:

FORMULA 15.3

$$b = \frac{N\Sigma\, XY - (\Sigma\, X)(\Sigma\, Y)}{N\Sigma\, X^2 - (\Sigma\, X)^2}$$

where b = the slope
N = the number of cases
$\Sigma\, XY$ = the summation of the crossproducts of the scores
$\Sigma\, X$ = the summation of the scores on X
$\Sigma\, Y$ = the summation of the scores on Y
$\Sigma\, X^2$ = the summation of the squared scores on X

Admittedly, this formula appears formidable at first glance; but it can be solved without too much difficulty if computations are organized into table format. The computing table displayed in Table 15.3 has a column for each

TABLE 15.3   COMPUTATION OF THE SLOPE (b)

X	Y	X^2	Y^2*	XY
1	1	1	1	1
1	2	1	4	2
1	3	1	9	3
1	5	1	25	5
2	3	4	9	6
2	1	4	1	2
3	5	9	25	15
3	0	9	0	0
4	6	16	36	24
4	3	16	9	12
5	7	25	49	35
5	4	25	16	20
Totals = 32	40	112	184	125

$\Sigma\, X = 32$
$\Sigma\, Y = 40$
$\Sigma\, X^2 = 112$
$\Sigma\, Y^2 = 184$
$\Sigma\, XY = 125$

*The quantity $\Sigma\, Y^2$ is not used in the computation of b. We will need it later, however, when we compute Pearson's r (see Section 15.5).

of the four quantities needed to solve the formula. The data are from the dual-wage-earner family sample (see Table 15.1).

In Table 15.3, the first two columns list the original X and Y scores for each case. The third column contains the squared scores on X, and the fourth lists the squared scores on Y. The fifth column lists the crossproducts of the scores for each case. In other words, the entries in the last column are determined by multiplying both scores for each case. We can now replace the symbols in Formula 15.3 with the proper sums:

$$b = \frac{N\Sigma\, XY - (\Sigma\, X)(\Sigma\, Y)}{N\Sigma\, X^2 - (\Sigma\, X)^2}$$

$$b = \frac{(12)(125) - (32)(40)}{(12)(112) - (32)^2}$$

$$b = \frac{(1500 - 1280)}{(1344 - 1024)}$$

$$b = \frac{220}{320}$$

$$b = .69$$

A slope of .69 indicates that, for each unit change in X, there is an increase of .69 unit in Y. For our example, the addition of each child (an increase of one unit in X) results in an increase of .69 hour of housework being done by the husband (an increase of .69 unit—or hour—in Y).

Once the slope has been calculated, finding the intercept (a) is relatively easy. To compute the mean of X and the mean of Y, divide the sums of columns 1 and 2 of Table 15.3 by N and enter these figures into Formula 15.4:

FORMULA 15.4
$$a = \overline{Y} - b\overline{X}$$

For our sample problem, the value of a would be

$$a = \overline{Y} - b\overline{X}$$
$$a = 3.33 - (.69)(2.67)$$
$$a = 3.33 - 1.84$$
$$a = 1.49$$

Thus, the least-squares regression line will cross the Y axis at the point where Y equals 1.49.

The full least-squares regression line for our sample data can now be specified:

$$Y = a + bX$$
$$Y = (1.49) + (.69)X$$

This formula can be used to estimate or predict scores on Y for any value of X. In Section 15.3, we used the freehand regression line to predict a score on Y (husband's housework) for a family with six children ($X = 6$). Our prediction was that, in families of six children, husbands would contribute about five hours per week to housekeeping chores. By using the least-squares regression line, we can see how close our impressionistic, eyeball prediction was.

$$Y' = a + bX$$
$$Y' = (1.49) + (.69)(6)$$
$$Y' = (1.49) + (4.14)$$
$$Y' = 5.63$$

Based on the least-squares regression line, we would predict that in a dual-wage-earner family with six children, husbands would devote 5.63 hours a week to housework. What would our prediction of husband's housework be for a family of seven children ($X = 7$)?

Note that our predictions of Y scores are basically "educated guesses." We will be unlikely to predict values of Y exactly except in the (relatively rare) case where the bivariate relationship is perfect and perfectly linear. Note also, however, that the accuracy of our predictions will increase as relationships become stronger because in stronger relationships, the dots are more clustered around the least-squares regression line. *(The slope and Y intercept may be computed for any problem at the end of this chapter, but see problems 15.1 to 15.5 in particular. These problems have smaller data sets and will provide good practice until you are comfortable with these calculations.)*

15.5 THE CORRELATION COEFFICIENT (PEARSON'S r)

I pointed out in Section 15.4 that the slope of the least-squares regression line (b) is a measure of the effect of X on Y. Since the slope is the amount of change produced in Y by a unit change in X, b will increase in value as the relationship increases in strength. However, b does not vary between zero and one and is therefore awkward to use as a measure of association. Instead, researchers rely heavily (almost exclusively) on a statistic called **Pearson's r,** or the correlation coefficient, to measure association between interval-ratio variables. Like the ordinal measures of association discussed in Chapter 14, Pearson's r varies from 0.00 to ± 1.00, with 0.00 indicating no association and $+1.00$ and -1.00 indicating perfect positive and perfect negative relationships, respectively. The definitional formula for Pearson's r is

FORMULA 15.5

$$r = \frac{\Sigma(X - \overline{X})(Y - \overline{Y})}{\sqrt{[\Sigma(X - \overline{X})^2][\Sigma(Y - \overline{Y})^2]}}$$

Note that the numerator of this formula is the covariation of X and Y, as was the case with Formula 15.2. This formula is awkward to use, and the computational Formula 15.6 is usually preferred.

FORMULA 15.6

$$r = \frac{N\Sigma\, XY - (\Sigma\, X)(\Sigma\, Y)}{\sqrt{[N\Sigma\, X^2 - (\Sigma\, X)^2][N\Sigma\, Y^2 - (\Sigma\, Y)^2]}}$$

A computing table such as Table 15.3 is strongly recommended as a way of organizing the quantities needed to solve this equation. For our sample problem involving dual-wage-earner families, the quantities displayed in Table 15.3 can be substituted directly into Formula 15.6:

$$r = \frac{(12)(125) - (32)(40)}{\sqrt{[(12)(112) - (32)^2][(12)(184) - (40)^2]}}$$

$$r = \frac{1500 - 1280}{\sqrt{(1344 - 1024)(2208 - 1600)}}$$

$$r = \frac{220}{\sqrt{194,560}}$$

$$r = \frac{220}{441.09}$$

$$r = .50$$

An r value of .50 indicates a moderately strong, positive linear relationship between the variables. As the number of children in the family increases, the hourly contribution of husbands to housekeeping duties also increases. *(Every problem at the end of this chapter requires the computation of Pearson's r. It is probably a good idea to practice with smaller data sets and easier computations first—see problem 15.1 in particular.)*

15.6 INTERPRETING THE CORRELATION COEFFICIENT: r^2

Pearson's r is an index of the strength of the linear relationship between two variables. While a value of 0.00 indicates no linear relationship and a value of ± 1.00 indicates a perfect linear relationship, values between these extremes have no direct interpretation. We can, of course, describe relationships in terms of how closely they approach the extremes (for example, coefficients approaching 0.00 can be described as "weak" and those approaching ± 1.00 as "strong"), but this description is somewhat subjective.

Fortunately, a more direct interpretation is provided by calculating an additional statistic called the **coefficient of determination.** This statistic, which is simply the square of Pearson's r (r^2), can be interpreted with a logic akin to proportional reduction in error (PRE). As you recall, the logic of PRE measures of association is to predict the value of the dependent variable under two different conditions. First, Y is predicted while ignoring the information supplied by X and, second, the independent variable is taken into account. With r^2, both the method of prediction and the construction of the final statistic are somewhat different and require the introduction of some new concepts.

When working with variables measured at the interval-ratio level, the predictions of Y under the first condition (while ignoring X) will be the mean of the Y scores ($\overline{Y}$) for every case. Given no information on X, this prediction strategy will be optimal because we know that the mean of any distribution is closer to all the scores than any other point in the distribution. I remind you of the principle of minimized variation introduced in Chapter 3 and expressed as

$$\Sigma(Y - \overline{Y})^2 = \text{minimum}$$

The scores of any variable vary less around the mean than around any other point. If we predict the mean of Y for every case, we will make fewer errors of prediction than if we predict any other value for Y.

Of course, we will still make many errors in predicting Y even if we faithfully follow this strategy. The amount of error is represented in Figure 15.6, which displays the relationship between number of children and husband's housework with the mean of Y ($\overline{Y}$) noted. The vertical lines from the actual scores to the predicted score ($\overline{Y}$) represent the amount of error we would make when predicting Y while ignoring X.

We can define the extent of our prediction error under the first condition (while ignoring X) by subtracting the mean of Y from each actual Y score and squaring and summing these deviations. The resultant figure, which can be noted as $\Sigma(Y - \overline{Y})^2$, is called the **total variation** in Y. We now have a visual representation (Figure 15.6) and a method for calculating the error we incur by predicting Y without knowledge of X. As we shall see below, we do not need to actually calculate the total variation to find the value of the coefficient of determination, r^2.

Our next step will be to determine the extent to which knowledge of X improves our ability to predict Y. If the two variables have a linear relationship, then predicting scores on Y from the least-squares regression equation will incorporate knowledge of X and reduce our errors of prediction. So, under the second condition, our predicted Y score for each value of X will be

$$Y' = a + bX$$

FIGURE 15.6   PREDICTING Y WITHOUT X (dual-career families)

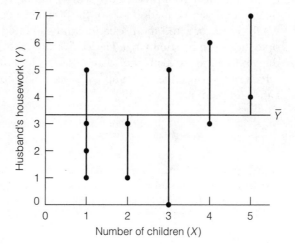

FIGURE 15.7 PREDICTING Y WITH X (dual-career families)

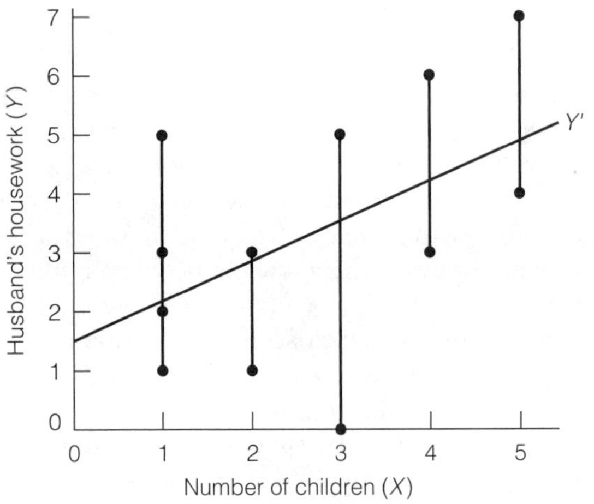

Figure 15.7 displays the data from the dual-career families with the regression line, as determined by the above formula, drawn in. The vertical lines from each data point to the regression line represent the amount of error in predicting Y that remains even after X has been taken into account.

As was the case under the first condition, we can precisely define the reduction in error that results from taking X into account. Specifically, two different sums can be found and then compared with the total variation of Y to construct a statistic that will indicate the improvement in prediction.

The first sum, called the **explained variation,** represents the improvement in our ability to predict Y when taking X into account. This sum is found by subtracting $\overline{Y}$ (our predicted Y score without X) from the score predicted by the regression equation (Y', or the Y score predicted with knowledge of X) for each case and then squaring and summing these differences. These operations can be summarized as $\Sigma(Y' - \overline{Y})^2$, and the resultant figure could then be compared with the total variation in Y to ascertain the extent to which our knowledge of X improves our ability to predict Y. Specifically, it can be shown mathematically that

FORMULA 15.7

$$r^2 = \frac{\Sigma(Y' - \overline{Y})^2}{\Sigma(Y - \overline{Y})^2} = \frac{\text{Explained variation}}{\text{Total variation}}$$

Thus, the coefficient of determination, or r^2, is the proportion of the total variation in Y attributable to or explained by X. Like other PRE measures, r^2 indicates precisely the extent to which X helps us predict, understand, or explain Y.

Application 15.1

For five cities, information has been collected on number of civil disturbances (riots, strikes, and so forth) over the past year and on unemployment rate. Are these variables associated?

The data are presented in the table below. Columns have been added for all necessary sums.

The slope (b) is

$$b = \frac{N\Sigma\,XY - (\Sigma\,X)(\Sigma\,Y)}{N\Sigma\,X^2 - (\Sigma\,X)^2}$$

$$b = \frac{(5)(985) - (76)(53)}{(5)(1290) - (76)^2}$$

$$b = \frac{897}{674}$$

$$b = 1.33$$

A slope of 1.33 means that for every unit change in X (for every increase of 1 in the unemployment rate), there was a change of 1.33 units in Y (the number of civil disturbances increased by 1.33). The Y intercept (a) is

$$a = \overline{Y} - b\overline{X}$$

$$a = \frac{53}{5} - (1.33)\left(\frac{76}{5}\right)$$

$$a = 10.6 - (1.33)(15.20)$$

$$a = 10.6 - 20.2$$

$$a = -9.6$$

The least-squares regression equation is

$$Y = a + bX = -9.6 + (1.33)X$$

The correlation coefficient is

$$r = \frac{N\Sigma\,XY - (\Sigma\,X)(\Sigma\,Y)}{\sqrt{[N\Sigma\,X^2 - (\Sigma\,X)^2][N\Sigma\,Y^2 - (\Sigma\,Y)^2]}}$$

$$r = \frac{(5)(985) - (76)(53)}{\sqrt{[(5)(1290) - (76)^2][(5)(919) - (53)^2]}}$$

$$r = \frac{897}{\sqrt{(674)(1786)}}$$

$$r = \frac{897}{\sqrt{1203764}}$$

$$r = \frac{897}{1097.16}$$

$$r = 0.82$$

These variables have a strong, positive association. The number of civil disturbances increases as the unemployment rate increases. The coefficient of determination, r^2, is $(0.82)^2$, or 0.67. This indicates that 67% of the variance in civil disturbances is explained by the unemployment rate.

City	Unemployment Rate (X)	Civil Disturbances (Y)	X^2	Y^2	XY
A	22	25	484	625	550
B	20	13	400	169	260
C	10	10	100	100	100
D	15	5	225	25	75
E	9	0	81	0	0
	76	53	1290	919	985

Above, we refer to the improvement in predicting Y with X as the explained variation. The use of this term suggests that some of the variation in Y will be "unexplained" or not attributable to the influence of X. In fact, the vertical lines in Figure 15.7 represent the **unexplained variation,** or the difference between our best prediction of Y with X and the actual scores.

The unexplained variation is thus the scattering of the actual scores around the regression line and can be found by subtracting the predicted Y scores from the actual Y scores for each case and then squaring and summing the differences. These operations can be summarized as $\Sigma(Y - Y')^2$, and the resultant sum would measure the amount of error in predicting Y that remains even after X has been taken into account. The proportion of the total variation in Y unexplained by X can be found by subtracting the value of r^2 from 1.00. Unexplained variation is usually attributed to the influence of some combination of other variables, measurement error, and random chance.

As you may have recognized by this time, the explained and unexplained variations bear a reciprocal relationship with each other. As one of these sums increases in value, the other decreases. Furthermore, the stronger the linear relationship between X and Y, the greater the value of the explained variation and the lower the unexplained variation. In the case of a perfect relationship ($r = \pm 1.00$), the unexplained variation would be 0 and r^2 would be 1.00. This would indicate that X explains or accounts for all the variation in Y and that we could predict Y from X without error. On the other hand, when X and Y are not linearly related ($r = 0.00$), the explained variation would be 0 and r^2 would be 0.00. In such a case, we would conclude that X explains none of the variation in Y and does not improve our ability to predict Y.

Relationships intermediate between these two extremes can be interpreted in terms of how much X increases our ability to predict or explain Y. For the dual-career families, we calculated an r of 0.50. Squaring this value yields a coefficient of determination of 0.25 ($r^2 = 0.25$), which indicates that number of children (X) explains 25% of the total variation in husband's housework (Y). When predicting the number of hours per week that husbands in such families would devote to housework, we will make 25% fewer errors by basing the predictions on number of children and predicting from the regression line, as opposed to ignoring this variable and predicting the mean of Y for every case. Also, 75% of the variation in Y is unexplained by X and presumably due to some combination of the influence of other variables, measurement error, and random chance. *(For practice in the interpretation of r^2, see any of the problems at the end of this chapter.)*

15.7 TESTING PEARSON'S r FOR SIGNIFICANCE

When the relationship measured by Pearson's r is based on data from a random sample, it will usually be necessary to test r for its statistical significance. That is, we will need to know if a relationship between the variables can be assumed to exist in the population from which the sample was drawn. To illustrate this test, the r of .50 from the dual-wage-earner family sample will be used. As was the case when testing gamma and Spearman's rho, the null hypothesis states that there is no linear association between the two variables in the population from which the sample was drawn. The population parameter is symbolized as ρ (rho), and the appropriate sampling distribution is the t distribution.

To conduct this test, we need to make a number of assumptions in step 1. Most should be quite familiar, but several are new. First, we must assume that both variables are normal in distribution (**bivariate normal distributions**). Second, we must assume that the relationship between the two variables is roughly linear in form.

The third assumption involves a new concept: **homoscedasticity.** Basically, a homoscedastistic relationship is one where the variance of the Y scores is uniform for all values of X. That is, if the Y scores are evenly spread above and below the regression line for the entire length of the line, the relationship is homoscedastistic.

A visual inspection of the scattergram will usually be sufficient to appraise the extent to which the relationship conforms to the assumptions of linearity and homoscedasticity. As a rule of thumb, if the data points fall in a roughly symmetrical, cigar-shaped pattern, whose shape can be approximated with a straight line, then it is appropriate to proceed with this test of significance. Any significant evidence of nonlinearity or marked departures from homoscedasticity may indicate the need for an alternative measure of association and thus a different test of significance.

Step 1. Making Assumptions.

> Model: Random sampling
> Level of measurement is interval-ratio
> Bivariate normal distributions
> Linear relationship
> Homoscedasticity
> Sampling distribution is normal

Step 2. Stating the Null Hypothesis.

$$H_0: \rho = 0.0$$
$$(H_1: \rho \neq 0.0)$$

Step 3. Selecting the Sampling Distribution and Establishing the Critical Region. With the null of "no relationship" in the population, the sampling distribution of all possible sample r's is approximated by the t distribution. Degrees of freedom are equal to $(N - 2)$.

$$\text{Sampling distribution} = t \text{ distribution}$$
$$\text{Alpha} = .05$$
$$\text{Degrees of freedom} = N - 2 = 10$$
$$t\,(\text{critical}) = -2.228$$

Step 4. Computing the Test Statistic.

$$t \text{ (obtained)} = r\sqrt{\frac{N-2}{1-r^2}}$$

$$t \text{ (obtained)} = (.50)\sqrt{\frac{12-2}{1-(.50)^2}}$$

$$t \text{ (obtained)} = (.50)\sqrt{\frac{10}{.75}}$$

$$t \text{ (obtained)} = (.50)(\sqrt{13.33})$$

$$t \text{ (obtained)} = (.50)(3.65)$$

$$t \text{ (obtained)} = 1.83$$

Step 5. **Making a Decision.** Since the test statistic does not fall into the critical region as marked by t (critical), we fail to reject the null hypothesis. Even though the variables are substantially related in the sample, we do not have sufficient evidence to conclude that the variables are also related in the population. The test indicates that the sample value of $r = .50$ could have occurred by chance alone if the null hypothesis is true and the variables are unrelated in the population. *(For practice in conducting and interpreting tests of significance with Pearson's r, see problems 15.1, 15.2, 15.4, 15.6, 15.8, and 15.9.)*

15.8 INTERPRETING STATISTICS: IS THE U.S. ECONOMY DOMINATED BY INTERLOCKED CORPORATIONS?

Do a few giant corporations control the U.S. economy? More specifically, is the economy controlled by a small number of people who serve on the boards of directors of several different corporations, especially the largest and most powerful? The view that a small but powerful elite control the economy would be supported if we found that U.S. corporations are networked together by individuals who concurrently sit on two or more corporate boards, a phenomenon called interlocking directorates. The view that the U.S. economy is dominated by a few corporations and powerful individuals would be even more supported if a positive correlation were found between the size of corporations and the number of ties or interlocks with other corporations. This would indicate that the most powerful corporations had the greatest number of linkages to other corporations and could exert the most control over the economy. Does size matter? Are the most important corporations the most interconnected? In this section, we'll examine the associations between the number of interlocking directorates a corporation has and its revenues using a randomly selected sample of 125 U.S. corporations for the year 1995.*

*Barnes, Roy C. 2000. "Patterns of Corporate Interlocking in the United States: 1962–1995," Paper presented at the Annual Meeting of the Southern Sociological Society, New Orleans, LA.

The companies included in the sample represent many different types of businesses, including transportation, financial, and utility corporations among others. Corporations can have as many as 30 directors, so it is possible for two corporations to have multiple ties. Our dependent variable will be the total number of ties a corporation has with all other corporations. Our independent variable is the total revenues a corporation reported for 1995 measured in billions of dollars. Table 15.4 presents the basic descriptive statistics for these two variables.

The first step in assessing an association between interval-ratio level variables is to produce a scatterplot. Figure 15.8 plots the total number of interlocks on the vertical or *Y* axis and corporate revenue along the horizontal or *X* axis. Perhaps the most striking feature of the scatterplot is the absence of a "textbook" relationship between the variables. While it is important that you be able to conceptualize idealized or perfect positive and negative relationships, you also need to realize that such plots are uncommon. Actual research is much more likely to produce "messy" bivariate relationships as in this example.

TABLE 15.4   BASIC DESCRIPTIVE STATISTICS

Variable	N of Cases	Mean	Standard Dev.
Total Number of Interlocks	125	6.9920	5.7788
Revenues in 1995 ($ billions)	125	15.6259	20.4317

FIGURE 15.8   TOTAL INTERLOCKS BY REVENUES IN 1995

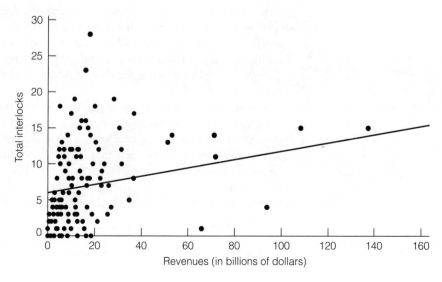

What can we say about this relationship? The regression line is not horizontal, so there is some relationship between these variables. Given the absence of a tight clustering along the regression line, we conclude that the association is somewhat weak. The fact that the regression line slopes up from left to right means that the linear relationship is generally positive: corporations with larger revenues do have more interlocks with other corporations. Note also that most cases are in the lower ranges of the X variable (i.e., they had lower revenues and are on the left-hand side of the horizontal axis), and these cases seem highly variable in the number of directors they share with other corporations. Some corporations with lower revenues shared few or no directors with other corporations, but others had many ties with other corporations. The larger corporations, on the other hand, do tend to have more interlocks, making the overall relationship positive.

The second step in assessing this relationship is to specify the regression line for the total number of interlocks and revenues. We'll skip the mechanics of computation here and simply report the values for the regression coefficients (a and b).

$$Y = a + bX$$
$$Y = 5.52 + (.14)X$$

The regression line will cross the Y axis at the point where $Y = 5.52$, and every unit increase in X (revenue) produces an increase of .14 interlock.

Next, we will calculate and interpret Pearson's r. Again, we'll skip the details of the computation and report that $r = 0.33$, which suggests that there is, at best, a modest relationship between revenues and number of interlocks. A test of significance (remember that we are working with a randomly selected sample of corporations) shows that the relationship is statistically significant at the .05 level ($t = 3.88$, df = 123). We reject the null hypothesis of "no association" and infer that there is a positive association between these variables in the population of all corporations. The coefficient of determination (r^2) is .11, which means that corporate revenue, by itself, explains 11% of the variation in the number of directors that corporations share with each other.

In sum, the results of the linear regression equation, as well as our assessment of the correlation coefficient, suggest that size does matter in predicting the total number of interlocks a company has with other corporations. These results support the view that larger corporations are more likely to be networked with other corporations, but the relationship, while statistically significant, is at best moderate in strength. Other variables besides revenues have an important influence on the number of corporate interlocks.

SUMMARY

This summary is based on the example used throughout the chapter.

1. We began with a question: Is the number of children in dual-wage-earner families related to the number of hours per week husbands devote to housework? We presented the observations in a scattergram (Figure 15.1), and our visual impression was that the variables were associated in a positive direction. The pattern formed by the observation points in the scattergram could be approximated with a straight line; thus, the relationship was roughly linear.
2. Values of Y can be predicted with the freehand regression line, but predictions are more accurate if the least-squares regression line is used. The least-squares regression line is the line that best fits the data by minimizing the variation in Y. Using the formula that defines the least-squares regression line ($Y = a + bX$), we found a slope (b) of .69, which indicates that each additional child (a unit change in X) is accompanied by an increase of .69 hour of housework per week for the husbands. We also predicted, based on this formula, that in a dual-wage-earner family with six children ($X = 6$), husbands would contribute 5.63 hours of housework a week ($Y' = 5.63$ for $X = 6$).

3. Pearson's r is a statistic that measures the overall linear association between X and Y. Our impression from the scattergram of a substantial positive relationship was confirmed by the computed r of .50. We also saw that this relationship yields an r^2 of .25, which indicates that 25% of the total variation in Y (husband's housework) is accounted for or explained by X (number of children).
4. Assuming that the 12 families represented a random sample, we tested the Pearson's r for its statistical significance and found that, at the .05 level, we could not assume that these two variables were also related in the population.
5. We acquired a great deal of information about this bivariate relationship. We know the strength and direction of the relationship and have also identified the regression line that best summarizes the effect of X on Y. We know the amount of change we can expect in Y for a unit change in X. In short, we have a greater volume of more precise information about this association between interval-ratio variables than we ever did about associations between ordinal or nominal variables. This is possible, of course, because the data generated by interval-ratio measurement are more precise and flexible than those produced by ordinal or nominal measurement techniques.

SUMMARY OF FORMULAS

Least-squares regression line:

15.1 $Y = a + bX$

Definitional formula for the slope:

15.2 $b = \dfrac{\Sigma(X - \overline{X})(Y - \overline{Y})}{\Sigma(X - \overline{X})^2}$

Computational formula for the slope:

15.3 $b = \dfrac{N\Sigma\,XY - (\Sigma\,X)(\Sigma\,Y)}{N\Sigma\,X^2 - (\Sigma\,X)^2}$

Y intercept:

15.4 $a = \overline{Y} - b\overline{X}$

Definitional formula for Pearson's r:

15.5 $r = \dfrac{\Sigma(X - \overline{X})(Y - \overline{Y})}{\sqrt{[\Sigma(X - \overline{X})^2][\Sigma(Y - \overline{Y})^2]}}$

Computational formula for Pearson's r:

15.6 $r = \dfrac{N\Sigma\,XY - (\Sigma\,X)(\Sigma\,Y)}{\sqrt{[N\Sigma\,X^2 - (\Sigma\,X)^2][N\Sigma\,Y^2 - (\Sigma\,Y)^2]}}$

Coefficient of determination:

15.7 $r^2 = \dfrac{\Sigma(Y' - \overline{Y})^2}{\Sigma(Y - \overline{Y})^2} = \dfrac{\text{Explained variation}}{\text{Total variation}}$

GLOSSARY

Bivariate normal distributions. The model assumption in the test of significance for Pearson's r that both variables are normally distributed.

Coefficient of determination (r^2). The proportion of all variation in Y that is explained by X. Found by squaring the value of Pearson's r.

Conditional means of Y. The mean of all scores on Y for each value of X.

Explained variation. The proportion of all variation in Y that is attributed to the effect of X. Equal to $\Sigma(Y' - \overline{Y})^2$.

Homoscedasticity. The model assumption in the test of significance for Pearson's r that the variance of the Y scores is uniform across all values of X.

Linear relationship. A relationship between two variables in which the observation points (dots) in the scattergram can be approximated with a straight line.

Pearson's r (r). A measure of association for variables that have been measured at the interval-ratio

level; ρ (Greek letter rho) is the symbol for the *population* value of Pearson's r.

Regression line. The single, best-fitting straight line that summarizes the relationship between two variables. Regression lines are fitted to the data points by the least-squares criterion, whereby the line touches all conditional means of Y or comes as close to doing so as possible.

Scattergram. Graphic display device that depicts the relationship between two variables.

Slope (b). The amount of change in one variable per unit change in the other; b is the symbol for the slope of a regression line.

Total variation. The spread of the Y scores around the mean of Y. Equal to $\Sigma(Y - \overline{Y})^2$.

Unexplained variation. The proportion of the total variation in Y that is not accounted for by X. Equal to $\Sigma(Y - Y')$.

Y intercept (a). The point where the regression line crosses the Y axis.

Y'. Symbol for predicted score on Y.

MULTIMEDIA RESOURCES

The Wadsworth Sociology Resource Center: Virtual Society
http://sociology.wadsworth.com/

Visit the companion web site for the sixth edition of *Statistics: A Tool for Social Research* to access a wide range of student resources. Begin by clicking on the Student Resources section of the book's web site to access the following study tools:

- Basic math review
- Statistics review

- Flash cards
- Internet links
- Additional chapter problems
- Table of random numbers
- MicroCase and SPSS examples and exercises
- "Find the text" flowcharts
- Hypothesis testing for variables measured at the ordinal level

PROBLEMS

15.1 PS Why does voter turnout vary from election to election? For municipal elections in five different cities, information has been gathered on the percent of registered voters who actually voted, unemployment rate, average years of education for the city, and the percentage of all political ads

that used "negative campaigning" (personal attacks, negative portrayals of the opponent's record, etc.). For each relationship:

a. Draw a scattergram and a freehand regression line.

b. Compute the slope (b) and find the Y intercept (*a*). (*HINT: remember to compute b before computing a. A computing table such as Table 15.3 is highly recommended.*)

c. State the least-squares regression line and predict the voter turnout for a city in which the unemployment rate was 12, a city in which the average years of schooling was 11, and an election in which 90% of the ads were negative.

d. Compute *r* and *r²*. (*HINT: a computing table such as Table 15.3 is highly recommended. If you constructed one for computing b, you already have most of the quantities you will need to solve for r.*)

e. Assume these cities are a random sample and conduct a test of significance for each relationship.

f. Describe the strength and direction of the relationships in a sentence or two. Which (if any) relationships were significant? Which factor had the strongest effect on turnout?

TURNOUT AND UNEMPLOYMENT

City	Turnout	Unemployment Rate
A	55	5
B	60	8
C	65	9
D	68	9
E	70	10

TURNOUT AND LEVEL OF EDUCATION

City	Turnout	Average Years of School
A	55	11.9
B	60	12.1
C	65	12.7
D	68	12.8
E	70	13.0

TURNOUT AND NEGATIVE CAMPAIGNING

City	Turnout	% of Negative Advertisements
A	55	60
B	60	63
C	65	55
D	68	53
E	70	48

15.25 SOC Occupational prestige scores for a sample of fathers and their oldest son and oldest daughter are shown in the table.

Family	Father's Prestige	Son's Prestige	Daughter's Prestige
A	80	85	82
B	78	80	77
C	75	70	68
D	70	75	77
E	69	72	60
F	66	60	52
G	64	48	48
H	52	55	57

Analyze the relationship between father's and son's prestige and the relationship between father's and daughter's prestige. For each relationship:

a. Draw a scattergram and a freehand regression line.

b. Compute the slope (*b*) and find the *Y* intercept (*a*).

c. State the least-squares regression line. What prestige score would you predict for a son whose father had a prestige score of 72? What prestige score would you predict for a daughter whose father had a prestige score of 72?

d. Compute *r* and *r²*.

e. Assume these families are a random sample and conduct a test of significance for both relationships.

f. Describe the strength and direction of the relationships in a sentence or two. Does the occupational prestige of the father have an impact on his children? Does it have the same impact for daughters as it does for sons?

15.3 GER The residents of a housing development for senior citizens have completed a survey whereon they indicated how physically active they are and how many visitors they receive each week. Are these two variables related for the 10 cases reported on page 392? Draw a scattergram and compute *r* and *r²*. Find the least-squares regression line. What would be the predicted number of visitors for a person whose level of activity was a 5? How about a person who scored 18 on level of activity?

Case	Level of Activity	Number of Visitors
A	10	14
B	11	12
C	12	10
D	10	9
E	15	8
F	9	7
G	7	10
H	3	15
I	10	12
J	9	2

15.4 PS The variables below were collected for a random sample of 10 precincts during the last national election. Draw scattergrams and compute r and r^2 for each combination of variables and test the correlations for their significance. Write a paragraph interpreting the relationship between these variables.

Precinct	Percent Democrat	Percent Minority	Voter Turnout
A	50	10	56
B	45	12	55
C	56	8	52
D	78	15	60
E	13	5	89
F	85	20	25
G	62	18	64
H	33	9	88
I	25	0	42
J	49	9	36

15.5 SOC/CJ The table presents the scores of 10 states on each of six variables: three measures of criminal activity and three measures of population structure. Crime rates are number of incidents per 100,000 population as of 1997.

For each combination of crime rate and population characteristic:

a. Draw a scattergram and a freehand regression line.

b. Compute the slope (b) and find the Y intercept (a).

c. State the least-squares regression line. What homicide rate would you predict for a state with a growth rate of -1? What robbery rate would you predict for a state with a population density of 250? What auto theft rate would you predict for a state in which 50% of the population lived in urban areas?

d. Compute r and r^2.

e. Assume these states are a random sample and conduct a test of significance for both relationships.

f. Describe the strength and direction of each of these relationships in a sentence or two.

15.6 PS The city of Shinbone recently had an election for city council. For the 15 randomly selected precincts listed below, was voter turnout related to the percentage of minority-group population?

Precinct	Registered Voters Who Voted (%)	Minority-group Voters (%)
A	23	0
B	47	15
C	65	75
D	17	10
E	9	5
F	64	25
G	50	97

State	Homicide	Robbery	Auto Theft	Growth[1]	Population Density[2]	Urbanization[3]
Maine	2	21	132	.9	40	36
New York	6	309	439	.2	385	92
Ohio	5	159	406	.6	274	81
Iowa	2	56	234	.8	51	44
Virginia	7	125	281	2.9	172	78
Kentucky	6	91	248	2.1	99	48
Texas	7	157	523	5.7	75	84
Arizona	8	166	970	8.4	41	88
Washington	4	120	568	4.7	85	83
California	8	253	709	3.8	209	97

[1]Percentage change in population from 1990 to 1998.
[2]Population per square mile of land area, 1998.
[3]Percent of population living in metropolitan areas, 1996.
Source: United States Bureau of the Census, *Statistical Abstracts of the United States: 1999* (119th edition). Washington, D.C., 1999.

Precinct	Registered Voters Who Voted (%)	Minority-group Voters (%)
H	49	45
I	45	7
J	43	16
K	27	17
L	25	8
M	33	23
N	34	24
O	25	25

a. Compute r and r^2.

b. Is r statistically significant?

c. Summarize the relationship in terms of its strength, direction, and statistical significance.

15.7 | SOC | The basketball coach at a small local college believes that his team plays better and scores more points in front of larger crowds. The number of points scored and attendance for all home games last season are reported below. Do these data support the coach's argument?

Game	Points Scored	Attendance
1	54	378
2	57	350
3	59	320
4	80	478
5	82	451
6	75	250
7	73	489
8	53	451
9	67	410
10	78	215
11	67	113
12	56	250
13	85	450
14	101	489
15	99	472

15.8 The table below presents the scores of 15 states on three variables. Compute r and r^2 for each combination of variables. Assume that these 15 states are a random sample of all states and test the correlations for their significance. Write a paragraph interpreting the relationship among these three variables.

State	Per Capita Expenditures on Education	Percent High School Graduates[1]	Rank in per Capita Income
Arkansas	1008	77	46
Colorado	1136	90	9

State	Per Capita Expenditures on Education	Percent High School Graduates[1]	Rank in Per Capita Income
Connecticut	1566	84	1
Florida	1074	82	19
Illinois	1147	84	8
Kansas	1141	89	24
Louisiana	951	79	41
Maryland	1267	85	5
Michigan	1370	85	18
Mississippi	939	77	50
Nebraska	1151	88	27
New Hampshire	1142	84	7
North Carolina	1013	81	31
Pennsylvania	1140	84	16
Wyoming	1347	90	34

[1] Based on percentage of population age 25 and older.
Source: United States Bureau of the Census, *Statistical Abstracts of the United States: 1999* (119th edition). Washington, D.C., 1999.

15.9 Fifteen individuals were randomly selected from the 1998 General Social Survey data set, and their scores on five variables are reproduced below. Is there a relationship between occupational prestige and age? Between church attendance and number of children? Between number of children and hours of TV watching? Between age and hours of TV watching? Between age and number of children? Between hours of TV watching and occupational prestige? Which of these relationships are significant? Variables and codes are explained in detail in Appendix G.

Occupational Prestige	Number of Children	Age	Church Attendance	Hours of TV per Day
32	3	34	3	1
50	0	41	0	3
17	0	52	7	2
69	3	67	0	5
17	0	40	0	5
52	0	22	2	3
32	3	31	0	4
50	0	23	8	4
19	9	64	1	6
37	4	55	0	2
14	3	66	5	5
51	0	22	6	0
45	0	19	3	7
44	0	21	4	1
46	4	58	2	0

Using *SPSS for Windows* to Produce Pearson's *r*

SPSS DEMONSTRATION 15.1
What Are the Correlates of Occupational Prestige?

To what extent is a person's social class, as measured by the prestige of her or his occupation, a result of personal efforts and talents (or lack thereof)? How does the social class of our parents shape our ability to rise in the class structure? We can investigate these issues by analyzing the relationships between three variables: *prestg80,* or the prestige of the respondent's occupation; *educ,* the respondent's years of schooling; and *papres80,* the prestige of the respondent's father's occupation.

If a person's social class position reflects her or his own preparation for the job market, there should be a strong positive correlation between *educ* and *prestg80.* On the other hand, if it's *who* you know and not *what* you know—if class position is greatly affected by the social class of one's family of origin—then there should be a strong positive relationship between *papres80* and *prestg80.* By comparing the strength of these bivariate correlations, we may be able to make some judgment about the relative importance of these factors in determining occupational prestige.

To request Pearson's *r,* click **Analyze, Correlate,** and **Bivariate.** The **Bivariate Correlations** window appears with the variable list on the left. Find *educ, papres80,* and *prestg80* in the list and click the arrow to move them into the **Variables** box. If you wish, you can get descriptive statistics for each variable by clicking the **Options** button and requesting means and standard deviations. Unless you request otherwise, the program will conduct two-tailed tests of significance on the correlations. Note that Spearman's rho (see Chapter 14) is also an option. Click **OK,** and your output should look like this:

Correlations

		HIGHEST YEAR OF SCHOOL COMPLETED	FATHERS OCCUPATIONAL PRESTIGE SCORE (1980)	RS OCCUPATIONAL PRESTIGE SCORE (1980)
HIGHEST YEAR OF SCHOOL COMPLETED	Pearson Correlation	1.000	.288	.505
	Sig. (2-tailed)		.000	.000
	N	1381	1098	1313
FATHERS OCCUPATIONAL PRESTIGE SCORE (1980)	Pearson Correlation	.2881	.000	.189
	Sig. (2-tailed)	.000		.000
	N	1098	1100	1053
RS OCCUPATIONAL PRESTIGE SCORE (1980)	Pearson Correlation	.505	.1891	.000
	Sig. (2-tailed)	.000	.000	
	N	1313	1053	1318

The output is in the form of a correlation matrix showing the relationships between all variables, including the correlation of a variable with itself. For each possible relationship, Pearson's r is reported in the top row, the results of the test of significance in the second row, and sample size in the third row (N varies because not everyone answered every question).

For our questions, we need to focus on the correlations between *prestg80, educ,* and *papres80*. As expected, the relationships are statistically significant ($p < .000$), positive, and moderate-to-strong. The relationship between *prestg80* and *educ* is stronger (.505) than the relationship between *prestg80* and *papres80* (.189). This suggests that preparation for the job market (*educ*) is more important than the relative advantage or disadvantage conferred by one's family of origin (*papres80*). Success may come more easily to people from an advantaged family background, but preparation (education) matters more. To further disentangle these relationships would require more sophisticated statistics and, fortunately, some will become available in Chapter 17.

SPSS DEMONSTRATION 15.2 Are the Correlates of Occupational Prestige Affected by Gender?

Are the pathways to success and higher prestige open to all equally? We can begin to address this question by analyzing the relationships reported in Demonstration 15.1 for men and women separately. If gender has no effect on a person's opportunities for success, *prestg80, papres80,* and *educ* should be related in the same way for both men and women.

To observe the effect of sex, we will split the GSS sample into two subfiles. Men and women will then be processed separately, and SPSS will produce one correlation matrix for men and another for women.

Click **Data** from the main menu and then click **Split File.** On the **Split File** window, click the button next to "organize output by groups". This will generate separate outputs for men and women. Select *sex* from the variable list and click the arrow to move the variable name into the **Groups Based On** window. Click **OK,** and all procedures requested will be done separately for men and women. To restore the full sample, call up the **Split File** window again and click the **Reset** button. For now, click **Statistics, Correlate,** and **Bivariate** and rerun the correlation matrix for *prestg80, papres80,* and *educ*. The output will look like this:

FOR MALES: Correlations

		HIGHEST YEAR OF SCHOOL COMPLETED	FATHERS OCCUPATIONAL PRESTIGE SCORE (1980)	RS OCCUPATIONAL PRESTIGE SCORE (1980)
HIGHEST YEAR OF SCHOOL COMPLETED	Pearson Correlation	1.000	.303	.463
	Sig. (2-tailed)		.000	.000
	N	620	497	603
FATHERS OCCUPATIONAL PRESTIGE SCORE (1980)	Pearson Correlation	.303	1.000	.173
	Sig. (2-tailed)	.000		.000
	N	497	498	487
RS OCCUPATIONAL PRESTIGE SCORE (1980)	Pearson Correlation	.463	.173	1.000
	Sig. (2-tailed)	.000	.000	
	N	603	487	605

FOR FEMALES: Correlations

		HIGHEST YEAR OF SCHOOL COMPLETED	FATHERS OCCUPATIONAL PRESTIGE SCORE (1980)	RS OCCUPATIONAL PRESTIGE SCORE (1980)
HIGHEST YEAR OF SCHOOL COMPLETED	Pearson Correlation Sig. (2-tailed) *N*	1.000 761	.271 .000 601	.542 .000 710
FATHERS OCCUPATIONAL PRESTIGE SCORE (1980)	Pearson Correlation Sig. (2-tailed) *N*	.271 .000 601	1.000 602	.207 .000 566
RS OCCUPATIONAL PRESTIGE SCORE (1980)	Pearson Correlation Sig. (2-tailed) *N*	.542 .000 710	.207 .000 566	1.000 713

The correlations for both groups are significant, positive, moderately strong, and roughly the same value for both genders. The social class of females is a little more affected by father's prestige (social class of family of origin) and education, but these results suggest that gender makes little difference for these relationships.

MicroCase

Using MicroCase to Produce Pearson's *r*

MICROCASE DEMONSTRATION 15.1
What Are the Correlates of Occupational Prestige?

To what extent is a person's social class, as measured by the prestige of her or his occupation, a result of personal efforts and talents (or lack thereof)? How does the social class of our parents shape our ability to rise in the class structure? We can investigate these issues by analyzing the relationships between three variables: *prestg80,* or the prestige of the respondent's occupation; *educ,* the respondent's years of schooling; and *papres80,* the prestige of the respondent's father's occupation.

If a person's social class position reflects her or his own preparation for the job market, there should be a strong positive correlation between *educ* and *prestg80.* On the other hand, if it's *who* you know and not *what* you know—if class position is greatly affected by the social class of one's family of origin—then there should be a strong positive relationship between *papres80* and *prestg80.* By comparing the strength of these bivariate correlations, we may be able to make some judgment about the relative importance of these factors in determining occupational prestige.

To request Pearson's *r,* click **Correlation** from the Statistics menu. The **Correlation** window appears with the variable list on the left. You can request correlations for only two variables at a time. First, find *papres80* (variable #4) and *prestg80* (variable #2) in the list and click the arrow to move them into the **Select Variables** box. Click **OK,** and your output should look like this:

```
Correlation Coefficients
PAIRWISE deletion (2-tailed test)  Significance Levels: ** = .01, * = .05

           PAPRES80     PRESTG80
PAPRES80    1.000         0.189**
           (1100)        (1053)
PRESTG80    0.189**       1.000
           (1053)        (1318)
```

Next, return to the **Correlation** window, delete *papres80* from the **Select Variable** window and choose *educ* (variable #7) instead. Click **OK,** and the following output will appear:

```
Correlation Coefficients
PAIRWISE deletion (2-tailed test)  Significance Levels: ** = .01, * = .05

           PRESTG80     EDUC
PRESTG80    1.000        0.505**
           (1318)       (1313)
EDUC        0.505**      1.000
           (1313)       (1381)
```

The output is in the form of a correlation matrix showing the relationships between all variables, including the correlation of a variable with itself. For each possible relationship, Pearson's *r* is reported in the top row, and sample size is reported in parentheses in the second row (*N* varies because not everyone answered every question). The significance of the Pearson's *r* is indicated by asterisks. As expected, the relationships between *prestg80, educ,* and *papres80* are statistically significant ($p < .01$), positive, and moderate-to-strong. The relationship between *prestg80* and *educ* is stronger (.505) than the relationship between *prestg80* and *papres80* (.189). This suggests that preparation for the job market (*educ*) is more important than the relative advantage or disadvantage conferred by one's family of origin (*papres80*). Success may come more easily to people from an advantaged family background, but preparation (education) matters more. To further disentangle these relationships would require more sophisticated statistics and, fortunately, some will become available in Chapter 17.

MICROCASE DEMONSTRATION 15.2 Are the Correlates of Occupational Prestige Affected by Gender?

Are the pathways to success and higher prestige open to all equally? We can begin to address this question by analyzing the relationships reported in Demonstration 15.1 for men and women separately. If gender has no effect on a person's opportunities for success, *prestg80, papres80,* and *educ* should be related in the same way for both men and women.

To observe the effect of sex, we will use the **Subset Variables** window under **Options** on the **Correlation** window. From the variable list, highlight *sex* (variable #10) and click the arrow pointing to the **Subset Variables** window. Select men first and rerun the **Correlation** command for *prestg80* and *educ* and then *papres80* and *educ*. The output will look like this:

```
Correlation Coefficients
Subset: SEX Categories: MALE
PAIRWISE deletion (1-tailed test)  Significance Levels: ** = .01, * = .05

            PRESTG80    EDUC
PRESTG80    1.000     0.463**
            (605)      (603)
EDUC        0.463**    1.000
            (603)      (620)
```

```
Correlation Coefficients
Subset: SEX Categories: MALE
PAIRWISE deletion (1-tailed test)  Significance Levels: ** = .01, * = .05

            PRESTG80   PAPRES80
PRESTG80    1.000      0.173**
            (605)       (487)
PAPRES80    0.173**    1.000
            (487)       (498)
```

Now, return to the **Correlation** window, select females rather than males, and repeat the **Correlation** commands. The output will look like this:

```
Correlation Coefficients
Subset: SEX Categories: FEMALE
PAIRWISE deletion (1-tailed test)  Significance Levels: ** = .01, * = .05

            PRESTG80    EDUC
PRESTG80    1.000     0.542**
            (713)      (710)
EDUC        0.542**    1.000
            (710)      (761)
```

```
Correlation Coefficients
Subset: SEX Categories: FEMALE
PAIRWISE deletion (1-tailed test)  Significance Levels: ** = .01, * = .05

            PRESTG80   PAPRES80
PRESTG80    1.000      0.207**
            (713)       (566)
PAPRES80    0.207**    1.000
            (566)       (602)
```

The correlations for both groups are significant, positive, moderately strong and roughly the same value for both genders. The social class of females is a little more affected by father's prestige (social class of family of origin) and education but these results suggest that gender makes little difference for these relationships.

Exercises

15.1 Run the analysis in Demonstration 15.1 again with *income98* rather than *prestg80* as the indicator of social class standing. Use the same independent variables (*papres80, educ*) and see if the patterns are similar to those identified in Demonstration 15.1. (Ignore the fact that *income98* is only ordinal in level of measurement.) Write up your conclusions.

15.2 Run the analysis in Demonstration 15.2 again but substitute *income98* for *prestg80*. Divide the sample by sex as before. Is there a gender difference in the correlates of *income98* as there was with *prestg80?*

15.3 Switch topics entirely and see if you can explain the variation in television viewing habits. What are the correlates of *tvhours?* Choose three or four possible independent variables (perhaps *age* and *educ*). Write up your results.

1. For each situation, compute and interpret the appropriate measure of association. Also, compute and interpret percentages for bivariate tables. Describe relationships in terms of the strength and pattern or direction.

   a. For 10 cities, data has been gathered on total crime rate (major felonies per 100,000 population) and the percentage of people who are new immigrants (arrived in the United States within the past five years). Are the variables related?

City	Total Crime Rate	Percent Immigrants
A	1500	10
B	1200	18
C	2000	9
D	1700	11
E	1600	15
F	1000	20
G	1700	9
H	1300	22
I	900	10
J	700	15

   b. There is some evidence that people's involvement in their communities (membership in voluntary organizations, participation in local politics, and so forth) has been declining, and television has been cited as the cause. Do the data below support the idea that TV is responsible for the decline?

Hours of Community Service:	Television Viewing			Totals
	Low	Moderate	High	
Low	5	10	18	33
Moderate	10	12	10	32
High	15	8	7	30
Totals	30	30	35	95

   c. A national magazine has rated the states in terms of "quality of life" (a scale that includes health care, availability of leisure facilities, unemployment rate, and a number of other variables) and the quality of the state system of higher education. Both scales range from 1 (low) to 20 (high). Is there a correlation between these two variables for the 10 states listed below?

State	Quality of Life	Quality of Higher Education
A	10	10
B	12	13
C	15	18

State	Quality of Life	Quality of Higher Education
D	18	20
E	10	15
F	9	11
G	11	12
H	8	6
I	13	9
J	5	8

d. Racial intermarriages are increasing in the United States, and the number of mixed-race people is growing. The table below shows the relationship between racial mixture of parents and the racial category, if any, with which a sample of people of mixed race identifies. Is there a relationship between these variables?

Do You Consider Yourself to Be:	Racial Mixture of Parents			Totals
	Black/White	White/Asian	Black/Asian	
Black	2	0	3	5
White	8	4	0	12
Asian	0	3	3	6
None of the above	10	8	4	22
	20	15	10	45

2. A number of research questions are stated below. Each can be answered by at least one of the techniques presented in Chapters 12 through 15. For each research situation, compute the most appropriate measure of association and write a sentence or two of response to the question. The questions are presented in random order. In selecting a measure of association, you need to consider the number of possible values and the level of measurement of the variables.

The research questions refer to the database below, which is taken from the 1998 General Social Survey (GSS). The actual questions asked and the complete response codes for the GSS are presented in Appendix G. Abbreviated codes are listed below. Some variables have been recoded for this exercise.

a. Are scores on "frequency of sex during the last year" associated with income or age? Compute a measure of association, assuming that the variables are interval-ratio in level of measurement.
b. Is fear associated with sex? Is it associated with marital status?
c. Is support for sex education in the public schools associated with church attendance? Is it associated with marital status?

Survey Items:

1. Marital status of respondent (MARITAL):
   1. Married
   2. Not married (includes widowed, divorced, etc.)

2. How often do you attend church? (ATTEND) See Appendix G for original codes:
   0. Never
   1. Rarely
   2. Often
3. Age of respondent (AGE) (Values are actual numbers.).
4. Respondent's total family income (INCOME98) See Appendix G for codes.
5. Sex:
   1. Male
   2. Female
6. Support for sex education in the public schools (SEXEDUC):
   1. Favor
   2. Oppose
7. Fear of walking alone at night (FEAR):
   1. Yes
   2. No
8. Frequency of sex during past year. See Appendix G for codes.

Case	Marital Status	Church Attendance	Age	Income	Sex	Sex Educ.	Fear	Frequency of Sex
1	2	0	22	12	2	1	2	5
2	1	0	52	13	1	1	2	2
3	1	2	44	16	1	2	1	3
4	1	0	56	10	1	1	2	1
5	1	2	61	8	2	2	2	0
6	1	0	28	19	2	1	2	5
7	2	1	59	9	1	2	1	0
8	1	1	69	11	1	2	2	5
9	2	0	23	4	1	1	2	6
10	2	2	31	20	2	1	1	6
11	1	2	67	21	2	2	1	1
12	1	1	46	9	1	1	1	2
13	2	0	19	10	2	1	1	5
14	1	2	34	11	1	2	2	2
15	2	0	29	18	2	1	1	4
16	1	1	31	16	1	1	1	5
17	2	1	88	6	2	2	2	1
18	1	0	24	15	2	1	1	4
19	2	2	69	11	1	1	1	2
20	1	2	60	14	1	2	2	3
21	1	1	29	12	2	1	2	4
22	1	2	43	13	2	2	1	2
23	1	0	35	20	1	1	1	4
24	2	2	38	9	2	2	2	4
25	2	0	83	19	1	2	1	0

16 ELABORATING BIVARIATE TABLES

LEARNING OBJECTIVES

By the end of this chapter, you will be able to

1. Explain the purpose of multivariate analysis in terms of observing the effect of a control variable.
2. Construct and interpret partial tables.
3. Compute and interpret partial measures of association.
4. Recognize and interpret direct, spurious or intervening, and interactive relationships.
5. Compute and interpret partial gamma.
6. Cite and explain the limitations of elaborating bivariate tables.

16.1 INTRODUCTION

Few questions can be answered by a statistical analysis of only two variables and, therefore, the typical research project will include many variables. In Chapters 12–15, we saw how various statistical techniques can be applied to bivariate relationships. In this chapter and Chapter 17, we will see how some of these techniques can be extended to probe the relationships among three or more variables. This chapter will present some multivariate techniques appropriate for variables that have been measured at the nominal or ordinal level and organized into table format. Chapter 17 will present some techniques that can be used when the variables have been measured at the interval-ratio level.

Before considering the techniques themselves, we should consider why they are important and what they might be able to tell us. There are two general reasons for utilizing multivariate techniques. First, and most fundamental, is the goal of simply gathering additional information about a specific bivariate relationship by observing how that relationship is affected (if at all) by the presence of a third variable (or a fourth or a fifth variable). Multivariate techniques will increase the amount of information we have on the basic bivariate relationship and (we hope) enhance our understanding of that relationship.

A second and very much related rationale for multivariate statistics involves the issue of causation. While multivariate statistical techniques cannot prove the existence of causal connections between variables, they can provide valuable evidence in support of causal arguments and are very important tools for testing and revising theory.

16.2 CONTROLLING FOR A THIRD VARIABLE

For variables arrayed in tables, multivariate analysis proceeds by observing the effects of other variables on the bivariate relationship. That is, we observe the relationship between the independent (X) and dependent (Y) variables after a third variable (which we will call **Z**) has been controlled. We do this by reconstructing the relationship between X and Y for each value or score of Z. If the control variable has an effect, the relationship between X and Y will change under the various conditions of Z.*

To illustrate, suppose that a researcher wishes to analyze the relationship between how well an individual is integrated into a group or organization and that individual's level of satisfaction with the group. The researcher has decided to focus on students and their level of satisfaction with the college as a whole. The necessary information on satisfaction (Y) is gathered from a sample of 270 students, and all are asked to list the student clubs or organizations to which they belong. Integration (X) is measured by dividing the students into two categories: the first includes students who are not members of any organizations (nonmembers), and the second includes students who are members of at least one organization (members). The researcher suspects that membership and satisfaction are positively related and that students who are members of at least one organization will report higher levels of satisfaction than students who are nonmembers. The relationship between these two variables is displayed in Table 16.1.

Inspection of the table suggests that these two variables are associated. The conditional distributions of satisfaction (Y) change across the two conditions of membership (X), and membership in at least one organization is associated with high satisfaction, whereas nonmembership is associated with low satisfaction. The existence and direction of the relationship are confirmed by the computation of a gamma of $+.40$ for this table.[†] In short, Table 16.1 strongly suggests the existence of a relationship between integration and morale, and these results can be taken as evidence for a causal or direct relationship between the two variables. The causal relationship is summarized symbolically in Figure 16.1, where the arrow represents the effect of X on Y.

---

*For the sake of brevity, we will focus on the simplest application of multivariate statistical analysis, the case where the relationship between an independent (X) and dependent (Y) variable is analyzed in the presence of a single control variable (Z). Once the basic techniques are grasped, they can be readily expanded to situations involving more than one control variable.

[†]We will not display the computation of gamma here. See Chapter 14 for a review of this measure of association.

TABLE 16.1   SATISFACTION WITH COLLEGE BY NUMBER OF MEMBERSHIPS
IN STUDENT ORGANIZATIONS

| | Memberships (X) | | |
Satisfaction (Y)	None	At Least One	Totals
Low	57 (54.3%)	56 (33.9%)	113
High	48 (45.7%)	109 (66.1%)	157
Totals	105 (100%)	165 (100%)	270
	Gamma = .40		

FIGURE 16.1   A DIRECT RELATIONSHIP BETWEEN TWO VARIABLES

$X \longrightarrow Y$

The researcher recognizes, of course, that membership is not the sole factor associated with satisfaction (if it were, the gamma noted above would be +1.00). Other variables may alter this relationship, and the researcher will need to consider the effects of these third variables (Z's) in a systematic and logical way. By doing so, the researcher will accumulate further information and detail about the bivariate relationship and can further probe the possible causal connection between X and Y.

For example, perhaps levels of satisfaction are affected by the rewards received by students from the university. Perhaps the most significant reward the university can dispense is recognition of academic achievements (that is, grades). Perhaps students who are more highly rewarded (have higher grade-point averages) are more satisfied with the university and are also more likely to participate actively in the life of the campus by joining student organizations. How can the effects of this third variable on the bivariate relationship be investigated?

Consider Table 16.1 again. This table provides information about the distribution of 270 cases on two variables. By analyzing the table, we can see that 165 students hold at least one membership, that 48 students are both nonmembers and highly satisfied, that the majority of the students are highly satisfied, and so forth. What the bivariate table cannot show us, of course, is the distribution of these students on GPA. For all we know at this point, the 109 highly satisfied members could all have high GPA's, low GPA's, or any combination of scores on this third variable. GPA is "free to vary" in this table, since the distribution of the cases on this variable is not accounted for.

We control for the effect (if any) of third variables by fixing their distributions so that they are no longer free to vary. We do this by sorting all cases in the sample according to their score on the third variable (Z) and then observing the relationship between X and Y for each value (or score) of Z. In

TABLE 16.2   SATISFACTION BY MEMBERSHIP, CONTROLLING FOR GPA

A. High GPA

| Satisfaction (Y) | Memberships (X) | | Totals |
	None	At Least One	
Low	29 (54.7%)	28 (34.2%)	57
High	24 (45.3%)	54 (65.9%)	78
Totals	53 (100%)	82 (100%)	135

Gamma = .40

B. Low GPA

| Satisfaction (Y) | Memberships (X) | | Totals |
	None	At Least One	
Low	28 (53.9%)	28 (33.7%)	56
High	24 (46.2%)	55 (66.3%)	79
Totals	52 (100%)	83 (100%)	135

Gamma = .39

the example at hand, we will construct separate tables displaying the relationship between membership and satisfaction for each category of GPA. The tables so produced are called **partial tables** and are displayed in Table 16.2. The students have been classified as "high" or "low" on GPA, and there are exactly 135 students in each category. Within each category of GPA, students have been reclassified by membership (X) and satisfaction (Y) in order to produce the two partial tables.

This type of multivariate analysis is called **elaboration** because the partial tables present the original bivariate relationship in a more detailed or elaborate form. The cell frequencies in the partial tables are subdivisions of the cell frequencies reported in Table 16.1. For example, if the frequencies of the cells in the partial tables are added together, the original frequencies of Table 16.1 will be reproduced.*

Also note how this method of controlling for other variables can be extended. In our example, the control variable (grades) had two categories and, thus, there were two partial tables, one for each value of the control variable. Had the control variable had more than two categories, we would have had more partial tables. By the same token, we can control for more than one variable at a time by sorting the cases on all scores of all control variables

---

*The total of the cell frequencies in the partial tables will always equal the corresponding cell frequencies in the bivariate table except when, as often happens in "real life" research situations, the researcher is missing scores on the third variable for some cases. These cases must be deleted from the analysis and, as a consequence, the partial tables will have fewer cases than the bivariate table.

and producing partial tables for each combination of scores on the control variables. Thus, if we had controlled for both GPA and gender, we would have had four partial tables to consider. There would have been one partial table for males with low GPA's, a second for males with high GPA's, and two partial tables for females with high and low GPA's.

To summarize, for nominal and ordinal variables, multivariate analysis begins by constructing partial tables or tables that display the relationship between X and Y for each value of Z. The next step is to trace the effect of Z by comparing the partial tables with each other and with the original bivariate table. By analyzing the partial tables, we can observe the effects (if any) of the control variable on the original bivariate relationship.

16.3 INTERPRETING PARTIAL TABLES

The cell frequencies in the partial tables may follow a variety of forms, but we will concentrate on three basic patterns. The patterns, determined by comparing the partial tables with each other and with the original bivariate table, are

1. **Direct relationships:** The relationship between X and Y is the same in all partial tables and in the bivariate table.
2. **Spurious relationships** or **intervening relationships:** The relationship between X and Y is the same in all partial tables but much weaker than in the bivariate table.
3. **Interaction:** Each partial table and the bivariate table all show different relationships between X and Y.

Each pattern has different implications for the relationships among the variables and for the subsequent course of the statistical analysis. I will now describe each pattern in detail and then summarize our discussion in Table 16.5.

Direct Relationships.   This pattern is often called **replication** because the partial tables reproduce (or replicate) the bivariate table; the cell frequencies are the same in the partial tables and the bivariate table. Measures of association calculated on the partial tables have the same value as the measure of association calculated for the bivariate table.

This outcome indicates that the control variable (Z) has no effect on the relationship between X and Y. Table 16.2 provides an example of this outcome. In this table, the relationship between membership (X) and satisfaction (Y) was investigated with GPA (Z) controlled. Table 16.2A shows the relationship for high-GPA students, and Table 16.2B shows the relationship for low-GPA students. The partial tables show the same conditional distributions of Y: for both high- and low-GPA students, about 45% of the nonmembers are highly satisfied, versus about 66% of the members. This same pattern was observed in the bivariate table (Table 16.1). Thus, the conditional distributions of Y are the same in each partial table as they were in the bivariate table.

The pattern of the cell frequencies will be easier to identify if you calculate measures of association. Working from the cell frequencies presented in Table 16.2, the gamma for high-GPA students is .40, and the gamma for low-GPA students is .39. The bivariate gamma (from Table 16.1) is .40, and the essential equivalence of these gammas reinforces our finding that the relationship between X and Y is essentially the same in the partial tables and the bivariate table.

This pattern of outcomes indicates that the control variable has no important impact on the bivariate relationship (if it did, the pattern in the partial tables would be different from the bivariate table) and may be ignored in any further analysis. In terms of the original research problem, the researcher may conclude that students who are members are more likely to express high satisfaction with the university regardless of their GPA. The level of rewards dispensed to students by the institution (as measured by GPA) has no effect on the relationship between membership and satisfaction. Low-GPA students who are members of at least one organization are just as likely to report high satisfaction as high-GPA students who are members of at least one organization. *(For practice in dealing with direct relationships, see Problems 16.1 and 16.2.)*

Spurious or Intervening Relationships. In this pattern, the relationship between X and Y is much weaker in the partial tables than in the bivariate table but the same across all partials. Measures of association for the partial tables are much lower in value (perhaps even dropping to 0.00) than the measure computed for the bivariate table. This outcome is consistent with two different causal relationships among the three variables. The first is called a spurious relationship or **explanation;** in this situation, Z is conceptualized as being antecedent to both X and Y (that is, Z is thought to occur before the other two variables in time). In this pattern, Z is a common cause of both X and Y, and the original bivariate relationship is said to be spurious. The apparent bivariate relationship between X and Y is explained by or due to the effect of Z. Once Z is controlled, the association between X and Y disappears, and the value of the measures of association for the partial tables is dramatically lower than the measure of association for the bivariate table.

To illustrate, suppose that the researcher in our example had also controlled for class standing by dividing the sample into upperclasses (seniors and juniors) and underclasses (sophomores and freshmen). The reasoning of the researcher might be that upperclass students, as a function of simple longevity, display higher levels of satisfaction with the college than do underclass students. Self-selection processes may be operating, and dissatisfied students may have transferred to another college or dropped out before they attain upperclass standing. Students who have been on campus longer will

Application 16.1

Seventy-eight juvenile males in a sample have been classified as high or low on a scale that measures involvement in delinquency. Also, each subject has been classified, using school records, as either a good or poor student. The following table displays a strong relationship between these two variables for this sample ($G = -.69$).

Delin-quency	Academic Record Poor	Good	Totals
Low	13 (27.1%)	20 (66.7%)	33 (42.3%)
High	35 (72.9%)	10 (33.3%)	45 (57.7%)
Totals	48 (100.0%)	30 (100.0%)	78 (100.0%)

Gamma = −.69

Judging by the column percentages and the gamma, it is clear that juvenile males with poor academic records are especially prone to delinquency. Is this relationship between delinquency and academic record affected by whether the subject resides in an urban or nonurban area?

Urban areas:

Delin-quency	Academic Record Poor	Good	Totals
Low	10 (27.8%)	3 (30.0%)	13 (28.3%)
High	26 (72.2%)	7 (70.0%)	33 (71.7%)
Totals	36 (100.0%)	10 (100.0%)	46 (100.0%)

Gamma = −.05

Nonurban areas:

Delin-quency	Academic Record Poor	Good	Totals
Low	3 (25.0%)	17 (85.0%)	20 (62.%)
High	9 (75.0%)	3 (15.0%)	12 (37.5%)
Totals	12 (100.0%)	20 (100.0%)	32 (100.0%)

Gamma = −.89

For urban juvenile males, the relationship between academic record and delinquency disappears. The gamma for this table is −0.05, and the column percentages are very similar. For this group, delinquency is not associated with experience in school.

For nonurban males, on the other hand, there is a very strong relationship between the two variables. Gamma is −.89, and there is a dramatic difference in the column percentages. Poor students living in nonurban areas are especially prone to delinquency.

Comparing the partial tables with each other and with the bivariate table reveals an interactive relationship. Although urban juvenile males are more delinquent than nonurban males (71.7%, of the urban males were highly delinquent, as compared to 37.5% of the nonurban males), their delinquency is not associated with academic record. For nonurban males, academic record is very strongly associated with delinquency. Urban males are more delinquent than nonurban males but not because of their experience in school. Nonurban males who are also poor students are especially likely to become involved in delinquency.

be more likely to locate an organization of sufficient appeal or interest to join. This might especially be the case for organizations based on major field (such as the Accounting Club), which underclass students are less likely to join, or honorary organizations for which underclass students are unlikely to qualify.

FIGURE 16.2 A SPURIOUS RELATIONSHIP

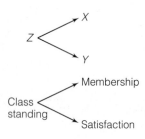

These thoughts about the possible relationships among these variables are expressed in diagram form in Figure 16.2. The absence of an arrow from membership (X) to satisfaction (Y) indicates that these variables are not truly associated with each other but rather are both caused by class standing (Z). If this causal diagram is a correct description of the relationship between these three variables (if the association between X and Y is spurious), then the association between membership and satisfaction should disappear once class standing has been controlled. That is, even though the bivariate gamma was .40 (Table 16.1), the gammas computed on the partial tables will approach 0. Table 16.3 displays the partial tables generated by controlling for class standing.

The partial tables indicate that, once class standing is controlled, membership is no longer related to satisfaction. Upperclass students are likely to express high satisfaction and underclass students are likely to express low satisfaction regardless of their number of memberships. In the partial tables, the distributions of Y no longer vary by the conditions of X, and the gammas computed on the partial tables are virtually 0. These results indicate that the bivariate association is spurious and that X and Y have no direct relationship. Class standing (Z) is a key factor in accounting for varying levels of satisfaction, and the analysis must be reoriented with class standing as an independent variable.

This outcome (partial measures much weaker than the original measure but equal to each other) is consistent with another conception of the causal links among the variables. In addition to a causal scheme wherein Z is ante-

TABLE 16.3 SATISFACTION BY MEMBERSHIP, CONTROLLING FOR CLASS STANDING

A. Upperclass Students

| Satisfaction (Y) | Memberships (X) | | Totals |
	None	At Least One	
Low	8 (25.0%)	32 (24.8%)	40
High	24 (75.0%)	97 (75.2%)	121
Totals	32 (100%)	129 (100%)	161

Gamma = .01

B. Underclass Students

| Satisfaction (Y) | Memberships (X) | | Totals |
	None	At Least One	
Low	49 (67.1%)	24 (66.7%)	73
High	24 (32.9%)	12 (33.3%)	36
Totals	73 (100%)	36 (100%)	109

Gamma = .01

FIGURE 16.3 AN INTERVENING RELATIONSHIP

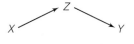

cedent to both X and Y, Z may also intervene between the two variables. This pattern is also called **interpretation** and is illustrated in Figure 16.3, where X is causally linked to Z, which is in turn linked to Y. This pattern indicates that, although X and Y are related, they are associated primarily through the control variable Z. This pattern of outcomes does not allow the researcher to distinguish between spurious relationships (Figure 16.2) and intervening relationships (Figure 16.3). The differentiation between these two types of causal patterns may be made on temporal or theoretical grounds, but not on statistical grounds. *(For practice in dealing with spurious or intervening relationships see problem 16.6b.)*

Interaction.   In this pattern, also called **specification,** the relationship between X and Y changes markedly for the different values of the control variable. The partial tables differ from each other and from the bivariate table. Interaction can be manifested in various ways in the partial tables. One possible pattern, for example, is for one partial table to display a stronger relationship between X and Y than that displayed in the bivariate table, while the relationship between X and Y drops to 0 in a second partial table. Symbolically, this outcome could be represented as in Figure 16.4, which would indicate that X and the first category of Z (Z_1) have strong effects on Y but, for the second category of Z (Z_2), there is no association between X and Y.

FIGURE 16.4 AN INTERACTIVE RELATIONSHIP

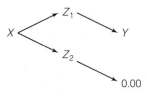

Interaction might be found, for example, in a situation in which *all* employees of a corporation were required to attend a program (X) designed to reduce anti-black prejudice (Y). Such a program would be likely to have stronger effects on white employees (Z_1) than African American employees (Z_2). That is, the program would have an effect on prejudice only for certain categories of subjects. The partial tables for white employees (Z_1) might show a strong relationship between program attendance and reduction in prejudice, whereas the partial tables for African American employees showed no relationship. African American employees, being unprejudiced against themselves in the first place, would be less affected by the program.

Interaction can take other forms. For example, the relationship between X and Y can vary not only in strength but also in direction between the partial tables. This causal relationship is symbolically represented in Figure 16.5, which indicates a situation where X and Y are positively related for the first category of Z (Z_1) and negatively related for the second category of Z (Z_2).

FIGURE 16.5 AN INTERACTIVE RELATIONSHIP

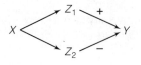

The researcher investigating the relationship between club membership and satisfaction with college life establishes a final control for race and divides the sample into white and black students. The partial tables are displayed in Table 16.4 and show an interactive relationship. The relationship between membership and satisfaction is different for white students (Z_1) and for black students (Z_2), and each partial table is different from the bivariate table. For white students, the relationship is positive and stronger than in the bivariate table. White students who are also members are much more likely

TABLE 16.4   SATISFACTION BY MEMBERSHIP, CONTROLLING FOR RACE

A. White Students

Satisfaction (Y)	Memberships (X)		Totals
	None	At Least One	
Low	40 (50.0%)	20 (16.7%)	60
High	40 (50.0%)	100 (83.3%)	140
Totals	80 (100%)	120 (100%)	200

Gamma = .67

B. Black Students

Satisfaction (Y)	Memberships (X)		Totals
	None	At Least One	
Low	17 (68.0%)	36 (80.0%)	53
High	8 (32.0%)	9 (20.0%)	17
Totals	25 (100%)	45 (100%)	70

Gamma = −.31

to express high overall satisfaction with the university, as indicated by both the percentage distribution (83% of the white students who are members are highly satisfied) and the measure of association (gamma = +.67 for white students).

For black students, the relationship between membership and satisfaction is very different—nearly the reverse of the pattern shown by white students. The great majority (80%) of the black students who are members report low satisfaction, and the gamma for this partial table is negative (−.31). Thus, for white students, satisfaction increases with membership, whereas, for black students, satisfaction decreases with membership. Based on these results, one might conclude that the social meanings and implications of joining student clubs and organizations vary by race and that the function of joining has different effects for black students. It may be, for example, that black students join different kinds of organizations than do white students. If the black students belonged primarily to a black student association that had an antagonistic relationship with the university, then belonging to such an organization could increase dissatisfaction with the university as it increased awareness of racial problems. *(For practice in dealing with interaction, see problem 16.4.)*

In closing this section, let me stress that, for the sake of clarity, the examples present here have been unusually "clean." In any given research project, the results of controlling for third variables will probably be considerably more ambiguous and open to varying interpretation than the examples presented here. In the case of spurious relationships, for example, the measures of association computed for the partial tables will probably not actu-

TABLE 16.5   A SUMMARY OF THE POSSIBLE RESULTS OF CONTROLLING FOR THIRD VARIABLES

Partial Tables (compared with bivariate table) Show	Pattern	Implications for Further Analysis	Likely Next Step in Statistical Analysis	Theoretical Implications
Same relationship between X and Y	Direct relationship, replication	Disregard Z	Analyze another control variable	Theory that X causes Y is supported
Weaker relationship between X and Y	Spurious relationship	Incorporate Z	Focus on relationship between Z and Y	Theory that X causes Y is not supported
	Intervening relationship	Incorporate Z	Focus on relationships between X, Y, and Z	Theory that X causes Y is partially supported but must be revised to take Z into account
Mixed	Interaction	Incorporate Z	Analyze subgroups (categories of Z) separately	Theory that X causes Y is partially supported but must be revised to take Z into account

ally drop to zero, even though they may be dramatically lower than the bivariate measure. (The pattern where the partial measures are roughly equivalent and much lower than the bivariate measure, but not zero, is sometimes called attenuation.) It is probably best to consider the examples above as ideal types against which any given empirical result can be compared.

Table 16.5 summarizes this discussion by outlining guidelines for decision making for each of the three outcomes discussed in this section. Since your own results will probably be more ambiguous than the ideal types presented here, you should regard this table as a set of suggestions and not as a substitute for your own creativity and sensitivity to the problem under consideration.

16.4 PARTIAL GAMMA (G_p)

When the results of controlling for a third variable indicate a direct, spurious, or intervening relationship, it is often useful to compute an additional measure that indicates the overall strength of the association between X and Y after the effects of the control variable **(Z)** have been removed. This statistic, called **partial gamma (G_p),** is somewhat easier to compare with the bivariate gamma than are the gammas computed on the partial tables separately. G_p is computed across all partial tables by Formula 16.1:

FORMULA 16.1

$$G_p = \frac{\Sigma \, N_s - \Sigma \, N_d}{\Sigma \, N_s + \Sigma \, N_d}$$

where $\Sigma \, N_s$ = the number of pairs of cases ranked the same across all partial tables
$\Sigma \, N_d$ = the number of pairs of cases ranked differently across all partial tables

In words, $\Sigma \, N_s$ is the total of all N_s's from the partial tables, and $\Sigma \, N_d$ is the total of all N_d's from all partial tables. (See Chapter 14 to review the computation of N_s and N_d.)

To illustrate the computation of G_p, let us return to Table 16.2, which showed the relationship between satisfaction and membership while controlling for GPA. The gammas computed on the partial tables were essentially equal to each other and to the bivariate gamma. Our conclusion was that the control variable had no effect on the relationship, and this conclusion can be confirmed by also computing partial gamma.

From Table 16.2A High GPA	From Table 16.2B Low GPA
$N_s = (29)(54) = 1566$	$N_s = (28)(55) = 1540$
$N_d = (28)(24) = 672$	$N_d = (28)(24) = 672$

$$\Sigma N_s = 1566 + 1540 = 3106$$
$$\Sigma N_d = 672 + 672 = 1344$$

$$G_p = \frac{\Sigma N_s - \Sigma N_d}{\Sigma N_s + \Sigma N_d}$$

$$G_p = \frac{3106 - 1344}{3106 + 1344}$$

$$G_p = \frac{1762}{4450}$$

$$G_p = .40$$

The partial gamma measures the strength of the association between X and Y once the effects of Z have been removed. In this instance, the partial gamma is the same value as the bivariate gamma ($G_p = G = .40$) and indicates that GPA has no effect on the relationship between satisfaction and membership.

When class standing was controlled (Table 16.3), clear evidence of a spurious relationship was found, since the gammas computed on the partial tables dropped almost to zero. Let us see what the value of partial gamma (G_p) would be for this second control:

From Table 16.3A Upperclassmen	From Table 16.3B Underclassmen
$N_s = (8)(97) = 776$	$N_s = (49)(12) = 588$
$N_d = (32)(24) = 768$	$N_d = (24)(24) = 576$

$$\Sigma N_s = 776 + 588 = 1364$$
$$\Sigma N_d = 768 + 576 = 1344$$

$$G_p = \frac{\Sigma N_s - \Sigma N_d}{\Sigma N_s + \Sigma N_d}$$

$$G_p = \frac{1364 - 1344}{1364 + 1344}$$

$$G_p = \frac{20}{2708}$$

$$G_p = .01$$

Once the effects of the control variable are removed, there is no relationship between X and Y. The very low value of G_p confirms our previous conclusion that the bivariate relationship between membership and satisfaction is spurious and actually due to the effects of class standing.

In a sense, G_p tells us no more about the relationships than we can see for ourselves from a careful analysis of the percentage distributions of Y in the partial tables or by a comparison of the measures of association computed on the partial tables. The advantage of G_p is that it tells us, in a single number (that is, in a compact and convenient way), the precise effects of Z on the relationship between X and Y. While G_p is no substitute for the analysis of the partial tables per se, it is a convenient way of stating our results and conclusions when working with direct or spurious relationships.

Although G_p can be calculated in cases of interactive relationships (see Table 16.4), it is rather difficult to interpret in these instances. If substantial interaction is found in the partial tables, this indicates that the control variable has a profound effect on the bivariate relationship. Thus, we should not attempt to separate the effects of Z from the bivariate relationship and, since G_p involves exactly this kind of separation, it should not be computed.

16.5 WHERE DO CONTROL VARIABLES COME FROM?

In one sense, this question is quite easy to answer. Although control variables can arise from various sources, they arise especially from theory. Social research proceeds in many different ways and is begun in response to a variety of problems. However, virtually all research projects are guided by a more-or-less-explicit theory or by some question about the relationship between two or more variables. The ultimate goal of social research is to develop defensible generalizations that improve our understanding of the variables under consideration and link back to theory at some level.

Thus, research projects are anchored in theory; and the concepts of interest, which will later be operationalized as variables, are first identified and their interrelationships first probed at the theoretical level. Since the social world is exceedingly complex, very few theories attempt to encompass that world in only two variables. Theories are, almost by definition, multivariate even when they focus on a bivariate relationship. Thus, to the extent that a research project is anchored in theory, the theory itself will suggest the control variables that need to be incorporated into the analysis. In the example used throughout this chapter, I tried to suggest that any researcher attempting to probe the relationship between involvement in an organization and satisfaction would, in the course of thinking over the possibilities, identify a number of additional variables that needed to be explicitly incorporated into the analysis.

Of course, textbook descriptions of the research process are oversimplified and tend to imply that research flows smoothly from conceptualization to operationalization to quantification to generalization. In reality, research is full of surprises, unexpected outcomes, and unanticipated results. Research in "real life" is more loosely structured and requires more imagination and creativity than textbooks can fully convey. My point is that the control variables that might be appropriate to incorporate in the data-analysis phase will be suggested or implied in the theoretical backdrop of the research project. They will flow from the researcher's imagination and sensitivity to the problem being addressed as much as from any other source.

These considerations have taken us well beyond the narrow realm of statistics and back to the planning stages of the research project. At this early time the researcher must make decisions about which variables to measure during the data-gathering phase and, thus, which variables might be incorporated as potential controls. Careful thinking and an extended consideration of possible outcomes at the planning stage will pay significant dividends during the data-analysis phase. Ideally, all relevant control variables will be incorporated and readily available for statistical analysis.

Thus, control variables come from the theory underlying the research project and from creative and imaginative thinking and planning during the early phases of the project. Nonetheless, it is not unheard of for a researcher to realize during data analysis that the control variable now so obviously relevant was never measured during data gathering and is thus unavailable for statistical analysis.

16.6 THE LIMITATIONS OF ELABORATING BIVARIATE TABLES

The basic limitation of this technique involves sample size. Elaboration is a relatively inefficient technique for multivariate analysis because it requires that the researcher divide the sample into a series of partial tables. If the control variable has more than two or three possible values, or if we attempt to control for more than one variable at a time, many partial tables will be produced. The greater the number of partial tables, the more likely we are to run out of cases to fill all of the cells. Empty or small cells, in turn, can create serious problems in terms of generalizability and confidence in our findings.

To illustrate, the example used throughout this chapter began with two dichotomized variables and a four-cell table (Table 16.1). Each of the control variables was also dichotomized, and we never confronted more than two partial tables with four cells each, or eight cells for each control variable. If we had used a control variable with three values, we would have had 12 cells to fill up, and if we had attempted to control for two dichotomized variables, we would have had 16 cells in four different partial tables. Clearly, as control variables become more elaborate and/or as the process of controlling becomes more complex, the phenomenon of empty or small cells will increasingly become a problem.

Two potential solutions to this dilemma immediately suggest themselves. The easy solution is to reduce the number of cells in the partial tables by collapsing categories within variables. If all variables are dichotomized, for example, the number of cells will be kept to a minimum. The best solution is to work with only very large samples so that there will be plenty of cases to fill up the cells. Unfortunately, the easy solution will often violate common sense (collapsed categories are more likely to group dissimilar elements together); and the best solution is not always feasible (mundane matters of time and money rear their ugly heads).

A third solution to the problem of empty cells requires the (sometimes risky) assumption that the variables of interest are measured at the interval-ratio level. At that level the techniques of partial and multiple correlation and regression, to be introduced in Chapter 17, are available. These multivariate techniques are more efficient than elaboration because they utilize all cases simultaneously and do not require that the sample be divided among the various partial tables.

16.7 INTERPRETING STATISTICS: ANALYZING POLITICAL PARTICIPATION

In Chapter 14, we analyzed the sources of civic engagement in the United States. In this section, we'll consider a related issue: what are the correlates of political participation? Our database will be the 1996 General Social Survey, a nationally representative sample. The dependent variable is measured by the response to the question: Would you attend a public meeting? This variable had four response categories: definitely would not, probably would not, probably would, and definitely would. To keep the number of cells at a minimum, these categories were collapsed into a simple dichotomy: yes and no. The independent variable is social class, and the bivariate relationship is presented in Table 16.6.

The bivariate gamma of .12 suggests that there is a positive but weak association between social class and willingness to participate. Lower- (60.6%) and working-class (59.1%) respondents are about equally willing to attend, and the percentage increases for middle- (65.7%) and upper- (68.4%) class people, but the differences in the column percentages are relatively small. What would happen to this relationship if we control for sex and race?

TABLE 16.6   PARTICIPATION BY SOCIAL CLASS

Would You Attend a Public Meeting?	Social Class				TOTALS
	Lower	Working	Middle	Upper	
No	28 (39.4%)	235 (40.9%)	196 (34.3%)	18 (31.6%)	477 (37.4%)
Yes	43 (60.6%)	339 (59.1%)	376 (65.7%)	39 (68.4%)	797 (62.6%)
Totals	71 (100.0%)	574 (100.0%)	572 (100.0%)	57 (100.0%)	1274 (100.0%)
			Gamma = .13		

TABLE 16.7 PARTICIPATION BY SOCIAL CLASS CONTROLLING FOR SEX (percentages)

A. MALES

Would You Attend a Public Meeting?	Social Class			
	Lower	Working	Middle	Upper
No	50.0%	42.3%	28.3%	22.2%
Yes	50.0%	57.7%	71.7%	77.8%
Totals	100.0%	100.0%	100.0%	100.0%
	(24)	(272)	(247)	(27)

Gamma = .30

B. FEMALES

Would You Attend a Public Meeting?	Social Class			
	Lower	Working	Middle	Upper
No	34.0%	39.7%	38.8%	40.0%
Yes	64.0%	60.3%	61.2%	60.0%
Totals	100.0%	100.0%	100.0%	100.0%
	(47)	(302)	(325)	(30)

Gamma = −.01

Table 16.7 presents the partial tables and gammas after controlling for sex. For male respondents (Part A of the table), there is a positive and stronger association between class and willingness to participate. Fifty percent of the lower-class males and about 58% of the working-class males indicated that they would attend a public meeting versus 70% or more of the middle- and upper-class males. The gamma of 0.30 confirms that this is a moderate, positive relationship. The partial table for females (Part B), on the other hand, shows essentially no relation between political participation and social class. The column percentages are very similar from class to class, and the gamma is −0.01. Comparing the partial tables with each other and with the bivariate table reveals an interactive relationship. Willingness to participate increases moderately with social class for males but not for females.

Table 16.8 presents results of controlling for race. The relationship for white respondents, like the relationship in the bivariate table, is positive but relatively weak, as indicated by the gamma of 0.14. For nonwhites, the relationship is still positive but weaker than the bivarate relationship ($G = 0.07$).

How can these relationships be summarized? While there is an association between willingness to attend a public meeting and social class for men, the association essentially disappears among women. The relationship is also affected by race: it is weaker for nonwhites than for whites. Since there is interaction in both these relationships, we can dispense with computing partial gamma (see Section 16.4).

TABLE 16.8   PARTICIPATION BY SOCIAL CLASS CONTROLLING FOR RACE (percentages)

A. WHITES

Would You Attend a Public Meeting?	Social Class			
	Lower	Working	Middle	Upper
No	38.2%	42.4%	34.2%	32.7%
Yes	61.8%	57.6%	65.8%	67.3%
Totals	100.0%	100.0%	100.0%	100.0%
	(55)	(455)	(483)	(52)

Gamma = .14

B. NONWHITES*

Would You Attend a Public Meeting?	Social Class			
	Lower	Working	Middle	Upper
No	43.8%	35.3%	34.8%	20.0%
Yes	56.2%	64.7%	65.2%	80.0%
Totals	100.0%	100.0%	100.0%	100.0%
	(16)	(119)	(89)	(5)

Gamma = .07

*The original version of the variable measuring race distinguished between whites, blacks, and others. Since very few respondents selected the last category, I collapsed blacks and others together.

Sex and race do affect the bivariate association, and the multivariate analysis reveals that these interrelations are complex. We found interaction in both relationships, and our results are consistent with the general recognition in the social sciences that class, gender, and race—some of the most consequential lines of conflict in our society—are complexly interrelated in their effects on political participation.

SUMMARY

1. Most research questions require the analysis of the interrelationship among many variables, even when the researcher is primarily concerned with a specific bivariate relationship. Multivariate statistical techniques provide the researcher with a set of tools by which additional information can be gathered about the variables of interest and by which causal interrelationships can be probed.

2. When variables have been organized in bivariate tables, multivariate analysis proceeds by controlling for a third variable. Partial tables are constructed and compared with each other and with the original bivariate table. Comparisons are made easier if appropriate measures of association are computed for all tables.

3. A direct relationship exists between the independent (X) and dependent (Y) variables if, after controlling for the third variable (Z), the relationship

between X and Y is the same across all partial tables and the same as in the bivariate table. This pattern suggests a causal relationship between X and Y.

4. If the relationship between X and Y is the same across all partial tables but much weaker than in the bivariate table, the relationship is either spurious (Z causes both X and Y) or intervening (X causes Z, which causes Y). Either pattern suggests that Z must be explicitly incorporated into the analysis.

5. Interaction exists if the relationship between X and Y varies across the partial tables and between each partial and the bivariate table. This pattern suggests that no simple or direct causal relationship exists between X and Y and that Z must be explicitly incorporated into the causal scheme.

6. Partial gamma (G_p) is a useful summary statistic that measures the strength of the association between X and Y after the effects of the control variable (Z) have been removed. Partial gamma should not be computed when an analysis of the partial tables shows substantial interaction.

7. Potential control variables must be identified before the data-gathering phase of the research project. The theoretical backdrop of the research project, along with creative thinking and some imagination, will suggest the variables that should be controlled for and measured.

8. Controlling for third variables by constructing partial tables is inefficient in that the cases must be spread out across many cells. If the variables have many categories and/or the researcher attempts to control for more than one variable simultaneously, "empty cells" may become a problem. It may be possible to deal with this problem by either collapsing categories or gathering very large samples. If interval-ratio level of measurement can be assumed, the multivariate techniques presented in the next chapter will be preferred, since they do not require the partitioning of the sample.

SUMMARY OF FORMULAS

Partial gamma   16.1

$$G_p = \frac{\Sigma N_s - \Sigma N_d}{\Sigma N_s + \Sigma N_d}$$

GLOSSARY

Direct relationship. A multivariate relationship in which the control variable has no effect on the bivariate relationship.

Elaboration. The basic multivariate technique for analyzing variables arrayed in tables. Partial tables are constructed to observe the bivariate relationship in a more detailed or elaborated format.

Explanation. See **spurious relationship.**

Interaction. A multivariate relationship wherein a bivariate relationship changes across the categories of the control variable.

Interpretation. See **intervening relationship.**

Intervening relationship. A multivariate relationship wherein a bivariate relationship becomes substantially weaker after a third variable is controlled for. The independent and dependent variables are linked primarily through the control variable.

Partial gamma (G_p). A statistic that indicates the strength of the association between two variables after the effects of a third variable have been removed.

Partial tables. Tables produced when controlling for a third variable.

Replication. See **direct relationship.**

Specification. See **interaction.**

Spurious relationship. A multivariate relationship in which a bivariate relationship becomes substantially weaker after a third variable is controlled for. The independent and dependent variables are not causally linked. Rather, both are caused by the control variable.

Z. Symbol for any control variable.

MULTIMEDIA RESOURCES

The Wadsworth Sociology Resource Center: Virtual Society
http://sociology.wadsworth.com/

Visit the companion web site for the sixth edition of *Statistics: A Tool for Social Research* to access a wide range of student resources. Begin by clicking on the Student Resources section of the book's web site to access the following study tools:

- Basic math review
- Statistics review

- Flash cards
- Internet links
- Additional chapter problems
- Table of random numbers
- MicroCase and SPSS examples and exercises
- "Find the text" flowcharts
- Hypothesis testing for variables measured at the ordinal level

PROBLEMS

16.1 SOC Problem 14.1 concerned the assimilation of a small sample of immigrants. One table showed the relationship between length of residence in the U.S. and facility with the English language:

	Length of Residence		
English Facility	Less than Five Years	More than Five Years	Totals
Low	20	10	30
High	5	15	20
Totals	25	25	50

a. If necessary, find the column percentages and compute gamma for this table. Describe the bivariate relationship in terms of strength and direction.

Is the relationship between residence and language affected by the gender of the immigrant? The partial tables showing the bivariate relationship for males and females are

A. Males:

	Length of Residence		
English Facility	Less than Five Years	More than Five Years	Totals
Low	10	5	15
High	2	8	10
Totals	12	13	25

B. Females:

	Length of Residence		
English Facility	Less than Five Years	More than Five Years	Totals
Low	10	5	15
High	3	7	10
Totals	13	12	25

b. Find the column percentages and compute gamma for each of the partial tables. Compare these percentages and gammas with each other and with the bivariate table. Does controlling for gender change the bivariate relationship?

Is the relationship between residence and language affected by the origin of the immigrant? The partial tables showing the bivariate relationship for Asian and Hispanic immigrants are

A. Asian immigrants:

	Length of Residence		
English Facility	Less than Five Years	More than Five Years	Totals
Low	8	4	12
High	2	5	7
Totals	10	9	19

B. Hispanic immigrants:

| English | Length of Residence | | |
Facility	Less than Five Years	More than Five Years	Totals
Low	12	6	18
High	3	10	13
Totals	15	16	31

c. Find the column percentages and compute gamma for each of the partial tables. Compare these percentages and gammas with each other and with the bivariate table. Does controlling for the origin of the immigrants change the bivariate relationship?

16.2 SOC Data on suicide rates, age structure, and unemployment rates have been gathered for 100 census tracts. Suicide and unemployment rates have been dichotomized at the median so that each tract could be rated as high or low. Age structure is measured in terms of the percentage of the population age 65 and older. This variable has also been dichotomized, and tracts have been rated as high or low. The tables below display the bivariate relationship between suicide rate and age structure and the same relationship controlling for unemployment.

| Suicide Rate | Population 65 and Older (%) | | |
	Low	High	Totals
Low	45	20	65
High	10	25	35
Totals	55	45	100

a. Calculate percentages and gamma for the bivariate table. Describe the bivariate relationship in terms of strength and direction.

Suicide rate by age, controlling for unemployment:

A. High unemployment:

| Suicide Rate | Population 65 and Older (%) | | |
	Low	High	Totals
Low	23	10	33
High	5	12	17
Totals	28	22	50

B. Low unemployment:

| Suicide Rate | Population 65 and Older (%) | | |
	Low	High	Totals
Low	22	10	32
High	5	13	18
Totals	27	23	50

b. Calculate percentages and gamma for each partial table. Compare the partial tables with each other and with the bivariate table. Compute partial gamma (G_p).

c. Summarize the results of the control. Does unemployment rate have any effect on the relationship between age and suicide rate? Describe the effect of the control variable in terms of the pattern of percentages, the value of the gammas, and the possible causal relationships among these variables.

16.3 SW Do long-term patients in mental health facilities become more withdrawn and reclusive over time? A sample of 608 institutionalized patients was rated by a standard "reality orientation scale." Is there a relationship with length of institutionalization? Does gender have any effect on the relationship?

Reality orientation by length of institutionalization:

| Reality Orientation | Length of Institutionalization | | |
	Less than 5 Years	More than 5 Years	Totals
Low	200	213	413
High	117	78	195
Totals	317	291	608

Reality orientation by length of institutionalization, controlling for gender:

A. Females:

| Reality Orientation | Length of Institutionalization | | |
	Less than 5 Years	More than 5 Years	Totals
Low	95	120	215
High	60	37	97
Totals	155	157	312

B. Males:

Length of Institutionalization

Reality Orientation	Less than 5 Years	More than 5 Years	Totals
Low	105	93	198
High	57	41	98
Totals	162	134	296

16.4 SOC Is there a relationship between attitudes on sexuality and age? Are older people more conservative with respect to questions of sexual morality? A national sample of 925 respondents has been questioned about attitudes on premarital sex. Responses have been collapsed into two categories: those who believe that premarital sex is "always wrong" and those who believe it is not wrong under certain conditions ("sometimes wrong"). These responses have been crosstabulated by age, and the results are reported below:

Age

Premarital Sex Is	Younger than 35	35 or Older	Totals
Always wrong	90	235	325
Sometimes wrong	420	180	600
Totals	510	415	925

a. Calculate percentages and gamma for the table. Is there a relationship between these two variables? Describe the bivariate relationship in terms of its strength and direction. Below, the bivariate relationship is reproduced after controlling for the sex of the respondent. Does gender have any effect on the relationship?

Attitude toward premarital sex by age, controlling for gender:
A. Male respondents:

Age

Premarital Sex Is	Younger than 35	35 or Older	Totals
Always wrong	70	55	125
Sometimes wrong	190	80	270
Totals	260	135	395

B. Female respondents:

Age

Premarital Sex Is	Younger than 35	35 or Older	Totals
Always wrong	20	180	200
Sometimes wrong	230	100	330
Totals	250	280	530

c. Summarize the results of this control. Does gender have any effect on the relationship between age and attitude? If so, describe the effect of the control variable in terms of the pattern of percentages, the value of the gammas, and the possible causal interrelationships among these variables.

16.5 SOC A job-training center is trying to justify its existence to its funding agency. To this end, data on four variables have been collected for each of the 403 trainees served over the past three years: (1) whether or not the trainee completed the program, (2) whether or not the trainee got and held a job for at least a year after training, (3) the sex of the trainee, and (4) the race of the trainee. Is employment related to completion of the program? Is the relationship between completion and employment affected by race or sex?

Employment by training:

Held Job for at Least One Year?	Training Completed?		Totals
	Yes	No	
Yes	145	60	205
No	72	126	198
Totals	217	186	403

Employment by training, controlling for race:
A. Whites:

Held Job for at Least One Year?	Training Completed?		Totals
	Yes	No	
Yes	85	33	118
No	38	47	85
Totals	123	80	203

B. Blacks:

Held Job for at Least One Year?	Training Completed?		Totals
	Yes	No	
Yes	60	27	87
No	34	79	113
Totals	94	106	200

Employment by training, controlling for sex:

C. Males:

Held Job for at Least One Year?	Training Completed?		Totals
	Yes	No	
Yes	73	30	103
No	36	62	98
Totals	109	92	201

D. Females:

Held Job for at Least One Year?	Training Completed?		Totals
	Yes	No	
Yes	72	30	102
No	36	64	100
Totals	108	94	202

16.6 SOC What are the social sources of support for the environmental movement? A recent survey gathered information on level of concern for such issues as global warming and acid rain. Is concern for the environment related to level of education? What effects do the control variables have? Write a paragraph summarizing your conclusions.

Concern for the environment by level of education:

Concern for the Environment	Level of Education		Totals
	Low	High	
Low	27	35	62
High	22	48	70
Totals	49	83	132

a. Concern for the environment by level of education, controlling for gender:

A. Males:

Concern for the Environment	Level of Education		Totals
	Low	High	
Low	14	17	31
High	11	22	33
Totals	25	39	64

B. Females:

Concern for the Environment	Level of Education		Totals
	Low	High	
Low	13	18	31
High	11	26	37
Totals	24	44	68

b. Concern for the environment by level of education, controlling for "level of trust in the nation's leadership":

A. Low levels of trust:

Concern for the Environment	Level of Education		Totals
	Low	High	
Low	6	22	28
High	10	40	50
Totals	16	62	78

B. High levels of trust:

Concern for the Environment	Level of Education		Totals
	Low	High	
Low	21	13	34
High	12	8	20
Totals	33	21	54

c. Concern for the environment by level of education, controlling for race:

A. White:

Concern for the Environment	Level of Education		Totals
	Low	High	
Low	19	11	30
High	18	44	62
Totals	37	55	92

B. Black:

Concern for the Environment	Level of Education		Totals
	Low	High	
Low	8	24	32
High	4	4	8
Totals	12	28	40

16.7 SOC In problem 14.15, we used the 1998 General Social Survey data set to investigate the relationships between income and five dependent variables. Let's return to two of those relationships and see if sex has any effect on the bivariate relationships. The bivariate and partial tables are presented below along with the bivariate gammas that were computed in the previous exercise. Compute percentages and gammas for each of the partial tables and state your conclusions. Does controlling for sex have any effect? Bivariate table (from problem 14.15):

Right to Abortion?	Income			Totals
	Less than $24,900	$24,900 to $50,000	More than $50,000	
Yes	220	218	226	664
No	366	299	250	915
Totals	586	517	476	1579

Gamma = −.14

(HINT: note that the higher score on ABANY is associated with "No." The negative sign on the gamma means that as score on income increases, score on ABANY—opposition to abortion—decreases. In other words, people with lower incomes were more opposed to the legal right to abortion.)

a. Support for the legal right to abortion by income, controlling for sex:

A. Males:

Right to Abortion?	Income			Totals
	Less than $24,900	$24,900 to $50,000	More than $50,000	
Yes	89	96	100	285
No	130	141	122	393
Totals	219	237	222	678

B. Females:

Right to Abortion?	Income			Totals
	Less than $24,900	$24,900 to $50,000	More than $50,000	
Yes	131	122	126	379
No	236	158	128	522
Totals	367	280	254	901

Bivariate table (from problem 14.15):

Right to Suicide?	Income			Totals
	Less than $24,900	$24,900 to $50,000	More than $50,000	
Favor	343	341	338	1022
Oppose	227	194	147	568
Totals	570	535	485	1590

Gamma = −.22

b. Approval of suicide for people with an incurable disease (SUICIDE1) by income controlling for sex:

A. Males:

Right to Suicide?	Income			Totals
	Less than $24,900	$24,900 to $50,000	More than $50,000	
Favor	140	165	170	475
Oppose	82	80	61	223
Totals	222	245	231	698

B. Females:

Right to Suicide?	Income			Totals
	Less than $24,900	$24,900 to $50,000	More than $50,000	
Favor	203	176	168	547
Oppose	145	114	86	345
Totals	348	290	254	892

16.8 SOC Does support for traditional gender roles vary by social class? Is this relationship affected by sex? By political ideology? By age? The 1998 General Social Survey includes an item (FEFAM) that asked respondents to agree or disagree that it is better if men work and women tend house. Responses to this item have been collapsed for this exercise into (1) agree (support for traditional gender roles) and (2) disagree (opposition to traditional gender roles).

a. Compute percentages and gamma for the bivariate table. Describe the relationship in terms of strength and direction. *(HINT: be careful when interpreting direction. Remember that a higher score on FEFAM means the respondent opposes traditional gender roles.)*

b. Compute percentages and gamma for each set of partial tables. What effect did each control variable have on the bivariate relationship? *(NOTE: since some respondents did not answer all questions, there are fewer cases in the partial tables than in the bivariate table.)*

c. Summarize the results of the analysis in a paragraph.

The bivariate table:

| Support for Traditional Gender Roles | Income | | | |
	Less than $22,500 (1)	$22,500 to $50,000 (2)	More than $50,000 (3)	Totals
Agree (1)	229	190	133	552
Disagree (2)	350	358	363	1071
Totals	579	548	496	1623

Support for traditional gender roles by income, controlling for sex:

A. Males:

| Support for Traditional Gender Roles | Income | | | |
	Less than $22,500 (1)	$22,500 to $50,000 (2)	More than $50,000 (3)	Totals
Agree (1)	87	93	67	247
Disagree (2)	138	156	164	458
Totals	225	249	231	705

B. Females:

| Support for Traditional Gender Roles | Income | | | |
	Less than $22,500 (1)	$22,500 to $50,000 (2)	More than $50,000 (3)	Totals
Agree (1)	142	97	66	305
Disagree (2)	212	202	199	613
Totals	354	299	265	918

Support for traditional gender toles by income, controlling for political ideology.*

A. Liberals:

| Support for Traditional Gender Roles | Income | | | |
	Less than $22,500 (1)	$22,500 to $50,000 (2)	More than $50,000 (3)	Totals
Agree (1)	58	38	15	111
Disagree (2)	143	121	109	373
Totals	201	159	124	484

B. Moderates:

| Support for Traditional Gender Roles | Income | | | |
	Less than $22,500 (1)	$22,500 to $50,000 (2)	More than $50,000 (3)	Totals
Agree (1)	79	60	38	177
Disagree (2)	114	133	131	378
Totals	193	193	169	555

C. Conservatives:

| Support for Traditional Gender Roles | Income | | | |
	Less than $22,500 (1)	$22,500 to $50,000 (2)	More than $50,000 (3)	Totals
Agree (1)	72	84	78	234
Disagree (2)	79	95	119	293
Totals	151	179	197	527

Support for traditional gender roles by income, controlling for age:

A. Ages 18–30:

| Support for Traditional Gender Roles | Income | | | |
	Less than $22,500 (1)	$22,500 to $50,000 (2)	More than $50,000 (3)	Totals
Agree (1)	36	23	6	65
Disagree (2)	125	68	48	241
Totals	161	91	54	306

*Partial tables have fewer cases than the bivariate table because of missing scores. Not all respondents answered all questions.

B. Ages 30–49:

Support for Traditional Gender Roles	Income			
	Less than $22,500 (1)	$22,500 to $50,000 (2)	More than $50,000 (3)	Totals
Agree (1)	62	83	70	215
Disagree (2)	133	208	215	556
Totals	195	291	285	771

C. Ages 50–89:

Support for Traditional Gender Roles	Income			
	Less than $22,500 (1)	$22,500 to $50,000 (2)	More than $50,000 (3)	Totals
Agree (1)	131	84	57	272
Disagree (2)	91	82	99	272
Totals	222	166	156	544

SPSS for Windows

Using *SPSS for Windows* to Elaborate Bivariate Tables

SPSS DEMONSTRATION 16.1 Analyzing the Effects of Age and Sex on Sexual Attitudes

Since we have been concerned with bivariate tables in this chapter, it will come as no surprise to find that we will once again make use of the **Crosstabs** procedure. To control for a third variable with **Crosstabs,** add the name of the control variable to the box at the bottom of the **Crosstabs** dialog window. In other words, on the **Crosstabs** window, place the name of your dependent variable(s) in the top (row) box, the name of your independent variable(s) in the middle (column) box, and the name of your control variable(s) in the box at the bottom of the screen.

To illustrate, in Demonstration 12.3 and again in Demonstration 14.1, we looked at the relationships between *premarsx* and recoded *age* or *ager* (see Demonstrations 12.3 and 14.1 for the recoding scheme) and found moderate, negative relationships (gamma = −.27). That is, approval of premarital sex decreased as age increased. Now, let's see if gender has any effect on these bivariate relationships. Are older women more opposed to premarital sex than older men?

If necessary, recode *age* as in Demonstration 12.3. On the **Crosstabs** dialog window, *premarsx* should be in the top (row) box, recoded *age* in the middle (column) box, and *sex* in the box at the bottom. Make sure you request gamma and column percentages. The output will consist of two partial tables, one for males and one for females, and gammas computed for each of these partial tables.

The bivariate relationships and statistics were displayed in Demonstration 14.1. To conserve space, we will concentrate on gamma and not reproduce the partial tables here. Gamma is −.24 for males and −.27 for females. If you compare the percentages in the partial tables, you will see that males are generally less opposed to premarital sex than females. Specifically, compare the top row of the two partial tables, and you'll see that males are less likely to say that premarital sex is "always wrong" regardless of age group. Overall, however, this is a direct relationship: opposition to premarital sex increases with age regardless of sex (although males are slightly more permissive in every age group).

SPSS DEMONSTRATION 16.2 Are Prestigious Jobs More Satisfying? Another Look

In Demonstration 14.3, we looked at the relationship between recoded *satjob* and *prestg80*. We found a bivariate gamma of −.26, indicating that people with higher-prestige jobs had higher levels of satisfaction with their jobs. While not exactly surprising, we might be able to learn a little more about the relationship with some additional controls. In this demonstration, the bivariate relationship is reexamined with *sex* and *degree* as control variables. Does the relationship between satisfaction and prestige vary by sex? Does the educational level of the respondent affect the relationship between job satisfaction and prestige?

The variable *degree* has a total of five values—too many to permit the variable to be used as a control. To simplify the variable, I grouped all respondents with any level of college education into a single category. The recoding scheme I used was

$$0 \rightarrow 0 \text{ (less than high school)}$$
$$1 \rightarrow 1 \text{ (high school diploma)}$$
$$2 \text{ thru } 4 \rightarrow 2 \text{ (at least some college)}$$

I called the recoded version of this variable *rdeg*. If necessary, see Demonstration 14.3 for the recode schemes for *satjob* (*rsat*) and *prestg80* (*rprest*).

Run the **Crosstabs** procedure and move *rsat* into the **Row Variable** box, *rprest* into the **Column Variable** box, and *rdeg* and *sex* into the bottom box. Don't forget gamma and column percents. To conserve space, the partial tables are not reproduced here. Instead, we will concentrate on the gammas:

	Gamma
For the bivariate table	−.26
For people with less than a high school education	−.13
For high school graduates	−.29
For people with at least some college	−.26

There appears to be some interaction between the variables. Satisfaction increases with prestige for all three levels of education but the relationship is weaker for the lower level of education. As you can see from the column percentages, prestige has less impact on job satisfaction for less educated respondents.

The control for gender also shows some interaction:

	Gamma
For the bivariate table	−.26
For males	−.33
For females	−.20

For both males and females, higher-prestige jobs are associated with greater satisfaction, but the effect is stronger for males. The column percentages also show that the prestige of the job makes more difference for men's satisfaction than for women's satisfaction.

Using MicroCase to Elaborate Bivariate Tables

MICROCASE DEMONSTRATION 16.1 Analyzing the Effects of Age and Sex on Sexual Attitudes

Since we have been concerned with bivariate tables in this chapter, it will come as no surprise to find that we will once again make use of the **Cross-tabulation** procedure. To control for a third variable with **Cross-tabulation,** click on the name of the control variable and move it to the **Control Variables** box at the bottom of the **Cross-tabulation** dialog window. In other words, on the **Cross-tabulation** window, place the name of your dependent variable in the **Row Variable** box, the name of your independent variable in the **Column Variable** box, and the name of your control variable(s) in the **Control Variables** box.

To illustrate, in Demonstration 12.3 and again in Demonstration 14.1, we looked at the relationships between *premarsx* and recoded *age* or *ager* (see Demonstrations 12.3 and 14.1 for the recoding scheme) and found moderate, negative relationships (gamma = −.27). That is, approval of premarital sex decreased as age increased. Now, let's see if gender has any effect on these bivariate relationships. Are older women more opposed to premarital sex than older men?

If necessary, recode *age* as in Demonstration 12.3. On the **Cross-tabulation** dialog window, *premarsx* should be in the **Row Variable** box, recoded *age* in the **Column Variable** box, and *sex* in the **Control Variable** box. The output will consist of two partial tables, one for males and one for females. Click **Column %** and get gammas for each table by clicking **Summary** in the **Statistics** box on the left of the screen.

The bivariate relationships and statistics were displayed in Demonstration 14.1. To conserve space, we will concentrate on gamma and not reproduce the partial tables here. Gamma is −.24 for males and −.27 for females. If you compare the percentages in the partial tables, you will see that males are generally less opposed to premarital sex than females. Specifically, compare the top row of the two partial tables, and you'll see that males are less likely to say that premarital sex is "always wrong" regardless of age group. Overall, however, this is a direct relationship: opposition to premarital sex increases with age regardless of sex (although males are slightly more permissive in every age group).

MICROCASE DEMONSTRATION 16.2 Are Prestigious Jobs More Satisfying? Another Look

In Demonstration 14.3, we looked at the relationship between recoded *satjob* and *prestg80*. We found a bivariate gamma of −.26, indicating that people with higher-prestige jobs had higher levels of satisfaction with their jobs. While not exactly surprising, we might be able to learn a little more about the relationship with some additional controls. In this demonstration, the bivariate relationship is reexamined with *sex* and *degree* as control variables. Does the relationship between satisfaction and prestige vary by sex? Does the educational level of the respondent affect the relationship between job satisfaction and prestige?

The variable *degree* has a total of five values—too many to permit the variable to be used as a control. To simplify the variable, I grouped all respondents with any level of college education into a single category. The recoding scheme I used was

$$0 \rightarrow 0 \text{ (less than high school)}$$
$$1 \rightarrow 1 \text{ (high school diploma)}$$
$$2 \text{ thru } 4 \rightarrow 2 \text{ (at least some college)}$$

I called the collapsed version of this variable *rdeg*. If necessary, see Demonstration 14.3 for the schemes for *satjob* (*rsat*) and *prestg80* (*rprest*).

Run the **Cross-tabulation** procedure and move *rsat* into the **Row Variable** box, *rprest* into the **Column Variable** box, and *rdeg* and *sex* into the **Control Variable** box. To conserve space, the partial tables are not reproduced here. Instead, we will concentrate on the gammas:

	Gamma
For the bivariate table	−.26
For people with less than a high school education	−.13
For high school graduates	−.29
For people with at least some college	−.26

There appears to be some interaction between the variables. Satisfaction increases with prestige for all three levels of education, but the relationship is weaker for the lower level of education. As you can see from the column percentages, prestige has less impact on job satisfaction for less educated respondents.

The control for gender also shows some interaction:

	Gamma
For the bivariate table	−.26
For males	−.33
For females	−.20

For both males and females, higher-prestige jobs are associated with greater satisfaction, but the effect is stronger for males. The column percentages also show that the prestige of the job makes more difference for men's satisfaction than for women's satisfaction.

Exercises

16.1 Use Demonstration 16.1 as a guide and analyze the relationship of recoded *age* on *cappun, grass,* and *gunlaw,* using *sex* as the control variable. Summarize the results in a paragraph or two. Be sure to characterize the relationships as direct, spurious, or interactive.

16.2 Using Demonstration 16.2 as a guide, analyze the relationships between income (*income91*) and job satisfaction, controlling for *sex* and recoded *degree*. Recode *income91* at the median before conducting the analysis. Summarize the results in a paragraph or two. Be sure to characterize the relationships as direct, spurious, or interactive.

17 PARTIAL CORRELATION AND MULTIPLE REGRESSION AND CORRELATION

LEARNING OBJECTIVES

By the end of this chapter, you will be able to

1. Compute and interpret partial correlation coefficients.
2. Find and interpret the least-squares multiple regression equation with partial slopes.
3. Calculate and interpret the multiple correlation coefficient (R^2).
4. Explain the limitations of partial and multiple regression analysis.

17.1 INTRODUCTION

In Chapter 16, I made the point that very few (if any) worthwhile research questions can be answered through a statistical analysis of only two variables. Social science research is, by nature, multivariate and often involves the simultaneous analysis of scores of variables. Some of the most powerful and widely used statistical tools for multivariate analysis are introduced in this chapter. We will cover techniques that are used to analyze causal relationships and to make predictions—crucial endeavors in any science.

These techniques are based on Pearson's r (see Chapter 15). They are generally more flexible than the techniques presented in Chapter 16, produce more information, and provide a wider variety of ways of disentangling the interrelationships among the variables.

We will first consider partial correlation analysis, a technique analogous to controlling for a third variable by constructing partial tables (see Chapter 16). The second technique involves multiple regression and correlation and allows the researcher to assess the effects, separately and in combination, of more than one independent variable on the dependent variable.

Throughout this chapter, we will focus our attention on research situations involving only three variables. This is the least complex application of these techniques but extensions to situations involving four or more variables are relatively straightforward. To deal efficiently with the computations required by the more complex applications, I refer you to any of the computerized statistical packages probably available at your local computer center.

17.2 PARTIAL CORRELATION

The technique of **partial correlation** can be used when a researcher wishes to observe how a specific bivariate relationship behaves in the presence of a third variable. By observing the partial correlation coefficients, direct, spurious and intervening relationships can be detected (see Chapter 16 and Table 16.5), and the causal relationships among the variables can be probed. As opposed to the techniques considered in Chapter 16, however, partial correlation analysis does not involve the inspection of the bivariate relationship for each category of the control variable. For this reason, partial correlation techniques will not reveal interactive relationships among the variables (that is, relationships where the nature of the bivariate relationship changes across the categories of the control variable).

Before dealing with partial correlation per se, let me introduce some new terminology. Since we will be dealing with more than one bivariate relationship, we will use subscripts to identify the various correlations. Thus, the symbol r_{yx} will refer to the correlation coefficient between variable Y and variable X. By the same token, r_{yz} will refer to the correlation coefficient between Y and Z, and r_{xz} to the correlation coefficient between X and Z. Correlation coefficients calculated for bivariate relationships are often referred to as **zero-order correlations.**

Partial correlation coefficients, when controlling for a single variable, are called first-order partials and can be symbolized as $r_{yx.z}$. The variable to the right of the dot is the control variable. Thus, $r_{yx.z}$ refers to the partial correlation coefficient that measures the relationship between variables X and Y while controlling for variable Z. The formula for the first-order partial is

FORMULA 17.1

$$r_{yx.z} = \frac{r_{yx} - (r_{yz})(r_{xz})}{\sqrt{1 - r_{yz}^2} \sqrt{1 - r_{xz}^2}}$$

Note that you must first calculate the zero-order coefficients between all possible pairs of variables (variables X and Y, X and Z, and Y and X) before solving this formula.

To illustrate the computation of a first-order partial, let us return to the problem introduced in Chapter 15 that examined the relationship between number of children (X) and husband's contribution to housework (Y) for 12 dual-career families. The zero-order r between these two variables (r_{yx} = .50) indicated a moderate, positive relationship. Suppose the researcher wished to investigate the possible effects of socioeconomic status (SES) on the bivariate relationship. The original data (from Table 15.1) and the scores of the 12 families on the new variable of SES (measured by the years of education completed by the husband) are presented in Table 17.1.

The three correlations indicate that the husband's contribution to housework is related to number of children (r_{yx} = .50), that husbands in higher-SES families tend to do less housework (r_{yz} = −.30), and that higher-SES families have fewer children (r_{xz} = −.47). Is the relationship between husband's

TABLE 17.1   SCORES ON THREE VARIABLES FOR 12 DUAL-WAGE-EARNER
FAMILIES AND ZERO-ORDER CORRELATIONS

Family	Husband's Housework (Y)	Number of Children (X)	Husband's Years of Education (Z)
A	1	1	12
B	2	1	14
C	3	1	16
D	5	1	16
E	3	2	18
F	1	2	16
G	5	3	12
H	0	3	12
I	6	4	10
J	3	4	12
K	7	5	10
L	4	5	16
	$r_{yx} = .50$	$r_{yz} = -.30$	$r_{xz} = -.47$

housework and number of children affected by SES? Substituting the above correlations into Formula 17.1, we would have

$$r_{yx.z} = \frac{r_{yx} - (r_{yz})(r_{xz})}{\sqrt{1 - r_{yz}^2}\,\sqrt{1 - r_{xz}^2}}$$

$$r_{yx.z} = \frac{.50 - (-.30)(-.47)}{\sqrt{1 - (-.30)^2}\,\sqrt{1 - (-.47)^2}}$$

$$r_{yx.z} = \frac{.50 - (.14)}{\sqrt{1 - .09}\,\sqrt{1 - .22}}$$

$$r_{yx.z} = \frac{.36}{\sqrt{.91}\,\sqrt{.78}}$$

$$r_{yx.z} = \frac{.36}{(.95)(.88)}$$

$$r_{yx.z} = \frac{.36}{.84}$$

$$r_{yx.z} = .43$$

The first-order partial ($r_{yx.z} = .43$) is lower in value than the zero-order coefficient ($r_{yx} = .50$), but the difference in the two values is not great. This result suggests a direct relationship between variables X and Y (see Section 16.3). That is, when controlling for SES, the statistical relationship between husband's housework and number of children is essentially unchanged. Regardless of SES, husband's hours of housework increase with the number of children. The next step in statistical analysis would probably be to discard this control variable (SES) and select another (see Table 16.5). The more the bivariate relationship retains its strength across a series of controls for the third variables (Z's), the stronger the evidence for a direct relationship between X and Y.

In addition to direct relationships, there are two other possible relationships between the partial and zero-order correlation coefficients. If the partial is much lower in value than the zero-order coefficient ($r_{yx.z} < r_{yx}$), the bivariate relationship between X and Y is spurious or intervening. That is, the association between X and Y is due mainly to the effects of Z. In a spurious relationship, X and Y are both caused by Z, whereas an intervening relationship means that X and Y are linked by Z (see Figures 16.2 and 16.3). In either case, the relationship between X and Y disappears once the effect of Z is controlled, and the next step in the statistical analysis would be to incorporate the control variable into the analysis as an independent (see Table 16.5). Let me stress that partial correlation cannot distinguish between the case where the control variable is antecedent (where Z causes both X and Y) and the case where the control variable is intervening (when X causes Z, which in turn causes Y). Judgments about possible causal relationships must be made on the grounds of temporal ordering among the variables and/or theory.

A third possible outcome of the application of partial correlation is for the partial correlation coefficient to be greater in value than the zero-order coefficient ($r_{yx.z} > r_{yx}$). This outcome would be consistent with a causal model wherein the variable taken as independent and the control variable each had a separate effect on the dependent variable and were uncorrelated with each other. This relationship is depicted in Figure 17.1. The absence of an arrow between X and Z indicates that they have no mutual relationship. This pattern means that both X and Z should be treated as independent variables, and the next step in the statistical analysis would probably involve multiple correlation and regression. As we shall see in Sections 17.3 and 17.4, these techniques enable the researcher to isolate the separate effects of several independent variables on the dependent variable and thus to make judgments about which independent has the stronger effect on the dependent. *(For practice in computing and interpreting partial correlation coefficients, see problems 17.1 to 17.3.)*

FIGURE 17.1 A POSSIBLE CAUSAL RELATIONSHIP AMONG THREE VARIABLES

17.3 MULTIPLE REGRESSION: PREDICTING THE DEPENDENT VARIABLE

In Chapter 15, the least-squares regression line was introduced as a way of describing the overall linear relationship between two interval-ratio variables and of predicting scores on Y from scores on X. This line was the best-fitting line to summarize the bivariate relationship and was defined by the formula

FORMULA 17.2
$$Y = a + bX$$

where a = the Y intercept
b = the slope

The least-squares regression line can be modified to include (theoretically) any number of independent variables. This technique is called **multiple regression.** For ease of explication, we will confine our attention to the case involving two independent variables. The least-squares multiple regression equation is

FORMULA 17.3
$$Y = a + b_1X_1 + b_2X_2$$

where b_1 = the partial slope of the linear relationship between the first independent variable and Y
b_2 = the partial slope of the linear relationship between the second independent variable and Y

Some new notation and some new concepts are introduced in this formula. First, while the dependent variable is still symbolized as Y, the independent variables are differentiated by subscripts. Thus, X_1 identifies the first independent variable and X_2 the second. The symbol for the slope (b) is also subscripted to identify the independent with which it is associated.

A major difference between the multiple and bivariate regression equations concerns the slopes (b's). In the case of multiple regression, the b's are called **partial slopes,** and they show the amount of change in Y for a unit change in the independent while controlling for the effects of the other independents in the equation. The partial slopes are thus analogous to partial correlation coefficients and represent the direct effect of the associated independent variable on Y.

The partial slopes for the independent variables are determined by Formula 17.4 and Formula 17.5:*

FORMULA 17.4
$$b_1 = \left(\frac{s_y}{s_1}\right)\left(\frac{r_{y1} - r_{y2}r_{12}}{1 - r_{12}^2}\right)$$

FORMULA 17.5
$$b_2 = \left(\frac{s_y}{s_2}\right)\left(\frac{r_{y2} - r_{y1}r_{12}}{1 - r_{12}^2}\right)$$

where b_1 = the partial slope of X_1 on Y
b_2 = the partial slope of X_2 on Y
s_y = the standard deviation of Y
s_1 = the standard deviation of the first independent variable (X_1)
s_2 = the standard deviation of the second independent variable (X_2)

---

*Partial slopes can be computed from zero-order slopes but Formulas 17.4 and 17.5 are somewhat easier to use.

r_{y1} = the bivariate correlation between Y and X_1
r_{y2} = the bivariate correlation between Y and X_2
r_{12} = the bivariate correlation between X_1 and X_2

To illustrate the computation of the partial slopes, we will assess the combined effects of number of children (X_1) and SES (X_2) on husband's contribution to housework. All the relevant information can be calculated from Table 17.1 and is reproduced below:

Husband's Housework	Number of Children	SES
$\overline{Y} = 3.3$	$\overline{X}_1 = 2.7$	$\overline{X}_1 = 13.7$
$s_y = 2.1$	$s_1 = 1.5$	$s_2 = 2.6$

Zero-order correlations
$r_{y1} = .50$
$r_{y2} = -.30$
$r_{12} = -.47$

The partial slope for the first independent variable (X_1) is

$$b_1 = \left(\frac{s_y}{s_1}\right)\left(\frac{r_{y1} - r_{y2}r_{12}}{1 - r_{12}^2}\right)$$

$$b_1 = \left(\frac{2.1}{1.5}\right)\left(\frac{.50 - (-.30)(-.47)}{1 - (-.47)^2}\right)$$

$$b_1 = (1.4)\left(\frac{.50 - .14}{1 - .22}\right)$$

$$b_1 = (1.4)\left(\frac{.36}{.78}\right)$$

$$b_1 = (1.4)(.46)$$

$$b_1 = .65$$

For the second independent variable, SES or X_2, the partial slope is

$$b_1 = \left(\frac{s_y}{s_2}\right)\left(\frac{r_{y2} - r_{y1}r_{12}}{1 - r_{12}^2}\right)$$

$$b_1 = \left(\frac{2.1}{2.6}\right)\left(\frac{-.30 - (-.24)}{1 - .22}\right)$$

$$b_1 = (.81)\left(\frac{-.30 + .24}{.78}\right)$$

$$b_1 = (.81)\left(\frac{-.06}{.78}\right)$$

$$b_1 = (.81)(-.08)$$

$$b_1 = -.07$$

Now that partial slopes have been determined for both independent variables, the Y intercept (a) can be found. Note that a is calculated from the mean of the dependent variable (symbolized as $\overline{Y}$) and the means of the two independent variables ($\overline{X}_1$ and $\overline{X}_2$).

FORMULA 17.6
$$a = \overline{Y} - b_1\overline{X}_1 - b_2\overline{X}_2$$

Substituting the proper values for the example problem at hand, we would have

$$a = \overline{Y} - b_1\overline{X}_1 - b_2\overline{X}_2$$
$$a = 3.3 - (.65)(2.7) - (-.07)(13.7)$$
$$a = 3.3 - (1.8) - (-1.0)$$
$$a = 3.3 - 1.8 + 1.0$$
$$a = 2.5$$

For our example problem, the full least-squares multiple regression equation would be

$$Y = a + b_1X_1 + b_2X_2$$
$$Y = 2.5 + (.65)X_1 + (-.07)X_2$$

As was the case with the bivariate regression line, this formula can be used to predict scores on the dependent variable from scores on the independent variables. For example, what would be our best prediction of husband's housework (Y') for a family of four children ($X_1 = 4$) where the husband had completed 11 years of schooling ($X_2 = 11$)? Substituting these values into the least-squares formula, we would have

$$Y' = 2.5 + (.65)(4) + (-.07)(11)$$
$$Y' = 2.5 + 2.6 - .8$$
$$Y' = 4.3$$

Our prediction would be that this husband would contribute 4.3 hours per week to housework. This prediction is, of course, a kind of "educated guess," which is unlikely to be perfectly accurate. However, we will make fewer errors of prediction using the least-squares line (and, thus, incorporating information from the independent variables) than we would using any other method of prediction (assuming, of course, that there is a linear association between the independent and the dependent variables). *(For practice in predicting Y scores and in computing slopes and the Y intercept, see problems 17.1 to 17.6.)*

17.4 MULTIPLE REGRESSION: ASSESSING THE EFFECTS OF THE INDEPENDENT VARIABLES

The least-squares multiple regression equation (Formula 17.3) is used to isolate the separate effects of the independents and to predict scores on the dependent variable. However, in many situations, using this formula to determine the relative importance of the various independents will be awkward

—especially when the independent variables differ in terms of units of measurement (e.g., number of children vs. years of education). When the units of measurement differ, a comparison of the partial slopes will not necessarily tell us which independent variable has the strongest effect and is thus the most important. Comparing the partial slopes of variables that differ in units of measurement is a little like comparing apples and oranges.

The comparability of the independent variables can be increased by converting all variables in the equation to a common scale and thereby eliminating variations in the values of the partial slopes that are solely a function of differences in units of measurement. We can, for example, standardize all distributions by changing all scores to Z scores. Each distribution of scores would then have a mean of 0 and a standard deviation of 1 (see Chapter 5), and comparisons between the independent variables would be much more meaningful. To standardize the variables to the normal curve, we could actually convert all scores into the equivalent Z scores and then recompute the slopes and the Y intercept. This would require a good deal of work and, fortunately, a shortcut is available for directly computing the slopes of the standardized scores. These **standardized partial slopes** are called **beta-weights** and are symbolized b^*. The beta-weights show the amount of change in the standardized scores of Y for a one-unit change in the standardized scores of each independent variable while controlling for the effects of all other independent variables. For the case where we have two independents, the beta-weight for each is found by using Formula 17.7 and Formula 17.8:

FORMULA 17.7
$$b_1^* = b_1\left(\frac{s_1}{s_y}\right)$$

FORMULA 17.8
$$b_2^* = b_2\left(\frac{s_2}{s_y}\right)$$

where b_1^* = the standardized partial slope of X_1 on Y
b_2^* = the standardized partial slope of X_2 on Y

Using standardized scores, the least-squares regression equation can be written as

FORMULA 17.9
$$Z_y = a_z + b_1^* Z_1 + b_2^* Z_2$$

where the symbol Z indicates that all scores
have been standardized to the normal curve

The standardized regression equation can be further simplified by dropping the term for the Y intercept, since this term will always be zero when scores have been standardized. This value is the point where the regression line crosses the Y axis and is equal to the mean of Y when all independents equal 0. This relationship can be seen by substituting 0 for all independent variables in Formula 17.6:

$$a = \overline{Y} - b_1\overline{X_1} - b_2\overline{X_2}$$
$$a = \overline{Y} - b_1(0) - b_2(0)$$
$$a = \overline{Y}$$

Since the mean of any standardized distribution of scores is zero, the mean of the standardized Y scores will be zero and the Y intercept will also be zero ($a = \overline{Y} = 0$). Thus, Formula 17.9 simplifies to

FORMULA 17.10

$$Z_y = b_1^* Z_1 + b_2^* Z_2$$

We can now compute the beta-weights for our sample problem to see which of the two independents has the stronger effect on the dependent. For the first independent variable, number of children (X_1):

$$b_1^* = b_1\left(\frac{s_1}{s_y}\right)$$

$$b_1^* = (.65)\left(\frac{1.5}{2.1}\right)$$

$$b_1^* = (.65)(.71)$$

$$b_1^* = .46$$

For the second independent variable, SES (X_2):

$$b_2^* = b_2\left(\frac{s_2}{s_y}\right)$$

$$b_2^* = (-.07)\left(\frac{2.6}{2.1}\right)$$

$$b_2^* = (-.07)(1.24)$$

$$b_2^* = -.09$$

Thus, the standardized regression equation, with beta-weights noted, would be

$$Z_y = (.46)Z_1 + (-.09)Z_2$$

and it is immediately obvious that the first independent variable has a much stronger direct effect on Y than the second independent variable.

In summary, multiple regression analysis permits the researcher to summarize the linear relationship among two or more independents and a dependent variable. The unstandardized regression equation (Formula 17.2) permits values of Y to be predicted from the independent variables in the original units of the variables. The standardized regression equation (Formula 17.10) allows the researcher to easily assess the relative importance of the various independent variables by comparing the beta-weights. *(For practice in computing and interpreting beta-weights, see any of the problems at the end of this chapter. It is probably a good idea to start with problem 17.1 since it has the smallest data set and the least complex computations.)*

READING STATISTICS 12: Regression and Correlation

Research projects that analyze the interrelationships among many variables are particularly likely to employ regression and correlation as central statistical techniques. The results of these projects will typically be presented in summary tables that report the multiple correlations, slopes, and, if applicable, the significance of the results. The zero-order correlations are often presented in the form of a matrix that displays the value of *Pearson's r* for every possible bivariate relationship in the data set. An example of such a matrix can be found in Section 17.6.

Usually, tables that summarize the multivariate analysis will report R^2 and the slope for each independent variable in the regression equation. An example of this kind of summary table would look like this:

Independents	Multiple R^2	Beta-weights
X_1	.17	.47
X_2	.23	.32
X_3	.27	.16
.	.	.
.	.	.

This table reports that the first independent variable, X_1, has the strongest direct relationship with the dependent variable and explains 17% of the variance in the dependent by itself ($R^2 = 0.17$). The second independent, X_2, adds 6% to the explained variance ($R^2 = 0.23$ after X_2 is entered into the equation). The third independent, X_3, adds 4% to the explained variance of the dependent ($R^2 = 0.27$ after X_3 is entered into the equation).

Statistics in the Professional Literature

Researchers have consistently reported a relationship between various measures of poverty and homicide for whites but not for blacks. Sociologist Matthew Lee wondered if this was because traditional measures of poverty do not take account of the way in which poverty is spatially concentrated in central cities for minority group members. In other words, the effects of poverty for urban black Americans are compounded by residential segregation and housing discrimination. Is it this combination of factors that leads to higher homicide rates? Lee gathered data on homicide, black disadvantage (a composite scale that includes measures of unemployment, poverty, housing quality, and other variables), the extent to which poverty is concentrated residentially for blacks, and other control variables for a sample of 121 central cities for 1990. His results are reported in two models in Table 1.

TABLE 1 MULTIPLE REGRESSION RESULTS[†]

	Model 1		Model 2	
	Unstandardized Coefficient	Beta-weight	Unstandardized Coefficient	Beta-weight
Black city disadvantage	0.046**	0.347	0.009	0.065
Black poverty concentration	—	—	2.416*	0.345
Blacks age 15–24	−0.019	−0.069	−0.20	−0.70
Residential density	0.200	0.067	0.026	.087
Vacant housing	2.371	0.126	2.311	0.123
South	−0.149	−0.172	−0.254*	−0.226
Intercept (a)	3.844**		3.115**	
R^2	0.154		0.181	

*$p < .05$ **$p < .01$
[†]Adapted from Table 2

The first model looks at the effect of black disadvantage and the control variables on homicide rates for blacks and shows a positive, statistically significant relationship. As the levels of disadvantage increase, so does the black homicide rate even when the control variables are included in the equation. Model 2 adds the measure of black poverty concentration to the other variables and produces a number of interesting results.

[T]he concentration of black poverty has a strong positive and statistically significant relationship with the black homicide rate. . . . What is particularly important is that the inclusion of this variable completely accounts for the relationship between [black] disadvantage and . . .

homicide rate observed in the previous model. Thus, . . . [the relationship between disadvantage and homicide rate is] due to . . . the spatial concentration of poor people.

In other words, the standardized coefficient (beta-weight) for black disadvantage was .347 in Model 1 but drops to .065 in Model 2 (after the measure of poverty concentration has been entered). This indicates that the relationship in Model 1 is spurious and due to the effects of the spatial concentration of poverty rather than poverty or disadvantage itself.

Matthew R. Lee. "Concentrated Poverty, Race, and Homicide." *The Sociological Quarterly,* (41), pp. 189–210.

17.5 MULTIPLE CORRELATION

We use the multiple regression equations to disentangle the separate direct effects of each independent variable on the dependent. Using **multiple correlation** techniques, we can also ascertain the combined effects of all independents on the dependent variable. We do so by computing the **multiple correlation coefficient (R)** and the **coefficient of multiple determination (R)^2**. The value of the latter statistic represents the proportion of the variance in Y that is explained by all the independent variables combined. In terms of zero-order correlation, we have seen that "number of children" (X_1) explains a proportion of .25 of the variance in Y ($r_{y1}^2 = (.50)^2 = .25$) by itself and that SES explains a proportion of .09 of the variance in Y ($r_{y2}^2 = (-.30)^2 = .09$). The two zero-order correlations cannot be simply added together to ascertain their combined effect on Y, because the two independents are also correlated with each other and, therefore, they will "overlap" in their effects on Y and explain some of the same variance. This overlap is eliminated in Formula 17.11:

FORMULA 17.11

$$R^2 = r_{y1}^2 + r_{y2.1}^2(1 - r_{y1}^2)$$

where R^2 = the multiple correlation coefficient
r_{y1}^2 = the zero-order correlation between Y and X_1, the quantity squared
$r_{y2.1}^2$ = the partial correlation of Y and X_2 while controlling for X_1, the quantity squared

The first term in this formula (r_{y1}^2) is the coefficient of determination for the relationship between Y and X_1. It represents the amount of variation in Y explained by X_1 by itself. To this quantity we add the amount of the variation remaining in Y (given by $1 - r_{y1}^2$) that can be explained by X_2 after the effect of X_1 is controlled ($r_{y2.1}^2$). Basically, Formula 17.11 allows X_1 to explain as much of Y as it can and then adds in the effect of X_2 after X_1 is controlled (thus eliminating the "overlap" in the variance of Y that X_1 and X_2 have in common).

To observe the combined effects of number of children (X_1) and SES (X_2) on husband's housework (Y), we need two quantities. The correlation between X_1 and Y ($r_{y1} = .50$) has already been found; but, before we can solve Formula 17.11, we must first calculate the partial correlation of Y and X_2 while controlling for X_1 ($r_{y2.1}$):

$$r_{y2.1} = \frac{r_{y2} - (r_{y1})(r_{12})}{\sqrt{1 - r_{y1}^2}\ \sqrt{1 - r_{12}^2}}$$

$$r_{y2.1} = \frac{(-.30) - (.50)(-.47)}{\sqrt{1 - (.50)^2}\ \sqrt{1 - (-.47)^2}}$$

$$r_{y2.1} = \frac{(-.30) - (-.24)}{\sqrt{.75}\ \sqrt{.78}}$$

$$r_{y2.1} = \frac{-.06}{.77}$$

$$r_{y2.1} = -.08$$

Formula 17.11 can now be solved for our sample problem:

$$R^2 = r_{y1}^2 + r_{y2.1}^2(1 - r_{y1}^2)$$
$$R^2 = (.5)^2 + (-.08)^2(1 - .50^2)$$
$$R^2 = .25 + (.006)(1 - .25)$$
$$R^2 = .25 + .005$$
$$R^2 = .255$$

The first independent variable (X_1), number of children, explains 25% of the variance in Y by itself. To this total, the second independent (X_2), SES, adds only a half a percent, for a total explained variance of 25.5%. In combination, the two independents explain a total of 25.5% of the variation in the dependent variable. *(For practice in computing and interpreting R and R^2, see any of the problems at the end of this chapter. It is probably a good idea to start with problem 17.1 since it has the smallest data set and the least complex computations.)*

17.6 INTERPRETING STATISTICS: DOES SIZE *REALLY* MATTER FOR CORPORATE INTERLOCKS?

In Chapter 15, we assessed the association between the number of interlocking directorates a corporation has and its revenues in 1995. In this section, we will apply some of our new statistical tools to introduce a third variable, the size of the board (i.e., the number of seats), into the analysis. This is a logical control variable because corporations with larger boards would have potentially more interlocks with other corporations as a simple function of the number of people involved. A multivariate analysis of total interlocks, revenues, and board size should tell us if size *really* matters.

TABLE 17.2   ZERO-ORDER CORRELATION

Coefficient (2-tail Probability)	Revenues in 1995	Board Size
Total interlocks	0.33 (0.00)	.36 (0.00)
Revenues in 1995		0.20 (0.03)

As you recall, the data for this analysis come from 125 randomly selected corporations in 1995.[*] The basic descriptive statistics for total number of interlocking directorates and revenue were reported in Table 15.4, and to this we will add that the boards of the corporations in the sample averaged 12.28 members with a standard deviation of 3.75. The minimum number of directors was 3, and the largest board had 30 members.

Before beginning the multivariate analysis, we should examine the correlation coefficients between all possible pairs of variables. Table 17.2 reports the zero-order correlations in a matrix. As we saw in Chapter 15, there is a moderate, positive relationship between number of interlocking directorates and revenues ($r = .33$), and the relationship is statistically significant at less than 0.001. Companies with larger revenues tend to have more ties through shared directors than companies with smaller revenues.

Turning next to the control variable, board size, we observe a significant, moderately strong, positive association ($r = .36$) with total interlocks. As we anticipated when selecting this control variable, corporations with larger boards of directors do tend to network more than companies with smaller boards of directors. Finally, there is also an association between the size of a corporation—as measured by its revenues in 1995—and the number of individuals who sit on its corporate board. This relationship is positive in direction, significant at less than .05, but weaker than the other relationships. Larger corporations tend to have larger boards.

Let's begin the multivariate analysis by asking if the original relation between total interlocks and revenues exists when controlling for the effects of board size. Remember that the partial correlation coefficient indicates the strength and direction of a bivariate relation when controlling for the effects of a third variable. In this case, the zero-order correlation was .33 and the partial correlation coefficient between the total number of interlocks and a corporation's revenues controlling for board size is .29 ($r_{yx.z} = .29$). Since the partial correlation is only slightly smaller than the original association, we

---

[*]Barnes, Roy C. 2000. "Patterns of Corporate Interlocking in the United States: 1962–1995," Paper presented at the Annual Meeting of the Southern Sociological Society, New Orleans, LA.

TABLE 17.3 MULTIPLE REGRESSION RESULTS

	Model 1			Model 2		
	Unstan-dardized Coefficient	Beta-weight	t-value	Unstan-dardized Coefficient	Beta-weight	t-value
Constant (a)	0.19	0.11	0.01	0.01		
Board size	0.55	.36	4.23***	0.47	0.30	3.66***
Revenues (in billions)	—	—	—	0.08	0.27	3.31**
R^2 or Coefficient of multiple determination	0.13			0.20		

$**p < 0.01$; $***p < 0.001$

can conclude that interlocks and revenues are associated even when controlling for board size. Total number of interlocks tends to increase as revenue increases regardless of the size of the board.

Next, we can assess the relative importance of revenues and board size in predicting the total number of interlocks and see what their combined effects are on the dependent variable. One way to conduct this analysis, commonly used by researchers, is to compare the results of two regression equations as reported in Table 17.3. The first is a simple (bivariate) linear regression model in which board size is used to predict the total number of interlocks. The second model adds revenues, the original independent variable and the more important for assessing the argument that the economy is dominated by a few, densely interlocked corporations, as a second independent variable. By comparing the two models, we can assess the relative importance of revenues and board size in predicting total interlocks (Model 2) and the added explanatory power of revenues, once the effects of board size have been accounted for (comparing Model 2 with Model 1).

Model 1 shows that the size of the board has a substantial and significant impact on the number of interlocks. The slope (b or "unstandardized coefficient") is highly significant, and board size alone explains 13% of the variation in total interlocks. These results reinforce the logical and analytic importance of this control variable.

Turning our attention to Model 2, let's begin with the unstandardized regression coefficients or partial slopes (b), which, as you recall, show the change in the dependent variable caused by each independent variable while controlling for the other independent variable. These coefficients could be substituted into the multiple regression equation (see Formula 17.3) and used to predict scores on the dependent variable for various combinations of scores in the independent variables. For example, how many interlocks would we predict for a corporation with a board of 10 members (X_1) and revenues of 15 billion dollars (X_2)?

Application 17.1

Five recently divorced men have been asked to rate subjectively the success of their adjustment to single life on a scale ranging from 5 (very successful adjustment) to 1 (very poor adjustment). Is adjustment related to the length of time married? Is adjustment related to socioeconomic status as measured by yearly income?

Case	Adjustment (Y)	Years Married (X_1)	Income (dollars) (X_2)
A	5	5	30,000
B	4	7	45,000
C	4	10	25,000
D	3	2	27,000
E	1	15	17,000

$$\overline{Y} = 3.4 \quad \overline{X}_1 = 7.8 \quad \overline{X}_2 = 28,800.00$$
$$s = 1.4 \quad s = 4.5 \quad s = 9,173.88$$

The zero-order correlations among these three variables are

	Years Married (X_1)	Income (X_2)
Adjustment (Y)	$-.62$	$.62$
Years married (X_1)		$-.49$

These results suggest strong but opposite relationships between each independent and adjustment. Adjustment decreases as years married increases, and increases as income increases.

To find the multiple regression equation, we must find the partial slopes.

For years married (X_1):

$$b_1 = \left(\frac{s_y}{s_1}\right)\left(\frac{r_{y1} - r_{y2}r_{12}}{1 - r_{12}^2}\right)$$

$$b_1 = \frac{1.4}{4.5}\left(\frac{(-.62) - (.62)(-.49)}{1 - (-.49)^2}\right)$$

$$b_1 = (.31)\left(\frac{(-.62) - (-.30)}{1 - .24}\right)$$

$$b_1 = (.31)\left(\frac{-.32}{.76}\right)$$

$$b_1 = (.31)(-.42)$$

$$b_1 = -.13$$

Partial slope for income (X_2):

$$b_2 = \left(\frac{s_y}{s_2}\right)\left(\frac{r_{y2} - r_{y1}r_{12}}{1 - r_{12}^2}\right)$$

$$b_2 = \frac{1.4}{9173.88}\left(\frac{(.62) - (.62)(-.49)}{1 - (-.49)^2}\right)$$

$$b_2 = (.00015)\left(\frac{(.62) - (.30)}{1 - .24}\right)$$

$$b_2 = (.00015)\left(\frac{.32}{.76}\right)$$

$$b_2 = (.00015)(.42)$$

$$b_2 = -.000063$$

The Y intercept would be

$$a = \overline{Y} - b_1\overline{X}_1 - b_2\overline{X}_2$$
$$a = 3.4 - (-.13)(7.8) - (.000063)(28,800)$$
$$a = 3.4 - (-1.01) - (1.81)$$
$$a = 3.4 + 1.01 - 1.81$$
$$a = 2.60$$

The multiple regression equation is

$$Y = a + b_1X_1 + b_2X_2$$
$$Y = 2.60 + (-.13)X_1 + (.000063)X_2$$

What adjustment score could we predict for a male who had been married 30 years $(X_1 = 30)$ and had an income of $50,000 $(X_2 = 50,000)$?

$$Y' = 2.60 + (-.13)(30) + (.000063)(50,000)$$
$$Y' = 2.60 + (-3.9) + (3.15)$$
$$Y' = 1.85$$

To assess which of the two independents has the stronger effect on adjustment, the standardized

(continued)

partial slopes must be computed. For years married (X_1):

$$b_1^* = b_1\left(\frac{s_1}{s_y}\right)$$

$$b_1^* = (-.13)\left(\frac{4.5}{1.4}\right)$$

$$b_1^* = -.42$$

For income (X_2):

$$b_2^* = b_2\left(\frac{s_2}{s_y}\right)$$

$$b_2^* = (.000063)\left(\frac{9173.88}{1.4}\right)$$

$$b_2^* = .41$$

The standardized regression equation is

$$Z_y = b_1^*Z_1 + b_2^*Z_2 = (-0.42)Z_1 + (0.41)Z_2$$

and the independents have nearly equal but opposite effects on adjustment. To assess the combined effects of the two independents on adjustment, the coefficient of multiple determination must be computed.

$$R^2 = r_{y1}^2 + r_{y2.1}^2(1 - r_{y1}^2)$$
$$R^2 = (-.62)^2 + (.46)^2(1 - (-.62)^2)$$
$$R^2 = .38 + (.21)(1 - .38)$$
$$R^2 = .38 + (.21)(.62)$$
$$R^2 = .38 + .13$$
$$R^2 = .51$$

The first independent, years married, explains 38% of the variation in adjustment by itself. To this quantity, income explains an additional 13% of the variation in adjustment. Taken together, the two independents explain a total of 51% of the variation in adjustment.

$$Y' = a + b_1X_1 + b_2X_2$$
$$Y' = .01 + (4.7)(10) + (.08)(15)$$
$$Y' = 0.1 + 4.7 + 1.2$$
$$Y' = 5.91$$

Model 2 also presents the standardized regression coefficients or beta-weights (b^*). Recall that these coefficients make it easier to compare the effects of the independent variables by standardizing all scores to Z scores and thus eliminating differences in scale (number of members vs. revenue in billions of dollars). The beta-weights show that the effects of the two independent variables are nearly identical with board size having a slightly greater impact on total interlocks than revenues. However, revenue still has a substantial effect on total interlocks even after controlling for the effects of board size.

Finally, it is important to note the magnitude of R^2. In Model 2, board size and revenues account for one-fifth of the variation in the total number of interlocks in the sample. More interesting is the change in R^2 between Model 1 and Model 2. In Model 1, board size accounts for 13% of the variation and in Model 2, both independent variables together account for 20% of the variation. By entering revenues last in the equation, we can observe its effect on the dependent variable even after the effects of the control variable have been taken into consideration. In other words, the difference between R^2 in Model 1 and 2 demonstrates that even after the amount of variation explained by board size is accounted for, we gain an *additional* 7% of explanatory power by adding revenues.

To conclude, let's restate some of the key multivariate findings. The partial correlation coefficient between total interlocks and revenues, controlling for the effects of board size, remains fairly strong. The multiple regression analysis shows that both revenues and board size have important relationships with total number of corporate interlocks. Finally, when considering the change in R^2 from Model 1 to Model 2, we see an improvement in explanatory power of 7%. Thus, even taking the effects of board size into account, size (measured by revenues) *really does* matter in understanding the extent to which corporations interlock with each other through shared board members.

17.7 THE LIMITATIONS OF MULTIPLE REGRESSION AND CORRELATION

Multiple regression and correlation are very powerful tools for analyzing the interrelationships among three or more variables. The techniques presented in this chapter permit the researcher to predict scores on one variable from two or more other variables, to distinguish between independent variables in terms of the importance of their direct effects on a dependent, and to ascertain the total effect of a set of independent variables on a dependent variable. In terms of the flexibility of the techniques and the volume of information they can supply, multiple regression and correlation represent some of the most powerful statistical techniques available to social science researchers.

Powerful tools are not cheap. They demand high-quality data, and measurement at the interval-ratio level is difficult to accomplish at this stage in the development of the social sciences. Furthermore, these techniques assume that the interrelationships among the variables follow a particular form. First, they assume that each independent variable has a linear relationship with the dependent variable. How well a given set of variables meets this assumption can be quickly checked with scattergrams.

Second, the techniques presented in this chapter assume that there is no interaction among the variables in the equation. If there is interaction among the variables, it will not be possible to accurately estimate or predict the dependent variable by simply adding the effects of the independents. There are techniques for handling interaction among the variables in the set, but these techniques are beyond the scope of this text.

Third, the techniques of multiple regression and correlation assume that the independent variables are uncorrelated with each other. Strictly speaking, this condition means that the zero-order correlation among all pairs of independents should be zero; but, practically, we act as if this assumption has been met if the intercorrelations among the independents are low.

To the extent that these assumptions are violated, the regression coefficients (especially partial and standardized slopes) and the coefficient of multiple determination (R^2) become less and less trustworthy and the techniques less and less useful. If the assumptions of the model cannot be met, the alternative might be to turn to the multivariate techniques described in the

READING STATISTICS 13: The Causes of Juvenile Delinquency

Professor Siu Kwong Wong recently conducted a study of delinquency among Chinese-Canadian youth living in Winnipeg, Manitoba. Along with other theories of delinquency, Professor Wong examined assimilation as a possible cause of delinquency for this group. Chinese culture is often said to emphasize conformity and strict adherence to authority and convention. Does delinquency increase as people assimilate from Chinese to Canadian culture? Do more acculturated Chinese-Canadian young people have higher rates of delinquency? Table A seems to suggest that there is a relationship between these variables.

The measure of acculturation (the Behavior Acculturation Scale) had a positive and significant effect on delinquency, suggesting that ". . . the more the respondent was acculturated into Canadian society, the more likely they (sic) had committed delinquency." This finding is quite consistent with the idea that Chinese culture inhibits deviance, but note that, when the dependent variable is divided into minor and serious offenses, the effect of acculturation is much stronger for the former. Also, the effect of having Chinese friends is quite weak and suggests that integration into the Chinese-Canadian community (as opposed to adherence to Chinese culture) is not a particular insulation against delinquency. Incidentally, after examining other results, Professor Wong concludes that the most significant cause of delinquency for this group is not acculturation

TABLE A REGRESSION OF DELINQUENCY ON ACCULTURATION

Independent	Beta-weight	Serious Offenses	Minor Offenses
Sex[1]	$-.25^{**}$	$-.11$	$-.28^{**}$
Age	$.25^{**}$	$.03$	$.30^{**}$
Behavior Acculturation Scale	$.21^{*}$	$.10$	$.22^{*}$
Association with Chinese friends	$-.01$	$-.08$	$.02$
N	285	285	285
R^2	.17	.04	.22

NOTE: All regression coefficients are standardized.
$^{*}p < .01$ $^{**}p < .001$
[1]It may surprise you to see a nominal-level variable like sex in a regression equation. Actually, it is quite common to include nominal-level variables—called "dummy" variables—as long as they have only two categories. In this case, sex was coded so that males = 1 and females = 2. The negative sign of the beta-weight thus indicates that higher rates of delinquency are associated with male respondents.

per se but conflict with less acculturated immigrant parents.

Siu Kwong Wong. "Delinquency of Chinese-Canadian Youth: A Test of Opportunity, Control and Intergenerational Conflict Theories." *Youth and Society*. (29) pp. 112–113, copyright © 1997 by Sage Publications. Reprinted by permission of Sage Publications, Inc.

previous chapter. Unfortunately, those techniques, in general, supply a lower volume of less precise information about the interrelationships among the variables.

Finally, we should note that we have covered only the simplest applications of partial correlation and multiple regression and correlation. In terms of logic and interpretation, the extensions to situations involving more than one control variable or more than two independent variables are relatively straightforward. However, the computations for these situations are extremely complex. If you are faced with a situation involving more than three variables, turn to one of the computerized statistical packages that are commonly available on college campuses (e.g., SPSS and Microcase). These programs require minimal computer literacy and can handle complex calculations in, literally, the blink of an eye. Efficient use of these packages will enable you to avoid drudgery and will free you to do what social scientists everywhere enjoy doing most: pondering the meaning of your results and, by extension, the nature of social life.

SUMMARY

1. Partial correlation involves controlling for third variables in a manner analogous to that introduced in the previous chapter. Partial correlations permit the detection of direct and spurious or intervening relationships between X and Y.

2. Multiple regression includes statistical techniques by which predictions of the dependent variable from more than one independent variable can be made (by partial slopes and the multiple regression equation) and by which we can disentangle the relative importance of the independent variables (by standardized partial slopes).

3. The multiple correlation coefficient (R^2) summarizes the combined effects of all independents on the dependent variable in terms of the proportion of the total variation in Y that is explained by all of the independents.

4. Partial correlation and multiple regression and correlation are some of the most powerful tools available to the researcher and demand high-quality measurement and relationships among the variables that are linear and noninteractive. Further, correlations among the independents must be low (preferably zero). Although the price is high, these techniques pay considerable dividends in the volume of precise and detailed information they generate about the interrelationships among the variables.

SUMMARY OF FORMULAS

Partial correlation coefficient:

17.1 $$r_{yx.z} = \frac{r_{yx} - (r_{yz})(r_{xz})}{\sqrt{1 - r_{yz}^2}\,\sqrt{1 - r_{xz}^2}}$$

Least-squares regression line (bivariate):

17.2 $$Y = a + bX$$

Least-squares multiple regression line:

17.3 $$Y = a + b_1 X_1 + b_2 X_2$$

Partial slope for X_1:

17.4 $$b_1 = \left(\frac{s_y}{s_1}\right)\left(\frac{r_{y1} - r_{y2}r_{12}}{1 - r_{12}^2}\right)$$

Partial slope for X_2:

17.5 $$b_2 = \left(\frac{s_y}{s_2}\right)\left(\frac{r_{y2} - r_{y1}r_{12}}{1 - r_{12}^2}\right)$$

Y intercept:

17.6 $$a = \overline{Y} - b_1\overline{X_1} - b_2\overline{X_2}$$

Standardized partial slope (beta-weight) for X_1:

17.7 $$b_1^* = b_1\left(\frac{s_1}{s_y}\right)$$

Standardized partial slope (beta-weight) for X_2:

17.8 $$b_2^* = b_2\left(\frac{s_2}{s_y}\right)$$

Standardized least-squares regression line:

17.9 $$Z_y = a_z + b_1^* Z_1 + b_2^* Z_2$$

Standardized least-squares regression line (simplified):

17.10 $$Z_y = b_1^* Z_1 + b_2^* Z_2$$

Coefficient of multiple determination:

17.11 $$R^2 = r_{y1}^2 + r_{y2.1}^2(1 - r_{y1}^2)$$

GLOSSARY

Beta-weights (b^*). Standardized partial slopes.

Coefficient of multiple determination (R^2). A statistic that equals the total variation explained in the dependent variable by all independent variables combined.

Multiple correlation. A multivariate technique for examining the combined effects of more than one independent variable on a dependent variable.

Multiple correlation coefficient (R). A statistic that indicates the strength of the correlation between a dependent variable and two or more independent variables.

Multiple regression. A multivariate technique that breaks down the separate effects of the independent variables on the dependent variable; used to make predictions of the dependent variable.

Partial correlation. A multivariate technique for examining a bivariate relationship while controlling for other variables.

Partial correlation coefficient. A statistic that shows the relationship between two variables while controlling for other variables; $r_{yx.z}$ is the symbol for the partial correlation coefficient when controlling for one variable.

Partial slopes. In a multiple regression equation, the slope of the relationship between a particular independent variable and the dependent variable while controlling for all other independents in the equation.

Standardized partial slopes (beta-weights). The slope of the relationship between a particular independent variable and the dependent when all scores have been normalized.

Zero-order correlations. Correlation coefficients for bivariate realtionships.

MULTIMEDIA RESOURCES

The Wadsworth Sociology Resource Center: Virtual Society
http://sociology.wadsworth.com/

Visit the companion web site for the sixth edition of *Statistics: A Tool for Social Research* to access a wide range of student resources. Begin by clicking on the Student Resources section of the book's web site to access the following study tools:

- Basic math review
- Statistics review

- Flash cards
- Internet links
- Additional chapter problems
- Table of random numbers
- MicroCase and SPSS examples and exercises
- "Find the text" flowcharts
- Hypothesis testing for variables measured at the ordinal level

PROBLEMS

17.1 PS In problem 15.1 data regarding voter turnout in five cities was presented. For the sake of convenience, the data for three of the variables are presented again here along with descriptive statistics and zero-order correlations.

City	Turnout	Rate of Unemployment	% of Negative Advertisements
A	55	5	60
B	60	8	63
C	65	9	55
D	68	9	53
E	70	10	48

Univariate statistics:

	Turnout	Unemployment	Negative Ads
Mean	63.6	8.2	55.8
Standard deviation	5.5	1.7	5.3

Bivariate correlations:

	Unemployment	Negative Ads
Turnout	.95	−.87
Unemployment		−.70

a. Compute the partial correlation coefficient for the relationship between turnout (Y) and unemployment (X) while controlling for the effect of negative advertising (Z). What effect does this control variable have on the bivariate relationship? Is the relationship between turnout and unemployment direct? *(HINT: use Formula 17.1 and see Section 17.2.)*

b. Compute the partial correlation coefficient for the relationship between turnout (Y) and negative advertising (X) while controlling for the effect of unemployment (Z). What effect does this have on the bivariate relationship? Is the relationship between turnout and negative advertising direct? *(HINT: use Formula 17.1 and see Section 17.2. You will need this partial correlation to compute the multiple correlation coefficient.)*

c. Find the unstandardized multiple regression equation with unemployment (X_1) and negative ads (X_2) as the independent variables. What turnout would be expected in a city in which the unemployment rate was 10% and 75% of the campaign ads were negative? *(HINT: use Formulas 17.4 and 17.5 to compute the partial slopes and then use Formula 17.6 to find a, the Y intercept. The regression line is stated in Formula 17.3. Substitute 10 for X_1 and 75 for X_2 to compute predicted Y.)*

d. Compute beta-weights for each independent variable. Which has the stronger impact on turnout? *(HINT: use Formulas 17.7 and 17.8 to calculate the beta-weights.)*

e. Compute the multiple correlation coefficient (R) and the coefficient of multiple determina-

tion (R^2). How much of the variance in voter turnout is explained by the two independent variables? *(HINT: use Formula 17.11. You calculated $r^2_{y2.1}$ in part b of this problem.)*

f. Write a paragraph summarizing your conclusions about the relationships among these three variables.

17.2 SOC A scale measuring support for increases in the national defense budget has been administered to a sample. The respondents have also been asked to indicate how many years of school they have completed and how many years, if any, they served in the military. Take "support" as the dependent variable.

Case	Support	Years of School	Years of Service
A	20	12	2
B	15	12	4
C	20	16	20
D	10	10	10
E	10	16	20
F	5	8	0
G	8	14	2
H	20	12	20
I	10	10	4
J	20	16	0

a. Compute the partial correlation coefficient for the relationship between support (Y) and years of school (X) while controlling for the effect of years of service (Z). What effect does this have on the bivariate relationship? Is the relationship between support and years of school direct?

b. Compute the partial correlation coefficient for the relationship between support (Y) and years of service (X) while controlling for the effect of years of school (Z). What effect does this have on the bivariate relationship? Is the relationship between support and years of service direct? *(HINT: you will need this partial correlation to compute the multiple correlation coefficient.)*

c. Find the unstandardized multiple regression equation with school (X_1) and service (X_2) as the independent variables. What level of support would be expected in a person with 13 years of school and 15 years of service?

d. Compute beta-weights for each independent variable. Which has the stronger impact on turnout?

e. Compute the multiple correlation coefficient (R) and the coefficient of multiple determination (R^2). How much of the variance in support is explained by the two independent variables? (*HINT: you calculated $r^2_{y2.1}$ in part b of this problem.*)

f. Write a paragraph summarizing your conclusions about the relationships among these three variables.

17.3 SOC Data on civil strife (number of incidents), unemployment, and urbanization have been gathered for 10 nations. Take civil strife as the dependent variable. Compute the zero-order correlations among all three variables.

Number of Incidents of Civil Strife	Unemployment Rate	Percentage of Population Living in Urban Areas
0	5.3	60
1	1.0	65
5	2.7	55
7	2.8	68
10	3.0	69
23	2.5	70
25	6.0	45
26	5.2	40
30	7.8	75
53	9.2	80

a. Compute the partial correlation coefficient for the relationship between strife (Y) and unemployment (X) while controlling for the effect of urbanization (Z). What effect does this have on the bivariate relationship? Is the relationship between strife and unemployment direct?

b. Compute the partial correlation coefficient for the relationship between strife (Y) and urbanization (X) while controlling for the effect of unemployment (Z). What effect does this have on the bivariate relationship? Is the relationship between strife and urbanization direct? (*HINT: you will need this partial correlation to compute the multiple correlation coefficient.*)

c. Find the unstandardized multiple regression equation with unemployment (X_1) and urbanization (X_2) as the independent variables. What level of strife would be expected in a nation in which the unemployment rate was 10% and 90% of the population lived in urban areas?

d. Compute beta-weights for each independent variable. Which has the stronger impact on turnout?

e. Compute the multiple correlation coefficient (R) and the coefficient of multiple determination (R^2). How much of the variance in strife is explained by the two independent variables?

f. Write a paragraph summarizing your conclusions about the relationships among these three variables.

17.4 SOC/CJ In problem 15.5, crime and population data were presented for each of 10 states. For each of the three crime variables:

a. Find the multiple regression equations (unstandardized) with growth and urbanization as the independent variables.

b. Make a prediction for each crime variable for a state with a 5% growth rate and a population that is 90% urbanized.

c. Compute beta-weights for each independent variable in each equation and compare their relative effect on each dependent.

d. Compute R and R^2 for each crime variable, using the population variables as independents.

e. Write a paragraph summarizing your findings.

17.5 PS Problem 15.4 presented some data on 10 precincts. Take voter turnout as the dependent variable.

a. Find the multiple regression equations (unstandardized).

b. What turnout would you expect for a precinct in which 0% of the voters were Democrats and 5% were minorities?

c. Compute beta-weights for each independent variable and compare their relative effect on turnout. Which was the more important factor?

d. Compute R and R^2.

e. Write a paragraph summarizing your findings.

17.6 SW Twelve families have been referred to a counselor, and she has rated each of them on a cohesiveness scale. Also, she has information on family income and number of children currently living at home. Take family cohesion as the dependent variable.

Family	Cohesion Score	Family Income	Number of Children
A	10	30,000	5
B	10	70,000	4
C	9	35,000	4
D	5	25,000	0
E	1	55,000	3
F	7	40,000	0
G	2	60,000	2
H	5	30,000	3
I	8	50,000	5
J	3	25,000	4
K	2	45,000	3
L	4	50,000	0

a. Find the multiple regression equations (unstandardized).

b. What level of cohesion would be expected in a family with an income of $20,000 and 6 children?

c. Compute beta-weights for each independent variable and compare their relative effect on cohesion. Which was the more important factor?

d. Compute R and R^2.

e. Write a paragraph summarizing your findings.

17.7 Problem 15.8 presented per capita expenditures on education for 15 states, along with rank on income per capita and the percentage of the population that has graduated from high school. Take educational expenditures as the dependent variable.

a. Compute beta-weights for each independent variable and compare their relative effect on expenditures. Which was the more important factor?

b. Compute R and R^2.

c. Write a paragraph summarizing your findings.

17.8 SOC Twenty individuals were randomly selected from the 1998 General Social Survey data set, and their scores on four variables are reported below. Take "Hours of TV Watching per Day" as the dependent variable and select two of the remaining variables as independents.

Hours of TV Watching per Day (TVHOURS)	Occupational Prestige (PRESTG80)	Number of Children (CHILDS)	Age (AGE)
4	50	2	43
3	36	3	58
3	36	1	34
4	50	2	42
2	45	2	27
3	50	5	60
4	50	0	28
7	40	3	55
1	57	2	46
3	33	2	65
1	46	3	56
3	31	1	29
1	19	2	41
0	52	0	50
2	48	1	62
4	36	1	24
3	48	0	25
1	62	1	87
5	50	0	45
1	27	3	62

a. Compute beta-weights for each of the independent variables you selected and compare their relative effect on the hours of television watching. Which was the more important factor?

b. Compute R and R^2.

c. Write a paragraph summarizing your findings.

Using *SPSS for Windows* for Regression Analysis

SPSS DEMONSTRATION 17.1 What Are the Correlates of Occupational Prestige? Another Look

In Demonstration 15.1, we used the **Correlate** procedure to calculate zero-order correlation coefficients between *prestg80, papres80,* and *educ*. In this section, we will use a significantly more complex and flexible procedure called **Regression** to analyze the effects of *papres80* and *educ* on *prestg80*. The **Regression** procedure permits the user to control many aspects of the regression formula, and it can produce a much greater volume of output than the **Correlate** procedure. Among other things, **Regression** displays the slope (*b*) and the *Y* intercept (*a*), so we can use this procedure to find least-squares regression lines. This demonstration represents a very sparing use of the power of this command and an extremely economical use of all the options available. I urge you to explore some of the variations and capabilities of this powerful data-analysis procedure.

With the 1998 GSS loaded, click **Analyze, Regression,** and **Linear,** and the **Linear Regression** window will appear. Move *prestg80* into the **Dependent** box and *educ* and *papres80* into the **Independent(s)** box. If you wish, you can click the **Statistics** button and then click **Descriptives** to get zero-order correlations, means, and standard deviations for the variables. Click **Continue** and **OK,** and the following output will appear (descriptive information about the variables and the zero-order correlations are omitted here to conserve space).

Model Summary

Model	*R*	*R* Square	Adjusted *R* Square	Std. Error of the Estimate
1	.518	.269	.267	11.78

ANOVA

Model		Sum of Squares	df	Mean Square	*F*	Sig.
1	Regression	53419.148	2	26709.574	192.464	.000
	Residual	145438.110	1048	138.777		
	Total	198857.258	1050			

Coefficients

Model		Unstandardized Coefficients *B*	Std. Error	Standardized Coefficients Beta	*t*	Sig.
1	(Constant)	9.436	1.906		4.952	.000
	FATHERS OCCUPATIONAL PRESTIGE SCORE (1980)	4.097E-02	.028	.040	1.445	.149
	HIGHEST YEAR OF SCHOOL COMPLETED	2.426	.133	.505	18.276	.000

The Model Summary block reports the multiple R (.518) and R square (.269). The ANOVA output block shows that significance of the relationship (Sig. = .000). So far, we know that the independent variables explain about 27% of the variance in *prestg80* and that this result is statistically significant. In the last output block, we see the slopes (B) of the independent variables on *prestg80,* the standardized partial slopes (Beta), and the Y intercept (reported as a constant of 9.436). From this information, we can build a regression equation to predict scores on *prestg80.* The beta for *educ* (.505) is greater than the beta for *papres80* (.040), so education is the more important independent variable.

What does all this mean? At least for this sample, a person's occupational prestige is more affected by education than by the social class of his or her family of origin.

SPSS DEMONSTRATION 17.2 What Are the Correlates of Sexual Activity?

The 1998 GSS includes an item (*sexfreq*) that asks respondents about the frequency of their sexual activity over the past year. What causes variations in sexual activity? Some obvious correlates include gender and age, since it is commonly supposed that men are more sexually active than women and younger people more active than older. What are the effects of social class? Does the level of sexual activity—like so many other aspects of the social world—vary by a person's economic status?

Gender is a nominal-level variable, and its inclusion in a regression procedure may be surprising. Actually, nominal variables with only two categories—sometimes called "dummy variables"—are commonly used in regression (see also Reading Statistics 13). Click **Analyze, Regression,** and **Linear** and name *sexfreq* as the dependent variable and *age, income91,* and *sex* as the independent variables. Request descriptive statistics, if you wish, by clicking the **Statistics** button and making the appropriate selection. The output will look like this (descriptive information about the variables and the zero-order correlations are omitted to conserve space):

Model Summary

Model	R	R Square	Adjusted R Square	Std. Error of the Estimate
1	.447	.200	.197	1.72

ANOVA

Model		Sum of Squares	df	Mean Square	F	Sig.
1	Regression	757.205	3	252.402	84.913	.000
	Residual	3037.853	1022	2.972		
	Total	3795.058	1025			

Coefficients

Model		Unstandardized Coefficients B	Std. Error	Standardized Coefficients Beta	t	Sig.
1	(Constant)	4.617	.283		16.325	.000
	RESPONDENTS SEX	-.335	.108	-.087	-3.096	.002
	AGE OF RESPONDENT	-4.847E-02	.003	-.410	-14.654	.000
	TOTAL FAMILY INCOME	5.754E-02	.010	.154	5.499	.000

The R and R^2 indicate that the independent variables account for about 20% of the variation in frequency of sexual activity. The slopes (B) indicate that the dependent variable increases with income (.05754) and decreases with age (-.04847). The slope for gender is also negative (-.335). Since a male is coded as 1 and a female as 2, this indicates that the frequency of sex is greater for men. In other words, the frequency of sex increases as gender "decreases"—its rate is higher for people with the "lower" score on gender. The beta-weights (Beta) suggest that age has the greatest influence on frequency of sexual activity, followed by income and gender. The rate of sexual activity increases as affluence increases, decreases as age increases, and is more associated with males than females.

<hr>

MicroCase

Using MicroCase for Regression Analysis

MICROCASE DEMONSTRATION 17.1 What Are the Correlates of Occupational Prestige? Another Look

In Demonstration 15.1, we used the **Correlation** procedure to calculate zero-order correlation coefficients between *prestg80, papres80,* and *educ.* In this section, we will use a more complex procedure called **Regression** to analyze the effects of *papres80* and *educ* on *prestg80.* **Regression** produces a greater volume of output than the **Correlate** procedure. Among other things, **Regression** displays the slope (b) and the Y intercept (a), so we can use this procedure to find least-squares regression lines.

With the 1998 GSS loaded, click **Regression** and move *prestg80* into the **Dependent Variable** box and *educ* and *papres80* into the **Independent Variables** box. Click **OK,** and a graph displaying regression information will appear. The R^2 at the top of the screen indicates that these two independent variables together explain about 27% of the variance in *prestg80.* The independent variables are on the left with lines connecting them to the dependent variable. The value above the line is the beta weight (β), and the value below the line is the zero-order correlation with the *prestg80.* The beta for *educ* (.505) is greater than the beta for *papres80* (.040), so education is the more important independent variable and, at least for this sample, a person's occupational prestige is more affected by education than by the social class of his or her family of origin.

The **Regression** procedure can produce a great deal of additional information about the relationship between these variables. For example, if you click ANOVA in the Statistics box to the left of the screen, the following output will be produced:

```
Analysis of Variance
Dependent Variable: PRESTG80
N: 1051        Missing: 336
Multiple R-Square = 0.269        Y-Intercept = 9.436
Standard error of the estimate = 11.780
LISTWISE deletion (1-tailed test)        Significance Levels: ** = .01,  * = .05

Source      Sum of Squares    DF   Mean Square       F   Prob.
REGRESSION       53419.148     2    26709.574  192.464  0.000
RESIDUAL        145438.110  1048      138.777
TOTAL           198857.258  1050

            Unstand.b  Stand.Beta  Std.Err.b      t
PAPRES80       0.041       0.040      0.028    1.445
EDUC           2.426       0.505      0.133   18.276**
```

Some of this output repeats the information provided on the graph (e.g., R^2 and beta) but note that the values of a ("Y intercept") and b ("Unstand. b") are also reported. From this information, we can build a regression equation to predict scores on *prestg80*.

MICROCASE DEMONSTRATION 17.2
What Are the Correlates of Sexual Activity?

The 1998 GSS includes an item (*sexfreq*) that asks respondents about the frequency of their sexual activity over the past year. What causes variations in sexual activity? Some obvious correlates include gender and age, since it is commonly supposed that men are more sexually active than women and younger people more active than older. What are the effects of social class? Does the level of sexual activity—like so many other aspects of the social world—vary by a person's economic status?

Gender is a nominal-level variable, and its inclusion in a regression procedure may be surprising. Actually, nominal variables with only two categories—sometimes called "dummy variables"—are commonly used in regression (see also Reading Statistics 13). Click **Regression** and name *sexfreq* (variable #48) as the dependent variable and *age* (#6), *income91* (#12), and *sex* (#10) as the independent variables. The first output you will see is the graph with R^2, betas, and zero-order correlations. From this screen, we can see that the independent variables account for about 20% of the variation in frequency of sexual activity. The beta-weights show that *age* has the strongest effect on sexual activity ($\beta = -0.410$), followed by social class ($\beta = 0.154$) and gender ($\beta = -0.087$). The signs of the betas tell us that sexual activity decreases with age and increases with income. The beta for gender is negative. Male is coded as 1 and female as 2, so the negative beta indicates that the frequency of sex is greater for men. In other words, the frequency of sex increases as gender "decreases"—its rate is higher for people with the "lower" score on gender. Click ANOVA to get the coefficients (*a* and the slopes or *b*s) for the regression equation.

Exercises

17.1 Conduct the analysis in Demonstration 17.1 again with *income91* as the dependent variable. Compare your conclusions with those that you made in Demonstration 15.1. Write a paragraph summarizing the relationships.

17.2 Conduct the analysis in Demonstration 17.2 again with *partnrs5* (number of different sexual partners over the past five years) as the dependent variable. Compare and contrast with the analysis of *sexfreq*.

17.3 Conduct a regression analysis of *tvhours, childs,* or *attend*. Choose two or three potential independent variables and follow the instructions in Demonstrations 17.1 and 17.2. Ideally, your independent variables should be interval-ratio in level of measurement, but ordinal variables with a broad range of scores and nominal variables with only two scores will work as well.

PART IV CUMULATIVE EXERCISES

1. Two research questions that can be answered by one of the techniques presented in Chapters 16 and 17 are stated below. For each research situation, choose either the elaboration technique or regression analysis. The level of measurement of the variables should have a great deal of influence on your decision.

   a. For a sample of college graduates, did having a job during the school year interfere with academic success? For 20 students, data have been gathered on college GPA, the average number of hours the student worked each week, and College Board scores (a measure of preparedness for college-level work). Take GPA as the dependent variable.

   b. Only about half of this sample graduated within four years (1 = yes, 2 = no). Was their progress affected by their level of social activity (1 = low, 2 = high)? Is the relationship between these variables the same for both males (1) and females (2)?

GPA	Hours of Work Each Week	Average College Board Scores	Graduated in 4 Years	Social Activity	Sex
3.14	13	550	1	1	1
2.00	20	375	1	2	2
2.11	22	450	1	1	2
3.00	10	575	1	2	1
3.75	0	600	1	1	1
3.11	21	650	1	1	2
3.22	7	605	1	2	1
2.75	25	630	1	1	2
2.50	30	680	1	1	1
2.10	32	610	1	1	2
2.45	20	580	2	2	2
2.01	40	590	2	1	1
3.90	0	675	2	2	2
3.45	0	650	2	2	1
2.30	25	550	2	2	1
2.20	18	470	2	1	2
2.60	25	600	2	2	1
3.10	15	525	2	2	1
2.60	27	480	2	1	2
2.20	20	500	2	2	1

2. A research project has been conducted on the audiences of Christian religious television programs. Several research questions have been developed with regard to the following variables:

   How many hours per week do you watch religious programs on television? _____ (actual hours)

   What is your age? _____ (years)

How many years of formal schooling have you completed? _____ (years)

Have you ever donated money to a religious televison program?

_____ 1. Yes

_____ 2. No

What is your religious preference?

_____ 1. Protestant

_____ 2. Catholic

What is your sex?

_____ 1. Female

_____ 2. Male

a. The amount of time devoted to religious television will increase with age and decrease with education, but age will have the strongest effect.

b. Protestants will be more likely to donate money. Protestant females will be especially likely to donate money.

Does the data support these hypotheses?

Case	Hours	Age	Education	Ever Donate?	Denomi- nation	Sex
1	14	52	10	1	1	1
2	10	45	12	1	1	2
3	8	18	12	1	1	1
4	5	45	12	1	1	1
5	10	57	10	1	1	2
6	14	65	8	2	1	1
7	3	23	12	1	1	1
8	12	47	11	2	1	1
9	2	30	14	2	1	2
10	1	20	16	2	1	2
11	21	60	16	1	2	1
12	15	55	12	1	2	2
13	12	47	9	2	2	1
14	8	32	14	1	2	2
15	15	50	12	1	2	1
16	10	45	10	2	2	2
17	20	72	16	1	2	1
18	12	40	16	2	2	2
19	14	42	14	2	2	2
20	10	38	12	1	2	1

Appendix A   Area Under the Normal Curve

Column (a) lists Z scores from 0.00 to 4.00. Only positive scores are displayed, but, since the normal curve is symmetrical, the areas for negative scores will be exactly the same as areas for positive scores. Column (b) lists the proportion of the total area between the Z score and the mean. Figure A.1 displays areas of this type. Column (c) lists the proportion of the area beyond the Z score, and Figure A.2 displays this type of area.

FIGURE A.1   AREA BETWEEN MEAN AND Z

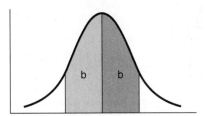

FIGURE A.2   AREA BEYOND Z

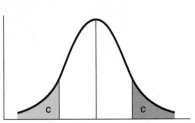

(a) Z	(b) Area Between Mean and Z	(c) Area Beyond Z	(a) Z	(b) Area Between Mean and Z	(c) Area Beyond Z
0.00	0.0000	0.5000	0.27	0.1064	0.3936
0.01	0.0040	0.4960	0.28	0.1103	0.3897
0.02	0.0080	0.4920	0.29	0.1141	0.3859
0.03	0.0120	0.4880	0.30	0.1179	0.3821
0.04	0.0160	0.4840			
0.05	0.0199	0.4801	0.31	0.1217	0.3783
0.06	0.0239	0.4761	0.32	0.1255	0.3745
0.07	0.0279	0.4721	0.33	0.1293	0.3707
0.08	0.0319	0.4681	0.34	0.1331	0.3669
0.09	0.0359	0.4641	0.35	0.1368	0.3632
0.10	0.0398	0.4602	0.36	0.1406	0.3594
			0.37	0.1443	0.3557
0.11	0.0438	0.4562	0.38	0.1480	0.3520
0.12	0.0478	0.4522	0.39	0.1517	0.3483
0.13	0.0517	0.4483	0.40	0.1554	0.3446
0.14	0.0557	0.4443			
0.15	0.0596	0.4404	0.41	0.1591	0.3409
0.16	0.0636	0.4364	0.42	0.1628	0.3372
0.17	0.0675	0.4325	0.43	0.1664	0.3336
0.18	0.0714	0.4286	0.44	0.1700	0.3300
0.19	0.0753	0.4247	0.45	0.1736	0.3264
0.20	0.0793	0.4207	0.46	0.1772	0.3228
			0.47	0.1808	0.3192
0.21	0.0832	0.4168	0.48	0.1844	0.3156
0.22	0.0871	0.4129	0.49	0.1879	0.3121
0.23	0.0910	0.4090	0.50	0.1915	0.3085
0.24	0.0948	0.4052			
0.25	0.0987	0.4013	0.51	0.1950	0.3050
0.26	0.1026	0.3974	0.52	0.1985	0.3015

(a)	(b) Area Between Mean and Z	(c) Area Beyond Z	(a)	(b) Area Between Mean and Z	(c) Area Beyond Z
Z			Z		
0.53	0.2019	0.2981	1.05	0.3531	0.1469
0.54	0.2054	0.2946	1.06	0.3554	0.1446
0.55	0.2088	0.2912	1.07	0.3577	0.1423
0.56	0.2123	0.2877	1.08	0.3599	0.1401
0.57	0.2157	0.2843	1.09	0.3621	0.1379
0.58	0.2190	0.2810	1.10	0.3643	0.1357
0.59	0.2224	0.2776	1.11	0.3665	0.1335
0.60	0.2257	0.2743	1.12	0.3686	0.1314
0.61	0.2291	0.2709	1.13	0.3708	0.1292
0.62	0.2324	0.2676	1.14	0.3729	0.1271
0.63	0.2357	0.2643	1.15	0.3749	0.1251
0.64	0.2389	0.2611	1.16	0.3770	0.1230
0.65	0.2422	0.2578	1.17	0.3790	0.1210
0.66	0.2454	0.2546	1.18	0.3810	0.1190
0.67	0.2486	0.2514	1.19	0.3830	0.1170
0.68	0.2517	0.2483	1.20	0.3849	0.1151
0.69	0.2549	0.2451	1.21	0.3869	0.1131
0.70	0.2580	0.2420	1.22	0.3888	0.1112
0.71	0.2611	0.2389	1.23	0.3907	0.1093
0.72	0.2642	0.2358	1.24	0.3925	0.1075
0.73	0.2673	0.2327	1.25	0.3944	0.1056
0.74	0.2703	0.2297	1.26	0.3962	0.1038
0.75	0.2734	0.2266	1.27	0.3980	0.1020
0.76	0.2764	0.2236	1.28	0.3997	0.1003
0.77	0.2794	0.2206	1.29	0.4015	0.0985
0.78	0.2823	0.2177	1.30	0.4032	0.0968
0.79	0.2852	0.2148	1.31	0.4049	0.0951
0.80	0.2881	0.2119	1.32	0.4066	0.0934
0.81	0.2910	0.2090	1.33	0.4082	0.0918
0.82	0.2939	0.2061	1.34	0.4099	0.0901
0.83	0.2967	0.2033	1.35	0.4115	0.0885
0.84	0.2995	0.2005	1.36	0.4131	0.0869
0.85	0.3023	0.1977	1.37	0.4147	0.0853
0.86	0.3051	0.1949	1.38	0.4162	0.0838
0.87	0.3078	0.1922	1.39	0.4177	0.0823
0.88	0.3106	0.1894	1.40	0.4192	0.0808
0.89	0.3133	0.1867	1.41	0.4207	0.0793
0.90	0.3159	0.1841	1.42	0.4222	0.0778
0.91	0.3186	0.1814	1.43	0.4236	0.0764
0.92	0.3212	0.1788	1.44	0.4251	0.0749
0.93	0.3238	0.1762	1.45	0.4265	0.0735
0.94	0.3264	0.1736	1.46	0.4279	0.0721
0.95	0.3289	0.1711	1.47	0.4292	0.0708
0.96	0.3315	0.1685	1.48	0.4306	0.0694
0.97	0.3340	0.1660	1.49	0.4319	0.0681
0.98	0.3365	0.1635	1.50	0.4332	0.0668
0.99	0.3389	0.1611	1.51	0.4345	0.0655
1.00	0.3413	0.1587	1.52	0.4357	0.0643
1.01	0.3438	0.1562	1.53	0.4370	0.0630
1.02	0.3461	0.1539	1.54	0.4382	0.0618
1.03	0.3485	0.1515	1.55	0.4394	0.0606
1.04	0.3508	0.1492	1.56	0.4406	0.0594

(a)	(b) Area Between Mean and Z	(c) Area Beyond Z	(a)	(b) Area Between Mean and Z	(c) Area Beyond Z
Z			Z		
1.57	0.4418	0.0582	2.09	0.4817	0.0183
1.58	0.4429	0.0571	2.10	0.4821	0.0179
1.59	0.4441	0.0559	2.11	0.4826	0.0174
1.60	0.4452	0.0548	2.12	0.4830	0.0170
1.61	0.4463	0.0537	2.13	0.4834	0.0166
1.62	0.4474	0.0526	2.14	0.4838	0.0162
1.63	0.4484	0.0516	2.15	0.4842	0.0158
1.64	0.4495	0.0505	2.16	0.4846	0.0154
1.65	0.4505	0.0495	2.17	0.4850	0.0150
1.66	0.4515	0.0485	2.18	0.4854	0.0146
1.67	0.4525	0.0475	2.19	0.4857	0.0143
1.68	0.4535	0.0465	2.20	0.4861	0.0139
1.69	0.4545	0.0455	2.21	0.4864	0.0136
1.70	0.4554	0.0446	2.22	0.4868	0.0132
1.71	0.4564	0.0436	2.23	0.4871	0.0129
1.72	0.4573	0.0427	2.24	0.4875	0.0125
1.73	0.4582	0.0418	2.25	0.4878	0.0122
1.74	0.4591	0.0409	2.26	0.4881	0.0119
1.75	0.4599	0.0401	2.27	0.4884	0.0116
1.76	0.4608	0.0392	2.28	0.4887	0.0113
1.77	0.4616	0.0384	2.29	0.4890	0.0110
1.78	0.4625	0.0375	2.30	0.4893	0.0107
1.79	0.4633	0.0367	2.31	0.4896	0.0104
1.80	0.4641	0.0359	2.32	0.4898	0.0102
1.81	0.4649	0.0351	2.33	0.4901	0.0099
1.82	0.4656	0.0344	2.34	0.4904	0.0096
1.83	0.4664	0.0336	2.35	0.4906	0.0094
1.84	0.4671	0.0329	2.36	0.4909	0.0091
1.85	0.4678	0.0322	2.37	0.4911	0.0089
1.86	0.4686	0.0314	2.38	0.4913	0.0087
1.87	0.4693	0.0307	2.39	0.4916	0.0084
1.88	0.4699	0.0301	2.40	0.4918	0.0082
1.89	0.4706	0.0294	2.41	0.4920	0.0080
1.90	0.4713	0.0287	2.42	0.4922	0.0078
1.91	0.4719	0.0281	2.43	0.4925	0.0075
1.92	0.4726	0.0274	2.44	0.4927	0.0073
1.93	0.4732	0.0268	2.45	0.4929	0.0071
1.94	0.4738	0.0262	2.46	0.4931	0.0069
1.95	0.4744	0.0256	2.47	0.4932	0.0068
1.96	0.4750	0.0250	2.48	0.4934	0.0066
1.97	0.4756	0.0244	2.49	0.4936	0.0064
1.98	0.4761	0.0239	2.50	0.4938	0.0062
1.99	0.4767	0.0233	2.51	0.4940	0.0060
2.00	0.4772	0.0228	2.52	0.4941	0.0059
2.01	0.4778	0.0222	2.53	0.4943	0.0057
2.02	0.4783	0.0217	2.54	0.4945	0.0055
2.03	0.4788	0.0212	2.55	0.4946	0.0054
2.04	0.4793	0.0207	2.56	0.4948	0.0052
2.05	0.4798	0.0202	2.57	0.4949	0.0051
2.06	0.4803	0.0197	2.58	0.4951	0.0049
2.07	0.4808	0.0192	2.59	0.4952	0.0048
2.08	0.4812	0.0188	2.60	0.4953	0.0047

(a) Z	(b) Area Between Mean and Z	(c) Area Beyond Z	(a) Z	(b) Area Between Mean and Z	(c) Area Beyond Z
2.61	0.4955	0.0045	3.11	0.4991	0.0009
2.62	0.4956	0.0044	3.12	0.4991	0.0009
2.63	0.4957	0.0043	3.13	0.4991	0.0009
2.64	0.4959	0.0041	3.14	0.4992	0.0008
2.65	0.4960	0.0040	3.15	0.4992	0.0008
2.66	0.4961	0.0039	3.16	0.4992	0.0008
2.67	0.4962	0.0038	3.17	0.4992	0.0008
2.68	0.4963	0.0037	3.18	0.4993	0.0007
2.69	0.4964	0.0036	3.19	0.4993	0.0007
2.70	0.4965	0.0035	3.20	0.4993	0.0007
2.71	0.4966	0.0034	3.21	0.4993	0.0007
2.72	0.4967	0.0033	3.22	0.4994	0.0006
2.73	0.4968	0.0032	3.23	0.4994	0.0006
2.74	0.4969	0.0031	3.24	0.4994	0.0006
2.75	0.4970	0.0030	3.25	0.4994	0.0006
2.76	0.4971	0.0029	3.26	0.4994	0.0006
2.77	0.4972	0.0028	3.27	0.4995	0.0005
2.78	0.4973	0.0027	3.28	0.4995	0.0005
2.79	0.4974	0.0026	3.29	0.4995	0.0005
2.80	0.4974	0.0026	3.30	0.4995	0.0005
2.81	0.4975	0.0025	3.31	0.4995	0.0005
2.82	0.4976	0.0024	3.32	0.4995	0.0005
2.83	0.4977	0.0023	3.33	0.4996	0.0004
2.84	0.4977	0.0023	3.34	0.4996	0.0004
2.85	0.4978	0.0022	3.35	0.4996	0.0004
2.86	0.4979	0.0021	3.36	0.4996	0.0004
2.87	0.4979	0.0021	3.37	0.4996	0.0004
2.88	0.4980	0.0020	3.38	0.4996	0.0004
2.89	0.4981	0.0019	3.39	0.4997	0.0003
2.90	0.4981	0.0019	3.40	0.4997	0.0003
2.91	0.4982	0.0018	3.41	0.4997	0.0003
2.92	0.4982	0.0018	3.42	0.4997	0.0003
2.93	0.4983	0.0017	3.43	0.4997	0.0003
2.94	0.4984	0.0016	3.44	0.4997	0.0003
2.95	0.4984	0.0016	3.45	0.4997	0.0003
2.96	0.4985	0.0015	3.46	0.4997	0.0003
2.97	0.4985	0.0015	3.47	0.4997	0.0003
2.98	0.4986	0.0014	3.48	0.4997	0.0003
2.99	0.4986	0.0014	3.49	0.4998	0.0002
3.00	0.4986	0.0014	3.50	0.4998	0.0002
3.01	0.4987	0.0013	3.60	0.4998	0.0002
3.02	0.4987	0.0013	3.70	0.4999	0.0001
3.03	0.4988	0.0012	3.70	0.4999	0.0001
3.04	0.4988	0.0012	3.80	0.4999	0.0001
3.05	0.4989	0.0011	3.90	0.4999	<0.0001
3.06	0.4989	0.0011	3.90	0.4999	<0.0001
3.07	0.4989	0.0011	4.00	0.4999	<0.0001
3.08	0.4990	0.0010			
3.09	0.4990	0.0010			
3.10	0.4990	0.0010			

Appendix B   Distribution of *t*

Degrees of Freedom (df)	Level of Significance for One-tailed Test					
	.10	.05	.025	.01	.005	.0005
	Level of Significance for Two-tailed Test					
	.20	.10	.05	.02	.01	.001
1	3.078	6.314	12.706	31.821	63.657	636.619
2	1.886	2.920	4.303	6.965	9.925	31.598
3	1.638	2.353	3.182	4.541	5.841	12.941
4	1.533	2.132	2.776	3.747	4.604	8.610
5	1.476	2.015	2.571	3.365	4.032	6.859
6	1.440	1.943	2.447	3.143	3.707	5.959
7	1.415	1.895	2.365	2.998	3.499	5.405
8	1.397	1.860	2.306	2.896	3.355	5.041
9	1.383	1.833	2.262	2.821	3.250	4.781
10	1.372	1.812	2.228	2.764	3.169	4.587
11	1.363	1.796	2.201	2.718	3.106	4.437
12	1.356	1.782	2.179	2.681	3.055	4.318
13	1.350	1.771	2.160	2.650	3.012	4.221
14	1.345	1.761	2.145	2.624	2.977	4.140
15	1.341	1.753	2.131	2.602	2.947	4.073
16	1.337	1.746	2.120	2.583	2.921	4.015
17	1.333	1.740	2.110	2.567	2.898	3.965
18	1.330	1.734	2.101	2.552	2.878	3.922
19	1.328	1.729	2.093	2.539	2.861	3.883
20	1.325	1.725	2.086	2.528	2.845	3.850
21	1.323	1.721	2.080	2.518	2.831	3.819
22	1.321	1.717	2.074	2.508	2.819	3.792
23	1.319	1.714	2.069	2.500	2.807	3.767
24	1.318	1.711	2.064	2.492	2.797	3.745
25	1.316	1.708	2.060	2.485	2.787	3.725
26	1.315	1.706	2.056	2.479	2.779	3.707
27	1.314	1.703	2.052	2.473	2.771	3.690
28	1.313	1.701	2.048	2.467	2.763	3.674
29	1.311	1.699	2.045	2.462	2.756	3.659
30	1.310	1.697	2.042	2.457	2.750	3.646
40	1.303	1.684	2.021	2.423	2.704	3.551
60	1.296	1.671	2.000	2.390	2.660	3.460
120	1.289	1.658	1.980	2.358	2.617	3.373
∞	1.282	1.645	1.960	2.326	2.576	3.291

Source: Table III of Fisher & Yates: *Statistical Tables for Biological, Agricultural and Medical Research*, published by Longman Group Ltd., London (1974), 6th edition (previously published by Oliver & Boyd Ltd., Edinburgh).

Appendix C   Distribution of Chi Square

df	.99	.98	.95	.90	.80	.70	.50	.30	.20	.10	.05	.02	.01	.001
1	.0³157	.0³628	.00393	.0158	.0642	.148	.455	1.074	1.642	2.706	3.841	5.412	6.635	10.827
2	.0201	.0404	.103	.211	.446	.713	1.386	2.408	3.219	4.605	5.991	7.824	9.210	13.815
3	.115	.185	.352	.584	1.005	1.424	2.366	3.665	4.642	6.251	7.815	9.837	11.341	16.268
4	.297	.429	.711	1.064	1.649	2.195	3.357	4.878	5.989	7.779	9.488	11.668	13.277	18.465
5	.554	.752	1.145	1.610	2.343	3.000	4.351	6.064	7.289	9.236	11.070	13.388	15.086	20.517
6	.872	1.134	1.635	2.204	3.070	3.828	5.348	7.231	8.558	10.645	12.592	15.033	16.812	22.457
7	1.239	1.564	2.167	2.833	3.822	4.671	6.346	8.383	9.803	12.017	14.067	16.622	18.475	24.322
8	1.646	2.032	2.733	3.490	4.594	5.527	7.344	9.524	11.030	13.362	15.507	18.168	20.090	26.125
9	2.088	2.532	3.325	4.168	5.380	6.393	8.343	10.656	12.242	14.684	16.919	19.679	21.666	27.877
10	2.558	3.059	3.940	4.865	6.179	7.267	9.342	11.781	13.442	15.987	18.307	21.161	23.209	29.588
11	3.053	3.609	4.575	5.578	6.989	8.148	10.341	12.899	14.631	17.275	19.675	22.618	24.725	31.264
12	3.571	4.178	5.226	6.304	7.807	9.034	11.340	14.011	15.812	18.549	21.026	24.054	26.217	32.909
13	4.107	4.765	5.892	7.042	8.634	9.926	12.340	15.119	16.985	19.812	22.362	25.472	27.688	34.528
14	4.660	5.368	6.571	7.790	9.467	10.821	13.339	16.222	18.151	21.064	23.685	26.873	29.141	36.123
15	5.229	5.985	7.261	8.547	10.307	11.721	14.339	17.322	19.311	22.307	24.996	28.259	30.578	37.697
16	5.812	6.614	7.962	9.312	11.152	12.624	15.338	18.418	20.465	23.542	26.296	29.633	32.000	39.252
17	6.408	7.255	8.672	10.085	12.002	13.531	16.338	19.511	21.615	24.769	27.587	30.995	33.409	40.790
18	7.015	7.906	9.390	10.865	12.857	14.440	17.338	20.601	22.760	25.989	28.869	32.346	34.805	42.312
19	7.633	8.567	10.117	11.651	13.716	15.352	18.338	21.689	23.900	27.204	30.144	33.687	36.191	43.820
20	8.260	9.237	10.851	12.443	14.578	16.266	19.337	22.775	25.038	28.412	31.410	35.020	37.566	45.315
21	8.897	9.915	11.591	13.240	15.445	17.182	20.337	23.858	26.171	29.615	32.671	36.343	38.932	46.797
22	9.542	10.600	12.338	14.041	16.314	18.101	21.337	24.939	27.301	30.813	33.924	37.659	40.289	48.268
23	10.196	11.293	13.091	14.848	17.187	19.021	22.337	26.018	28.429	32.007	35.172	38.968	41.638	49.728
24	10.856	11.992	13.848	15.659	18.062	19.943	23.337	27.096	29.553	33.196	36.415	40.270	42.980	51.179
25	11.524	12.697	14.611	16.473	18.940	20.867	24.337	28.172	30.675	34.382	37.652	41.566	44.314	52.620
26	12.198	13.409	15.379	17.292	19.820	21.792	25.336	29.246	31.795	35.563	38.885	42.856	45.642	54.052
27	12.879	14.125	16.151	18.114	20.703	22.719	26.336	30.319	32.912	36.741	40.113	44.140	46.963	55.476
28	13.565	14.847	16.928	18.939	21.588	23.647	27.336	31.391	34.027	37.916	41.337	45.419	48.278	56.893
29	14.256	15.574	17.708	19.768	22.475	24.577	28.336	32.461	35.139	39.087	42.557	46.693	49.588	58.302
30	14.953	16.306	18.493	20.599	23.364	25.508	29.336	33.530	36.250	40.256	43.773	47.962	50.892	59.703

Source: Table IV of Fisher & Yates: *Statistical Tables for Biological, Agricultural and Medical Research*, published by Longman Group Ltd., London (1974), 6th edition (previously published by Oliver & Boyd Ltd., Edinburgh). Reprinted by permission of Addison Wesley Longman Ltd.

Appendix D   Distribution of *F*

$$p = .05$$

n_1 n_2	1	2	3	4	5	6	8	12	24	∞
1	161.4	199.5	215.7	224.6	230.2	234.0	238.9	243.9	249.0	254.3
2	18.51	19.00	19.16	19.25	19.30	19.33	19.37	19.41	19.45	19.50
3	10.13	9.55	9.28	9.12	9.01	8.94	8.84	8.74	8.64	8.53
4	7.71	6.94	6.59	6.39	6.26	6.16	6.04	5.91	5.77	5.63
5	6.61	5.79	5.41	5.19	5.05	4.95	4.82	4.68	4.53	4.36
6	5.99	5.14	4.76	4.53	4.39	4.28	4.15	4.00	3.84	3.67
7	5.59	4.74	4.35	4.12	3.97	3.87	3.73	3.57	3.41	3.23
8	5.32	4.46	4.07	3.84	3.69	3.58	3.44	3.28	3.12	2.93
9	5.12	4.26	3.86	3.63	3.48	3.37	3.23	3.07	2.90	2.71
10	4.96	4.10	3.71	3.48	3.33	3.22	3.07	2.91	2.74	2.54
11	4.84	3.98	3.59	3.36	3.20	3.09	2.95	2.79	2.61	2.40
12	4.75	3.88	3.49	3.26	3.11	3.00	2.85	2.69	2.50	2.30
13	4.67	3.80	3.41	3.18	3.02	2.92	2.77	2.60	2.42	2.21
14	4.60	3.74	3.34	3.11	2.96	2.85	2.70	2.53	2.35	2.13
15	4.54	3.68	3.29	3.06	2.90	2.79	2.64	2.48	2.29	2.07
16	4.49	3.63	3.24	3.01	2.85	2.74	2.59	2.42	2.24	2.01
17	4.45	3.59	3.20	2.96	2.81	2.70	2.55	2.38	2.19	1.96
18	4.41	3.55	3.16	2.93	2.77	2.66	2.51	2.34	2.15	1.92
19	4.38	3.52	3.13	2.90	2.74	2.63	2.48	2.31	2.11	1.88
20	4.35	3.49	3.10	2.87	2.71	2.60	2.45	2.28	2.08	1.84
21	4.32	3.47	3.07	2.84	2.68	2.57	2.42	2.25	2.05	1.81
22	4.30	3.44	3.05	2.82	2.66	2.55	2.40	2.23	2.03	1.78
23	4.28	3.42	3.03	2.80	2.64	2.53	2.38	2.20	2.00	1.76
24	4.26	3.40	3.01	2.78	2.62	2.51	2.36	2.18	1.98	1.73
25	4.24	3.38	2.99	2.76	2.60	2.49	2.34	2.16	1.96	1.71
26	4.22	3.37	2.98	2.74	2.59	2.47	2.32	2.15	1.95	1.69
27	4.21	3.35	2.96	2.73	2.57	2.46	2.30	2.13	1.93	1.67
28	4.20	3.34	2.95	2.71	2.56	2.44	2.29	2.12	1.91	1.65
29	4.18	3.33	2.93	2.70	2.54	2.43	2.28	2.10	1.90	1.64
30	4.17	3.32	2.92	2.69	2.53	2.42	2.27	2.09	1.89	1.62
40	4.08	3.23	2.84	2.61	2.45	2.34	2.18	2.00	1.79	1.51
60	4.00	3.15	2.76	2.52	2.37	2.25	2.10	1.92	1.70	1.39
120	3.92	3.07	2.68	2.45	2.29	2.17	2.02	1.83	1.61	1.25
∞	3.84	2.99	2.60	2.37	2.21	2.09	1.94	1.75	1.52	1.00

Values of n_1 and n_2 represent the degrees of freedom associated with the between and within estimates of variance, respectively.

Source: Table V of Fisher and Yates: *Statistical Tables for Biological, Agricultural and Medical Research*, published by Longman Group Ltd., London (1974), 6th edition (previously published by Oliver and Boyd Ltd., Edinburgh). Reprinted by permission of Addison Wesley Longman Ltd.

$p = .01$

n_1 n_2	1	2	3	4	5	6	8	12	24	∞
1	4052	4999	5403	5625	5764	5859	5981	6106	6234	6366
2	98.49	99.01	99.17	99.25	99.30	99.33	99.36	99.42	99.46	99.50
3	34.12	30.81	29.46	28.71	28.24	27.91	27.49	27.05	26.60	26.12
4	21.20	18.00	16.69	15.98	15.52	15.21	14.80	14.37	13.93	13.46
5	16.26	13.27	12.06	11.39	10.97	10.67	10.27	9.89	9.47	9.02
6	13.74	10.92	9.78	9.15	8.75	8.47	8.10	7.72	7.31	6.88
7	12.25	9.55	8.45	7.85	7.46	7.19	6.84	6.47	6.07	5.65
8	11.26	8.65	7.59	7.01	6.63	6.37	6.03	5.67	5.28	4.86
9	10.56	8.02	6.99	6.42	6.06	5.80	5.47	5.11	4.73	4.31
10	10.04	7.56	6.55	5.99	5.64	5.39	5.06	4.71	4.33	3.91
11	9.65	7.20	6.22	5.67	5.32	5.07	4.74	4.40	4.02	3.60
12	9.33	6.93	5.95	5.41	5.06	4.82	4.50	4.16	3.78	3.36
13	9.07	6.70	5.74	5.20	4.86	4.62	4.30	3.96	3.59	3.16
14	8.86	6.51	5.56	5.03	4.69	4.46	4.14	3.80	3.43	3.00
15	8.68	6.36	5.42	4.89	4.56	4.32	4.00	3.67	3.29	2.87
16	8.53	6.23	5.29	4.77	4.44	4.20	3.89	3.55	3.18	2.75
17	8.40	6.11	5.18	4.67	4.34	4.10	3.79	3.45	3.08	2.65
18	8.28	6.01	5.09	4.58	4.25	4.01	3.71	3.37	3.00	2.57
19	8.18	5.93	5.01	4.50	4.17	3.94	3.63	3.30	2.92	2.49
20	8.10	5.85	4.94	4.43	4.10	3.87	3.56	3.23	2.86	2.42
21	8.02	5.78	4.87	4.37	4.04	3.81	3.51	3.17	2.80	2.36
22	7.94	5.72	4.82	4.31	3.99	3.76	3.45	3.12	2.75	2.31
23	7.88	5.66	4.76	4.26	3.94	3.71	3.41	3.07	2.70	2.26
24	7.82	5.61	4.72	4.22	3.90	3.67	3.36	3.03	2.66	2.21
25	7.77	5.57	4.68	4.18	3.86	3.63	3.32	2.99	2.62	2.17
26	7.72	5.53	4.64	4.14	3.82	3.59	3.29	2.96	2.58	2.13
27	7.68	5.49	4.60	4.11	3.78	3.56	3.26	2.93	2.55	2.10
28	7.64	5.45	4.57	4.07	3.75	3.53	3.23	2.90	2.52	2.06
29	7.60	5.42	4.54	4.04	3.73	3.50	3.20	2.87	2.49	2.03
30	7.56	5.39	4.51	4.02	3.70	3.47	3.17	2.84	2.47	2.01
40	7.31	5.18	4.31	3.83	3.51	3.29	2.99	2.66	2.29	1.80
60	7.08	4.98	4.13	3.65	3.34	3.12	2.82	2.50	2.12	1.60
120	6.85	4.79	3.95	3.48	3.17	2.96	2.66	2.34	1.95	1.38
∞	6.64	4.60	3.78	3.32	3.02	2.80	2.51	2.18	1.79	1.00

Values of n_1 and n_2 represent the degrees of freedom associated with the between and within estimates of variance, respectively.

Appendix E

Using Statistics: Ideas for Research Projects

This appendix presents outlines for four research projects, each of which requires the use of SPSS or MicroCase to analyze a data set (usually the 1998 General Social Survey). The research projects should be completed at various intervals during the course, and each project permits a great deal of choice on the part of the student. The first project stresses description and should be done after completing Chapters 2–4. The second involves estimation and should be completed in conjunction with Chapter 7. The third project uses inferential statistics and should be done after completing Part II, and the fourth combines inferential statistics with measures of association (with an option for multivariate analysis) and should be done after Part III (or IV). Most of the projects can be done with any of the databases available at the web site for this text.

PROJECT #1—DESCRIPTIVE STATISTICS

1. Choose one of the databases available at the web site for this text. Select five variables from the database *(NOTE: your instructor may specify a different number of variables)* and use the **Frequencies** (SPSS) or **Univariate** (MicroCase) command to get frequency distributions and summary statistics. (For SPSS, click the **Statistics** button and request the mean, median, mode, standard deviation, and range.) See Demonstrations 3.1, 3.2, and 4.1 for guidelines and examples. Make a note of all relevant information when it appears on screen or make a hard copy. See Appendix G for a list of variables available in the 1998 GSS and the code libraries at the web site for this text for lists of variables in the other data sets.

2. For each variable, get bar or line charts to summarize the overall shape of the distribution of the variable. See Demonstration 2.3 for guidelines and examples.

3. Inspect the frequency distributions and graphs and choose appropriate measures of central tendency and, for ordinal- and interval-ratio-level variables, dispersion. Also, for interval-ratio and ordinal variables with many scores, check for skew both by using the line chart and by comparing the mean and median (see Sections 3.6 and 3.7). Write a sentence or two of description for each variable, being careful to include a description of the overall shape of the distribution (see Chapter 2), the central tendency (Chapter 3), and the dispersion (Chapter 4). For nominal- and ordinal-level variables, be sure to explain any arbitrary numerical codes. For example, on the variable *class* in the 1998 GSS (see Appendix G), a 1 is coded as 'lower class,' a 2 indicates 'working class,' and so forth. This is an ordinal-level variable, so you might choose to report the

median as a measure of central tendency. If the median score on *class* were 2.45, for example, you might place that value in context by reporting that "the median is 2.45, about halfway between 'working class' and 'middle class'."

4. Below are examples of *minimal* summary sentences, using fictitious data:

For a nominal-level variable (e.g., gender), report the mode and some detail about the overall distribution. For example:

"Most respondents were married (57.5%), but divorced (17.4%) and single (21.3%) individuals were also common."

For an ordinal-level variable (e.g., occupational prestige), use the median (and, perhaps, the mode) and the range.

"The median prestige score was 44.3, and the range extended from 34 to 87. The most common score was 42."

For an interval-ratio level variable (e.g., population density for 124 nations), use the mean (and, perhaps, the median or mode) and the standard deviation (and, perhaps, the range).

"For these nations, population density (population per square mile) ranged from a low of 4.00 to a high of 12,000. The standard deviation was 1000. The nations averaged 351 people per square mile, and the median density was 193. The distribution is positively skewed, and some nations have very high population density."

PROJECT #2—ESTIMATION In this exercise, you will use the 1998 GSS sample to estimate the characteristics of the U.S. population. You will use SPSS or MicroCase to generate the sample statistics and then use either Formula 7.2 or 7.3 to find the confidence interval and state each interval in words.

A. Estimating Means

1. There are relatively few interval-ratio variables in the 1998 GSS and, for this part of the project, you may use ordinal variables that have *at least* four categories or scores. Choose a total of three variables that fit this description *other than* the variables you used in Exercise 7.1. *(NOTE: your instructor may specify a different number of variables).*

2. SPSS: Use the **Descriptives** command to get means, standard deviations, and sample size (*N*), and use this information to construct 95% confidence intervals for the first of your variables. Make a note of the mean, standard deviation, and sample size or keep a hard copy. Use Formula 7.2

to compute the confidence intervals. Repeat this procedure for the remaining variables.

MicroCase: Use the **Univariate** command and click **Summary** in the **Statistics** window on the left of the screen. The 99% and 95% confidence intervals are produced automatically and printed just above the frequency distribution. Make a note of the intervals or get a hard copy. Repeat this procedure for the remaining variables.

3. Download the report form for this project from the web site for this text. Complete the form for each variable and, in the space provided, write a sentence reporting the interval, the confidence level, and sample size. Write in plain English, as if you were reporting results in a newspaper. Most importantly, you should make it clear that you are estimating characteristics of the population of the entire United States. For example, a summary sentence might look like this:

"Based on a random sample of 1231, I estimate at the 95% level that U.S. drivers average between 64.46 and 68.22 miles per hour when driving on interstate highways."

B. Estimating Proportions

4. Choose three variables that are nominal or ordinal with *two or three* categories *other than* the variables you used in Exercise 7.2. *(NOTE: your instructor may specify a different number of variables.)*

5. **SPSS:** Use the **Frequencies** command to get the percentage of the sample in the various categories of each variable. Change the percentages (remember to use the 'valid percents' column) to proportions and construct confidence intervals for one category of each variable (e.g., the % female for *sex*) using Formula 7.3.

   MicroCase: Use the **Univariate** command to get the percentages of the sample in the various categories of each variable. Change the percentages to proportions and construct confidence intervals for one category of each variable (e.g. the % female for *sex*) using Formula 7.3.

6. Download the report form for this project from the web site. Complete the form for each variable and, in the space provided, write a sentence reporting the interval, the confidence level, and sample size. Write in plain English, as if you were reporting results in a newspaper.

7. *OPTIONAL:* For any one of the confidence intervals you constructed, identify each of the concepts listed at the end of the Report Form for Project 2 and explain their role in estimation.

PROJECT # 3—
SIGNIFICANCE TESTING

Use the 1998 General Social Survey data set for this project.

A. Two Sample T Test (Chapter 9)

1. Choose two different dependent variables from the interval-ratio or ordinal variables that have three or more scores. Choose independent variables that might logically be a cause of your dependent variables. Remember that, for a *t* test, independent variables can have *only* two categories. Independent variables can be any level of measurement, and you may use the same independent variable for both tests.

2. SPSS: Click **Analyze, Compare Means,** and then **Independent Samples T Test.** Name your dependent variable(s) in the **Test Variable** window and your independent variable in the **Grouping Variable** window. You will also need to specify the scores used to define the groups on the independent variable. See SPSS Demonstration 9.1 for examples. Make a note of the test results (group means, obtained *t* score, significance, sample size) or keep a hard copy. Repeat the procedure for the second dependent variable.

MicroCase: Click **t Test** on the Statistics Menu, and the **t Test** dialog box will open with the variables listed on the left. Find your first dependent variable and click the arrow to move it to the **Dependent Variable** box. Next, find your independent variable and move it to the **Independent Variable** box. Click **OK,** and you will see the "box and whiskers" plot. To conduct the *t* test, click on the button next to **Means** in the **Statistics** box at the left of the screen. See MicroCase Demonstration 9.1 for examples. Make a note of the test results (group means, obtained *t* score, significance, sample size) or keep a hard copy. Repeat the procedure for the second dependent variable.

3. Write up the results of the test using the Report Form available at the web site for this text.

B. Analysis of Variance (Chapter 10)

1. Choose two different dependent variables from the interval-ratio or ordinal variables that have three or more scores. Choose independent variables that might logically be a cause of your dependent variables and that have between three and five categories. You may use the same independent variables for both tests.

2. SPSS: Click **Analyze, Compare Means,** and then **One-way Anova.** The **One-way Anova** window will appear. Find your dependent variable in the variable list on the left and click the arrow to move the variable name into the **Dependent List** box. Note that you can request more than one dependent variable at a time. Next, find the name of your independent variable and move it to the **Factor** box. Click **Options** and then click the box next to **Descriptive** in the **Statistics** box to request means and standard deviations. Click **Continue** and **OK.** Make a note of the test results or keep a hard copy. Repeat, if necessary, for your second dependent variable.

MicroCase: Click **Anova** from the **Statistics** menu. The **Anova** window appears. Find your dependent variable in the variable list and move it to the **Dependent Variable** box. Next, find your independent variable and move it into the **Independent Variable** box. Click **OK,** and the first screen will present a "box and whiskers" graph (see MicroCase Demonstration 9.1). From the **Statistics** box on the left of the screen, click **Means** to see the results of the ANOVA test. Make a note of the test results or keep a hard copy. Repeat for your second dependent variable.

3. Write up the results of the test using the Report Form available at the web site for this text.

C. Chi Square (Chapter 11)

1. Choose two different dependent variables of any level of measurement that have five or fewer (preferably 2–3) scores. For each dependent variable, choose an independent variable that might logically be a cause. Independent variables can be any level of measurement as long as they have five or fewer (preferably 2–3) categories. Output will be easier to analyze if you use variables with few categories. You may use the same independent variable for both tests.

2. SPSS: Click **Analyze, Descriptive Statistics,** and then **Crosstabs.** The **Crosstabs** dialog box will appear. Highlight your first dependent variable and move it into the **Rows** box. Next, highlight your independent variable and move it into the **Columns** box. Click the **Statistics** button at the bottom of the window and click the box next to chi square. Click **Continue** and **OK.** Make a note or a hard copy of the results. Repeat for your second dependent variable.

MicroCase: Begin by clicking **Cross-tabulation** from the **Statistics** menu. The **Cross-tabulation** dialog box will appear. Highlight your first dependent variable and move it into the **Row Variable** box and then move your independent variable into the **Column Variable** box. Click **OK.** Make a note or a hard copy of the results and repeat these procedures for your second dependent variable.

3. Write up the results of the test using the Report Form available at the web site for this text.

PROJECT 4—ANALYZING THE STRENGTH AND SIGNIFICANCE OF RELATIONSHIPS

A. Using Bivariate Tables

1. From the 1998 GSS data set, select either

> **a.** One dependent variable and 3 independent variables (possible causes)
>
> or **b.** One independent variable and 3 possible dependent variables (possible effects).

Variables can be from any level of measurement but must have only a few (2–5) categories or "scores." Develop research questions or hypotheses about the relationships between variables. Make sure that the causal links you suggest are sensible and logical.

2. SPSS: Use the **Crosstabs** procedure to generate bivariate tables. See any of the Demonstrations at the end of Chapters 12, 13, or 14 for examples. Click **Analyze, Descriptive Statistics,** and **Crosstabs** and place your dependent variable(s) in the rows and independent variable(s) in the columns. On the **Crosstabs** dialog box, click the **Statistics** button and choose chi square, phi or V, and gamma for every table you request. On the **Crosstabs** dialog box, click the **Cells** button and get column percentages for every table you request. Make a note of results as they appear on the screen or get hard copies.

MicroCase: Use the **Cross-tabulation** procedure to generate bivariate tables. See any of the Demonstrations at the end of Chapters 12, 13, or 14 for examples. Place your dependent variable(s) in the rows and independent variable(s) in the columns. Click "column %" in the **Tables** box to get column percentages and click "Summary" in the **Statistics** box to get chi square, Cramer's V, and gamma for every table you request. Make a note of results as they appear on the screen or get hard copies.

3. See the instructions at the web site for this text for writing up your results in a short paper.

4. *OPTIONAL MULTIVARIATE ANALYSIS:* Pick one of the bivariate relationships you produced in step 2 and find a logical control variable for this relationship. Run **Crosstabs** (SPSS) or **Cross-tabulation** (MicroCase) for the bivariate relationship again while controlling for the third variable. Compare the partial tables with each other and with the bivariate table. Is the original bivariate relationship direct? Is there evidence of a spurious or intervening relationship? Do the variables have an interactive relationship? Write up the results of this analysis and include them in your summary paper for this project.

B. Using Interval-Ratio Variables

1. From any data set, select either

   a. One dependent variable and 3 independent variables (possible causes)

or **b.** One independent variable and 3 possible dependent variables (possible effects).

Variables should be interval-ratio in level of measurement, but you may use ordinal-level variables as long as they have more than (preferably many more than) 5 scores. Develop research questions or hypotheses about the relationships between variables. Make sure that the causal links you suggest are sensible and logical.

2. SPSS: Use the **Regression** and **Scatterplot** (click Graphs and then Scatter) procedures to analyze the bivariate relationships. Make a note of results (including r, r^2, slope, beta-weights, and a) as they appear on the screen or get hard copies.

MicroCase: Use the **Scatterplot** and **Regression** procedures to analyze the bivariate relationships. Make a note of results (including r, r^2, beta-weights, and a) as they appear on the screen or get hard copies.

3. See the instructions at the web site for this text for writing up your results in a short paper.

4. *OPTIONAL MULTIVARIATE ANALYSIS:* Pick one of the bivariate relationships you produced in step 2 and find another logical independent variable. Run **Regression** again with both independent variables and analyze the results. How much improvement is there in the explained variance after the second independent variable is included? Write up the results of this analysis and include them in your summary paper for this project.

Appendix F   An Introduction to Computerized Statistics Programs (*SPSS for Windows* and *MicroCase*)

Computers have affected virtually every aspect of human society and, as you would expect, their impact on the conduct of social research has been profound. Researchers routinely use computers to organize data and compute statistics—activities that humans often find dull, tedious, and difficult but which computers accomplish with accuracy and ease. This division of labor allows social scientists to spend more time on analysis and interpretation— activities that humans typically enjoy but which are beyond the power of computers (so far, at least).

These days, the skills needed to use computers successfully are quite accessible, even for people with little or no experience. This appendix will prepare you to use one of the two statistics programs that are featured in this text: **SPSS for Windows** or **MicroCase.** Make sure you know which of the two is being used in your course and read only the sections of this appendix relevant to that program. Likewise, end-of-chapter exercises are organized in parallel but separate sections for the two programs. In either case, if you have used a mouse to "point and click" and run a computer program, you are ready to learn to use these programs. Even if you are completely unfamiliar with computers, you will find the programs accessible. After you finish this appendix, you will be ready to do the exercises found at the end of most chapters of this text.

A word of caution before we begin: This appendix is intended only as an *introduction* to SPSS and MicroCase. It will give you an overview of the program and enough information so that you can complete the assignments in the text. It is unlikely, however, that this appendix will answer all your questions or provide solutions to all the problems you might encounter. So, this is a good place to tell you that both programs have extensive and easy-to-use "help" facilities that will provide assistance as you request it. You should familiarize yourself with these features and use them as needed. To access them in either program, simply click on the **Help** command on the toolbar across the top of the screen.

Both SPSS and MicroCase are **statistical packages** (or **statpaks**): sets of computer programs that work with data and compute statistics as requested by the user (you). Once you have entered the data for a particular group of observations, you can easily and quickly produce an abundance of statistical information without doing any computations yourself.

Why bother to learn this technology? The truth is that the laborsaving capacity of computers is sometimes exaggerated, and there are research situations in which they are unnecessary. If you are working with a small number of observations or need only a few, uncomplicated statistics, then statistical packages are probably not going to be helpful. However, as the number of cases increases and as your requirements for statistics become more sophisticated, computers and statpaks will become more and more useful.

An example should make this point clearer. Suppose you have gathered a sample of 150 respondents and the *only* thing you want to know about these people is their average age. To compute an average, as you know, you add the scores and divide by the total number of cases. How long do you think it would take you to add 150 two-digit numbers (ages) with a hand calculator? (Don't even think about doing it by hand.) If you entered the scores at the (pretty fast) rate of one per second, or sixty scores a minute, it would take about 3 or 4 minutes to enter the ages and get the average. Even if you worked slowly and carefully and did the addition a second and third time to check your math, you could probably complete all calculations in less than 20 minutes. If this were all the information you needed, computers and statpaks would not save you any time.

Such a simple research project is not very realistic, however. Typically, researchers deal with not one but scores or even hundreds of variables, and samples have hundreds or thousands of cases. While you could add 150 numbers in perhaps 3 or 4 minutes, how long would it take to add the scores for 1500 cases? What are the chances of processing 1500 numbers without making significant errors of arithmetic? The more complex the research situation, the more valuable and useful statpaks become. SPSS and MicroCase can produce statistical information in a few keystrokes or clicks of the mouse that might take you minutes, hours, or even days to produce with a hand calculator.

Clearly, this is technology worth mastering by any social researcher. With statpaks, you can avoid the drudgery of mere computation, spend more time on analysis and interpretation, and conduct research projects with very large data sets. Mastery of this technology might be very handy indeed in your senior-level courses, in a wide variety of jobs, or in graduate school.

F.1 GETTING STARTED— DATABASES AND COMPUTER FILES

Before statistics can be calculated, SPSS or MicroCase must first have some data to process. A **database** is an organized collection of related information such as the responses to a survey. For purposes of computer analysis, a database is organized into a **file:** a collection of information that is stored under the same name in the memory of the computer, on a diskette, or in some other medium. Words as well as numbers can be saved in files. If you've ever used a word processing program to type a letter or term paper, you probably saved your work in a file so that you could update or make corrections at a later time. Data can be stored in files indefinitely. Since it can take months to conduct a thorough data analysis, the ability to save a database is another advantage of using computers.

For the SPSS and MicroCase exercises in this text, we will use a database that contains some of the results of the General Social Survey (GSS) for 1998. This database contains the responses of a sample of adult Americans to questions about a wide variety of social issues. The GSS has been conducted almost every year since 1972 and has been the basis for hundreds of research projects by professional social researchers. It is a rich source of information about public opinion in the United States and includes data on everything from attitudes about abortion to opinions on assisted suicide.

The GSS is especially valuable because the respondents are chosen so that the sample as a whole is representative of the entire U.S. population. A representative sample reproduces, in miniature form, the characteristics of the population from which it was taken (see Chapters 6 and 7). So, when you analyze the 1998 General Social Survey database, you are in effect analyzing U.S. society as of 1998. The data is real, and the relationships you will analyze reflect some of the most important and sensitive issues in American life.

The complete General Social Survey for 1998 includes hundreds of items of information (age, sex, opinion about such social issues as capital punishment, and so forth) for almost 3000 respondents. Some of you will be using student versions of MicroCase or SPSS for Windows, which are limited in the number of cases and variables they can process. To accommodate these limits, I have reduced the database to about 50 items of information and fewer than 1500 respondents.

The GSS data file is summarized in Appendix G. Please turn to this appendix and familiarize yourself with it. Note that the variables are listed alphabetically by their variable names. In SPSS, the names of variables must be no more than eight characters long. (MicroCase allows longer names, but we will use the SPSS format). In many cases, the resultant need for brevity is not a problem, and variable names (e.g., *age*) are easy to figure out. In other cases, the eight-character limit necessitates extreme abbreviation, and some variable names (like *abany* and *abhlth*) are not so obvious. Appendix G also shows the wording of the item that generated the variable. For example, the *abany* variable consists of responses to a question about legal abortion: should it be possible for a woman to have an abortion for "any reason"? Note that the variable name is formed from the question: should an abortion be possible for *any* reason? The *abhlth* variable consists of responses to a different scenario: should legal abortion be possible if the woman's health is in danger? Appendix G is an example of a code book for the database since it lists all the codes (or scores) for the survey items along with their meanings.

Notice in Appendix G that some of the possible responses to *abany* and *abhlth* and other variables are labeled 'Not applicable,' NA, or DK. The first of these responses means that the item in question was not given to a respondent. The full GSS is very long and, to keep the time frame for complet-

FIGURE F.1 THE DATA ANALYSIS PROCESS

DataBase	$\rightarrow$	**Statpak**	$\rightarrow$	**Output**	$\rightarrow$	**Analysis**
(raw information)		(computer programs)		(statistics and graphs)		(interpretation)

ing the survey reasonable, not all respondents are asked every question. NA stands for 'No Answer' and means that the respondent was asked the question but refused to answer. DK stands for 'Don't Know,' which means that the respondent did not have the requested information. All three of these scores are *Missing Values* and, as "noninformation," they are eliminated from statistical analysis. Missing values are common on surveys and, as long as they are not too numerous, they are not a particular problem.

It's important that you understand the difference between a statpak (SPSS or MicroCase) and a database (GSS) and what we are ultimately after here. A database consists of information. A statpak organizes the information in the database and produces statistics. Our goal is to apply the statpak to the database to produce output (for example, statistics and graphs) that we can analyze and use to answer questions. The process might be diagrammed as in Figure F.1.

Statpaks like SPSS and MicroCase are general research tools that can be used to analyze databases of all sorts: they are not limited to the 1998 GSS. In the same way, the 1998 GSS could be analyzed with statpaks other than those used in this text. Other widely used statpaks include SAS and Stata— each of which may be available on your campus.

Sections F.2 (SPSS) through F.6 (SPSS) introduce the SPSS program. If your class is using MicroCase, skip to Section F.7 (MicroCase).

F.2 (SPSS) STARTING *SPSS FOR WINDOWS* AND LOADING THE 1998 GSS

If you are using the complete, professional version of *SPSS for Windows,* you will probably be working in a computer lab, and you can begin running the program immediately. If you are using the student version of the program on your personal computer, the first thing you need to do is install the software. Follow the instructions that came with the program and return to this appendix when installation is complete.

To start *SPSS for Windows,* find the icon (or picture) on the screen of your monitor that has an "SPSS" label attached to it. Use the computer mouse to move the arrow on the monitor screen over this icon and then double-click the left button on the mouse. This will start up the SPSS program.

After a few seconds the *SPSS for Windows* screen will appear and ask, at the top of the screen, "What would you like to do?" Of the choices listed, the button next to 'Open an existing file' will be checked (or preselected), and there will probably be a number of data sets listed in the window at the bottom of the screen. Find the 1998 General Social Survey data set, probably labeled *GSS98.sav* or something similar. If you have the data set on a diskette, you will need to specify the correct drive. Check with your instructor to make sure that you know where to find the 1998 GSS.

Once you've located the data set, click on the name of the file with the left-hand button on the mouse, and SPSS will load the data. The next screen you will see is the SPSS Data Editor screen.

Note that there is a list of commands across the very top of the screen. These commands begin with **File** at the far left and end with **Help** at the far right. This is the main menu bar for SPSS. When you click any of these words, a **menu** of commands and choices will drop down. Basically, you tell SPSS what to do by clicking on your desired choices from these menus. Sometimes, submenus will appear, and you will need to specify your choices further.

SPSS provides the user with a variety of options for displaying information about the data file and output on the screen. I recommend that you tell the program to display lists of variables by name (e.g., *age, abany*) rather than labels (e.g, AGE OF RESPONDENT, ABORTION IF WOMAN WANTS FOR ANY REASON). Lists displayed this way will be easier to read and compare to Appendix G. To do this, click **Edit** on the main menu bar and then click **Options** from the drop-down submenu. A dialog box labeled "Options" will appear with a series of "tabs" along the top. The "General" options should be displayed but, if not, click on this tab. On the "General" screen, find the box labeled "Variable Lists" on the upper-right-hand side. Click "Display names" and "alphabetical" and then click **OK** if they are not already selected. If you make changes, a message will tell you that changes will take effect the next time a data file is opened.

In this section, you learned how to start up *SPSS for Windows,* load a data file, and set some of the display options for this program. These procedures are summarized in Table F.1.

TABLE F.1   SUMMARY OF COMMANDS (SPSS)

To start SPSS for Windows	Click the SPSS icon on the screen of the computer monitor
To open a data file	Double click on the data file name
To set display options for lists of variables	Click **Edit** from the main menu bar, then click **Options.** On the "General" tab make sure that "Display names" and "alphabetical" are selected and then click **OK.** You must restart SPSS before the change takes effect

F.3 (SPSS) WORKING WITH DATABASES

Note that in the SPSS Data Editor Window the data are organized into a two-dimensional grid with columns running up and down (vertically) and rows running across (horizontally). Each column is a variable or item of information from the survey. The names of the variables are listed at the tops of the columns. Remember that you can find the meaning of these variable names in the GSS 1998 code book in Appendix G.

Another way to decipher the meaning of variable names is to click **Utilities** on the menu bar and then click **Variables.** The Variables window opens. This window has two parts. On the left is a list of all variables in the database arranged in alphabetical order with the first variable highlighted. On the right is the Variable Information window with information about the highlighted variable. The first variable is listed as *abany*. The Variable Information window displays a fragment of the question that was actually asked during the survey ("ABORTION IF WOMAN WANTS FOR ANY REASON") and shows the possible scores on this variable (a score of 1 = yes and a score of 2 = no) along with some other information.

The same information can be displayed for any variable in the data set. For example, find the variable *wrkstat* in the list. You can do this by using the arrow keys on your keyboard or the slider bar on the right of the variable list window. You can also move through the list by typing the first letter of the variable name you are interested in. For example, type "w" and you will be moved to the first variable name in the list that begins with that letter. Now you can see that the variable measures labor force status and that a score of "1" indicates that the respondent was working full time, and so forth. What do *prestg80* and *marital* measure? Close this window by clicking the **Close** button at the bottom of the window.

Examine the window displaying the 1998 GSS a little more. Each row of the window (reading across or from left to right) contains the scores of a particular respondent on all the variables in the database. Note that the upper-left-hand cell is highlighted (outlined in a darker border than the other cells). This cell contains the score of respondent #1 on the first variable. The second row contains the scores of respondent #2, and so forth. You can move around in this window with the arrow keys on your keyboard. The highlight moves in the direction of the arrow, one cell at a time.

In this section, you learned to read information in the data display window and to decipher the meaning of variable names and scores. These commands are summarized in Table F.2, and we are now prepared to actually perform some statistical operations with the 1998 GSS database.

F.4 (SPSS) PUTTING SPSS TO WORK: PRODUCING STATISTICS

At this point, the database on the screen is just a mass of numbers with little meaning for you. That's okay because you will not have to actually read any information from this screen. Virtually all of the statistical operations you will conduct will begin by clicking the **Analyze** (or **Statistics** for version 8.0

TABLE F.2   SUMMARY OF COMMANDS (SPSS)

To move around in the Data window	1. Click the cell you want to highlight or 2. Use the arrow keys on your keyboard or 3. Move the slider buttons or 4. Click the arrows on the right-hand and bottom margins
To get information about a variable	1. From the menu bar, click Utilities and then click Variables 2. Scroll through the list of variable names until you highlight the name of the variable in which you are interested. Variable information will appear in the window on the right Or 3. See Appendix G

or earlier) command from the menu bar, selecting a procedure and statistics, and then naming the variable or variables you would like to process.

To illustrate, let's have *SPSS for Windows* produce a frequency distribution for the variable *sex*. Frequency distributions are tables that display the number of times each score of a variable occurred in the sample (see Chapter 2). So, when we complete this procedure, we will know the number of males and females in the 1998 GSS sample.

With the 1998 GSS loaded, begin by clicking the **Analyze** command on the menu bar. From the menu that drops down, click **Descriptive Statistics** and then **Frequencies.** The Frequencies window appears with the variables listed in alphabetical order in the box on the left. The first variable (*abany*) will be highlighted. Use the slider button or the arrow keys on the right-hand margin of this box to scroll through the variable list until you highlight the variable *sex*, or type 's' to move to the approximate location.

Once the variable you want to process has been highlighted, click the arrow button in the middle of the screen to move the variable name to the box on the right-hand side of the screen. SPSS will produce frequency distributions for all variables listed in this box, but, for now, we will confine our attention to sex. Click the **OK** button in the upper-right-hand corner of the Frequencies window and, in seconds, a frequency distribution will be produced.

SPSS sends all tables and statistics to the Output window or SPSS viewer. This window is now "closest" to you, and the Data Editor window is "behind" the Output window. If you wanted to return to the Data Editor, click on any part of it if it is visible, and it will move to the "front" and the Output window will be "behind" it. To display the Data Editor window if it is not visible, minimize the Output window by clicking the "-" box in the upper-right-hand corner.

Frequencies   What can we tell from this table? The score labels (male and female) are printed at the left with the number of cases (frequency) in each category of the variable one column to the right. As you can see, there are 622 males and 765 females in the sample. The next two columns give information about percentages, and the last column to the right displays cumulative percentages. We will defer a discussion of this last column until a later exercise.

One of the percentage columns is labeled Percent and the other is labeled Valid Percent. The difference between these two columns lies in the handling of missing values. The Percent column is based on all cases, including people who did not respond to the item (NA) and people who said they did not have the requested information (DK). The Valid Percent column excludes all missing scores. Since we will almost always want to ignore missing scores, we will pay attention only to the Valid Percent column. Note that for sex, there are no missing scores (gender was determined by the interviewer), and the two columns are identical.

F.5 (SPSS) PRINTING AND SAVING OUTPUT

Once you've gone to the trouble of producing statistics, a table, or a graph, you will probably want to keep a permanent record. There are two ways to do this. First, you can print a copy of the contents of the Output Window to take with you. To do this, click on **File** and then click **Print** from the **File** menu. Alternatively, find the icon of a printer (third from the left) in the row of icons just below the menu bar and click on it.

The other way to create a permanent record of SPSS output is to save the Output window to the computer's memory or to a diskette. To do this, click **Save** from the **File** menu. The Save dialog box opens. Give the output a name (some abbreviation such as 'freqsex' might do) and, if necessary, specify the name of the drive in which your diskette is located. Click **OK,** and the table will be permanently saved.

F.6 (SPSS) ENDING YOUR *SPSS FOR WINDOWS* SESSION

Once you have saved or printed your work, you may end your SPSS session. Click on **File** from the menu bar and then click **Exit.** If you haven't already done so, you will be asked if you want to save the contents of the Output window. You may save the frequency distribution at this point if you wish. Otherwise, click **NO.** The program will close, and you will be returned to the screen from which you began.

Sections F.7 (MicroCase) through F.11 (MicroCase) introduce the Micro-Case program. If your class is using SPSS, skip these sections.

F.7 (MicroCase) STARTING MICROCASE AND LOADING THE 1998 GSS

If you are using the complete, professional version of MicroCase, you will probably be working in a computer lab, and you can begin running the program immediately. If you are using the student version of the program on your personal computer, the first thing you need to do is install the software. Follow the instructions that came with the program and return to this appendix when installation is complete.

To start MicroCase, find the icon (or picture) on the screen of your monitor that is labeled "MC" and says "Student MicroCase" underneath. Use the computer mouse to move the arrow on the monitor screen over this icon and then double-click the left button on the mouse. This will start up the MicroCase program.

After a few seconds the "File and Data" window (Student version) or the "File Management" window (Professional version) will open and present you with a number of choices. The first thing you always need to do with Micro-Case is open a data file, so click on the **Open File** command at the top-left corner of the window. The "Open" window will appear and will probably list a number of data files. Find the 1998 General Social Survey data set, probably labeled GSS98 or something similar. If you have the data set on a disk, you will probably need to specify the correct drive. Check with your instructor to make sure that you know where to find the 1998 GSS. A "File Settings" window will open and present some summary information about the data file. Click **OK,** and you will be returned to the original window.

Note that there is a list of commands across the very top of the screen. These commands begin with **File** at the far left and end with **Help** at the far right. We won't be using these commands very often, but note the **Help** command. As we mentioned previously, MicroCase has an extensive, built-in, Help facility, and you should familiarize yourself with this command.

In this section, you learned how to start up MicroCase and load a data file. These procedures are summarized in Table F.3.

F.8 (MicroCase) WORKING WITH DATABASES

If you are working with the Student Version of MicroCase, click on the **List Data** command in the lower-right-hand corner of the "File and Data" screen. If you are working with the Professional Version of the program, click the "Data Management" command on the left of the screen and then click **List Data.** In either case, you will see a two-dimensional grid with columns running up and down (vertically) and rows running across (horizontally). This is how the data is organized for MicroCase. Each column is a variable or item of information from the General Social Survey. The names of the variables are listed at the tops of the columns, and you can see that the first column contains the scores for a variable named *wrkstat,* the second column is the variable *prestg90,* the third column is the variable *marital,* and so forth.

Recall that you can find the meaning of these variable names in the GSS 1998 code book in Appendix G. An alternate method for deciphering variable names is provided by MicroCase, which has a "built-in" or online code book—a duplicate of Appendix G in electronic form. To use this facility, click the **V** button at the top of the screen to open the Variables window.

TABLE F.3   SUMMARY OF COMMANDS (MicroCase)

To start MicroCase	Click the MicroCase icon on the screen of the computer monitor
To open a data file	Click **Open File** and then click on the data file name

This window has two parts. On the left is a list of all variables in the database with the first variable highlighted. On the right is the Variable Information box with information about the highlighted variable. The first variable is listed as *wrkstat.* The Variable Information window displays the question that was actually asked during the survey ("Last week, were you working full time, part time, going to school, keeping house, or what?) and shows the possible scores on this variable (a score of 1 = full time, a score of 2 = part time, and so forth) along with some other information.

The same information can be displayed for any variable in the data set. For example, find the variable *marital* in the list. You can do this by using the arrow keys on your keyboard or the slider bar on the right of the variable list window. You can also search for a particular variable by clicking the **Search** button and supplying the variable name. Once you have highlighted *marital,* you can see that the variable measures marital status and that a score of "1" indicates that the respondent was married. What does *prestg80* measure? Close this window by clicking the **Close** button at the bottom of the window.

Examine the window displaying the 1998 GSS a little more. Each row of the window (reading across or from left to right) contains the scores of a particular respondent on all the variables in the database. Note that the upper-left-hand cell is highlighted (outlined in a darker border than the other cells). This cell contains the score of respondent #1 on the first variable, the second row contains the scores of the second respondent, and so forth.

In this section, you learned to read information in the "List Data" window and to decipher the meaning of variable names and scores. These commands are summarized in Table F.4, and we are now prepared to actually perform some statistical operations with the 1998 GSS database.

F.9 (MicroCase) PUTTING MICROCASE TO WORK: PRODUCING STATISTICS

In MicroCase, statistics are produced from a separate screen that you can access by clicking the **Statistics Menu** command (Student version) or the **Basic Statistics** command (Professional version). When this screen appears, you will see a number of statistical commands. For now, let's have MicroCase produce a frequency distribution for the variable sex. Frequency distributions

TABLE F.4   SUMMARY OF COMMANDS (MicroCase)

To move around in the Data window	1. Click the cell you want to highlight or
	2. Use the arrow keys on your keyboard or
	3. Move the slider buttons or
	4. Click the arrows on the right-hand and bottom margins
To get information about a variable	1. Click the V button
	2. Scroll through the list of variable names until you highlight the name of the variable in which you are interested. Variable information will appear in the window on the right
	Or 3. See Appendix G

are tables that display the number of times each score of a variable occurred in the sample (see Chapter 2). So, when we complete this procedure, we will know the number of males and females in the 1998 GSS sample. Click the **Univariate** command, and a window will appear with the variables listed in the box on the left. Scroll down until you come to the variable *sex*. Highlight the variable you want to process by clicking on it and then click the arrow button pointing to the box labeled "Primary Variable," and the variable name will be copied to the box. Click **OK**, and a pie chart for this variable will be displayed. To see the frequency distribution, click on the **Summary** button in the box labeled **Statistics** on the left-hand side.

What can we tell from this table? Ignore the statistics at the top of the output for now and observe the table below. The score labels (male and female) are printed at the left with the number of cases (Freq. for frequency) in each category of the variable one column to the right. As you can see, there are 622 males and 765 females in the sample. The other columns display information about percentages (%), cumulative percentages (Cum. %), and Z scores. We will defer a discussion of these columns until a later exercise.

F.10 (MicroCase) PRINTING OUTPUT

Once you've gone to the trouble of producing statistics, a table, or a graph, you will probably want to keep a permanent record. You can print a copy of any statistical output by clicking the Print icon on the Menu bar or by clicking **File** and then **Print.**

F.11 (SPSS) ENDING YOUR MICROCASE SESSION

Once you have printed your work, you are ready to end your session. Return to the Statistics Menu by clicking the **Menu** button and then click **Exit Program.** The program will close, and you will be returned to the screen from which you began.

The General Social Survey (GSS) is a public opinion poll conducted yearly or biyearly by the National Opinion Research Council. A version of the 1998 GSS is available at the web site for this text and is used for all end-of-chapter exercises. Our version of the 1998 GSS includes about 50 variables for a randomly selected subsample of about half of the original respondents. This code book lists each item in the data set. The variable names are those used in the data files. The questions have been reproduced exactly as they were asked (with a few exceptions to conserve space), and the numbers beside each response are the scores recorded in the data file.

The data set includes variables that measure demographic or background characteristics of the respondents, including sex, age, race, religion, and several indicators of socioeconomic status. Also included are items that measure opinion on such current and controversial topics as abortion, capital punishment, and homosexuality.

Most variables in the data set have codes for "missing data." These refer to various situations in which the respondent does not or cannot answer the question. These codes are excluded from all statistical operations and are listed here in parentheses to the right of the variable name. The codes for missing data are: NAP or "not applicable" (the respondent was not asked the question), DK or "Don't Know" (the respondent didn't have the requested information), and NA or "No Answer" (the respondent refused to answer).

Please tell me if you think it should be possible for a woman to get a legal abortion if . . .

abany   She wants it for any reason.   (0. NAP, 8. DK, 9. NA)
      1. Yes
      2. No

abhlth   The woman's health is seriously endangered.
      (Same scoring as abany)

age   Age of respondent   (99. NA)
      18–89. Actual age in years

attend   How often do you attend religious services?   (9. DK or NA)

0. Never	1. Less than once per year
2. Once or twice a year	3. Several times per year
4. About once a month	5. 2–3 times a month
6. Nearly every week	7. Every week
8. Several times a week	

cappun — Do you favor or oppose the death penalty for persons convicted of murder?

(0. NAP, 8. DK, 9. NA)

    1. Favor   2. Oppose

childs — How many children have you ever had? Please count all that were born alive at any time (including any from a previous marriage).   (9. NA)

    0–7. Actual number

    8. Eight or more

class — Subjective class identification   (0. NAP, 8. DK, 9. NA)

    1. Lower class   2. Working class

    3. Middle class   4. Upper class

cohabit — Is it alright for a couple to live together without intending to get married?

(9. NA)

    1. Strongly agree

    2. Agree

    3. Neither agree nor disagree

    4. Disagree

    5. Strongly disagree

degree — Respondent's highest degree   (7. NAP, 8. DK, 9. NA)

    0. Less than HS   1. High school

    2. Assoc./Junior college   3. Bachelor's

    4. Graduate

educ — Highest year of school completed   (97. NAP, 98. DK, 99. NA)

    0–20. Actual number of years

eqwlth — Should the government reduce the income differences between the rich and the poor? If a score of 1 means the government SHOULD reduce income differences and a score of 7 means that the government SHOULD NOT do so, what score between 1 and 7 comes closest to the way you feel?

(8. DK, 9. NA, 10. NAP)

    1. Government should   2.

    3.   4.

    5.   6.

    7. Government should not

fear — Is there any area right around here—that is, within a mile—where you would be afraid to walk alone at night?   (0. NAP, 8. DK, 9. NA)

    1. Yes   2. No

fefam — It is much better for everyone involved if the man is the achiever outside the home and the woman takes care of the home and family.

(0. NAP, 8. DK, 9. NA)

    1. Strongly agree   2. Agree

    3. Disagree   4. Strongly disagree

fehelp   It is more important for a wife to help her husband's career than to have one herself.
   (Same scoring as fefam)

grass   Do you think the use of marijuana should be made legal or not?
   (0. NAP, 8. DK, 9. NA)
   1. Should
   2. Should not

gunlaw   Would you favor or oppose a law that requires a person to obtain a police permit before he or she could buy a gun?   (0. NAP, 8. DK, 9. NA)
   1. Favor
   2. Oppose

hapmar   Taking things all together, how would you describe your marriage? Would you say that your marriage is very happy, pretty happy, or not too happy?
   (0. NAP, 8. DK, 9. NA)
   1. Very happy
   2. Pretty happy
   3. Not too happy

happy   Taken all together, how would you say things are these days—would you say that you are very happy, pretty happy, or not too happy?
   (0. NAP, 8. DK, 9. NA)
   1. Very happy
   2. Pretty happy
   3. Not too happy

homosex   What about sexual relations between two adults of the same sex: Do you think it is always wrong, almost always wrong, wrong only sometimes, or not wrong at all?   (5. Other, 0. NAP, 8. DK, 9. NA)
   1. Always wrong   2. Almost always wrong
   3. Wrong only sometimes   4. Not wrong at all

income98   Respondent's total family income from all sources for 1998
   (98. DK, 99. NA)

1. Less than 1000	2. 1000 to 2999
3. 3000 to 3999	4. 4000 to 4999
5. 5000 to 5999	6. 6000 to 6999
7. 7000 to 7999	8. 8000 to 9999
9. 10,000 to 12,499	10. 12,500 to 14,999
11. 15,000 to 17,499	12. 17,500 to 19,999
13. 20,000 to 22,499	14. 22,500 to 24,999
15. 25,000 to 29,999	16. 30,000 to 34,999
17. 35,000 to 39,999	18. 40,000 to 49,999
19. 50,000 to 59,999	20. 60,000 to 74,999
21. 75,000 to 89,999	22. 90,000 to 109,999
23. 110,000 or over	

marital Are you currently married, widowed, divorced, separated, or have you never been married?   (9. NA)
> 1. Married   2. Widowed
> 3. Divorced   4. Separated
> 5. Never married

news How often do you read the newspaper? (0. NAP, 8. DK, 9. NA)
> 1. Every day   2. A few times a week
> 3. Once a week   4. Less than once a week
> 5. Never

paeduc Father's highest year of school completed (97. NAP, 98. DK, 99. NA)
> 0–20. Actual number of years

papres80 Prestige of father's occupation (0. NAP, DK, NA)
> 17–86. Actual score

partnrs5 How many sex partners have you had over the past five years? (9. 1 or more, don't know the number, 95. Several, 98. DK, 99. NA, −1. NAP)
> 0. No partners   1. 1 partner
> 2. 2 partners   3. 3 partners
> 4. 4 partners   5. 5–10 partners
> 6. 11–20 partners   7. 21–100 partners
> 8. More than 100 partners

polviews I'm going to show you a seven-point scale on which the political views that people might hold are arranged from extremely liberal to extremely conservative. Where would you place yourself on this scale?
                (0. NAP, 8. DK, 9. NA)
> 1. Extremely liberal   2. Liberal
> 3. Slightly liberal   4. Moderate
> 5. Slightly conservative   6. Conservative
> 7. Extremely conservative

postlife Do you believe there is life after death? (0. NAP, 8. DK, 9. NA)
> 1. Yes
> 2. No

premarsx There's been a lot of discussion about the way morals and attitudes about sex are changing in this country. If a man and a woman have sex relations before marriage, do you think it is always wrong, almost always wrong, wrong only sometimes, or not wrong at all? (0. NAP, 8. DK, 9. NA)
> 1. Always wrong   2. Almost always wrong
> 3. Wrong only sometimes   4. Not wrong at all

pres96 In 1996, you remember that Clinton ran for president on the Democratic ticket against Dole for the Republicans and Perot as an Independent. Did you vote

for Clinton, Dole, or Perot? (Includes only those who said they voted in this election)   (6. No presidential vote, 4. Other, 0. NAP, 8. DK, 9. NA)
   1. Clinton
   2. Dole
   3. Perot

prestg80   Prestige of respondent's occupation   (0. NAP, DK, NA)
   17–86. Actual score

race   Race of respondent
   1. White
   2. Black
   3. Other

racepush   African Americans shouldn't push themselves where they're not wanted.
    (0. NAP, 8. DK, 9. NA)
   1. Agree strongly   2. Agree slightly
   3. Disagree slightly   4. Disagree strongly

region   Region of interview
   1. New England   2. Mid-Atlantic
   3. East N. Cent.   4. West N. Cent.
   5. So. Atlantic   6. East So. Cent.
   7. West So. Cent.   8. Mountain
   9. Pacific

relig   What is your religious preference? Is it Protestant, Catholic, Jewish, some other religion, or no religion?   (8. DK, 9. NA)
   1. Protestant   2. Catholic
   3. Jewish   4. None
   5. Other

satfin   So far as you and your family are concerned, would you say that you are pretty well satisfied with your present financial situation, more or less satisfied, or not satisfied at all?   (8. DK, 9. NA)
   1. Pretty well satisfied
   2. More or less satisfied
   3. Not satisfied at all

satjob   On the whole, how satisfied are you with the work you do?
    (0. NAP, 8. DK, 9. NA)
   1. Very satisfied
   2. Moderately satisfied
   3. A little dissatisfied
   4. Very dissatisfied

sex   Respondent's gender
   1. Male
   2. Female

sexeduc

Would you be for or against sex education in your public schools?

(0. NAP, 3. Depends, 8. DK, 9. NA)

1. Favor
2. Oppose

sexfreq

About how many times did you have sex during the last 12 months?

(−1. NAP, 8. DK, 9. NA)

0. Not at all
1. Once or twice
2. About once a month
3. 2 or 3 times a month
4. About once a week
5. 2 or 3 times a week
6. More than 3 times a week

size

Size of place in thousands. Population figures from U.S. Census. Add three zeros to code for actual values.

spanking

Do you strongly agree, agree, disagree, or strongly disagree that it is sometimes necessary to discipline a child with a good, hard spanking?

(0. NAP, 8. DK, 9. NA)

1. Strongly agree
2. Agree
3. Disagree
4. Strongly disagree

spkath

There are always some people whose ideas are considered bad or dangerous by other people. For instance, somebody who is against all churches and religion. If such a person wanted to make a speech in your (city/town/community), . . . should he be allowed to speak or not?

(0. NAP, 8. DK, 9. NA)

1. Allowed
2. Not allowed

suicide1

Do you think a person has a right to end his or her own life if this person has an incurable disease? (0. NAP, 8. DK, 9. NA)

1. Yes
2. No

trust

Generally speaking, would you say that most people can be trusted or that you can't be too careful in dealing with people?

(3. Other, depends [volunteered], 0. NAP, 8. DK, 9. NA)

1. Most people can be trusted
2. Can't be too careful

tvhours

On the average day, about how many hours do you personally watch television? (−1. NAP, DK, NA)

00–24. Actual hours

tvrelig

About how much time per week do you spend watching religious shows on television? (−1, 99.9 NAP, 998. DK, 999, NA)

00–18.0 Actual hours

wrkstat

Last week, were you working full time, part time, going to school, keeping house, or what?
1. Working full time
2. Working part time
3. With a job but not at work (ill, vacation, etc.)
4. Unemployed
5. Retired
6. In school
7. Keeping house
8. Other

xmovie

Have you seen an X-rated movie in the last year?   (0. NAP, 8. DK, 9. NA)
1. Yes
2. No

Answers to Odd-Numbered Computational Problems

In addition to answers, this section suggests some problem-solving strategies and provides examples of how to interpret the numerical answers. You should try to solve and interpret the problems on your own before consulting this section.

In solving these problems, I let my calculator or computer do most of the work. I worked with whatever level of precision these devices permitted and, generally, didn't round off until the end or until I had to record an intermediate sum. I always rounded off to two places of accuracy (or, two places beyond the decimal point, or to 100ths). If you follow these same conventions, your answers will almost always match mine. However, there is no guarantee that our answers will always be exact matches, and you should be aware that small discrepancies might occur and that they are almost always trivial. If the difference between your answer and mine doesn't seem trivial, you should double-check to make sure you haven't made an error or solve the problem again using a greater degree of precision.

Finally, please allow me a brief disclaimer about mathematical errors in this section. Let me assure you, first of all, that I know how important this section is for most students and that I worked hard to be certain that these answers are correct. Human fallibility being what it is, however, I know that I cannot make absolute guarantees. Should you find any errors, please let me know so I can make corrections in the future.

Chapter 2

2.1 **a.** Complex A: $(5/20) \times 100 = 25.00\%$
Complex B: $(10/20) \times 100 = 50.00\%$
b. Complex A: $4{:}5 = 0.80$
Complex B: $6{:}10 = 0.60$
c. Complex A: $(0/20) = 0.00$
Complex B: $(1/20) = 0.05$
d. $(6/(4 + 6)) = (6/10) = 60.00\%$
e. Complex A: $8{:}5 = 1.60$
Complex B: $2{:}10 = 0.20$

2.3 Bank robbery rate =
$(47/211732) \times 100000 = 22.20$
Homicide rate = $(13/211732) \times 100000 = 6.14$
Auto theft rate = $(23/211732) \times 100000 = 10.86$

Chapter 3

3.1 "Region of birth" and "Religion" are nominal-level variables, "support for legalization" and "opinion of food" are ordinal, and "expenses" and "number of movies" are interval-ratio. The mode, the most common score, is the only measure of central tendency available for nominal-level variables. For the ordinal-level variables, don't forget to array the scores from high to low before locating the median. There are ten freshmen (N is even), so the median will be the score halfway between the scores of the two middle cases. There are 11 seniors (N is odd), so the median will be the score of the middle case. To find the mean for the interval-ratio variables, add the scores and divide by the number of cases.

	Freshmen:	Seniors:
Region of birth:	Mode = North	Mode = North
Legalization:	Median = 3	Median = 5
Expenses:	Mean = 38.50	Mean = 53.00
Movies:	Mean = 5.8	Mean = 5.18
Food:	Median = 6	Median = 4
Religion:	Mode = Protestant	Mode = Protestant and None (4 cases each)

3.3 For gender, the modal category is male. For social class, the median is "medium" (i.e., the middle case is in this category). The mean number of years in the party is 26.15. The median level of education is high school. The modal marital status is married. The average number of children is 2.39.

3.5 For marital status, the modal category is "married." For race, the mode is "white." For age, the mean is 27.53, and the median is 7 for attitude on abortion.

3.7 In Table 2.4, there are 10 males and 10 females, so sex is bimodal. The mode of "marital status" is "single," with 10 cases. The median of "satisfaction" is 3, and the mean age is 20.05.

3.9 Attitude and opinion scales almost always generate ordinal-level data, so the appropriate measure of central tendency would be the median. For the students, the median is 9 and, for the neighbors, the median is 2. Incidentally, the means are 7.80 for the students and 4.00 for the neighbors.

3.11 The mean for the class scores is 513.50, and the median is 502.50. The greater value for the mean indicates a positive skew (a few high scores). For the math test, the mean is 72.25 and the median is 72.50. Since the mean is lower, this distribution has a very slight negative skew. The mean on the verbal test is 78.19, and the median is 77.5. There is a very slight positive skew in this distribution. Also, the students did better on the verbal test than on the math test.

3.13 For freshmen, the median is 35, and the mean is 31.72. For seniors, the median is 30, and the mean is 28.6.

Chapter 4

4.1 The IQV is .89 for complex A, .99 for complex B, .71 for complex C, and .74 for complex D. Complex B is the most heterogeneous and complex C is the least.

4.3 The range is $50 - 10$ or 40. The standard deviation is 12.28.

4.5 For problem 2.8, the range is $21 - 0$, or 21. Q_1 is the score of the $(.25)(25) = 6.25$th or, rounding off, the 6th case. This value is 2. Q_3 is the score of the $(.75)(25) = 18.75$, or 19th case, or 7. $Q = Q_3 - Q_1 = 7 - 2 = 5$. The standard deviation is 5.49, and the variance is $(5.49)^2$, or 30.14.

4.7

		$\overline{X}$	s
Labor force participation:	Males	77.6	2.73
	Females	58.4	6.73
% High school grads:	Males	69.2	5.38
	Females	70.2	4.98
Mean income:	Males	33,896.60	4,443.16
	Females	29,462.40	4,597.93

Males and females are very similar in terms of educational level, but females are less involved in the labor force and, on the average, earn almost $4500 less than males per year. Females are much more variable in their labor force participation but are similar to males in dispersion on the other two variables. See Section 9.6 for more on this topic.

4.9

	$\overline{X}$	s
For students:	7.80	3.33
For neighbors:	4.00	4.15

4.11 $R = 62 - 5 = 57$, $s = 13.84$

4.13 A: $R = 18$, $s = 5.32$
B: $R = 8$, $s = 2.37$
C: $R = 3$, $s = 1.03$
D: $R = 15$, $s = 7.88$

4.15 Freshman: $s = 12.53$
Seniors: $s = 13.34$

Chapter 5

5.1

X_i	Z score	% Area Above	% Area Below
5	−1.67	95.25	4.75
6	−1.33	90.82	9.18
7	−1.00	84.13	15.87
8	−.67	74.86	25.14
9	−.33	62.93	37.07
11	.33	37.07	62.93
12	.67	25.14	74.86
14	1.33	9.18	90.82
15	1.67	4.75	95.25
16	2.00	2.28	97.72
18	2.67	.38	99.62

5.3

	Z scores	Area
a.	.10 & 1.10	32.45%
b.	.60 & 1.10	13.86%
c.	.60	27.43%
d.	.90	18.41%
e.	.60 & −.40	38.11%
f.	.10 & −.40	19.52%
g.	.10	53.98%
h.	.30	61.79%
i.	.60	72.57%
j.	1.10	86.43%

5.5

X_i	Z Score	Number of Students Above	Number of Students Below
60	−2.00	195	5
57	−2.50	199	1
55	−2.83	199	1
67	−0.83	159	41
70	−0.33	126	74
72	0.00	100	100
78	1.00	32	168
82	1.67	10	190
90	3.00	1	199
95	3.83	1	199

5.7

	Z score	Area
a.	−2.20	1.39%
b.	1.80	96.41%
c.	−0.20 & 1.80	54.34%
d.	0.80 & 2.80	20.93%
e.	−1.20	88.49%
f.	0.80	21.19%

5.9

	Z score	Area
a.	−1.00 & 1.50	.7745
b.	0.25 & 1.50	.3345
c.	1.50	.0668
d.	0.25 & −2.25	.5865
e.	−1.00 & −2.25	.1465
f.	−1.00	.1587

5.11 Yes. The raw score of 110 translates into a Z score of +2.88. 99.80% of the area lies below this score, so this individual was in the top 1% on this test.

5.13 For the first event, the probability is .0919 and, for the second, the probability is .0655. The first event is more likely.

Part I Cumulative Exercises

1. First, determine the level of measurement of each variable. This will help you to set up the frequency distributions and determine the appropriate measures of central tendency and dispersion. There are one nominal variable (religion), two ordinal variables (strength and comfort), and two interval-ratio-level variables (pray and age). For nominal-level variables, the mode is the category with the most cases. For this sample, there are more Protestants than any other religion. The index of qualitative variation (IQV), the only measure of dispersion presented in this text for nominal-level variables, is .86. This is a high value (the maximum would be 1.00) and indicates a great deal of dispersion.

To find the median for "Strength" and "Comfort," the scores must first be arrayed from high to low. There are fifteen cases in the sample, so the median is the score associated with the 8th case. The median for "Strength" is 7 and 1 for "Comfort." The ranges for these variables are 9 and 4, respectively. As an alternative, it would be reasonable to use the standard deviation as a measure of dispersion for "Strength." It is common in social science research for the standard deviation to be reported for ordinal-level variables with a broad range of scores. The standard deviation for this variable is 2.77. As opposed to "Strength," "Comfort" has only 5 categories, so an alternative to the range for this variable would be the IQV, which is .80.

For "Pray," the mean is 1.40, and the standard deviation is 1.62. For age, the mean is 41.53, and the standard deviation is 12.54.

3.

N of children:	Mean = 2.44,	s = 2.06
Yrs of school:	Median = 1	Range = 4
Race:	Mode = white,	IQV = .50
Birth control:	Median = 2	Range = 3
TV:	Mean = 3.08	s = 2.04
Religion:	Mode = Protestant,	IQV = .54

Chapter 7

7.1 **a.** 5.2 ± 0.11 **b.** 100 ± 0.71
c. 20 ± 0.40 **d.** 1020 ± 5.41
e. 7.3 ± 0.23 **f.** 33 ± 0.80

7.5 **a.** 2.30 ± 0.04 **b.** 2.10 ± 0.01, $0.78 \pm .07$
c. 6.00 ± 0.37

7.7 **a.** 178.23 ± 1.97 The estimate is that students spent between \$176.26 and \$180.20 on books.
b. $1.5 \pm .04$ The estimate is that students visited the clinic between 1.46 and 1.54 times on the average.
c. $2.8 \pm .13$ **d.** $3.5 \pm .19$

7.9. $14 \pm .07$ The estimate is that between 7% and 21% of the population consists of unmarried couples living together.

7.11 **a.** $P_s = (823/1496) = .55$
Confidence interval: $.55 \pm .02$
Between 53% and 57% of the population agrees with the statement.
b. $P_s = (650/1496) = .44$
Confidence interval: $.44 \pm .02$
c. $P_s = (375/1496) = .25$
Confidence interval: $.25 \pm .02$
d. $P_s = (1023/1496) = .68$
Confidence interval: $.68 \pm .02$
e. $P_s = (800/1496) = .54$
Confidence interval: $.54 \pm .02$

7.13 For alpha $= .10$: $100 \pm .74$
For alpha $= .05$: $100 \pm .88$
For alpha $= .01$: 100 ± 1.16
For alpha $= .001$: 100 ± 1.47

7.15 The confidence interval is $.51 \pm .05$. The estimate would be that between 46% and 56% of the population prefer candidate A. The population parameter (P_u) is equally likely to be anywhere in the interval (that is, it's just as likely to be 46% as it is to be 56%), so a winner cannot be predicted.

7.17 The confidence interval is 0.23 ± 0.08. At the 95% confidence level, the estimate would be that between 240 (15%) and 496 (31%) of the 1600 freshmen would be extremely interested. The estimated numbers are found by multiplying N (1600) by the upper (.31) and lower (.15) limits of the interval.

7.19 **a.** $43.87 \pm .73$ **b.** $2.86 \pm .13$
c. 1.81 ± 0.09
d. $27 \pm .03$ (About 27% of the population are Catholic.)
e. $24 \pm .03$ (About 24% of the population have never married.)
f. $53 \pm .03$ (About 49 % of the electorate voted for Clinton in 1996.)
g. $73 \pm .03$

Chapter 8

8.3 **a.** Z (obtained) $= -41.00$
b. Z (obtained) $= 29.09$

8.5 Z (obtained) $= 6.04$

8.7 **a.** Z (obtained) $= -13.66$
b. Z (obtained) $= 25.50$

8.9 t (obtained) $= 4.50$

8.11 Z (obtained) $= 3.06$

8.13 Z (obtained) $= -1.48$

8.15 **a.** Z (obtained) $= -0.74$
b. Z (obtained) $= 2.19$
c. Z (obtained) $= -8.55$
d. Z (obtained) $= -18.07$
e. Z (obtained) $= 2.09$
f. Z (obtained) $= -53.33$

8.17 t (obtained) $= -1.14$

Chapter 9

9.1 **a.** $\sigma = 1.39$, Z (obtained) $= -2.53$
b. $\sigma = 1.60$, Z (obtained) $= 2.49$

9.3 **a.** Z (obtained) $= 1.70$
b. Z (obtained) $= -2.49$

9.5   **a.** $\sigma = 0.08$ Z (obtained) $= 11.25$
   b. $\sigma = 0.12$ Z (obtained) $= -3.33$
   c. $\sigma = 0.15$ Z (obtained) $= 20.00$

9.7   These are small samples (combined N's of less than 100), so be sure to use Formulas 9.5 and 9.6 in step 4.
   a. $\sigma = 0.12$, t (obtained) $= -1.33$
   b. $\sigma = 0.13$, t (obtained) $= 14.85$

9.9 t (obtained) $= 3.52$

9.11 $P_u = .45$, $\sigma_p = .06$, Z (obtained) $= 0.67$

9.13   **a.** $P_u = .46$, $\sigma_p = 0.06$, Z (obtained) $= 2.17$
   b. $P_u = .80$, $\sigma_p = 0.07$, Z (obtained) $= 1.43$
   c. $P_u = .72$, $\sigma_p = 0.08$, Z (obtained) $= 0.75$

9.15   **a.** Z (obtained) $= -6.00$
   b. Z (obtained) $= -2.67$
   c. Z (obtained) $= -4.33$
   d. Z (obtained) $= -0.23$
   e. Z (obtained) $= -5.47$
   f. Z (obtained) $= -3.56$

Chapter 10

10.1   **a.** The overall mean is 12.17, SST $= 231.67$, SSB $= 173.17$, and SSW $= 58.5$. The F ratio is 13.32.
   b. The overall mean is 6.87, SST $= 455.73$, SSB $= 78.53$, and SSW $= 377.20$. The F ratio is 1.25.
   c. The overall mean is 31.65, SST $= 8362.55$, SSB $= 5053.35$, and SSW $= 3309.20$. The F ratio is 8.14.

10.3   **a.** The overall mean is 4.39, SST $= 86.28$, SSB $= 45.78$, and SSW $= 40.50$. The obtained F ratio is 8.48. With alpha set at .05 and 2 and 15 degrees of freedom, the critical F ratio would be 3.68, so the null may be rejected. Decision making does vary significantly by type of relationship. By inspection of the group means, it seems that the "cohabitational" category accounts for most of the differences.
   b. The overall mean is 16.44, SST $= 332.44$, SSB $= 65.44$, and SSW $= 267.00$. The obtained

F ratio is 1.84. At alpha $=.05$, the null hypothesis cannot be rejected.

10.5   SST $= 213.61$, SSB $= 2.11$, SSW $= 211.50$ F (obtained) $= .08$

10.7   SST $= 429.48$, SSB $= 305.42$, SSW $= 124.06$ F (obtained) $= 5.96$

Chapter 11

11.1   **a.** 1.11   **b.** 0.00   **c.** 1.52   **d.** 1.46

11.3   There is 1 degree of freedom in a 2 $\times$ 2 table. With alpha set at .05, the critical value for the chi square would be 3.841. The obtained chi square is 0.65, so we fail to reject the null hypothesis of independence between the variables. There is no statistically significant relationship between race and services received.

11.5   With 1 degree of freedom and alpha set at .05, the critical region will begin at 3.841. The obtained chi square of 2.15 does not fall within this area, so the null hypothesis cannot be rejected. No statistically significant relationship exists between unionization and salary.

11.7   The obtained chi square is 5.13, which is significant (df $= 1$, alpha $= .05$).

11.9   The obtained chi square is 6.67.

11.11   The obtained chi square is 12.59.

11.13   The obtained chi square is 19.34.

11.15   **a.** Chi square is 25.19
   b. Chi square is 1.80
   c. Chi square is 5.23
   d. Chi square is 28.43
   e. Chi square is 14.17

Part II Cumulative Exercises

1. One of the continuing challenges of empirical research is to make reasonable decisions about which statistical test to use in which situation. If you approach the decision systematically and consider the selection criteria carefully, the confusion and ambiguity can be minimized. Let's use this first problem of the exercise to consider some ways in which reasonable decisions can be made. The situation calls for a test of hypotheses ("is the difference significant?"), so our choice of procedures will be limited to Chapters 8–11. Next, determine the types of variables you are working with. Number of minutes is an interval-ratio-level variable, and gender is nominal and has only two categories. Which test should we use? The techniques in Chapter 8 (one-sample tests) and Chapter 10 (tests involving more than two samples or categories) are not relevant. Chi square (Chapter 11) won't work unless we collapse the scores on internet use into a few categories. This leaves Chapter 9. A test of sample means fits the situation, and we have a large sample (combined N's greater than 100), so it looks like we're going to wind up in Section 9.2.

Another way to approach test selection would be to use the flowcharts available at the web site for this text. The flowcharts would quickly lead to Chapter 9, where we would answer: YES, we want to test for the significance of the difference between population means, NO, the samples are not matched; and YES, sample size is large. These answers will also lead to Section 9.2.

a. $\sigma = .14$, Z (obtained) $= -35.51$

The difference in Internet minutes is significant. Men, on the average, use this technology more frequently.

b. The table format is a sure tip-off that the chi square test is appropriate. The obtained chi square is 0.87, which is not significant at the .05 level. There is no statistically significant relationship between involvement and social class.

c. "Number of partners" sounds like an interval-ratio-level variable, and education has three categories. Analysis of variance is an appropriate test for this situation. The F ratio is .13—not at all significant—so we must conclude that this dimension of sexuality does not vary by level of education.

d. The problem asks for a characteristic of a population ("how many times do adult Americans move?") but gives information only for a "random sample." The estimation procedures presented in Chapter 7 fit the situation and, since

the information is presented in the form of a mean, you should use Formula 7.2 to form the estimate. At an alpha level of .05, the confidence interval would be 3.5 ± 0.02.

e. The research question focuses on the difference between a single sample and a population, so Chapter 8 is relevant. Since the population standard deviation is unknown, and we have a large sample, Sections 8.1–8.5 and Formula 8.1 will be relevant. Z (obtained) is -2.50, which would be significant at alpha $= .05$. The sample is significantly different from the population—rural school districts are different from the universe of school districts in this state.

Chapter 12

Because of space constraints, I've provided answers to only two problems here.

12.1   Efficiency and Authoritarianism

	Authoritarianism	
Efficiency	Low	High
Low	37.04%	70.59%
	(10)	(12)
High	62.96%	29.41%
	(17)	(5)
Totals	100.00%	100.00%
	(27)	(17)

The conditional distributions change, so there is a relationship between the variables. The change from column to column is quite large, and the relationship is strong. Efficiency decreases as authoritarianism increases—workers with dictatorial bosses are less productive (or, maybe, bosses become more dictatorial when workers are inefficient), so this relationship is negative in direction.

12.9   GPA by Attractiveness

	Attractiveness		
GPA	Low	Moderate	High
Low	28.00%	26.67%	33.33%
Moderate	40.00%	33.33%	35.56%
High	32.00%	40.00%	31.11%
Totals	100.00%	100.00%	100.00%

The conditional distributions change, so there is a relationship between these two variables. The change from column to column is quite minimal, and the relationship is weak. There is a slight tendency for GPA to increase with attractiveness (I think—it's hard to tell when the relationship is so weak), but these patterns do not support the hypothesis.

Chapter 13

13.1 **a.** $\phi = 0.00, \lambda = 0.00$
b. $\phi = 0.09, \lambda = 0.00$
c. $\phi = 0.25, \lambda = 0.14$

13.3 $\phi = .17, \lambda = .00$

13.5 $\phi = .31, \lambda = 0.03$

13.7 $\phi = .00, \lambda = .00$

13.9 **a.** $\phi = .05, \lambda = .00$
b. $\phi = .54, \lambda = .33$
c. $\phi = .39, \lambda = .05$

13.11 Cramer's $V = .05, \lambda = .00$

Chapter 14

14.1 **a.** $G = 0.71$ **b.** $G = 0.71$ **c.** $G = -0.88$

These relationships are strong. Facility in English and income increase with length of residence $(+0.71)$. Use the percentages to help interpret the direction of a relationship. In the first table, 80% of the "newcomers" had "Low" facility in English, and 60% of the "oldtimers" had "High" facility. In this relationship, low scores on one variable are associated with low scores on the other, and scores increase together (as one increases, the other increases), so this is a positive relationship. Contact with the old country decreases with length of residence (-0.88). Most newcomers have higher levels of contact, and most oldtimers have lower levels.

14.3 $G = -0.61$

14.5 $G = 0.27$

14.7 $G = 0.22, Z$ (obtained) $= 0.93$

14.9 $G = -0.14$

14.11 $r_s = -0.46, t$ (obtained) $= -1.55$

14.13 $r_s = 0.70$

There is a strong positive relationship between ethnic heterogeneity and strife for this sample. The greater the diversity, the greater the strife.

14.15 a. $G = -.14$ **b.** $G = -.17$
c. $G = -.14$ **d.** $G = -.13$
e. $G = .39$

Chapter 15

15.1 (HINT: When finding the slope, remember that "Turnout" is the dependent or Y variable.)

Turnout and Unemployment:
$b = 3.00, \quad a = 39.00, \quad Y = (39) + (3)X$
$r = 0.95, \quad r^2 = 0.90, \quad t$ (obtained) $= 4.96$

Turnout and Education:
$b = 12.67, \quad a = -94.73,$
$Y = (-94.73) + (12.67)X$
$r = 0.98, \quad r^2 = 0.97, \quad t$ (obtained) $= 9.80$

Turnout and Negative Campaigning:
$b = -0.90, \quad a = 114.01$
$Y = (114.01) + (-0.90)X$
$r = -0.87, \quad r^2 = 0.76 \quad t$ (obtained) $= -3.08$

15.3 (HINT: When finding the slope, remember that "Number of visitors" is the dependent or Y variable.)

$b = -0.37, \quad a = 13.42,$
$r = -0.31, \quad r^2 = 0.10$

15.5 **b.**

		Growth Rate	Population Density	% Urban
Homicide	$b =$	.49	.01	.08
	$a =$	4.02	4.72	−0.11
Robbery	$b =$	3.20	.58	3.35
	$a =$	136.07	62.71	−99.31
Auto theft	$b =$	76.84	.00	8.75
	$a =$	219.71	450.49	−188.40

c. For a growth rate of -1, the predicted homicide rate would be 3.53.

For a population density of 250, the predicted robbery rate would be 207.71.

For a state with 50% urbanization, the predicted rate of auto theft would be 249.10.

d. Pearson r's and (r^2s) are

	Growth Rate	Population Density	% Urban
Homicide	.59 (.35)	.28 (.08)	.76 (.57)
Robbery	.10 (.01)	.78 (.61)	.85 (.73)
Auto theft	.81 (.65)	.00 (.00)	.76 (.57)

15.7 $b = 0.05$, $a = 53.18$, $r = 0.40$, $r^2 = 0.16$

15.9 Prestige by Age:
$a = 49.41$, $b = -.27$, $r = -.30$, $r^2 = .09$
Attend by Number of children:
$a = 3.53$, $b = -0.41$, $r = -0.39$,
$r^2 = .15$
TV by Number of children:
$a = 2.91$, $b = 0.15$, $r = 0.18$, $r^2 = .03$
TV by Age:
$a = 2.40$, $b = 0.02$, $r = 0.16$, $r^2 = .03$
Number of children by Age:
$a = -2.00$, $b = 0.10$, $r = 0.67$, $r^2 = .45$
TV by Prestige:
$a = 4.18$, $b = 0.03$, $r = -0.19$, $r^2 = .04$

Part III Cumulative Exercises

1. a. The choice of a measure of association is largely dependent on the level of measurement of the variables involved. Assuming that crime rates and "percent immigrant" are both measured at the interval-ratio level, Pearson's r would be the appropriate statistic for measuring the strength and direction of this relationship. This is a moderate to strong relationship ($r = -.47$), and percentage of immigrants explains 22% of the variation in crime rate for these ten cities. The relationship is negative, which means that as the percentage of immigrants increases, the crime rate decreases.

b. These variables are ordinal in level of measurement, so gamma would be the appropriate measure of association (see the web site for this text for other ordinal-level measures of association). Gamma is $-.43$, indicating a moderate to strong

negative relationship. As TV viewing increases, involvement decreases.

c. The appropriate measure for ordinal-level variables with many scores is Spearman's rho, which, for these variables, is .74. There is a strong positive relationship between these variables. States with higher quality of life also have superior systems of higher education.

d. Race is a nominal-level variable, and the table is larger than 2×2. Cramer's V is .40, indicating a moderate relationship between the variables. The column percentages show that this sample tends to reject all of the singular racial categories in favor of "None of the above." A second, weaker trend is for people with one white parent to identify with that group. Because of the uneven row totals, lambda is zero even though there is an association between these variables.

Chapter 16

16.1 a. $G = 0.71$
b. $G = 0.78$ (males)
$G = 0.65$ (females)
c. $G = 0.67$ (Asians)
$G = 0.74$ (Hispanics)

The bivariate relationship is strong and positive. The longer the residence, the greater the facility in English. The bivariate relationship is not affected by the sex or origin of the immigrants. These results would be taken as strong evidence of a direct (causal) relationship between length of residence and English facility.

16.3 $G = -.23$
$G = -.34$ (females)
$G = -.10$ (males)

This is an interactive relationship. Length of institutionalization has a greater effect on the reality orientation of females than of males.

16.5 $G = 0.62$ (bivariate table)
$G = 0.52$ (whites)
$G = 0.68$ (blacks)
$G_p = 0.60$ (controlling for race)
$G = 0.62$ (males)
$G = 0.62$ (females)
$G_p = 0.62$ (controlling for sex)

The bivariate relationship is strong and positive. Completion of the training program is closely associated with holding a job for at least one year. There is some interaction with race. The training has less impact for whites than for blacks. Blacks who did not complete the training were less likely to have held a job for at least one year. Gender has no impact at all on the relationship. Overall, there is a direct relationship between completion of the training and holding a job for at least a year, though there is some interaction with race.

16.7 For "Right to Abortion?" by Income by Sex

$G = -.14$

$G = -.06$ (males)

$G = .19$ (females)

$G_p = -.15$

For "Right to Suicide?" by Income by Sex

$G = -.22$

$G = -.16$ (males)

$G = -.11$ (females)

$G_p = -.13$

Chapter 17

17.1 a. For turnout (Y) and unemployment (X) while controlling for negative advertising (Z), $r_{yx.z} = 0.95$. The relationship between X and Y is not affected by the control variable Z.

b. For turnout (Y) and negative advertising (X) while controlling for unemployment (Z), $r_{yx.z} = -0.89$. The bivariate relationship is not affected by the control variable.

c. Turnout $(Y) = 70.25 + (2.09)$ unemployment $(X_1) + (-.43)$ negative advertising (X_2). For unemployment $(X_1) = 10$ and negative advertising $(X_2) = 75$, turnout $(Y) = 58.90$.

d. For unemployment (X_1): $b_1^* = .66$. For negative advertising (X_2): $b_2^* = -.41$.

Unemployment has a stronger effect on turnout than negative advertising. Note that the independent variables' effect on turnout is in opposite directions.

e. $R^2 = .98$

17.3 a. For strife (Y) and unemployment (X), controlling for urbanization (Z), $r_{yx.z} = .79$.

b. For strife (Y) and urbanization (X), controlling for unemployment (Z), $r_{yx.z} = .20$.

c. Strife $(Y) = (-14.60) + (4.94)$ unemployment $(X_1) + (.16)$ urbanization (X_2). With unemployment $= 10$ and urbanization $= 90$, strife (Y') would be 49.2.

d. For unemployment (X_1): $b_1^* = .78$. For urbanization (X_2): $b_2^* = .13$.

e. $R^2 = .65$

17.5 a. Turnout $(Y) = 83.80 + (-1.16)$ Democrat $(X_1) + (2.89)$ minority (X_2).

b. For $X_1 = 0$ and $X_2 = 5$, $Y' = 98.25$

c. $Z_y = (-1.27)Z_1 + (.84)Z_2$

d. $R^2 = 0.51$

17.7 a. $Z_y = (.30)$ HS grads $(Z_1) + (-.47)$ Rank (Z_2)

b. $R^2 = .46$

Part IV Cumulative Exercises

1. a. The choice of multivariate procedures will depend on level of measurement and the number of possible scores for each variable. Regression analysis is appropriate for interval-ratio, continuous variables. In this situation, we have three variables: GPA, hours of work, and College Board scores. College Board scores are arguably only ordinal in level of measurement, but the scores have a wide range and seem continuous. GPA is the dependent variable, and the zero-order correlation with "hours worked" is $-.83$, a strong and negative relationship, which indicates that having a job did interfere with academic success for this sample. The zero-order correlation between GPA and College Board scores is .55, a relationship that is consistent with the idea that College Board scores predict success in college.

The results of the regression analysis with "hours worked" (X_1) and College Board scores (X_2) as independents:

$$Y = 1.71 + (-.04)X_1 + (.003)X_2$$
$$Z_y = (-.75)Z_1 + (.43)Z_2$$
$$R^2 = .86$$

The beta-weights indicate that "hours worked" has a stronger direct effect on GPA than College Board scores. Even for students who were well prepared for college and had high College Board scores, having a job had a negative impact on GPA. The high value for R^2 means that only a small percentage of the variance in GPA (14%) is left unexplained by these two independent variables.

b. In this situation, we have three dichotomous variables, two ordinal-level and one nominal. With the limited variation possible, the elaboration technique is appropriate to analyze these relationships. The bivariate table shows that there is a relationship between graduating and level of social activity for this sample: 70% of the students with low levels of social activity graduated in four years vs. only 30% of the students with high levels. Bivariate gamma is .69, indicating a strong relationship.

Controlling for sex reveals some interaction. The gamma for males (.77) is stronger than the bivariate gamma, and the gamma for females (.60) is weaker. In other words, while level of social activity has an effect on graduation rates, the effect is stronger for males than for females. This suggests that sex should be incorporated into the analysis with a view to developing an understanding of why it would have different effects for males and females.

Glossary

Each entry includes a brief definition and notes the chapter in which the term was introduced.

Alpha error. See *Type I error.* Chapter 8

Alpha level (α). In inferential statistics, the probability of error. (1) In estimation, the probability that a confidence interval does not contain the population value. Chapter 7 (2) In hypothesis testing, the proportion of the area under the sampling distribution that contains unlikely sample outcomes if the null is true. The probability of Type I error. Chapter 8

Analysis of variance. A test of significance appropriate when testing for the differences among more than two sample means. Chapter 10

ANOVA. See *Analysis of variance.* Chapter 10

Arrow keys. When using SPSS, the keys used to move the cursor around the screen. Appendix F

Association. The relationship between two (or more) variables. Two variables are said to be associated if the distribution of one variable changes for the various categories or scores of the other variable. Chapter 12

Average deviation. The average of the absolute deviations of the scores around the mean. Chapter 4

Bar chart. A graphic display device for nominal- and ordinal-level variables. Chapter 2

Beta error. See *Type II error.* Chapter 8

Beta-weights (b^*). Standardized partial slopes. Chapter 17

Bias. A criterion used to select sample statistics for estimation procedures. A statistic is unbiased if the mean of its sampling distribution is equal to the population value of interest. Chapter 7

Bivariate normal distributions. The model assumption in the test of significance for Pearson's r that both variables are normally distributed. Chapter 15

Bivariate table. A table that displays the joint frequency distribution of two variables. Chapters 11 and 12

Cells. The cross-classification categories of the variables in a bivariate table. Chapter 11

Central Limit Theorem. A theorem that specifies the mean, standard deviation, and shape of the sampling distribution, given that the sample is large. Chapter 6

χ^2 (critical). The chi square score that marks the beginnings of the critical region of the chi square sampling distribution. Chapter 11

χ^2 (obtained). The test statistic computed in step 4 of the five-step model. Chapter 11

Chi square test. A nonparametric test of hypothesis for variables organized in a bivariate table. Chapter 11

Class intervals. The categories used in frequency distributions for interval-ratio-level variables. Chapter 2

Cluster sampling. A method of EPSEM sampling that is based on selecting groups, such as geographical areas, rather than cases for a list of the population. Chapter 6

Coefficient of determination (r^2). The proportion of all variation in Y that is explained by X. Chapter 15

Coefficient of multiple determination (R^2). A statistic that equals the total variation explained in the dependent variable by all independent variables combined. Chapter 17

Column. The vertical dimension of a bivariate table. Chapter 11

Conditional distribution of Y. The distribution of scores on the dependent variable for a specific score or category of the independent variable. Chapter 12

Conditional means of Y. The mean of all scores on Y for each value of X. Chapter 15

Confidence interval. An estimate of a population value in which a range of values is specified. Chapter 7

Confidence level. An alternative way to express alpha, the probability that a confidence interval will not contain the population value. Chapter 7

Continuous variable. A variable with a unit of measurement that can be subdivided infinitely. Chapter 1

Cramer's V. A chi square-based measure of association. Chapter 13

Critical region (region of rejection). The area under the sampling distribution that includes all unlikely sample outcomes. Chapter 8

Cumulative frequency. A column in a frequency distribution that displays the number of cases in an interval and all preceding intervals. Chapter 2

Cumulative percentage. A column in a frequency distribution that displays the percentage of cases in an interval and all preceding intervals. Chapter 2

Data. In social science research, information that is represented by numbers. Chapter 1

Database. An organized collection of related information. Appendix F

Data reduction. Summarizing many scores with a few statistics. Chapter 1

Deciles. The points that divide a distribution of scores into tenths. Chapter 3

Dependent variable. A variable that is identified as an effect, result, or outcome variable. The dependent variable is thought to be caused by the independent variable. Chapters 1 and 12

Descriptive statistics. The branch of statistics concerned with (1) summarizing the distribution of a single variable or (2) measuring the relationship between two or more variables. Chapter 1

Deviations. The distances between the scores and the mean. Chapter 4

Direct relationship. A multivariate relationship in which a control variable has no effect on the bivariate relationship. Chapter 16

Discrete variable. A variable with a basic unit of measurement that cannot be subdivided. Chapter 1

Dispersion. The amount of variety or heterogeneity in a distribution of scores. Chapter 4

E_1. For lambda, the number of errors of prediction made when predicting which category of the dependent variable cases will fall into while ignoring the independent variable. Chapter 13

E_2. For lambda, the number of errors of prediction made when predicting which category of the dependent variable cases will fall into while taking the independent variable into account. Chapter 13

Efficiency. The extent to which sample outcomes are clustered around the mean of the sampling distribution. Chapter 7

EPSEM. Equal probability of selection method. A technique for selecting samples in which every element or case in the population has an equal probability of being selected for the sample. Chapter 6

Expected frequency (f_e). The cell frequencies that would be expected in a bivariate table if the variables were independent. Chapter 11

Explained variation. The proportion of all variation in Y that is attributed to the effect of X. Chapter 15

Explanation. See *Spurious relationship*. Chapter 16

F ratio. For the analysis of variance, the test statistic computed in step 4 of the five-step model. Chapter 10

File. A database (or any other information) that is stored under the same name in the memory of the computer or on a floppy disk. Appendix F

Five-step model. A step-by-step guideline for conducting tests of hypotheses. Chapter 8

Frequency distribution. A table that displays the number of cases in each category of a variable. Chapter 2

Frequency polygon (line chart). A graphic display device for interval-ratio variables. Chapter 2

Gamma (G). A measure of association for ordinal variables organized in table format. Chapter 14

Histogram. A graphic display device for interval-ratio variables. Chapter 2

Homoscedasticity. The model assumption in the test of significance for Pearson's r that the variance of the Y scores is uniform across all values of X. Chapter 15

Hypothesis. A statement about the relationship between variables that is derived from a theory. Hypotheses are more specific than theories, and all terms and concepts are fully defined. Chapter 1

Independence. The null hypothesis in the chi square test. Two variables are independent if the classification of a case on one variable has no effect on the probability that the case will be classified in any particular category of the second variable. Chapter 11

Independent random samples. Random samples gathered so that the selection of a case for one sample has no effect on the probability that any particular case will be selected for the other samples. Chapter 9

Independent variable. A variable that is identified as a causal variable. The independent variable is thought to cause the dependent variable. Chapters 1 and 12

Index of qualitative variation (IQV). A measure of dispersion for variables that have been organized into frequency distributions. Chapter 4

Inferential statistics. The branch of statistics concerned with making generalizations from samples to populations. Chapter 1

Interaction. A multivariate relationship in which a bivariate relationship changes across the categories of the control variable. Chapter 16

Interpretation. See *Intervening relationship.* Chapter 16

Interquartile range (*Q*). The distance from the third quartile to the first. Chapter 4

Interval estimate. See *Confidence interval.* Chapter 7

Intervening relationship. A multivariate relationship in which the independent and dependent variables are linked primarily through the control variable. Chapter 16

Lambda (λ). A measure of association for nominal-level variables that have been organized into a bivariate table. Lambda is based on the logic of PRE. Chapter 13

Level of measurement. The mathematical characteristics of a variable as determined by the measurement process. A major criterion for selecting statistical techniques. Chapter 1

Linear relationship. A relationship between two variables in which the observation points (dots) in the scattergram can be approximated with a straight line. Chapter 15

Line chart. See *Frequency polygon.*

Marginals. The row and column totals of a bivariate table. Chapter 11

Mean ($\overline{X}$). The arithmetic average of a set of scores. Chapter 3

Mean square. In the analysis of variance, an estimate of the variance calculated by dividing the sum of squares within (SSW) or the sum of squares between (SSB) by the appropriate degrees of freedom. Chapter 10

Measures of association. Statistics that summarize the strength and direction of the relationship between variables. Chapters 1 and 12

Measures of central tendency. Statistics that summarize a distribution of scores by reporting the most typical or representative value of the distribution. Chapter 3

Measures of dispersion. Statistics that indicate the amount of variety or heterogeneity in a distribution of scores. Chapter 4

Median (Md). The point in a distribution of scores above and below which half of the cases fall. Chapter 3

Menu. A list of options in a statistical package. Appendix F

Microcase. A statistical package designed for use in social science research. Appendix F

Midpoint. The point halfway between the upper and lower limits of a class interval. Chapter 2

Mode. The most common value in a distribution or the largest category of a variable. Chapter 3

Multiple correlation. A multivariate technique for examining the combined effects of more than one independent variable on a dependent variable. Chapter 17

Multiple correlation coefficient (*R*). A statistic that indicates the strength of the correlation between a dependent variable and two or more independent variables. Chapter 17

Multiple regression. A multivariate technique that breaks down the separate effects of the independent variables on the dependent variable. Chapter 17

Negative association. A bivariate relationship in which the variables vary in opposite directions. As one variable increases, the other decreases, and high scores on one variable are associated with low scores on the other. Chapter 12

Nonparametric test. A type of significance test in which no assumptions about the shape of the sampling distribution are made. Chapter 11

Normal curve. A theoretical distribution of scores that is symmetrical, unimodal, and bell shaped. The standard normal curve always has a mean of 0 and a standard deviation of 1. Chapter 5

Normal curve table. Appendix A; a detailed description of the area between a Z score and the mean of a standardized normal distribution. Chapter 5

Null hypothesis (H_0). A statement of "no difference." The specific form varies from test to test. Chapter 8

Observed frequency (f_o). The cell frequencies actually observed in a bivariate table. Chapter 11

One-tailed test. A type of hypothesis test that can be used when (1) the direction of the difference can be predicted or (2) concern is focused on only one tail of the sampling distribution. Chapter 8

One-way analysis of variance. An application of ANOVA in which the effect of a single variable on another is observed. Chapter 10

Partial correlation. A multivariate technique for examining a bivariate relationship while controlling for other variables. Chapter 17

Partial correlation coefficient. A statistic that shows the relationship between two variables while controlling for other variables; $r_{yx.z}$ is the symbol for the partial correlation coefficient when controlling for one variable. Chapter 17

Partial gamma (G_p). A statistic that indicates the strength of the association between two variables after the effects of a third variable have been removed. Chapter 16

Partial slopes. In a multiple regression equation, the slope of the relationship between a particular independent variable and the dependent variable while controlling for all other independents in the equation. Chapter 17

Partial tables. Tables produced when controlling for a third variable. Chapter 16

Pearson's r (r). A measure of association for variables that have been measured at the interval-ratio level. Chapter 15

Percentage (%). The number of cases in a category divided by the number of cases in all categories, the entire quantity multiplied by 100. Chapter 2

Percentile. The point in a distribution of scores below which a specific percentage of the cases fall. Chapter 3

Phi (ϕ). A chi square–based measure of association. Chapter 13

Pie chart. A graphic display device for nominal- and ordinal-level variables. Chapter 2

Point estimate. An estimate of a population value in which a single value is specified. Chapter 7

Population. The total collection of all cases in which the researcher is interested. Chapter 1

Positive association. A bivariate relationship in which the variables vary in the same direction. As one variable increases, the other also increases, and high scores on one variable are associated with high scores on the other. Chapter 12

Proportion (p). The number of cases in a category divided by the number of cases in all categories. Chapter 2

Proportional reduction in error (PRE). The logic that underlies the definition and computation of several different measures of association. Statistics are derived by comparing the number of errors made in predicting the dependent variable while ignoring the independent variable with the number of errors made while taking the independent variable into account. Chapter 13

Quartiles. The points in a distribution of scores that divide the distribution into quarters. Chapter 3

Random samples. See **EPSEM.** Chapter 6

Range (R). The highest score minus the lowest score. Chapter 4

Rate. The number of actual occurrences divided by the number of possible occurrences per some unit of time. Chapter 2

Ratio. The number of cases in one category divided by the number of cases in another category. Chapter 2

Region of rejection. See **Critical region.** Chapter 8

Regression line. The best-fitting straight line that summarizes the relationship between two variables. The regression line is fitted to the data points by the least-squares criterion whereby the line touches all conditional means of Y or comes as close to doing so as possible. Chapter 15

Replication. See **Direct relationship.** Chapter 16

Representative. A characteristic of a random sample or a sample drawn according to the rule of EPSEM. A sample that accurately reproduces the major characteristics of the population from which it was drawn is said to be representative of the population. Chapter 6

Research hypothesis (H_1). A statement that contradicts the null hypothesis. The specific form varies from test to test. Chapter 8

Row. The horizontal dimension of a table. Chapter 11

Sample. A subset of a population. In inferential statistics, information is gathered from random or EPSEM samples and then generalized to populations. Chapter 1

Sampling distribution. The distribution of all possible sample outcomes of a given statistic. Chapter 6

Scattergram. A graphic display device that depicts the relationship between two variables. Chapter 15

Simple random sample. A sample drawn from a population so that every case has an equal chance of being included. Chapter 6

Skew. The extent to which a distribution of scores has a few cases that are extremely high (positive skew) or extremely low (negative skew). Chapter 3

Slope (b). The amount of change in a variable per unit change in the other variable. Chapter 15

SPSS for Windows. A statistical package designed for the analysis of social science data. Appendix F

Spearman's rho (r_s). A measure of association for

ordinal variables that are in "continuous" format. Chapter 14

Specification. See *Interaction.* Chapter 16

Spurious relationship. A multivariate relationship in which both the independent and dependent variables are actually caused by the control variable. The independent and dependent are not causally related. Chapter 16

Standard deviation (s or σ). The square root of the squared deviations of the scores around the mean, divided by N. The most commonly used measure of dispersion; s represents the standard deviation of a sample, and σ represents the standard deviation of a population. Chapter 4

Standard error of the mean. The standard deviation of a sampling distribution of sample means. Chapter 6

Standardized partial slopes (beta-weights). The slope of the relationship between a particular independent variable and the dependent when all scores are expressed as Z scores. Chapter 17

Statistical package (statpak). A set of computer programs designed to manipulate and statistically analyze data. Appendix F

Statistics. A set of mathematical techniques for organizing and analyzing data. Chapter 1

Stratified random sample. A random sample drawn from a population by selecting cases from sublists of groups in proportion to the representation of the groups in the population. Chapter 6

Sum of squares between (SSB). The sum of the squared deviations of the sample means from the overall mean, weighted by sample size. Chapter 10

Sum of squares total (SST). The sum of the squared deviations of the scores from the overall mean. Chapter 10

Sum of squares within (SSW). The sum of the squared deviations from the category means. Chapter 10

Systematic random sample. A sample selected by choosing the first case from a list of the population randomly and then choosing every kth case. Chapter 6

***t* (critical).** The t score that marks the beginnings of the critical region of a t distribution. Chapter 8

***t* distribution.** A distribution used to find the critical region for tests of sample means when N is small and is unknown. Chapter 8

***t* (obtained).** The test statistic computed in step 4 of the five-step model for tests of sample means when N is small and population standard deviation is unknown. Chapter 8

Test statistic. The value computed in step 4 of the five-step model that places the sample outcome on the sampling distribution. Chapter 8

Theory. A generalized explanation of the relationship between two or more variables. Chapter 1

Total variation. The spread of the Y scores around the mean of Y. Chapter 15

Two-tailed test. A type of hypothesis test that can be used when (1) the direction of the difference cannot be predicted or (2) concern is focused on both tails of the sampling distribution. Chapter 8

Type I error (alpha error). The probability of rejecting a null hypothesis that is true. Chapter 8

Type II error (beta error). The probability of failing to reject a null hypothesis that is false. Chapter 8

Unexplained variation. The proportion of the total variation in Y that is not accounted for by X. Chapter 15

Variable. Any trait that can change values from case to case. Chapter 1

Variance (s^2 or σ^2). The squared deviations of the scores around the mean, divided by N. A measure of dispersion used in inferential statistics and in regression techniques; s^2 represents the variance of a sample, and σ^2 represents the variance of a population. Chapter 4

***Y* intercept (a).** The point where the regression line crosses the Y axis. Chapter 15

***Z* (critical).** The Z score that marks the beginnings of the critical region of a Z distribution. Chapter 8

***Z* (obtained).** The test statistic computed in step 4 of the five-step model. Chapter 8

***Z* scores.** Standard scores; the way scores are expressed after they have been standardized to the theoretical normal curve. Chapter 5

Index

GLOSSARY OF SYMBOLS

The number in parentheses indicates the chapter in which the symbol is introduced.

a	Point at which the regression line crosses the Y axis (15)
AD	Average deviation (4)
ANOVA	The analysis of variance (10)
b	Slope of the regression line (15)
b_i	Partial slope of the linear relationship between the ith independent variable and the dependent variable (17)
b_i^*	Standardized partial slope of the linear relationship between the ith independent variable and the dependent variable (17)
df	Degrees of freedom (8)
f	Frequency (2)
F	The F ratio (10)
f_e	Expected frequency (11)
f_o	Observed frequency (11)
G	Gamma for a sample (14)
G_p	Partial gamma (16)
H_0	Null hypothesis (8)
H_1	Research or alternate hypothesis (8)
IQV	Index of qualitative variation (4)
Md	Median (3)
Mo	Mode (3)
N	Number of cases (2)
N_d	Number of pairs of cases ranked in different order on two variables (14)
N_s	Number of pairs of cases ranked in the same order on two variables (14)

%	Percentage (2)
P	Proportion (2)
P_s	A sample proportion (7)
P_u	A population proportion (7)
PRE	Proportional reduction in error (13)
Q	Interquartile range (4)
r	Pearson's correlation coefficient for a sample (15)
r^2	Coefficient of determination (15)
R	Range (4)
r_s	Spearman's rho for a sample (14)
$r_{xy.z}$	Partial correlation coefficient (17)
R^2	Multiple correlation coefficient (17)
s	Sample standard deviation (4)
SSB	The sum of squares between (10)
SST	The total sum of squares (10)
SSW	The sum of squares within (10)
s^2	Sample variance (4)
t	Student's t score (8)
V	Cramer's V (13)
X	Any independent variable (12)
$\overline{X}$	Mean of a sample (3)
X_i	Any score in a distribution (3)
Y	Any dependent variable (12)
Y'	A predicted score on Y (15)
Z scores	Standard scores (5)
Z	A control variable (16)